# Theory and Applications of Nanoparticles

# Theory and Applications of Nanoparticles

Edited by **Andrew Green**

NY RESEARCH PRESS

New York

Published by NY Research Press,
23 West, 55th Street, Suite 816,
New York, NY 10019, USA
www.nyresearchpress.com

**Theory and Applications of Nanoparticles**
Edited by Andrew Green



International Standard Book Number: 978-1-63238-446-1 (Hardback)



Printed in China.

# Contents

# Preface

Nanoparticles are ultrafine units which have a dimension measured in nanometers; its unit is nm; which stands for billionths of a meter. These nanoparticles are created by human activities and also exist in the natural world. Despite of their submicroscopic size, they exhibit unique material characteristics. Manufactured nanoparticles may exhibit properties that may be useful for practical applications in a variety of areas, which includes medicinal science, fields of engineering, catalysis, and environmental remediation.

Nanoparticles have three major physical properties and they all are interrelated:

1) Highly mobile in Free State,
2) Enormous specific surface area,
3) May or may not exhibit Quantum effects.

Also they can be very hard as well as very soft, thus nanoparticles have an enormous range of compositions but it depends on the product and on its use.

This book is a combined effort of lots of people working directly or indirectly in the production of this book, first of all I want to thank my teachers and my mentors for supporting me in the entire process, I want to thank the publishing team and my family, lastly, lots of gratitude towards all the authors for the contribution and their time.

It's my hope that this book will be a great success and it will help everyone in the field of nanoparticles, including students, scientists, researchers and every individual who seeks knowledge.

**Editor**

# 1

# The Impact of Nanochloroquine on Restoration of Hepatic and Splenic Mitochondrial Damage against Rodent Malaria

**Satyajit Tripathy, Sabyasachi Das, Sandeep Kumar Dash, Sourav Chattopadhyay, and Somenath Roy**

*Immunology and Microbiology Laboratory, Department of Human Physiology with Community Health, Vidyasagar University, Midnapore West Bengal 721102, India*

Correspondence should be addressed to Somenath Roy; sroy.vu@hotmail.com

Academic Editor: Vijaya Rangari

The applications of nanotechnology to pharmacology are the potential appliance of biodegradable polymers and convection-enhanced drug delivery in the diagnostics and treatment of diseases. Chitosan is a natural polysaccharide that has attracted significant scientific interest during the last two decades. The present study was to evaluate the possible effects of chitosan tripolyphosphate conjugated nanochloroquine against *Plasmodium berghei* infection on select makers of oxidative damage and antioxidant status in mitochondria of liver and spleen. *P. berghei* infection was developed in Swiss mice by intraperitoneal injection of 200 $\mu$L of infected blood. Parasite-infected mice were treated with chloroquine and nanoconjugated chloroquine. Superoxide radical generation, nitrate level, and oxidized glutathione were increased significantly ($P < 0.05$) in the mitochondria of infected group as compared to control group, and reduced glutathione level, activity of SOD, GPx, GR, and GST, and mitochondrial transmembrane potential were decreased significantly ($P < 0.05$), which were increased or decreased significantly ($P < 0.05$) near to normal in nanoconjugated chloroquine treated group than chloroquine treated group. So, the findings may suggest the advantageous role of nanoconjugated chloroquine against the *P. berghei* induced oxidative damage in hepatic and splenic mitochondria.

## 1. Introduction

Despite years of effort, the developments of effective drug against malaria are available. To date, antimalarial drugs remain the major way to treat the disease. Presently, the most effective way of dealing with malaria is the administration of chemotherapeutic agents. Although drug treatments of malaria are currently the best means of disease management, there is an urgent need for the development of structurally novel and effective antimalarial drugs because of increasing resistance to most presently available antimalarial drugs [1–3].

The multiplication of the malaria parasite in the blood causes the pathology such as anemia, hemolysis, and damage of the essential organs of the host by the parasite products. Like other microbes, *P. berghei* is also a causative agent that should change the normal condition of host by utilizing or protecting defence factors. It is generally accepted that ROS, NO, and peroxynitrite kill intraerythrocytic malarial parasites [4, 5], but the parasite protect itself by antioxidant enzyme [6]. Reactive oxygen species are considered mediators of the inflammatory response and tissue damage in malaria. After a short period of development and multiplication, these parasites leave the liver and invade erythrocytes where they multiply by schizogony, undergoing development through ring, trophozoite, and schizont stages that are responsible for malaria pathogenesis [7]. *Plasmodium* parasites ingest large quantities of hemoglobin and break it down in their food vacuole. The heam prosthetic group of hemoglobin is considered toxic, possibly due to the generation of oxygen radicals. Malaria infection damages several major organs, such as liver, kidney, brain, spleen, heart, and lungs, and severe malaria is characterized by multiorgan failure. Malarial infection induces mitochondrial pathology by inhibiting mitochondrial respiration, dehydrogenases, and transmembrane potential and damaging the ultrastructure as evident from transmission electron microscopic studies. Mitochondrial GSH depletion and formation of protein carbonyl indicate that mitochondrial pathology is associated with mitochondrial oxidative stress [8].

Although it is true that the pharmacokinetics of drugs differ between humans and mice, it is also true that all mammalian *Plasmodium* species have comparable life cycles and are sensitive to the same drugs [9]. The increasing resistance of malaria parasites to available drugs increases the burden of disease and the need to develop new and effective antimalarial agents [10]. The outbreed albino mouse inoculated with *P. berghei* is generally considered to be a valid model for the primary and large scale screening of drugs for eventual use against human malaria [11]. The use of strains of rodent malaria parasites can yield additional information concerning both its potential value against strains of human malaria and the mode of action of a compound.

Nanocarriers may also allow the use of potentially toxic antimalarials which are still under development. Therefore, chitosan, a natural linear biopolyaminosaccharide obtained by alkaline deacetylation of chitin, was used as a nanodrug carrier for the development of a potent antimalarial compound with conjugating the different age-old drug such as pyrimethamine [12], chloroquine [13], and chitosan have no toxic effect on liver, the most detoxifying organ, and has been recognized as biocompatible. It is also reported that chitosan nanoparticles reduced mitochondrial dehydrogenase activity and also inflicted extensive damage to the cell morphology.

Here, we aimed to isolate the mitochondria from liver and spleen and to study its oxidative stress, antioxidant enzymes level, and mitochondrial membrane potential of these vital organs and also compare the efficacy between only drug and nanoconjugated drug against *P. berghei* induced oxidative damage. This study seeks to demonstrate the potency of chitosan tripolyphosphate conjugated nanochloroquine on the mitochondrial oxidative status of *Plasmodium berghei* infected Swiss mice.

## 2. Materials and Methods

*2.1. Chemicals and Reagents.* Sodium chloride (NaCl), giemsa, Tris buffer, Triton-X 100, potassium dihydrogen phosphate ($KH_2PO_4$), dipotassium hydrogen phosphate ($K_2HPO_4$), ethylene diamine tetra acetate (EDTA), sodium hydroxide (NaOH), chloroform, sodium acetate, ammonium acetate, potassium hydroxide (KOH), methanol, Tris-HCl, formaldehyde, alcohol, and diphenylamine (DPA) were procured from Merck Ltd., SRL Pvt. Ltd., Mumbai, India. $5',5'$-dithio (bis)-2-nitrobenzoic acid (DTNB), standard reduced glutathione (GSH), glutathione reductase (GR), NADPH, Na4, NADPH, and oxidized glutathione (GSSG) were procured from Sigma (St. Louis, MO, USA). Commercially available histopaque 1077, HEPES, phorbol miresteds aceted (PMA), horse heart cytochrome-c were purchased from Sigma Chemical Co., USA. All other chemicals were from Merck Ltd., SRL Pvt., Ltd., Mumbai, India, and were of the highest purity grade available.

*2.2. Parasite.* The lethal rodent parasite, *Plasmodium berghei* NK65 strain, was maintained into sex- and age-matched wild type mice by weekly passage by intraperitoneal injection, and blood stage parasites were stored at −80°C.

*2.3. Animals.* Swiss albino mice, aged 6–8 weeks and weighing 20–25 g were kept up in accordance with the guidelines of the National Institute of Nutrition, Indian Council of Medical Research, Hyderabad, India, and approved by the ethical committee of Vidyasagar University. The animals were housed for a week for acclimatization in groups of five in standard polypropylene cages (Tarson) in the departmental animal house with stainless steel top grill. A clean paddy husk was used as a bedding material. The animals were fed on commercial pellet diet and water *ad libitum* in polypropylene bottles with stainless steel sipper tubes. The animals were maintained under standard conditions of humidity (55–65%), temperature (22 ± 3°C), and light (12 h light/12 h dark cycles). The animals used in this study did not show any sign of malignancy or other pathological processes.

*2.4. Preparation of Chitosan-Tripolyphosphate Nanoparticle and Loading of Chloroquine.* Chitosan-tripolyphosphate (CS-TPP) nanoparticle preparation by ionotropic gelation and drug loading was followed according to our previous laboratory report [14].

*2.5. Particle Characterization.* The drug loaded particle characterization was carried out in different techniques like FTIR, DLS, and zeta potential study, and also the *in vitro* drug toxicity, the actual drug content, and encapsulation efficiency was reported in our previous study [14].

*2.6. Parasite Density Determination and Inoculums Preparation.* Standard inoculum was prepared from a donor mouse with *P. berghei* parasitized erythrocytes. Infected blood from the donor mouse was obtained by cardiac puncture after anesthesia with chloroform. Parasitaemia was established by microscopic examination of a thin blood film under oil immersion at ×100 magnification and measured as a percentage of infected erythrocytes in fields of 1000 erythrocytes. Each mouse was infected with a standard inoculum of the $10^6$ parasitized erythrocyte suspension in phosphate buffered saline (0.2 mL) from a donor mouse that was prepared based on percentage parasitaemia.

*2.7. Experimental Design.* Mice were randomized into control, experimental, and treated groups of twenty four (24) animals and in each group six animals (6) were taken. Randomly selected eighteen (18) mice among twenty four mice were challenged by *P. berghei* (200 $\mu$L of infected blood containing $1 \times 10^5$ parasites) intraperitoneally (i.p.) for 10 days for successive infection development [15]. Other six mice (Group A) were injected with 200 $\mu$L 0.9gm% NaCl intraperitoneally (i.p.) and kept isolated from the infected group. After infection development for 10 days, the mice were isolated in three groups and marked as in the following.

Group "A": Control: Only 0.9 gm% Nacl;

Group "B": Infected control: *P. berghei* infection + 0.9 gm% NaCl;

Group "C": CQ treated: *P. berghei* infection + CQ drug with the concentration of actual drug content (27.36%) of 250 mg/Kg bw nanochloroquine [14];

Group "D": Nch treated: *P. berghei* infection + 250 mg/kg bw nanoconjugated chloroquine (effective dose) [14].

All the groups were treated for 15 days of the effective duration of Nch [14]. After the experiment was terminated at the end of schedule, all animals were sacrificed, and liver and spleen were collected to isolate mitochondria.

*2.8. Isolation of Mitochondria from Liver and Spleen.* After decapitation, tissues were excised from experimental mice of different experimental groups and washed with cold normal saline. Washed tissues of mice from each group were perfused with normal saline and were transferred for homogenize. Mitochondria of liver were isolated using the method described by Aprille and Austin [16]. Tissue was homogenized in the ice-cold buffer containing 0.25 M sucrose, 1 mM EDTA, 1 mM Tris-HCl, and pH 7.4. The homogenate was centrifuged at 600 ×g for 10 min at 4°C, and supernatant fraction were collected and further centrifuged 8000 ×g for 20 min at 4°C to pellet mitochondria and after that it was washed with sucrose buffer solution and then used for following parameters determination.

*2.9. Biochemical Estimation*

*2.9.1. Superoxide Radical Generation.* The superoxide production was measured according to the method of Boveris [17] by the SOD-inhibitable reduction of acetylated cytochrome c. Cytochrome c reduction by generated superoxide was monitored spectrophotometrically at 550 nm wavelength with or without the addition of 31 M SOD. The superoxide generation assay was performed in room air conditions.

*2.9.2. Nitrite (NO) Level.* 100 $\mu$L of Griess reagent (containing 1 part of 1% sulfanilamide in 5% phosphoric acid, and 1 part of 0.1% of N-C-1 napthyl ethylene diamine dihydrochloride) was added to 100 $\mu$L of mitochondria, incubated at room temperature for 10 minutes; readings were taken in a UV spectrophotometer at 550 nm and compared to a sodium nitrite standard curve (values ranging between 0.5 and 25 $\mu$M). The levels of NO were expressed as $\mu$M/mg protein [18].

*2.9.3. Determination of Superoxide Dismutase (SOD) Activity.* SOD activity of mitochondria of liver and spleen was determined from its ability to inhibit the auto-oxidation of pyrogallol according to KarMahapatra et al. [18]. The reaction mixture considered consisted of 50 mM Tris (hydroxymethyl) aminomethane (pH 8.2), 1 mM diethylen-etriamine penta acetic acid, and 20–50 $\mu$L of isolated mitochondria of liver and spleen. The reaction was initiated by the addition of 0.2 mM pyrogallol and the absorbance measured kinetically at 420 nm at 25°C for 3 min. SOD activity was expressed as unit/mg protein.

*2.9.4. Determination of Reduced Glutathione (GSH) Level.* Reduced glutathione estimation was performed by the method of KarMahapatra et al. [18]. The required amount of mitochondria of liver and spleen was mixed with 25% of TCA and centrifuged at 2,000 ×g for 15 min to settle the precipitated proteins. The supernatant was aspirated and diluted to 1 mL with 0.2 M sodium phosphate buffer (pH 8.0). Later, 2 mL of 0.6 mM DTNB was added. After 10 minutes, the optical density of the yellow-colored complex formed by the reaction of GSH and DTNB (Ellman's reagent) was measured at 405 nm. A standard curve was obtained with standard reduced glutathione. The levels of GSH were expressed as $\mu$g of GSH/mg protein.

*2.9.5. Determination of Oxidized Glutathione (GSSG) Level.* The oxidized glutathione level was measured after derivatization of GSH with 2-vinylpyidine according to the method of KarMahapatra et al. [18]. In brief, with 0.5 mL mitochondria, 2 $\mu$L of 2-vinylpyidine was added and incubated for 1 hr at 37°C. Then, the mixture was deproteinized with 4% sulfosalicylic acid and centrifuged at 1,000 ×g for 10 min to settle the precipitated proteins. The supernatant was aspirated and GSSG level was estimated with the reaction of DTNB at 412 nm in spectrophotometer and calculated with standard GSSG curve. The levels of GSSG were expressed as $\mu$g of GSSG/mg protein.

*2.9.6. Determination of Glutathione Peroxidase (GPx) Activity.* The GPx activity of was evaluated by the method of KarMahapatra et al. [18]. The reaction mixture contained 50 mM potassium phosphate buffer (pH 7.0), 1 mM EDTA, 1 mM sodium azide, 0.2 mM NADPH, 1 U glutathione reductase, and 1 mM reduced glutathione. The hepatic and splenic mitochondria, after its addition, were allowed to equilibrate for 5 min at 25°C. The reaction was initiated by adding 0.1 mL of 2.5 mM $H_2O_2$. Absorbance at 340 nm was recorded for 5 min. Values were expressed as nmol of NADPH oxidized to NADP by using the extinction coefficient of $6.2 \times 10^3\ M^{-1}\ cm^{-1}$ at 340 nm. The activity of GPx was expressed in terms of nmol NADPH consumed/min/mg protein.

*2.9.7. Determination of Glutathione Reductase (GR) Activity.* The GR activity was determined by the method of KarMahapatra et al. [18]. The tubes for enzyme assay were incubated at 37°C and contained 2.0 mL of 9 mM GSSG, 0.02 mL of 12 mM NADPH, Na4, 2.68 mL of 1/15 M phosphate buffer (pH 6.6), and 0.1 mL of mitochondria. The activity of this enzyme was determined by monitoring the decrease in absorbance at 340 nm. The activity of GR was expressed in terms of nmol NADPH consumed/min/mg protein.

*2.9.8. Determination of Glutathione-S-Transferase (GST) Activity.* The mitochondrial GST activity of liver and spleen was measured by the method of KarMahapatra et al. [18]. The tubes of enzyme assay were incubated at 25°C and contained 2.85 mL of 0.1 M potassium phosphate (pH 6.5) containing 1 mM of GSH, 0.05 mL of 60 mM 1-chloro-2, 4-dinitrobengene and 0.1 mL of mitochondria from liver and spleen. The activity of this enzyme was determined by monitoring the increase in absorbance at 340 nm. The activity of GST was expressed in terms of nmol NADPH consumed/min/mg protein.

*2.9.9. Protein Estimation.* Protein was determined using bovine serum albumin as a standard according to Lowry et al. [19].

*2.9.10. Determination of Mitochondrial Membrane Potential ($\Delta\psi_m$).* Mitochondrial membrane potential was determined according to Reungpatthanaphong and Mankhetkorn [20] with some modification. Briefly, the hepatic and splenic mitochondria were incubated separately with 40 nM rhodamine B in 1 cm quartz cuvettes containing HEPES-$Na^+$ buffer, pH 7.3 and vigorously stirred at 37°C. The rhodamine B fluorescence at 582 nm (excited at 553 nm) was monitored for 2 min using Hitachi F-7000 Fluorescence Spectrophotometer. Both excitation and emission slit width were set to 5.0.

*2.10. Statistical Analysis.* The data were expressed as mean ± SEM, $n = 6$. Comparisons between the means of control and infected group and also infected and treated group were made by Student's $t$-test (using a statistical package, Origin 6.1, Northampton, Mass, USA), $P < 0.05$, level of significance.

## 3. Results and Discussion

*3.1. Particle Characterization.* The detail of the test sample characterization was already described in our previous report [14]. Therefore, we showed only the brief information about the sample characterization in the present report. FTIR spectra showed the most important peaks at 1565 $cm^{-1}$, 2927 $cm^{-1}$, and 1406 $cm^{-1}$ which are assigned to N=O stretching of amine presence in drug, C–H stretching of methylene group, stretching vibrations of C=C, and C=N which are on indicate presence of chloroquine on chitosan NPs, respectively. The particles size, measured by DLS, were in the range of 150–225 nm, and the drug loaded nanoparticles size were in the range of 150–300 nm. The highest fraction of chloroquine NP present in the solution was of less than 250 nm, and the positive charge of Nch (+32.9 mV) is slightly higher than CS-TPP NPs. The cytotoxic activity of chitosan-TPP nanoparticle on HeLa cell lines showed no significant difference in cell viability between the cells treated with chitosan-TPP NPs, and our previous study revealed that the paprticle has no activity on parasitemia alteration, and in the drug loaded particle, the actual drug content was approximately 27.36% and the encapsulation efficiency was over 54% of the nanoparticles [14].

*3.2. Parasitemia.* The pathology of malaria depends not only on the strain of *Plasmodium* but also on the species and strain of the vertebrate host [21, 22]. Differences in the pathologies of *Plasmodia* strains appear to depend more on the strain of the parasite rather than on the type of mouse, indicating that the nature of the parasite and/or its antigens are more important than the mouse strain in determining the host defense response [22]. However, giemsa staining of tail vein blood smear was observed from control, infected, CQ treated, and Nch treated group to evaluate the parasitemia. In the

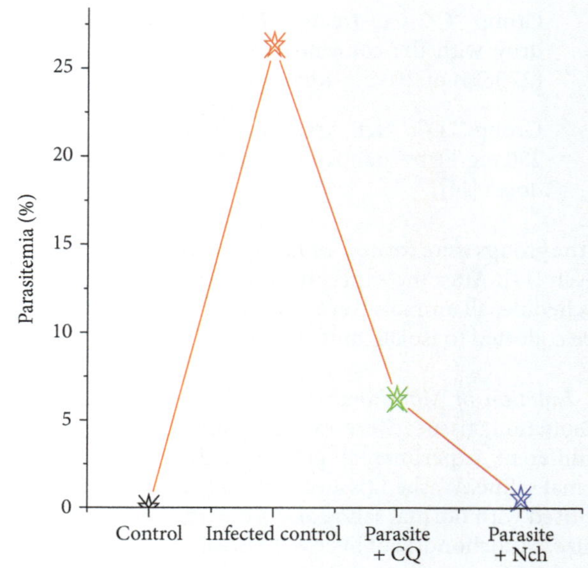


FIGURE 1: Graphical presentation of parasitemia level in different groups. Values are expressed as mean ± SEM, $n = 6$. *Indicates the significant difference ($P < 0.05$) compared with control. #Indicates the significant difference compared with infected group. Images are photographed at ×100 magnification.

infected group, parasitemia was 26.32%, whereas CQ treated group and Nch treated groups showed the parasitemia 6.21% and 0.56%, respectively (Figure 1). It may be due to free heme which is a powerful free radical generator, which can cause serious molecular damage to both host and parasite; the heme group contains $Fe^{2+}$ atoms that can catalyze Fenton and Haber-Weiss reactions, generating free radicals [23].

*3.3. Superoxide Radical Generation.* Alteration of the mitochondrial redox status by oxidative stress has been implicated in organ damage during malaria. Because mitochondria are the source and target of reactive oxygen species (ROS), we have investigated whether *P. berghei* is linked to mitochondrial pathology and mitochondrial ROS generation during malaria. It has been found that in the infected group the $O_2^{\bullet}$ level increased significantly by 41.27% in liver mitochondria as well as in spleen it also increased by 52.37% but Nch treatment it decreased by 23.86% in liver mitochondria and 43.14% in spleen as compared to infected control (Figure 2).

*3.4. Nitrate Generation.* Nitrate (NO) is an indicator of free radical generation. NO levels were significantly ($P < 0.05$) increased in mitochondria of liver by 22.92%, due to parasite infection as compared to control group, in which they were significantly ($P < 0.05$) decreased by 5.36% and 16.43% after treatment of CQ and Nch, respectively, as compared to infected control. In case of spleen mitochondria, the NO level in the infected group increased significantly ($P < 0.05$) by 38.48%, and CQ treatment reduced the NO generation by 3.5%, whereas Nch supplementation decreased by 10.5% as compared to infected control group. The results are shown in Figure 3.

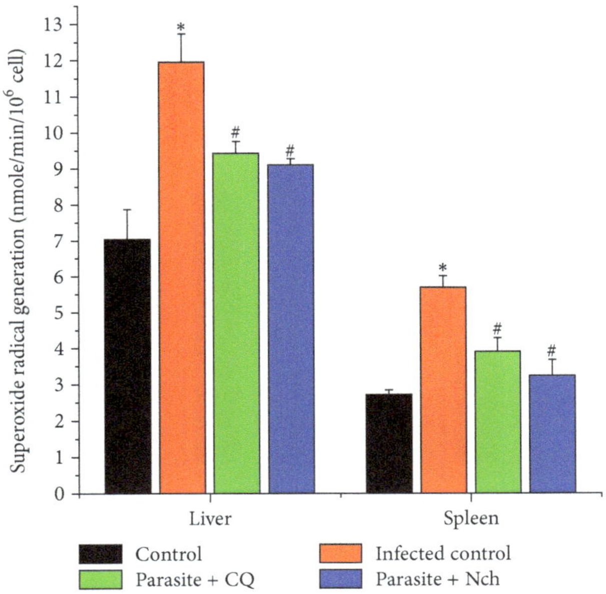


FIGURE 2: Graphical presentation of superoxide radical generation ($O_2^{\bullet}$) of hepatic and splenic mitochondria. Values are expressed as mean ± SEM, $n = 6$. *Indicates significant difference ($P < 0.05$) compared to control group. #Indicates significant difference ($P < 0.05$) compared to infected group.

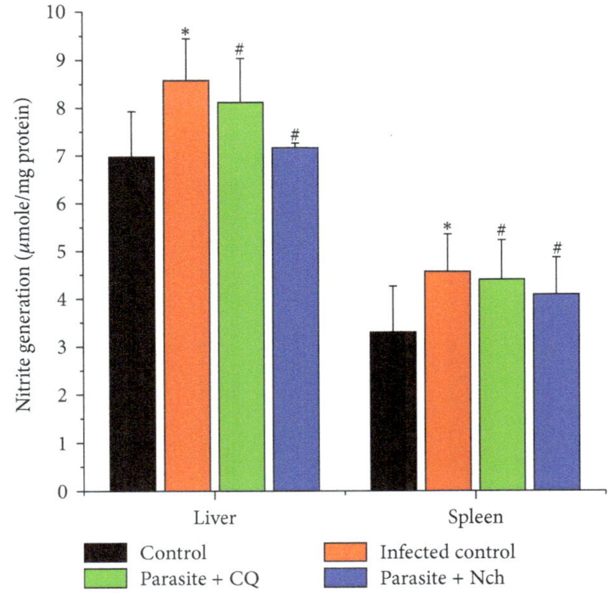


FIGURE 3: Graphical presentation of nitrite generation (NO) of hepatic and splenic mitochondria. Values are expressed as mean ± SEM, $n = 6$. *Indicates significant difference ($P < 0.05$) compared to control group. # indicates significant difference ($P < 0.05$) compared to infected group.

*3.5. Superoxide Dismutase (SOD) Activity.* SOD, a chain breaking antioxidant, plays an important role in protection against the deleterious effects of lipid peroxidation [24]. Free radicals such as partially reduced and highly reactive oxygen species ($O_2$), $H_2O_2$ or hydrogen peroxide, hydroxyl radical ($OH^-$), and singlet oxygen have been implicated in the etiology of various diseases. SOD is the major enzyme that protects cells from damage by catalyzing the dismutation of the superoxide into oxygen and hydrogen peroxide. From this study, the SOD generation in mitochondria, SOD level is decreased in liver mitochondria by 76.4% as compared to control due to infection of *P. berghei*, but in Nch treated group it has been increased significantly, similarly, the SOD level in splenic mitochondria also decreased in infected group by 47.59% and significantly ($P < 0.05$) increased in Nch treated group by 41.11% than CQ (39.16%) treated group as compared to infected control. The results are shown in Figure 4.

*3.6. Reduced Glutathione (GSH) and Oxidized Glutathione (GSSG).* Glutathione is an important antioxidant in cellular system. So, to understand glutathione level, we have measured both reduced and oxidized form of glutathione. The reduced glutathione level was decreased significantly ($P < 0.05$) by 53.24% in liver of infected group, as compared to control, whereas the oxidized glutathione level was increased significantly ($P < 0.05$) by 109.77%, as compared to control. Treatment of nanodrug significantly ($P < 0.05$) increased GSH level by 52.44% and decreased GSSG level significantly ($P < 0.05$) by 48.93% in liver. Similarly, the Nch supplementation increased the GSH level in splenic mitochondria as compared to infected control has been shown in Figures 5 and 6.

*3.7. Glutathione Reductase (GR) and Glutathione-S-Transferase (GST) Activity.* Many enzymes are important for synthesis, turnover, and maintenance of intracellular GSH concentrations [23, 24]. For the purpose of this paper, we term GSH reductase (GR), and glutathione-S-transferases (GSTs) was measured to understand the antioxidant enzyme activity of different experimental mitochondria. GR activity is significantly ($P < 0.05$) diminished in liver, in case of *P. berghei* infected group in Figure 7. It was recovered after nanoconjugated chloroquine treatment which increased antioxidant enzyme activity in mitochondria. Glutathione transferase (GST) is also measured to understand the antioxidant enzyme activity. Nanoconjugated chloroquine protects the mitochondria to elevate antioxidant enzyme activity to keep the organelles safe from oxidative damage due to malaria. The results are shown in Figure 8.

*3.8. Mitochondrial Membrane Potential ($\Delta\psi_m$).* The result in Figure 9 shows that the *P. berghei* infection decreased $\psi m$, as evidenced from 40.74% and 33.66% reduction in hepatic and splenic mitochondria then control in fluorescence at 582 nm due to rhodamine b uptake ($P < 0.05$), whereas in Nch treated group showed the membrane potential increment significantly ($P < 0.05$) then infected group.

Recently, biodegradable and biocompatible polymeric chitosan nanoparticles have attracted a considerable attention as potential carriers for the controlled and site-specific delivery of drugs [25]. Due to its unique polymeric cationic cha-

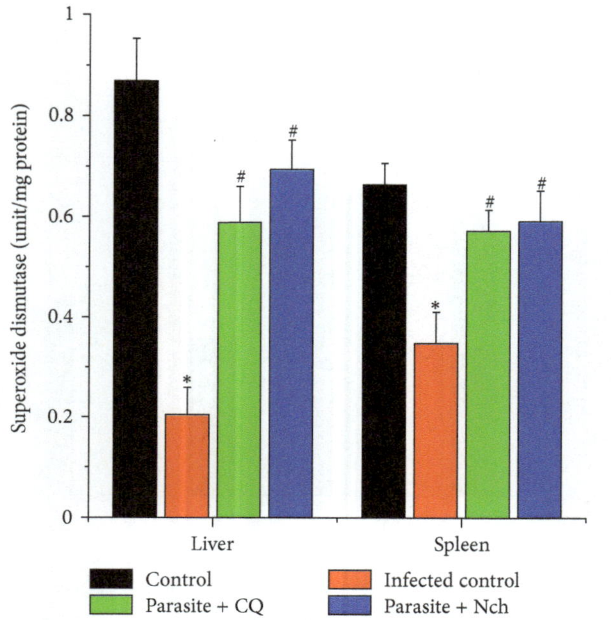


FIGURE 4: Graphical presentation of superoxide dismutase (SOD) of hepatic and splenic mitochondria. Values are expressed as mean ± SEM, $n = 6$. *Indicates significant difference ($P < 0.05$) compared to control group. #Indicates significant difference ($P < 0.05$) compared to infected group.

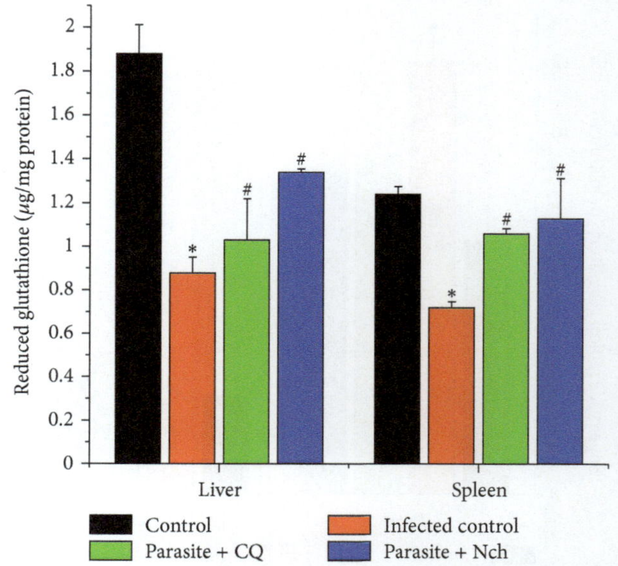


FIGURE 5: Graphical presentation of reduced glutathione (GSH) level of hepatic and splenic mitochondria. Values are expressed as mean ± SEM, $n = 6$. *Indicates significant difference ($P < 0.05$) compared to control group. #Indicates significant difference ($P < 0.05$) compared to infected group.

racter, CS has been extensively examined for the development of drug delivery systems in the pharmaceutical industry [26]. Due to the availability of free amino groups, it carries a positive charge and reacts with many negatively charged surfaces of the cell membrane, mucus lining (due to negatively charged sialic acid residues), and also with other anionic polymers. The nanodrug delivery systems seem to be a promising and viable approach for improving malaria treatment [27]. *Plasmodium* infection leads to increased oxidative stress in the vertebrate hosts. The high proliferation rate of parasites results in the production of large quantities of toxic redox-active by-products [28]. The role of oxidative stress during malarial infection needs clear illumination, as some have demonstrated a protective role while others suggested a relation with malarial pathophysiology and complication [29].

ROS are generated as by-products of aerobic respiration and various other catabolic and anabolic processes. Mitochondria are the major producer of ROS in cells, and the bulk of mitochondrial ROS is generated at the electron transport chain. Electrons leak from the electron transport chain directly to oxygen, producing short-lived free radicals such as superoxide anion ($O_2^-$). $O_2^-$ can be converted to nonradical derivatives such as hydrogen peroxide ($H_2O_2$) either spontaneously or catalyzed by superoxide dismutase (SOD). $H_2O_2$ is relatively stable and membrane permeable. It can be diffused within the cell and be removed by cytosolic antioxidant systems such as catalase, glutathione peroxidase. Because mitochondria are the major producer of ROS in mammalian cells, the close proximity to ROS places mitochondrial DNA (mtDNA) prone to oxidative damage. Consistently, many studies have shown that 8-oxo-dG, one of the common oxidative lesions, is detected at higher level in mtDNA than nuclear DNA, suggesting that mtDNA is more susceptible to oxidative damage [24].

In the study, it is reported that during the erythrocytic phase of malaria, red blood cell lysis and the release of hemozoin occur, which consists primarily of ferriprotoporphyrin IX dimmers and monomers (FP) and methemoglobin in plasmodial proteins. Free heme is a powerful free radical generator, which can cause serious molecular damage to both host and parasite; the heme group contains $Fe^{2+}$ atoms that can catalyze Fenton and Haber-Weiss reactions, generating free radicals [30]. This study demonstrates that infection of mice with *Plasmodium berghei* malaria produces demonstrable disturbance in the behavior of hepatic and splenic mitochondria and causes cell damage due to $O_2^{\bullet-}$ and NO production as well as antioxidant enzyme reduction where Nch treatment group showed the increment of $\psi m$. Malarial infection induces mitochondrial pathology by reducing transmembrane potential. Mitochondrial GSH depletion and formation of $O_2^{\bullet-}$ indicate that mitochondrial pathology is associated with mitochondrial oxidative stress intramitochondrial $O_2^{\bullet-}$ generation during malaria. $O_2^{\bullet-}$ inactivates mitochondrial aconitase to release iron from iron-sulfur clusters, which forms the hydroxyl radical. OH interacting with $H_2O_2$ produced concurrently [8].

Oxidative stress is known to induce apoptosis in many cellular systems, including liver [30–32] and spleen [33]. It has been reported that *P. yoelii* infection is responsible to induce opening of MPTP, as a result of that mitochondrial membrane permeability increases, leading to a decrease in $\psi m$. Mitochondrial anomalies and upregulation of apoptosis inducing

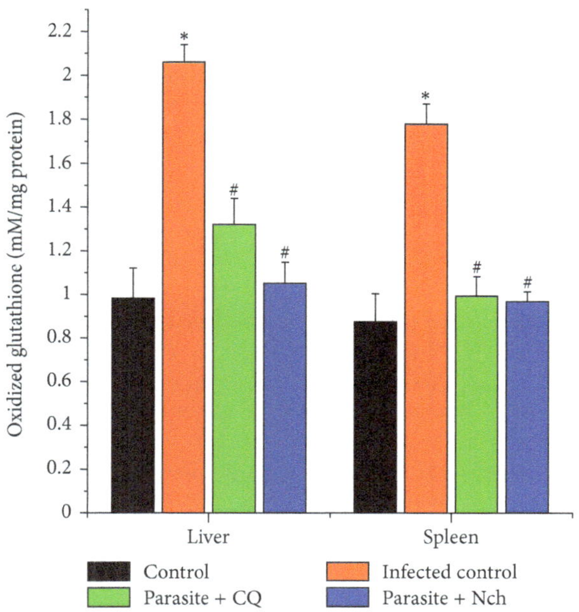


FIGURE 6: Graphical presentation of oxidized glutathione (GSSG) level of hepatic and splenic mitochondria. Values are expressed as mean ± SEM, $n = 6$. *Indicates significant difference ($P < 0.05$) compared to control group. #Indicates significant difference ($P < 0.05$) compared to infected group.

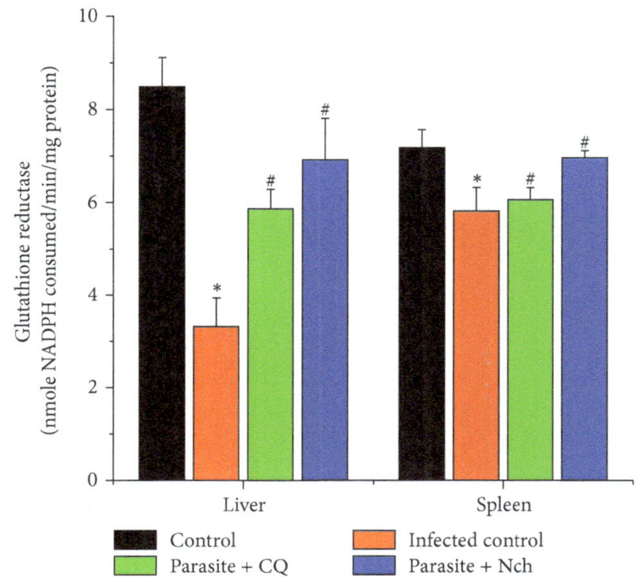


FIGURE 7: Graphical presentation of glutathione reductase (GR) activity of hepatic and splenic mitochondria. Values are expressed as mean ± SEM, $n = 6$. *Indicates significant difference ($P < 0.05$) compared to control group. #Indicates significant difference ($P < 0.05$) compared to infected group.

proteins such as Bax, p53 have also been reported in mice brain during experimental murine cerebral malaria. Opening of MPTP causes release of cytochrome c from mitochondria to cytosol [34, 35]. Generation of ROS and associated oxidative stress has been found to activate mitochondrial pathway of apoptosis [36]. We have found that *P. berghei* infection significantly induces OH generation in the liver and spleen of infected mice. Scavenging of $^{-}$OH also significantly inhibits the mitochondrial pathway, as evidenced from the near normal restoration of $\psi m$ [37]. The data suggest that administration of Nch attenuates *P. berghei* infection associated oxidative stress (assayed by lipid peroxidation) as well as apoptosis (as revealed from reduction of membrane potential), indicating a causative role of $^{-}$OH in the initiation of apoptosis. As mitochondria play a critical role in regulation of apoptosis, which is implicated in the pathogenic infection, the activation of the permeability transition pore in mitochondria is believed to play a critical role in cell necrosis and apoptosis [38]. Such selective oxidative modification may cause the cells to be more vulnerable to cell damage inducers.

## 4. Conclusion

From this study it may conclude that, chitosan-tripolyphosphate conjugated chloroquine nanoparticles potentially able to protect the mitochondrial damage of liver as well as defend the spleen from oxidative damage due to *P. berghei* infection by decreasing free radical generation and also by increasing the antioxidant enzymes level. The nanoconjugated drug may become the focus in future as a new therapeutic antimalarial agent as well as a potent free radical scavenger against such infection. However, further critical studies are necessary to clarify the mechanism of activity and also its role against chloroquine resistant parasite, including immune mechanisms in the course of medication.

## Abbreviations

| | |
|---|---|
| bw: | Body weight |
| DLS: | Dynamic light scattering |
| FTIR: | Fourier transform infrared spectroscopy |
| DTNB: | 5′, 5′-Dithio (bis)-2-nitrobenzoic acid |
| EDTA: | Ethylene diamine tetra acetate |
| GR: | Glutathione reductase |
| GSH: | Reduced glutathione |
| GSSG: | Oxidized glutathione |
| $H_2O_2$: | Hydrogen peroxide |
| MDA: | Malondialdehyde |
| MPO: | Myeloperoxidase |
| NADPH: | Nicotinamide adenine dinucleotide phosphate |
| Nch: | Nanochloroquine |
| OPD: | O-Dianisidine |
| PBS: | Phosphate buffer saline |
| ROS: | Reactive oxygen species |
| TBA: | Thiobarbituric acid |
| TCA: | Trichloroacetic acid. |

## Conflict of Interest

The authors declare that there is no conflicts of interests.

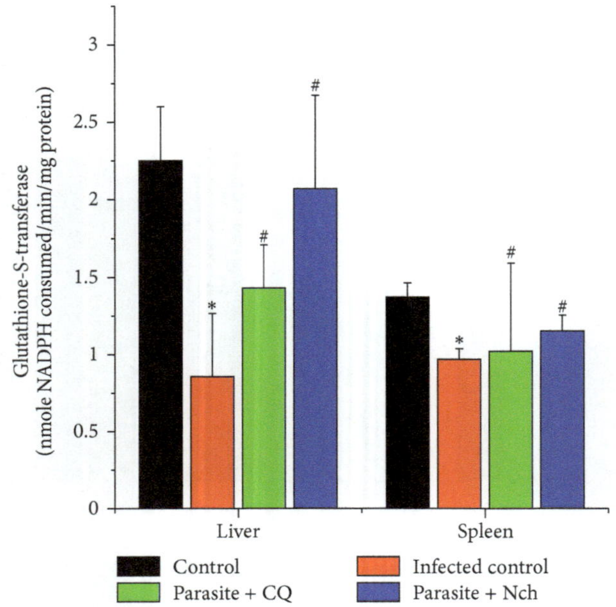


FIGURE 8: Graphical presentation of glutathione-s-transferase (GST) activity of hepatic and splenic mitochondria. Values are expressed as mean ± SEM, $n = 6$. *Indicates significant difference ($P < 0.05$) compared to control group. #Indicates significant difference ($P < 0.05$) compared to infected group.

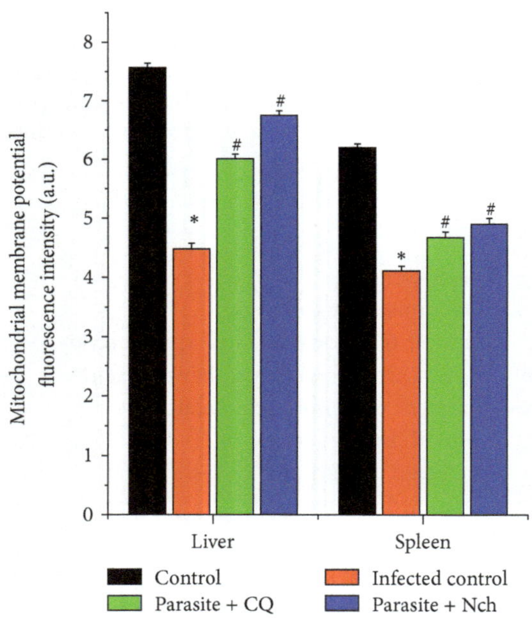


FIGURE 9: Graphical presentation of mitochondrial membrane potential alteration. Values are expressed as mean ± SEM, $n = 6$. *Indicates significant difference ($P < 0.05$) compared to control group. #Indicates significant difference ($P < 0.05$) compared to infected group.

## Acknowledgments

The authors thank Professor Pralhad C. Ghosh (Delhi University, South Campus, Delhi, India) for supplying the parasite strain and authors express gratitude to the Indian Institute of Technology, Kharagpur 721302, and Vidyasagar University, Midnapore 721102, for providing the facilities to execute these studies.

## References

[1] P. L. Olliaro and P. I. Trigg, "Status of antimalarial drugs under development," *Bulletin of the World Health Organization*, vol. 73, no. 5, pp. 565–571, 1995.

[2] W. Peters, "The problem of drug resistance in malaria," *Parasitology*, vol. 90, pp. 705–715, 1985.

[3] N. J. White, "Antimalarial drug resistance: the pace quickens," *Journal of Antimicrobial Chemotherapy*, vol. 30, no. 5, pp. 571–585, 1992.

[4] L. R. Brunet, "Nitric oxide in parasitic infections," *International Immunopharmacology*, vol. 1, no. 8, pp. 1457–1467, 2001.

[5] M. M. Stevenson and E. M. Riley, "Innate immunity to malaria," *Nature Reviews Immunology*, vol. 4, no. 3, pp. 169–180, 2004.

[6] A. Holmgren, "Antioxidant function of thioredoxin and glutaredoxin systems," *Antioxidants and Redox Signaling*, vol. 2, no. 4, pp. 811–820, 2000.

[7] P. Srivastava, S. K. Puri, G. P. Dutta, and V. C. Pandey, "Hepatic superoxide-scavenging system during *Plasmodium berghei* infection and chloroquine treatment," *Medical Science Research*, vol. 19, no. 10, pp. 307–308, 1991.

[8] C. Guinovart, M. M. Navia, M. Tanner, and P. L. Alonso, "Malaria: burden of disease," *Current Molecular Medicine*, vol. 6, no. 2, pp. 137–140, 2006.

[9] S. Dey, M. Guha, A. Alam et al., "Malarial infection develops mitochondrial pathology and mitochondrial oxidative stress to promote hepatocyte apoptosis," *Free Radical Biology and Medicine*, vol. 46, no. 2, pp. 271–281, 2009.

[10] I. Landau and P. Gautret, "Animal models: rodents," in *Malaria: Parasite Biology, Pathogenesis, and Protection*, I. W. Sherman, Ed., pp. 401–417, ASM Press, Washington, DC, USA, 1998.

[11] W. Peters, J. H. Portus, and B. L. Robinson, "The chemotherapy of rodent malaria, XXII. The value of drug resistant strains of *P. berghei* in screening for blood schizontocidal activity," *Annals of Tropical Medicine and Parasitology*, vol. 69, no. 2, pp. 155–171, 1975.

[12] E. C. Ibezim, C. T. Andrade, C. M. B. Barretto, D. C. Odimegwu, and F. F. D. Lima, "Ionically cross-linked chitosan/tripolyphosphate microparticles for the controlled delivery of pyrimethamine," *Ibinosina Journal of Medicine and Biological Sciences*, vol. 3, no. 3, pp. 77–87, 2011.

[13] J. W. Loh, M. Saunders, and L. Y. Lim, "Cytotoxicity of monodispersed chitosan nanoparticles against the Caco-2 cells," *Toxicology and Applied Pharmacology*, vol. 262, no. 3, pp. 273–282, 2012.

[14] S. Tripathy, S. Das, S. P. Chakraborty, S. K. Sahu, P. Pramanik, and S. Roy, "Synthesis, characterization of chitosan-tripolyphosphate conjugated chloroquine nanoparticle and its in vivo anti-malarial efficacy against rodent parasite: a dose and duration dependent approach,," *International Journal of Pharmaceutics*, vol. 434, pp. 292–305, 2012.

[15] S. Tripathy, S. P. Chakraborty, and S. Roy, "Superoxide radical generation mediated *Plasmodium berghei* infection in Swiss mice," *Al Ameen Journal of Medical Sciences*, vol. 5, pp. 69–81, 2012.

[16] J. R. Aprille and J. Austin, "Regulation of the mitochondrial adenine nucleotide pool size," *Archives of Biochemistry and Biophysics*, vol. 212, no. 2, pp. 689–699, 1981.

[17] A. Boveris, "Determination of the production of superoxide radicals and hydrogen peroxide in mitochondria," *Methods in Enzymology*, vol. 105, pp. 429–435, 1984.

[18] S. KarMahapatra, S. P. Chakraborty, S. Das, and S. Roy, "Methanol extract of *Ocimum gratissimum* protects murine peritoneal macrophages from nicotine toxicity by decreasing free radical generation, lipid and protein damage and enhances antioxidant protection," *Oxidative Medicine and Cellular Longevity*, vol. 2, no. 4, pp. 222–230, 2009.

[19] O. H. Lowry, N. J. Rosenbrough, A. L. Farr, and R. J. Randall, "Protein measurement with the Folin phenol reagent," *The Journal of Biological Chemistry*, vol. 193, no. 1, pp. 265–275, 1951.

[20] P. Reungpatthanaphong and S. Mankhetkorn, "Modulation of multidrug resistance by artemisinin, artesunate and dihydroartemisinin in K562/adr and GLC4/adr resistant cell lines," *Biological and Pharmaceutical Bulletin*, vol. 25, no. 12, pp. 1555–1561, 2002.

[21] B. R. Moore, J. D. Jago, and K. T. Batty, "*Plasmodium berghei*: parasite clearance after treatment with dihydroartemisinin in an asplenic murine malaria model," *Experimental Parasitology*, vol. 118, no. 4, pp. 458–467, 2008.

[22] R. Carter and D. Walliker, "New observations on the malaria parasites of rodents of the Central African Republic. *Plasmodium vinckei* petteri subsp. nov. and *Plasmodium chabaudi* Landau, 1965," *Annals of Tropical Medicine and Parasitology*, vol. 69, no. 2, pp. 187–196, 1975.

[23] B. S. Das and N. K. Nanda, "Evidence for erythrocyte lipid pepoxidation in acute falciparum malaria," *Transactions of the Royal Society of Tropical Medicine and Hygiene*, vol. 93, no. 1, pp. 58–62, 1999.

[24] J. H. Santos, B. S. Mandavilli, and B. Van Houten, "Measuring oxidative mtDNA damage and repair using quantitative PCR," *Methods in Molecular Biology*, vol. 197, pp. 159–176, 2002.

[25] N. Evelyne, P. Nathaline, F. Candau et al., "Biopolymer and polymer nanoparticles and their biomedical applications," *Handbook of Nanostructure and Nanotechnology*, vol. 5, pp. 577–635, 2000.

[26] K. A. Janes, P. Calvo, and M. J. Alonso, "Polysaccharide colloidal particles as delivery systems for macromolecules," *Advanced Drug Delivery Reviews*, vol. 47, no. 1, pp. 83–97, 2001.

[27] H. Swai, B. Semete, L. Kalombo, and P. Chelule, "Potential of treating tuberculosis with a polymeric nano-drug delivery system," *Journal of Controlled Release*, vol. 132, p. e48, 2008.

[28] I. Landau and A. G. Chabaud, "Natural infection by 2 plasmodia of the rodent *Thamnomys rutilans* in the Central African Republic," *Comptes Rendus Hebdomadaires des Seances de l'Academie des Sciences D*, vol. 261, no. 1, pp. 230–232, 1965.

[29] M. Yoeli, B. Hargreaves, R. Carter, and D. Walliker, "Sudden increase in virulence in a strain of Plasmodium yoelii," *Annals of Tropical Medicine and Parasitology*, vol. 69, no. 2, pp. 173–178, 1975.

[30] B. S. Das and N. K. Nanda, "Evidence for erythrocyte lipid pepoxidation in acute falciparum malaria," *Transactions of the Royal Society of Tropical Medicine and Hygiene*, vol. 93, no. 1, pp. 58–62, 1999.

[31] A. Pabon, J. Carmona, L. C. Burgos, and S. Blair, "Oxidative stress in patients with non-complicated malaria," *Clinical Biochemistry*, vol. 36, pp. 71–78, 2003.

[32] M. J. Czaja, "Induction and regulation of hepatocyte apoptosis by oxidative stress," *Antioxidants and Redox Signaling*, vol. 4, no. 5, pp. 759–767, 2002.

[33] T. A. Sarafian and D. E. Bredesen, "Is apoptosis mediated by reactive oxygen species?" *Free Radical Research*, vol. 21, no. 1, pp. 1–8, 1994.

[34] K. Biswas, U. Bandyopadhyay, I. Chattopadhyay, A. Varadaraj, E. Ali, and R. K. Banerjee, "A novel antioxidant and antiapoptotic role of omeprazole to block gastric ulcer through scavenging of hydroxyl radical," *Journal of Biological Chemistry*, vol. 278, no. 13, pp. 10993–11001, 2003.

[35] D. R. Green and J. C. Reed, "Mitochondria and apoptosis," *Science*, vol. 281, no. 5381, pp. 1309–1312, 1998.

[36] H. U. Simon, A. Haj-Yehia, and F. Levi-Schaffer, "Role of reactive oxygen species (ROS) in apoptosis induction," *Apoptosis*, vol. 5, no. 5, pp. 415–418, 2000.

[37] M. Guha, S. Kumar, V. Choubey, P. Maity, and U. Bandyopadhyay, "Apoptosis in liver during malaria: role of oxidative stress and implication of mitochondrial pathway," *The FASEB Journal*, vol. 20, no. 8, pp. E439–E449, 2006.

[38] L. Sanchez-Torres, A. Rodriguez-Ropon, M. Aguilar-Medina, and L. Favila-Castillo, "Mouse splenic $CD4^+$ and $CD8^+$ T cells undergo extensive apoptosis during a *Plasmodium chabaudi chabaudi* AS infection," *Parasite Immunology*, vol. 23, no. 12, pp. 617–626, 2001.

# Selective Hydrogenation of *p*-Chloronitrobenzene on Nanosized PdNiB Catalysts

**Yu-Wen Chen and Der-Shing Lee**

*Department of Chemical Engineering, National Central University, Chung-Li 320, Taiwan*

Correspondence should be addressed to Yu-Wen Chen; ywchen@ncu.edu.tw

Academic Editor: Mohammad H. Maneshian

A series of PdNiB bimetallic nanoalloy catalysts with various Pd contents was prepared. Pd was well dispersed in NiB. Even adding a small amount of Pd in NiB had a significant effect on activity and selectivity in hydrogenation of *p*-chloronitrobenzene (*p*-CNB) to *p*-chloroaniline (*p*-CAN). High activity and selectivity on PdNiB could be attributed to both ensemble effect and electronic effect. The particle size in PdNiB decreased with an increase in Pd content. Electron-enriched Ni could activate the polar-$NO_2$ groups of *p*-CNB and depress the dehalogenation of *p*-CAN.

## 1. Introduction

Bimetallic nanoparticle catalysts have received increasing attention [1–9]. The most remarkable features of nanoscopic materials are that their chemical and physical properties are quite different from those of bulk solids and those of atoms. This phenomenon has been known as a quantum effect [10–13]. Magnetic enhancement has been expected in Pd clusters owning to their finite size because Pd does not polarize magnetically in bulk metal state, but has giant magnetic moment in the present of ferromagnetic 3d transition metals [14–18]. One could successfully prepare chemically monodispersed ultrafine Pd and bimetallic PdNi particles [17–19].

The addition of the second metal plays a key role in controlling the activity, selectivity, and stability of catalysts in certain reaction. However, little is known about the electronic structures of bimetallic clusters of groups 8–10. Since both Ni and Pd belong to the same group 10 in the periodic table, the increase (mixing) between Ni $3d^8 4s^2$ electrons and Pd $4d^{10} 5s^0$ electrons is very likely involved in intermetallic Ni-Pd bonding [19]. Such effect on the catalysis by nanosized NiPd clusters is certainly interesting.

Aromatic chloramines are important intermediates in the synthesis of dyes, drugs, herbicides, and pesticides [20–34]. At present, these organic amines are generally produced through selective hydrogenation of the corresponding aromatic chloronitro compounds over transition metal catalysts such as Pt and N [35–39]. In this process, the hydrodechlorination of the aromatic chloramines often occurs over most metal catalysts because of the electron-donating effect of amino group in the aromatic ring. Keane [40] reported that in the continuous gas phase hydrogenation of *p*-chloronitrobenzene (*p*-CNB) over several supported Ni catalysts, *p*-chloroaniline (*p*-CAN) was produced as the main product at *p*-CNB conversion of about 15%, which is enlightening for the development of clean route to produce chloroanilines. It has been reported by this lab [41–43] that NiB is more active and selective than Raney nickel catalyst for hydrogenation of *p*-CNB to *p*-CAN. Many modifiers were added on NiB for hydrogenation of chloronitrobenzene [44–70]. Although intensive efforts have been made to solve the problem of hydrodechlorination in CNB hydrogenation, it is still a challenge to create novel catalysts for further improving the selectivity to CAN at complete conversion of the substrates and intermediates and simultaneously maintaining the high catalytic activity [20–32].

Usually the desire selectivity can be achieved by the addition of some inhibitors or some modifiers such as base or other electron-donating compounds to the catalyst or to the solution phase. By alloying a second metal to platinum, Coq et al. [71, 72] obtained good results in the hydrogenation of *p*-CNB over PtSn/$Al_2O_3$ catalysts, with 97.5% *p*-CAN selectivity at >98% conversion. Comparing with Pt catalysts, Pd catalysts exhibit low selectivity of CAN in the hydrogenation of CNB because the concomitant hydrodechlorination reactions are serious over Pd-based catalysts. The most attractive advantage of Pd-based catalysts over Pt-based ones is its low cost. In order to improve the selectivity to CAN in the hydrogenation of CNB over Pd catalysts, several strategies have been developed, such as introducing some promoting or inhabiting additives to the solution phase or to the catalysts, alloying Pd with other metals, applying partly poisoned catalysts, and modulating the metal/support interaction in supported Pd catalysts. PdB has been reported to be an active catalyst than Pd [14–19].

In a previous paper [9], one of the authors has reported the Pt-NiB catalyst for hydrogenation of *p*-CNB. However, Pt was too active to have high selectivity to the *p*-CAN product. Pd was chosen as the dopant of NiB in this study because its hydrogenation capability was not too high to have high selectivity to *p*-CAN and not too low to have low activity. By adding suitable amount of Pd in NiB, it was expected to have high conversion of *p*-CNB and high selectivity of *p*-CAN.

The objective of this study was to synthesize PdNiB bimetallic nanocatalysts and to investigate the effect of Pd content on the catalytic properties of PdNiB catalysts in liquid-phase hydrogenation of *p*-CNB to *p*-CAN.

## 2. Experimental

*2.1. Materials.* *p*-CNB, with a purity of >99%, was obtained from Acros (Belgium). High-purity hydrogen gas (>99.99% from Air Product) was used without further purification. Palladium nitrate and nickel acetate (>98%) were supplied by Showa Chemicals (Tokyo, Japan), and sodium borohydride was purchased from Lancaster (Morecambe, UK). Sodium hydroxide, methanol, and ammonia solutions were purchased from Tedia Co. (OH, USA). Ethanol was purchased from Showa Chemicals (Tokyo, Japan). Raney nickel catalyst was obtained from Merck. All other solvents and reagents were of analytical grade quality, purchased commercially, and used without any further purification. Water used in the whole process was doubly distilled water.

*2.2. Catalysts Preparation.* A series of PdNiB catalysts was prepared by the chemical reduction method. Pd/Ni atomic ratios were between 0.001 and 0.02 for investigation. Nickel acetate and palladium nitrate were added to the 50% methanol/water solution (both were 0.1 M) at room temperature under vigorous stirring and used nitrogen stream to remove air. The solution of reduction agent, sodium borohydride (1 M), was added with the microtubing pump into the solution drop by drop. The atomic ratio of (Ni + Pd) to B was fixed at 1 : 3. Excess amount of sodium borohydride was used to ensure that all nickel and palladium cations were reduced to metals. When no bubbles were released, the black powders were centrifuged and were washed to remove the impurities, that is, $Na^+$ ions and the excess amount boride species, by hot deionized water for three times and with methanol twice. Since all catalysts were easily oxidized in air, care must be taken to avoid their exposure to air. Generally, the catalysts were kept in methanol until further use.

*2.3. Catalyst Characterization.* The catalysts were characterized by XRD, TEM, and XPS. The crystalline structure of the catalysts was characterized by XRD using a Siemens D500 powder X-ray diffractometer. The XRD patterns were collected by using Cu $K_\alpha$ radiation ($\lambda$ = 0.15418 nm). The tube voltage and current were 40 kV and 40 mA, respectively. The scanning rate was 0.05° $s^{-1}$.

The morphologies of the catalysts were determined by TEM (Jeol JEM-2000 FX II). TEM was operated at 160 kV and the magnification was 200 K. A small amount of sample was put into the sample tube filled with a 95% ethanol solution. After agitation under ultrasonic environment for 90 min, one drop of the dispersed slurry was dipped onto a carbon coated copper mesh (300#) (Ted Pella Inc., CA, USA) and dried at room temperature in vacuum overnight.

The compositions and the electronic state of each element on the surface of the catalysts were studied by XPS on a Thermo VG Scientific Sigma Probe spectrometer. Al $K_\alpha$ radiation was used as the excitation source ($h\nu$ = 1486.6 eV) (20 kV, 30 mA). The sample was pressed as a self-supported plate and was mounted on the sample cell. It was degassed in the pretreatment chamber at 343 K for 2 h and then it was transferred into the analyzing chamber where the background pressure was lower than $10^{-7}$ Pa. Before the XPS test, the sample was sputtered using $Ar^+$ ions for 15 min to remove the oxidation parts, which were formed during the XPS operation. All the binding energies were calibrated by using the contaminant carbon (Cls = 284.6 eV). The XPS spectrum of each element was deconvoluted using a Gaussian-Lorentz curve-fitting program, the background was corrected using a Shirley-type baseline, and the peak type was Gaussian-Lorenz (7 : 3). The surface composition of each sample was calculated using the corresponding peak areas of Ni, Pd, and B, and the PHI sensitive factors were applied in the calculation. It should be noted that, for Ni, only the 2p3/2 subbands were studied instead of the 2p1/2 subbands, because the former have a much higher signal-to-noise ratio and provide the same information.

*2.4. Reaction Test.* All the experiments were carried out in the cylindrical stirred-tank reactor (Parr Instrument Model 4842). A four-bladed pitched impeller was placed for effective agitation, and the agitator was connected to an electric motor with variable speed up to 1700 rpm. A pressure transmitter and an automatic temperature controller were also provided. Hydrogen was introduced at the bottom of the reactor. A separate tube was used as sampling tube for the liquid phase. The reactor was charged by 0.002 mol Ni catalyst and 2.54 g *p*-CNB in 80 mL methanol; the concentration of *p*-CNB was 0.2 M. It was reported that methanol was a better reaction medium than ethanol for the hydrogenation reaction. The reaction was operated in reaction-controlled regime, as confirmed by using the different particle size and

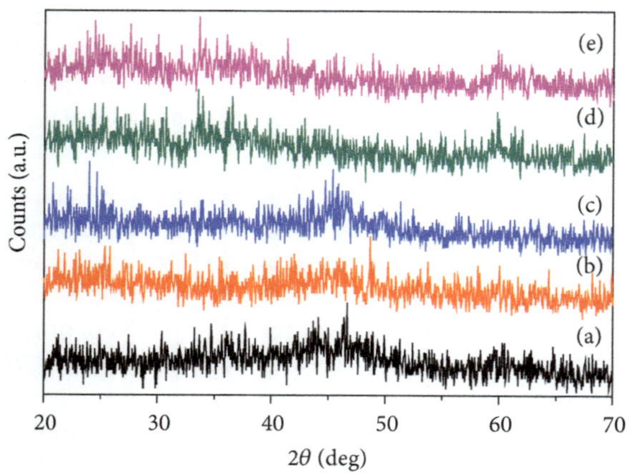


FIGURE 1: XRD patterns of catalyst samples, (a) NiB, (b) 0.001Pd-NiB, (c) 0.005Pd-NiB, (d) 0.01Pd-NiB, and (e) 0.02Pd-NiB.

stirring rate. The results showed that 400 rpm was enough to free from external diffusion. Therefore, 500 rpm stirring speed was used in this study. PdNiB sample was in nanosize range; one can conclude that the pore diffusion was not important in this study. During the reaction run, the samples were withdrawn periodically (every 10 min) and analyzed by a gas chromatograph equipped with a flame ionization detector and a 3 m × 1/8 inch stainless steel column packed with 5% OV-101 on Chromosorb WAW-DMSC (80–100 mesh). The conversion and the selectivity of each product were calculated as follows:

$$\text{conversion (\%)} = \left(1 - \frac{C_{p\text{-CNB}}}{C_{\text{AN}} + C_{\text{NB}} + C_{p\text{-CAN}} + C_{p\text{-CNB}}}\right) \times 100\%,$$

$$S_{\text{AN}}\ (\%) = \left(\frac{C_{\text{AN}}}{C_{\text{AN}} + C_{\text{NB}} + C_{p\text{-CAN}}}\right) \times 100\%,$$

$$S_{\text{NB}}\ (\%) = \left(\frac{C_{\text{NB}}}{C_{\text{AN}} + C_{\text{NB}} + C_{p\text{-CAN}}}\right) \times 100\%,$$

$$S_{p\text{-CAN}}\ (\%) = \left(\frac{C_{p\text{-CAN}}}{C_{\text{AN}} + C_{\text{NB}} + C_{p\text{-CAN}}}\right) \times 100\%, \quad (1)$$

where $C_{\text{AN}}$, $C_{\text{NB}}$, $C_{p\text{-CAN}}$, and $C_{p\text{-CNB}}$ represented the concentration of aniline, nitrobenzene, *p*-chloroaniline, and *p*-chloronitrobenzene, respectively.

## 3. Results and Discussion

*3.1. XRD.* XRD was used to characterize the structure of the Pd-NiB bimetallic nanoparticles. The XRD patterns of NiB and Pd-NiB samples with various contents of Pd are shown in Figure 1. The broad peak at $2\theta = 45^\circ$ in each pattern indicates the amorphous phase in all the as-prepared samples, in consistent with those in the literature [9, 11, 12]. It is further evidenced by a diffuse Debye ring rather than distinct dots in their SAED patterns. It has been reported that Ni can form several types of compounds with B such as NiB, $Ni_2B$, and $Ni_3B$. After adding the Pd, the intensities of this peak decreased. This shows that the presence of Pd could suppress the crystalline growth (long-range order) of NiB. No diffraction peak corresponding to crystalline Pd was found, indicating a high dispersion of Pd in NiB. No other crystalline phases (including palladium-related compounds, elemental Ni, B, and the corresponding oxides and hydroxides) were observed. One can conclude that PdNiB possessed short-range order and long range disorder, resulting in more surface coordinating unsaturated sites, more crystalline defects, and isotropic structure. Alloying Pd interrupts the long-range order of NiB, resulting in the characteristics of small cluster of NiB. Wang et al. [73] reported that a metal can disperse on the surface of another metal to form a highly dispersed bimetallic system due to the formation of bonds between the two metals. They noted that a highly dispersed Pd-Ni system occurred when <2.0 wt% Pd was loaded on a Pd/Ni catalyst. Our results are in accord.

*3.2. TEM.* Figure 2 shows the morphology of the as-prepared samples. With the addition of Pd, the particle size tends to decrease. Due to the high surface energy of the nanosized amorphous alloys, metal metalloids prepared by chemical reduction with borohydride are inclined to aggregate to form larger particles with diameters of several 10s to several 100s nanometers. Each sample was composed of many small particles, and the size of the aggregates was about 50 nm. It was difficult to obtain detailed information on the TEM photos because of the aggregation of the particles by the very strong van der Waals force and magnetization. One was not able to see any Pd particles due to very low concentration in the sample. After adding Pd on NiB, it became difficult to sediment in water, indicating that Pd-NiB particles were smaller than NiB particles. The primary particles of PdNiB were very fine. One could conclude that the size of primary particles of PdNiB was much smaller than 50 nm. Moreover, the magnetization of ultrafine particles remarkably increased with doping Pd into NiB.

*3.3. XPS.* The compositions and the electronic structure of each species on the surface of the samples were determined by XPS analysis. All of the spectra were deconvoluted, and the amount of each species in each state was calculated based on the corresponding peak area. The XPS spectra of $Ni_{2p}$ and $B_{1s}$ are shown in Figures 3 and 4. The binding energy of $Ni_{2p}$ level is ascribed to 856.8, 857.7, 856.0, 853.4, 856.3, and 856.7 eV, respectively, for the samples of NiB, 0.001Pd-NiB, 0.005Ni-PdB, 0.01Pd-NiB, and 0.02Pd-NiB. These are assigned to the metallic nickel. The peak around 862.2 eV is ascribed to nickel oxide (it was simply donated as $Ni^{2+}$ for the oxidized state of Ni; similar notations were used for the other elements). Tolman et al. [74] investigated a variety of $Ni^0$ organometallic compounds and found that in zero-valence state Ni complexes, the $Ni_{2p3/2}$ binding energies span a range of 853.6–856.0 eV, while higher binding energy peaks were observed in a range of 855.0–857.2 eV in $Ni^{2+}$ complexes, and even higher

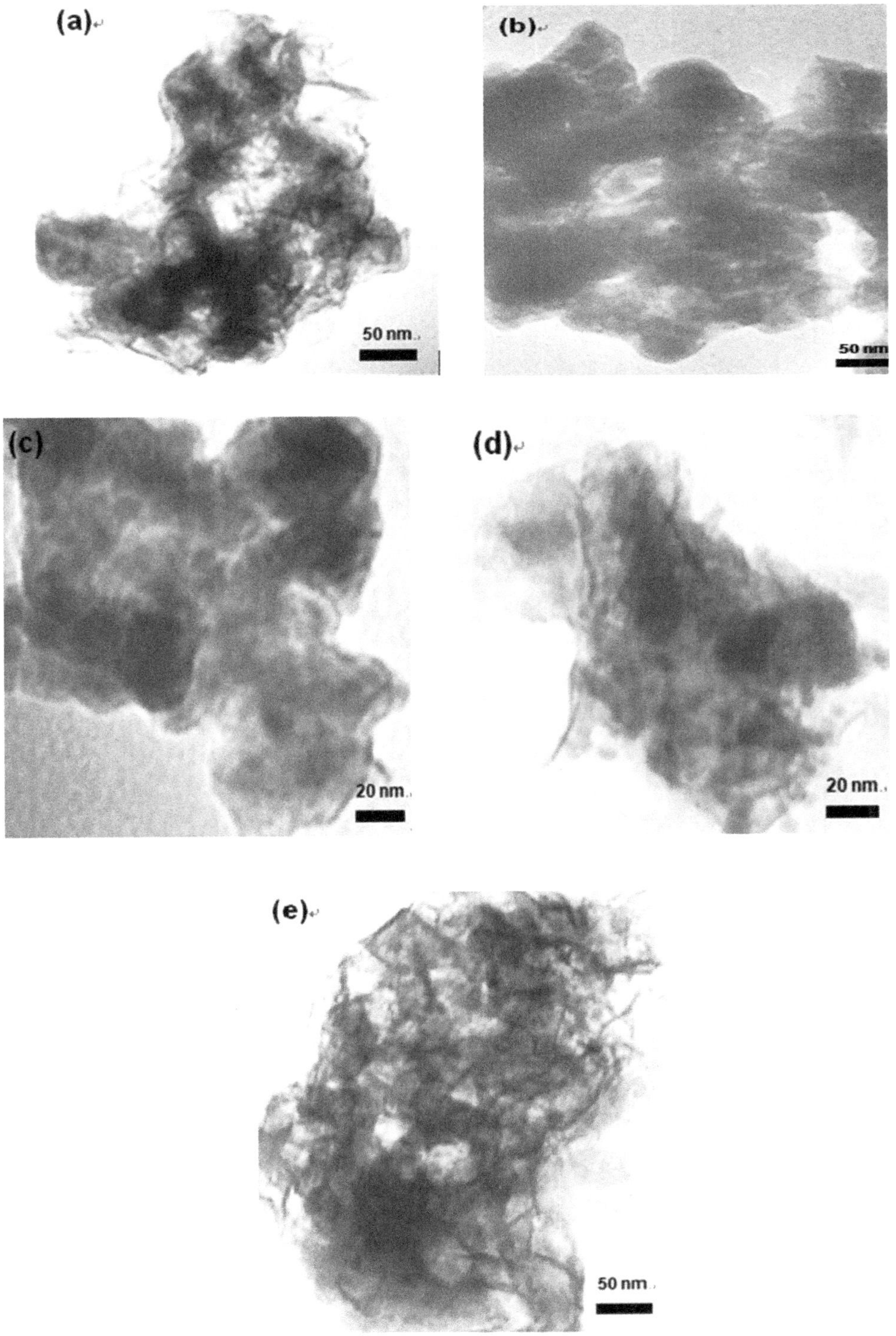


FIGURE 2: TEM images of (a) NiB, (b) 0.001Pd-NiB, (c) 0.005Pd-NiB, (d) 0.01Pd-NiB, and (e) 0.02Pd-NiB.

binding energy peaks were observed for $Ni^{4+}$ compounds [74]. Matienzo et al. [75] also extensively studied a series of nickel compounds in all of the oxidation states by XPS. In NiO, binding energy values of 854.0 and 872.0 eV were found for $Ni_{2p3/2}$ and $Ni_{2p1/2}$ peaks, respectively. Furthermore, in the case of NiO and tetrahedral $Ni^{2+}$ compounds, obvious shake-up (satellite) peaks can be found in the region at about 6 eV higher than a normal $Ni_{2p3/2}$ or $Ni_{2p1/2}$ band because such compounds are paramagnetic, whereas $Ni^0$ compounds and square-planar complexes of $Ni^{2+}$ do not produce satellite

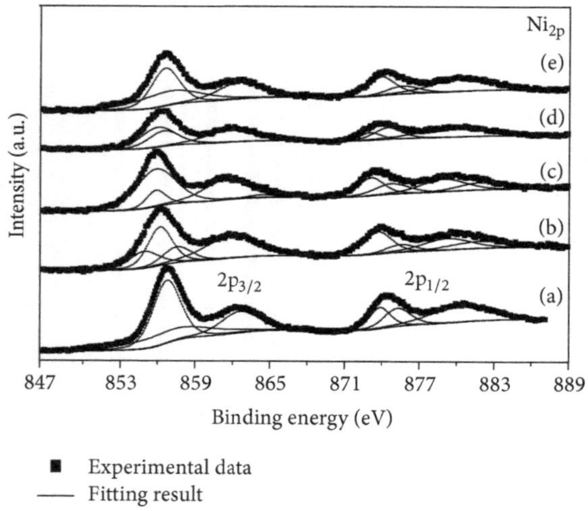


FIGURE 3: XPS spectra of $Ni_{2p}$ for (a) NiB, (b) 0.001Pd-NiB, (c) 0.005Pd-NiB, (d) 0.01Pd-NiB, and (e) 0.02Pd-NiB.

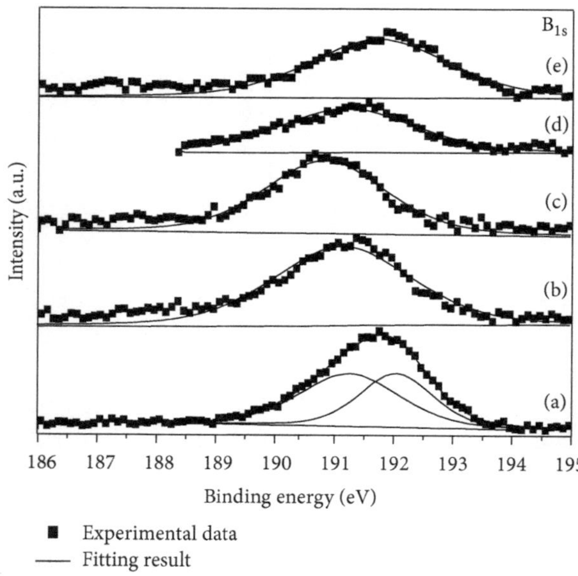


FIGURE 4: XPS spectra of $B_{1s}$ for (a) NiB, (b) 0.001Pd-NiB, (c) 0.005Pd-NiB, (d) 0.01Pd-NiB, and (e) 0.02Pd-NiB.

peaks, because they are diamagnetic. Therefore, the $Ni_{2p}$ bands in Figure 3 show low binding energy values and essentially no satellites features inferring that Ni exists mainly as zero-valence state inside PdNiB rather than in NiO-Pd-B form. The PdNiB catalysts exhibited electron-enriched state and effectively dispersed the Ni nanoparticles, resulting in more metallic Ni atoms on the surface. It is consistent with the results in the literature [76]. No obvious shift was presented in the XPS spectra, and those present in metallic state are not very close to the standard value 853.1 eV [77], which was attributed to the quantum size effect. It is known that hydrogen molecules adsorbed on the $Ni^0$ sites are activated to proceed hydrogenation reaction. Therefore, the amount of $Ni^0$ on the surface of catalyst is crucial for its catalytic activity in hydrogenation reaction.

There are two peaks in the $B_{1s}$ spectra. The peak near 187.1 eV is attributed to elemental B, and the peak near 192.5 eV is attributed to $B^{3+}$. The peak with the binding energy around 192.5 eV is related to the presence of the oxidized B species, which may be attributed to the interaction between $BH_4^-$ and $H_2O$ during the reduction of $Pd^{2+}$ ions by $BH_4^-$ in alkali solution or the surface boron oxide in the Pd-NiB alloy oxidized by air. It should be noted that most of B species is in oxidation state. The binding energies of $B_{1s}$ are 191.6, 191.2, 190.9, 191.2, and 191.8 eV for the samples of NiB, 0.001PdNiB, 0.005PdNiB, 0.01PdNiB, and 0.02PdNiB, respectively. The decreased BE's resulted from acceptance of electrons of B from other metal atoms. Strong interaction and interdiffusion between Pd and B substrate at room temperature have been reported [18, 78]. Here the change in BE's should be due to the interaction of Ni, Pd, and B. In conclusion, the amorphous structure of NiB and the presence of B promoted the surface-alloying effect of Pd and Ni. Pd peak was not observed due to the low loadings in these samples.

*3.4. Hydrogenation Reaction.* The catalytic activities of the as-prepared catalysts were investigated on the hydrogenation of *p*-CNB. There are two kinds of reactions in the hydrogenation process: selected hydrogenation of $-NO_2$ group and dehalogenation of –Cl group. In order to obtain high *p*-CAN yield, the dechlorination of *p*-CNB and *p*-CAN should be restrained; only the hydrogenation of $-NO_2$ group of *p*-CNB is desired [79]. The simplified reaction route is displayed in Scheme 1.

Figure 5 shows the conversion of *p*-CNB and the corresponding selectivity of *p*-CAN over these catalysts. The yields of *p*-CAN, NB, and AN are also shown in Figure 5. In each case, the concentration of NB increased at the first 30 min and then decreased and converted to AN and became zero eventually. There are two possible explanations the other one is that *p*-CNB only followed path (1) in reaction scheme in the first 30 min; is that the rate of the hydrogenation on the path (3) is faster than path (1). Since the increasing rate of aniline is not obvious, so the possibility of the latter case is low.

Figure 5 shows the *p*-CNB conversion and the selectivity of *p*-CAN versus reaction time. The reaction rate was zeroth order to *p*-CNB concentration. The activity of PdNiB was greater than NiB even with very small amount of Pd. The reaction results of the catalysts are listed in Table 1. The sample 0.02Pd-NiB had the highest reaction rate, which achieved 100% conversion within 30 min. The results showed that the activities increased with an increase in Pd content in the range of Pd/Ni atomic ratios between 0 and 0.01. Even adding small amount of Pd could enhance the activity and selectivity of *p*-CAN greatly. The hydrogenation of *p*-CNB on PdNiB catalysts was very selective to *p*-CAN. The results demonstrated that PdNiB can have the advantages of both metals, that is, high activity of Pd and high selectivity of Ni.

Both $Ni^0$ and $Pd^0$ are the active sites. Since only very small amount of Pd was added, the reaction was mainly on the surface of Ni. The surface concentration and electron

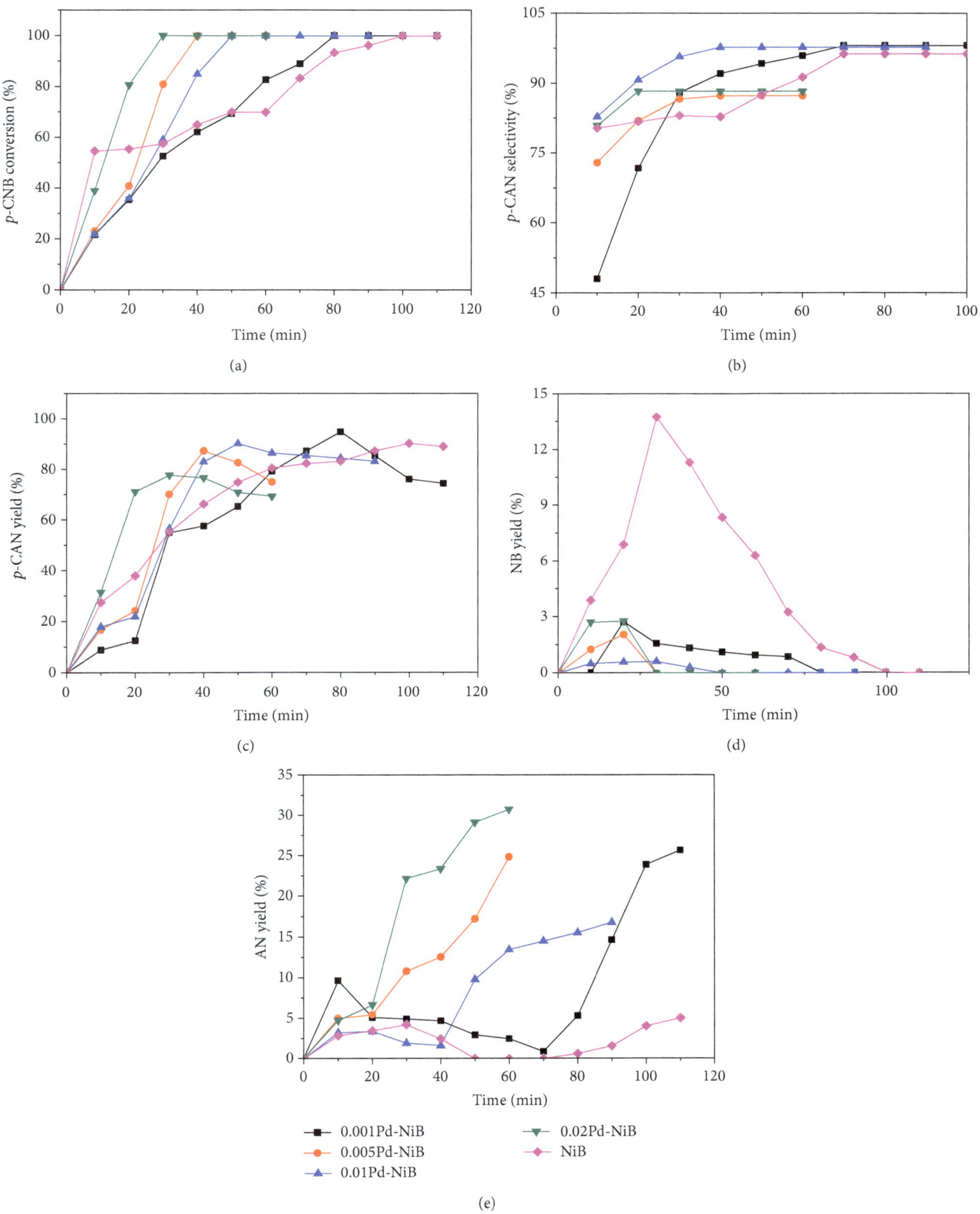


FIGURE 5: The relations of (a) *p*-CNB conversion, (b) *p*-CAN selectivity, (c) *p*-CAN yield, (d) NB yield, and (e) AN yield with reaction time on PdNiB catalysts. *Reaction conditions: 2 mmol NiPdB catalyst; 16 mmol *p*-CNB; 80 mL methanol; 80°C; 1.2 MPa hydrogen pressure; 500 rpm.

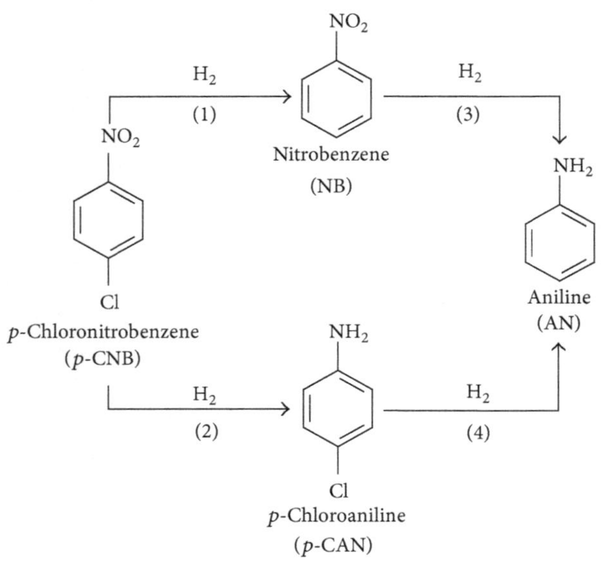


Scheme 1: Simplified reaction scheme of hydrogenation of *p*-CNB.

Table 1: The Effect of Pd content on the hydrogenation of *p*-CNB over PdNiB catalysts[a].

| Sample | Reaction time (min) to 100% conversion[a] | Selectivity (%) | | | Reaction rate constant ($s^{-1}$) |
|---|---|---|---|---|---|
| | | *p*-CAN[b] | AN[b] | NB[b] | |
| NiB | 95.6 | 96.9 | 5.5 | 0 | 0.0298 |
| 0.001Pd-NiB | 80.0 | 94.7 | 5.3 | 0 | 0.0167 |
| 0.005Pd-NiB | 40.0 | 75.1 | 24.9 | 0 | 0.0119 |
| 0.01Pd-NiB | 50.0 | 90.2 | 9.8 | 0 | 0.0127 |
| 0.02Pd-NiB | 28.6 | 77.8 | 22.2 | 0 | 0.0112 |

[a]Reaction condition: 1.2 MPa hydrogen pressure, 353 K reaction temperature, absolute methanol was medium, 500 rpm stirring speed, and 0.2 mmol Ni catalyst.
[b]*p*-CAN: *p*-chloroaniline; AN: aniline; NB: nitrobenzene.

density of $Ni^0$ are the important factors for the reaction. A higher $Ni^0$ content would favor the dissociative adsorption of $H_2$ molecules. According to the XPS analysis, one could find out that all the prepared samples had high densities of the elemental Ni. Thus, the $-NO_2$ group in the reactant *p*-CNB can be activated by the positive-charged Pd. The ability of the nearby dissociated H atoms to attack the O and N atoms of $-NO_2$ groups to form $-NH_2$ is enhanced, which benefits the high selectivity of *p*-CAN [33, 35, 80]. With the addition of Pd, the binding energy of Ni shifted to a lower value, indicating that Ni received more electrons from Pd. Combining the structure effect, palladium species could make the Ni particles smaller. This would increase the dispersion of Ni active sites and make the surface of the catalyst highly unsaturated, which favors the hydrogenation of *p*-CNB. The nitro group owned two different elements, N and O; both are highly electronegative. With the electrons donation from alloying B and Pd to Ni, the Ni on the catalyst surface became more electron enriched. Higher electronegativity of $-NO_2$ would be adsorbed on the surface easily, and the active sites would activate the N=O bond, which was polarized. The para-substituted nitro group had the higher electron negativity, resulting from the combination of both inductive and resonance effects. Since the $-NO_2$ group was more electronegative than $-Cl$, $-NO_2$ was supposed to occupy the active site on Ni surface at the start of the reaction. $-NO_2$ adsorbed on the catalyst surface is hydrogenated to form *p*-CAN, which is further desorbed. In addition, oxygen was more electronegative and alloying Pd and B could engage the oxygen to activate the polar $-NO_2$ group of *p*-CNB. The $-NH_2$ of *p*-CAN might adsorb on the surface alloying B and Pd and coordinate with each other. It would improve the selectivity of *p*-CAN by depressing the dehalogenation reaction. Pd-dopant in the NiB catalysts would increase the electron density of Ni and make the activity increase significantly. In addition, PdNiB is paramagnetic; it is easier to separate after reaction.

The results presented herein show that metal catalyst could be affected by the formation of the bimetallic colloid or by the addition of metal cations. The mechanism of the hydrogenation of *p*-CNB over the bimetallic collided system is very complicate and is affected by the electronic and conformation properties of the components as well as the composition on the mental surface.

*3.5. Reaction Kinetics.* One could calculate the reaction rate constant for each catalyst. Based on the conversion-time curves in the hydrogenation of *p*-CNB (Figure 5), it shows that the reaction was of first order with respect to the concentration of *p*-CNB [81]. The reaction rate can be expressed as

$$-r_A = kC_{A0} \cdot (1 - X_A), \quad (2)$$

where $k$ is the reaction rate constant, $s^{-1}$; $C_{A0}$ is the the initial concentration of reactant at $t = 0$; $X_A$ is the conversion of $A$.

Since the reaction was carried out in a constant-volume batch reactor, so

$$-r_A = C_{A0} \frac{dX_A}{dt}, \quad (3)$$

where $C_{A0}$ is the the initial concentration of reactant at $t = 0$; $X_A$ is the conversion of $A$; $t$ is the reaction time.

Combining the above two equations and using integral method of analysis of data, one could get

$$-\ln(1 - X_A) = kt. \quad (4)$$

A plot of $-\ln(1 - X_A)$ versus $t$ could derive the slope which represents the reaction rate constant. Because of the induction period appeared in the initial stage of reaction, the data in the initial stage were neglected. The results are tabulated in Table 1. As expected, the rate constant of NiPdB catalyst was higher than that of NiB.

## 4. Conclusion

A series of PdNiB nanoalloy catalysts with various Pd contents was prepared by chemical reduction method with

$NaBH_4$ as the reducing agent. The catalysts were characterized by XRD, TEM, and XPS. The catalytic properties of these catalysts were tested in the hydrogenation of *p*-CNB. NiB was amorphous as indicated by the broad peak around $2\theta = 45^\circ$. Upon modification with Pd, it could maintain NiB in the amorphous structure and decrease the crystallinity. With the addition of Pd, the particle size of NiB decreased. PdNiB formed nanoalloy and no discrete phase was found. The addition of Pd could reduce the particle size of NiB and improve the Ni dispersion. The binding energy of elemental Ni and B in the PdNiB is negatively shifted, indicating that Ni and B accepted partial electrons from Pd. The magnetization of PdNiB remarkably increased with doping Pd into NiB.

Since the $-NO_2$ was more electronegative than –Cl, $-NO_2$ was supposed to occupy the active site on Ni surface in the beginning of the reaction. $-NO_2$ adsorbed on the surface is hydrogenated to form *p*-CAN which is then desorbed. In addition, oxygen is more electronegative and the alloying B could engage the oxygen to activate the polar $-NO_2$ group of *p*-CNB. The $-NH_2$ of *p*-CAN might adsorb on the alloying B and coordinate with each other. Hence, it could improve the selectivity of *p*-CAN by suppressing the dehalogenation reaction.

High activity and selectivity on PdNiB in the hydrogenation of *p*-CNB could be attributed to both ensemble effect and electronic effect. Alloying Pd interrupts the long-range order of NiB, resulting in the characteristics of small cluster of NiB. On the other hand, B and Pd could partially donate electrons to Ni. Electron-enriched Ni could activate the polar $-NO_2$ groups of *p*-CNB and depress the dehalogenation of *p*-CAN. The results described here showed that PdNiB nanoalloy catalyst is a promising catalyst for industrial application.

## Acknowledgment

This research was supported by the Ministry of Economical Affairs, Taiwan, under Contract no. 101-EC-17-A-09-S1- 022.

## References

[1] J. Q. Du, Y. Zhang, T. Tian, S. C. Yan, and H. T. Wang, "Microwave irradiation assisted rapid synthesis of Fe–Ru bimetallic nanoparticles and their catalytic properties in water-gas shift reaction," *Materials Research Bulletin*, vol. 44, no. 6, pp. 1347-1351, 2009.

[2] J. C. S. Wu and H. C. Chou, "Bimetallic Rh–Ni/BN catalyst for methane reforming with $CO_2$," *Chemical Engineering Journal*, vol. 148, no. 2-3, pp. 539–545, 2009.

[3] N. He, P. Li, Y. Zhou, W. Ren, S. Fan, and V. A. Verkhozina, "Catalytic dechlorination of polychlorinated biphenyls in soil by palladium-iron bimetallic catalyst," *Journal of Hazardous Materials*, vol. 164, no. 1, pp. 126-132, 2009.

[4] G. Bernardotto, F. Menegazzo, F. Pinna, M. Signoretto, G. Cruciani, and G. Strukul, "New Pd–Pt and Pd–Au catalysts for an efficient synthesis of $H_2O_2$ from $H_2$ and $O_2$ under very mild conditions," *Applied Catalysis A*, vol. 358, no. 2, pp. 129–135, 2009.

[5] S. Penner, H. Lorenz, W. Jochum et al., "Pd/$Ga_2O_3$ methanol steam reforming catalysts, part I: morphology, composition and structural aspects," *Applied Catalysis A*, vol. 358, no. 2, pp. 193–202, 2009.

[6] N. M. Bertero, A. F. Trasarti, B. Moraweck, A. Borgna, and A. J. Marchi, "Selective liquid-phase hydrogenation of citral over supported bimetallic Pt–Co catalysts," *Applied Catalysis A*, vol. 358, no. 1, pp. 32–41, 2009.

[7] M. P. Latusek, R. M. Heimerl, B. P. Spigarelli, and J. H. Holles, "Synthesis and characterization of supported bimetallic overlayer catalysts," *Applied Catalysis A*, vol. 358, no. 1, pp. 79–87, 2009.

[8] N. Gyorffy, I. Bakos, S. Szabo et al., "Preparation, characterization and catalytic testing of GePt catalysts," *Journal of Catalysis*, vol. 263, no. 2, pp. 372–379, 2009.

[9] Y. W. Chen, N. Sasirekha, and Y. C. Liu, "Hydrogenation of *p*-chloronitrobenzene over NiPtB nanoalloy catalysts," *Journal of Non-Crystalline Solids*, vol. 355, no. 22-23, pp. 1193–1201, 2009.

[10] G. C. Bond and D. E. Webster, "Mixed metal catalysts in catalytic hydrogenation," *Annals of the New York Academy of Sciences*, vol. 158, pp. 540–559, 1969.

[11] L. F. Chen and Y. W. Chen, "Effect of additive (W, Mo, and Ru) on Ni–B amorphous alloy catalyst in hydrogenation of *p*-chloronitrobenzene," *Industrial & Engineering Chemistry Research*, vol. 45, no. 26, pp. 8866–8873, 2006.

[12] Y. Chen, "Chemical preparation and characterization of metal-metalloid ultrafine amorphous alloy particles," *Catalysis Today*, vol. 44, no. 1–4, pp. 3–16, 1998.

[13] X. F. Chen, H. X. Li, H. S. Luo, and M. H. Qiao, "Liquid phase hydrogenation of furfural to furfuryl alcohol over Mo-doped Co-B amorphous alloy catalysts," *Applied Catalysis A*, vol. 233, no. 1-2, pp. 13–20, 2002.

[14] T. Ding, Y. N. Qin, and M. A. Zhi, "Study of the Pd-B/$\gamma$-$Al_2O_3$ amorphous alloy catalyst," *Chinese Chemical Letters*, vol. 14, no. 3, pp. 319–322, 2003.

[15] X. Yu, M. Wang, and H. Li, "Study on the nitrobenzene hydrogenation over a Pd-B/$SiO_2$ amorphous catalyst," *Applied Catalysis A*, vol. 202, no. 1, pp. 17–22, 2000.

[16] W. X. Tu, S. J. Cao, L. P. Yang, and W. C. Wang, "Modification effects of magnetic supports and bimetallic structures on palladium nanocluster catalysts," *Chemical Engineering Journal*, vol. 143, no. 1–3, pp. 244–248, 2008.

[17] F. J. Urbano and J. M. Marinas, "Hydrogenolysis of organohalogen compounds over palladium supported catalysts," *Journal of Molecular Catalysis A*, vol. 173, no. 1-2, pp. 329–345, 2001.

[18] D. H. Shen, P. A. Kearney, J. M. Slaughter, and C. M. Falco, "Chemical interaction at the Pd-B interface," *Journal of Magnetism and Magnetic Materials*, vol. 126, no. 1–3, pp. 25–27, 1993.

[19] S. U. Son, Y. Jang, J. Park et al., "Designed synthesis of atom-economical Pd/Ni bimetallic nanoparticle-based catalysts for sonogashira coupling reactions," *Journal of the American Chemical Society*, vol. 126, no. 16, pp. 5026–5027, 2004.

[20] B. Coq, G. Ferrat, and F. Figueras, "Conversion of chlorobenzene over palladium and rhodium catalysts of widely varying dispersion," *Journal of Catalysis*, vol. 101, no. 2, pp. 434–445, 1986.

[21] B. Coq, A. Tijani, and F. Figuéras, "Hydrogenation of *para*-chloronitrobenzene over supported ruthenium-based catalysts," *Applied Catalysis*, vol. 76, no. 2, pp. 255–266, 1991.

[22] N. Lingaiah, N. S. Babu, R. Gopinath, P. S. S. Reddy, and P. S. S. Prasad, "Hydrodechlorination of chlorobenzene over supported metal catalysts," *Catalysis Surveys from Asia*, vol. 10, no. 1, pp. 29–39, 2006.

[23] L. M. Sikhwivhilu, N. J. Coville, B. M. Pulimaddi, J. Venkatreddy, and V. Vishwanathan, "Selective hydrogenation of o-chloronitrobenzene over palladium supported nanotubular titanium dioxide derived catalysts," *Catalysis Communications*, vol. 8, no. 12, pp. 1999–2006, 2007.

[24] M. Liu, W. Yu, H. Liu, and J. Zheng, "Preparation and characterization of polymer-stabilized ruthenium–platinum and ruthenium–palladium bimetallic colloids and their catalytic properties for hydrogenation of o-chloronitrobenzene," *Journal of Colloid and Interface Science*, vol. 214, no. 2, pp. 231–237, 1999.

[25] M. H. Liu, W. Y. Yu, and H. F. Liu, "Selective hydrogenation of o-chloronitrobenzene over polymer-stabilized ruthenium colloidal catalysts," *Journal of Molecular Catalysis A*, vol. 138, no. 2-3, pp. 295–303, 1999.

[26] Y. W. Chen and D. S. Lee, "Hydrogenation of p-chloronitrobenzene on nanosized modified NiMoB catalysts," *Catalysis Survey in Asia*, vol. 16, pp. 198–209, 2012.

[27] Y. Monguchi, A. Kume, K. Hattori, T. Maegawa, and H. Sajiki, "Pd/C–$Et_3$N-mediated catalytic hydrodechlorination of aromatic chlorides under mild conditions," *Tetrahedron*, vol. 62, no. 33, pp. 7926–7933, 2006.

[28] V. Vishwanathan, V. Jayasri, P. M. Basha, N. Mahata, L. Sikhwivhilu, and N. J. Coville, "Gas phase hydrogenation of *ortho*-chloronitrobenzene (O-CNB) to *ortho*-chloroaniline (O-CAN) over unpromoted and alkali metal promoted-alumina supported palladium catalysts," *Catalysis Communications*, vol. 9, no. 3, pp. 453–458, 2008.

[29] C. Y. Xi, H. Y. Cheng, J. M. Hao, S. X. Cai, and F. Y. Zhao, "Hydrogenation of o-chloronitrobenzene to o-chloroaniline over Pd/C in supercritical carbon dioxide," *Journal of Molecular Catalysis A*, vol. 282, no. 1-2, pp. 80–84, 2008.

[30] X. L. Yang, H. F. Liu, and H. Zhong, "Hydrogenation of o-chloronitrobenzene over polymer-stabilized palladium–platinum bimetallic colloidal clusters," *Journal of Molecular Catalysis A*, vol. 147, no. 1-2, pp. 55–62, 1999.

[31] Z. K. Yu, S. J. Liao, Y. Xu, B. Yang, and R. Yu, "A remarkable synergic effect of polymer-anchored bimetallic palladium–ruthenium catalysts in the selective hydrogenation of p-chloronitrobenzene," *Journal of the Chemical Society, Chemical Communications*, no. 11, pp. 1155–1156, 1995.

[32] Z. K. Yu, S. J. Liao, Y. Xu, B. Yang, and D. R. Yu, "Hydrogenation of nitroaromatics by polymer-anchored bimetallic palladium–ruthenium and palladium–platinum catalysts under mild conditions," *Journal of Molecular Catalysis A*, vol. 120, no. 1–3, pp. 247–255, 1997.

[33] X. X. Han, R. X. Zhou, G. H. Lai, B. H. Yue, and X. M. Zheng, "Effect of transition metal (Cr, Mn, Fe, Co, Ni and Cu) on the hydrogenation properties of chloronitrobenzene over Pt/$TiO_2$ catalysts," *Journal of Molecular Catalysis A*, vol. 209, no. 1-2, pp. 83–87, 2004.

[34] J. R. Kosak, "Catalytic hydrogenation of aromatic halonitro compounds," *Annals of the New York Academy of Sciences*, vol. 172, pp. 175–185, 1970.

[35] H. Li, Q. F. Zhao, and H. X. Li, "Selective hydrogenation of p-chloronitrobenzene over Ni–P–B amorphous catalyst and synergistic promoting effects of B and P ," *Journal of Molecular Catalysis A*, vol. 285, no. 1-2, pp. 29–35, 2008.

[36] N. Mahata, A. F. Cunha, J. J. M. Órfão, and J. L. Figueiredo, "Hydrogenation of chloronitrobenzenes over filamentous carbon stabilized nickel nanoparticles," *Catalysis Communications*, vol. 10, no. 8, pp. 1203–1206, 2009.

[37] M. H. Lin, B. Zhao, and Y. W. Chen, "Hydrogenation of p-chloronitrobenzene over Mo-modified NiCoB nanoalloy catalysts: effect of Mo content," *Industrial & Engineering Chemistry Research*, vol. 48, no. 15, pp. 7037–7043, 2009.

[38] B. Zhao, C. J. Chou, and Y. W. Chen, "Hydrogenation of p-chloronitrobenzene on tungsten-modified NiCoB catalyst," *Industrial & Engineering Chemistry Research*, vol. 49, no. 4, pp. 1669–1676, 2010.

[39] B. Zhao and Y. W. Chen, "Hydrogenation of p-chloronitrobenzene on Mo, La, Fe, and W-modified NiCoB nanoalloy catalysts," *Journal of Non-Crystalline Solids*, vol. 356, no. 18-19, pp. 839–847, 2010.

[40] M. A. Keane, "A review of catalytic approaches to waste minimization: case study-liquid-phase catalytic treatment of chlorophenols," *Journal of Chemical Technology and Biotechnology*, vol. 80, no. 11, pp. 1211–1222, 2005.

[41] Y. C. Liu, C. Y. Huang, and Y. W. Chen, "Hydrogenation of p-chloronitrobenzene on Ni–B nanometal catalysts," *Journal of Nanoparticle Research*, vol. 8, no. 2, pp. 223–234, 2006.

[42] Y. C. Liu, C. Y. Huang, and Y. W. Chen, "Liquid-phase selective hydrogenation of p-chloronitrobenzene on Ni–P–B nanocatalysts," *Industrial & Engineering Chemistry Research*, vol. 45, no. 1, pp. 62–69, 2006.

[43] Y. C. Liu and Y. W. Chen, "Hydrogenation of p-chloronitrobenzene on lanthanum-promoted NiB nanometal catalysts," *Industrial & Engineering Chemistry Research*, vol. 45, no. 9, pp. 2973–2980, 2006.

[44] J. F. Deng, J. Yang, S. Sheng, and H. Chen, "The study of ultrafine Ni–B and Ni–P amorphous alloy powders as catalysts," *Journal of Catalysis*, vol. 150, no. 2, pp. 434–438, 1994.

[45] J. F. Deng, H. Li, and W. J. Wang, "Progress in design of new amorphous alloy catalysts," *Catalysis Today*, vol. 51, no. 1, pp. 113–125, 1999.

[46] S. P. Lee and Y. W. Chen, "Nitrobenzene hydrogenation on Ni–P, Ni–B and Ni–P–B ultrafine materials," *Journal of Molecular Catalysis A*, vol. 152, no. 1-2, pp. 213–223, 2000.

[47] C. P. Li, Y. W. Chen, and W. J. Wang, "Nitrobenzene hydrogenation over aluminum borate-supported platinum catalyst," *Applied Catalysis A*, vol. 119, no. 2, pp. 185–194, 1994.

[48] H. Li, H. X. Li, W. L. Dai, W. J. Wang, and Z. G. Fang, "XPS studies on surface electronic characteristics of Ni–B and Ni–P amorphous alloy and its correlation to their catalytic properties," *Applied Surface Science*, vol. 152, no. 1-2, pp. 25–34, 1999.

[49] H. Li, H. X. Li, W. J. Wang, J. F. Deng, and J. F., "Excellent activity of ultrafine Co–B amorphous alloy catalyst in glucose hydrogenation," *Chemistry Letters*, no. 7, pp. 629–630, 1999.

[50] H. X. Li, H. S. Luo, L. Zhuang, W. L. Dai, and M. H. Qiao, "Liquid phase hydrogenation of furfural to furfuryl alcohol over the Fe-promoted Ni–B amorphous alloy catalysts," *Journal of Molecular Catalysis A*, vol. 203, no. 1-2, pp. 267–275, 2003.

[51] H. X. Li, H. Li, W. L. Dai, and M. H. Qiao, "Preparation of the Ni–B amorphous alloys with variable boron content and its correlation to the hydrogenation activity," *Applied Catalysis A*, vol. 238, no. 1, pp. 119–130, 2003.

[52] H. X. Li, Y. D. Wu, J. Zhang, W. L. Dai, and M. H. Qiao, "Liquid phase acetonitrile hydrogenation to ethylamine over a highly active and selective Ni–Co–B amorphous alloy catalyst," *Applied Catalysis A*, vol. 275, no. 1-2, pp. 199–206, 2004.

[53] H. Li, J. Zhang, and H. X. Li, "Ultrasound-assisted preparation of a novel Ni–B amorphous catalyst in uniform nanoparticles

for *p*-chloronitrobenzene hydrogenation," *Catalysis Communications*, vol. 8, no. 12, pp. 2211–2216, 2007.

[54] L. H. Lu, Y. H. Ma, F. Guo et al., "Structural characteristics of Mo modified skeletal Ni and its catalysis for liquid phase hydrogenation of nitro T-acid to T-acid," *Fine Chemicals*, vol. 25, no. 3, pp. 251–255, 2008.

[55] Y. F. Ma, W. Li, M. G. Zhang, Y. Zhou, and K. Y. Tao, "Preparation and catalytic properties of amorphous alloys in hydrogenation of sulfolene," *Applied Catalysis A*, vol. 246, no. 2, pp. 215–223, 2003.

[56] F. Cardenas-Lizana, S. Gomez-Quero, and M. A. Keane, "Clean production of chloroanilines by selective gas phase hydrogenation over supported Ni catalysts," *Applied Catalysis A*, vol. 334, no. 1-2, pp. 199–206, 2008.

[57] X. X. Han, R. X. Zhou, X. M. Zheng, and H. Jiang, "Effect of rare earths on the hydrogenation properties of *p*-chloronitrobenzene over polymer-anchored platinum catalysts," *Journal of Molecular Catalysis A*, vol. 193, no. 1-2, pp. 103–108, 2003.

[58] X. X. Han, R. X. Zhou, G. H. Lai, and X. M. Zheng, "Influence of support and transition metal (Cr, Mn, Fe, Co, Ni and Cu) on the hydrogenation of *p*-chloronitrobenzene over supported platinum catalysts," *Catalysis Today*, vol. 93–95, pp. 433–437, 2004.

[59] R. Hara, K. Sato, W. H. Sun, and T. Takahashi, "Catalytic dechlorination of aromatic chlorides using Grignard reagents in the presence of $(C_5H_5)_2TiCl_2$ ," *Chemical Communications*, vol. 12, no. 9, pp. 845–846, 1999.

[60] M. L. Hitchman, R. A. Spackman, N. C. Ross, and C. Agra, "Disposal methods for chlorinated aromatic waste," *Chemical Society Reviews*, vol. 24, no. 6, pp. 423–430, 1995.

[61] Y. J. Hou, Y. Q. Wang, F. He et al., "Liquid phase hydrogenation of 2-ethylanthraquinone over La-doped Ni–B amorphous alloy catalysts," *Materials Letters*, vol. 58, no. 7-8, pp. 1267–1271, 2004.

[62] G. Junfeng, A. J. David, and G. G. J. Brian, "Exploring the structural complexities of metal-metalloid nanoparticles: the case of Ni · B as catalyst," *Chemistry*, vol. 15, no. 5, pp. 1134–1143, 2009.

[63] V. Kratky, M. Kralik, M. Mecarova, M. Stolcova, L. Zalibera, and M. Hronec, "Effect of catalyst and substituents on the hydrogenation of chloronitrobenzenes," *Applied Catalysis A*, vol. 235, no. 1-2, pp. 225–231, 2002.

[64] I. H. Liu, C. Y. Chang, S. C. Liu, I. C. Chang, and S. M. Shih, "Absorption removal of sulfur dioxide by falling water droplets in the presence of inert solid particles," *Atmospheric Environment*, vol. 28, no. 21, pp. 3409–3415, 1994.

[65] Z. B. Yu, M. H. Qiao, H. X. Li, and J. F. Deng, "Preparation of amorphous Ni–Co–B alloys and the effect of cobalt on their hydrogenation activity," *Applied Catalysis A*, vol. 163, no. 1-2, pp. 1–13, 1997.

[66] H. Yamashita, H. Yoshikawa, T. Funabiki, and S. Yoshida, "Catalysis by amorphous metal alloys, part 4: structural modification towards metastable states and catalytic activity of amorphous $NI_{62}B_{38}$ ribbon alloy," *Journal of the Chemical Society, Faraday Transactions I*, vol. 82, no. 6, pp. 1771–1780, 1986.

[67] X. Yan, M. Liu, H. Liu, and K. Y. Liew, "Role of boron species in the hydrogenation of *o*-chloronitrobenzene over polymer-stabilized ruthenium colloidal catalysts," *Journal of Molecular Catalysis A*, vol. 169, no. 1-2, pp. 225–233, 2001.

[68] X. H. Yan, J. Q. Sun, Y. W. Wang, and J. F. Yang, "A Fe-promoted Ni–P amorphous alloy catalyst (Ni–Fe–P) for liquid phase hydrogenation of *m*- and *p*-chloronitrobenzene," *Journal of Molecular Catalysis A*, vol. 252, no. 1-2, pp. 17–22, 2006.

[69] X. H. Yan, J. Q. Sun, Y. H. Xu, and J. F. Yang, "Liquid-phase hydrogenation of chloronitrobenzene to chloroaniline over Ni–Co–B amorphous alloy catalyst," *Chinese Journal of Catalysis*, vol. 27, no. 2, pp. 119–123, 2006.

[70] J. Chen, N. Yao, R. Wang, and J. Zhang, "Hydrogenation of chloronitrobenzene to chloroaniline over $Ni/TiO_2$ catalysts prepared by sol-gel method," *Chemical Engineering Journal*, vol. 148, pp. 164–172, 2009.

[71] B. Coq, A. Tijani, and F. Figuéras, "Influence of alloying platinum for the hydrogenation of *p*-chloronitrobenzene over $PtM/Al_2O_3$ catalysts with M = Sn, Pb, Ge, Al, Zn," *Journal of Molecular Catalysis*, vol. 71, no. 3, pp. 317–333, 1992.

[72] B. Coq, A. Tijani, R. Dutartre, and F. Figueras, "Influence of support and metallic precursor on the hydrogenation of *p*-chloronitrobenzene over supported platinum catalysts," *Journal of Molecular Catalysis*, vol. 79, no. 1-3, pp. 253–264, 1993.

[73] M. G. Wang, H. X. Li, Y. D. Wu, and J. Zhang, "Comparative studies on the catalytic behaviors between the Ni–B amorphous alloy and other Ni-based catalysts during liquid phase hydrogenation of acetonitrile to ethylamine," *Materials Letters*, vol. 57, no. 19, pp. 2954–2964, 2003.

[74] C. A. Tolman, W. H. Riggs, W. J. Lin, C. M. King, and R. C. Wendt, "Electron spectroscopy for chemical analysis of nickel compounds," *Inorganic Chemistry*, vol. 12, no. 12, pp. 2770–2778, 1973.

[75] L. J. Matienzo, L. I. Yin, S. O. Grim, and W. E. Swartz Jr., "X-ray photoelectron spectroscopy of nickel compounds," *Inorganic Chemistry*, vol. 12, no. 12, pp. 2762–2769, 1973.

[76] W. L. Dai, M. H. Qiao, and J. F. Deng, "XPS studies on a novel amorphous Ni–Co–W–B alloy powder," *Applied Surface Science*, vol. 120, no. 1-2, pp. 119–124, 1997.

[77] J. F. Moulder, W. F. Stickle, P. E. Sobol, and K. D. Bomben, *Handbook of X-Ray Photoelectron Spectroscopy*, Physical Electronics, Chanhassen, Minn, USA, 1995.

[78] J. Tamaki, S. Nakayama, M. Yamamura, and T. Imanaka, "Preparation and catalytic properties of palladium-(boron, phosphorus) thin films using an RF sputtering method," *Journal of Materials Science*, vol. 24, no. 5, pp. 1582–1588, 1989.

[79] J. H. Shen and Y. W. Chen, "Catalytic properties of bimetallic NiCoB nanoalloy catalysts for hydrogenation of *p*-chloronitrobenzene," *Journal of Molecular Catalysis A*, vol. 273, no. 1-2, pp. 265–276, 2007.

[80] H. Noller and W. M. Lin, "Activity and selectivity of $Ni = Cu/Al_2O_3$ catalysts for hydrogenation of crotonaldehyde and mechanism of hydrogenation," *Journal of Catalysis*, vol. 85, no. 1, pp. 25–30, 1984.

[81] O. Levenspiel, *Chemical Reaction Engineering*, John Wiley & Sons, New York, NY, USA, 1999.

# Synthesis of Gold Nanoparticles Using Whole Cells of *Geotrichum candidum*

**Amit Kumar Mittal, Abhishek Kaler, Aparna Vasant Mulay, and Uttam Chand Banerjee**

*Department of Pharmaceutical Technology, National Institute of Pharmaceutical Education and Research, Sector-67, SAS Nagar, Punjab 160062, India*

Correspondence should be addressed to Uttam Chand Banerjee; ucbanerjee@niper.ac.in

Academic Editor: Hamed Bahmanpour

The synthesis of nanoparticles with desired size and shape is an important area of research in nanotechnology. Use of biological system is an alternative approach to chemical and physical procedures for the synthesis of metal nanoparticles. An efficient environment-friendly approach for the biosynthesis of rapid and stable Gold nanoparticles (AuNPs) using whole cells of *Geotrichum candidum* is discussed in this paper. The enzymes/proteins present in the microorganism might be responsible for the reduction of metal salts to nanoparticles. Various reaction parameters such as culture age, temperature, pH, metal salt, and cell mass concentrations were optimized. The AuNPs were characterized by UV-visible spectroscopy, dynamic light scattering (DLS), energy dispersive spectroscopy (EDS), scanning electron microscope (SEM), and Fourier transform infrared spectroscopy (FTIR). Nanoparticles were isolated by sonicating the whole cells after treatment with Tween 80. The whole cell mediated process showed the simplistic, feasible, easy to scale up, and low-cost approach for the synthesis of AuNPs.

## 1. Introduction

Metal nanoparticles have been an extensive area of research because of their unique chemical, physical, and optical properties [1]. These make them potential candidates in the field of catalysis, labeling, biosensing, drug delivery, antimicrobial, and so forth [2, 3]. AuNPs have wider ranges of applications in the biomedical field for biosensor development, drug delivery, imaging, photo diagnostics, and so forth [4]. Development of reliable processes for the synthesis of metal nanomaterials with excellent dispersity and stability with minimum harmful effects is the need of the day [5]. Traditional chemical and physical methods reported in the literature involve the use of hazardous chemicals and extreme reaction conditions [6–8]. The current research is directed towards the development of eco-friendly protocols for the synthesis of nanomaterials/nanostructures of desirable sizes and shapes [9, 10]. Considering applications of AuNPs in the fields of biology and medicine, environment, and technology, there is a growing need for the development of cost-effective method for the synthesis of new nanoparticles [11]. The biological methods fulfil all the requirements of a process to be green [12]. Synthesis of nanoparticles utilizing biological system such as bacteria, fungi, and several plant extracts have been widely reported in the literature [6, 13–15]. Microorganisms are able to produce metal nanomaterials either intra- or extracellularly. Intracellular synthesis of metal nanoparticles (*Pseudomonas stutzeri*, *Escherichia coli*, *Vibrio cholerae*, *Pseudomonas aeruginosa*, *Salmonells typlus*, and *Staphylococcus currens*) has been investigated by many workers [16, 17]. The formation of extracellular silver nanoparticles by photoautotrophic cyanobacteria, *Plectonema boryanum*, has been described in [18]. The intracellular synthesis of AuNPs by microbial reduction of $Au^{+3}$ using *Shewanella* algae has been investigated by Konishi et al. in 2006 and Konishi et al. in 2007 [19, 20]. The synthesis of AuNPs of different sizes and shapes has been reported in both aqueous and nonpolar organic solvents [21, 22]. Some well-known examples of plant mediated synthesis of gold nanotriangles using aloe-vera extract and tamarind leaf extract are also available in literature [23, 24]. In this paper, we report the intracellular synthesis of AuNPs using yeast, *Geotrichum candidum*, its downstream processing and characterization.

## 2. Materials and Methods

*2.1. Chemicals.* Gold (III) chloride trihydrate salt (99.99%) was purchased from Sigma-Aldrich (Steinheim, Germany). Different salts, buffer, and media components were purchased from Qualigens Chemicals (Mumbai, India), Hi Media (Mumbai, India), and Central Drug House (P) (New Delhi, India).

*2.2. Microorganism and Culture Conditions. Geotrichum candidum* NCIM 980 was procured from National Collection of Industrial Microorganisms, National Chemical Laboratory, Pune, India. The organism was grown in YPD medium (yeast extract 3 g/L, peptone 3 g/L and dextrose 10 g/L, pH 6.5) at 30°C in an incubator shaker (200 rpm).

*2.3. Synthesis of Gold Nanoparticle and Its Downstream Processing.* The cells were grown in specified medium for 24 h at 30°C in shaker incubator (200 rpm). The fermentation broth was centrifuged at 7,000 ×g for 15 min; cells were washed thoroughly and resuspended in deionised water. Suspension of wet cell mass (10%) in deionised water was prepared. Tetrachloroauric acid trihydrate ($AuCl_4{\cdot}3H_2O$) was added to the cell suspension (10%) to final concentration of 1 mM. The reaction was carried out at 30°C, 200 rpm for 72 h and observed for visual colour change. For the isolation of nanoparticles from the whole cells, various disruption methods were used. Cell disruption was performed using ultrasonication after treatment with surfactant (Tween 80). The supernatant was collected after centrifugation at 10,000 ×g for 15 min and lyophilized to concentrate the AuNPs.

*2.4. Characterization of Nanoparticles.* An essential part in the synthesis of nanoparticles is its characterization. The synthesised AuNPs were characterized by UV-visible spectrometry, dynamic light scattering (DLS), scanning electron microscopy (SEM), Fourier transform infrared spectroscopy (FTIR), and energy dispersive spectroscopy (EDS). Observation of strong broad surface plasmon peaks at visible regions (510–525 nm) has been well documented for AuNPs, with sizes ranging widely from 10 to 100 nm. The reduction of pure $Au^{+3}$ ions to $Au^{o}$ was monitored by measuring through UV-Vis spectroscopy by taking the samples of reaction mixture (gold nanoparticle solution). Dynamic light scattering (DLS or Zeta-sizer) was used to measure the size, size distribution, and potential (charge) of nanoparticles dispersed in a liquid. Scanning electron microscopic (SEM) analysis was done using the instrument (S-3400, Hitachi) operated at an excitation voltage of 25 kV. The nanoparticles or whole cell suspensions were mounted onto steel stage using double-sided adhesive tape on a glass cover slip by just dropping a very small amount of the sample on the cover slip, and the film was allowed to dry and coated with gold using ion sputter (E-1010, Hitachi). The SEM was used for the determination of size and morphology of the nanoparticles. The functional groups and capping of the synthesized AuNPs were confirmed by FTIR spectroscopy. Elemental analysis and chemical composition of NPs were determined using energy dispersive spectroscopy.

*2.5. Optimization of Various Reaction Parameters*

*2.5.1. Culture Age and Its Concentration.* Cell age has pronounced effect on the synthesis of nanoparticles. In order to determine the optimum culture cell age for nanoparticle production, the cells were harvested at different phases of growth. After harvesting, the cells were thoroughly washed, resuspended in deionised water, and tetrachloroauric acid trihydrate ($AuCl_4{\cdot}3H_2O$) was added to make final concentration of 1 mM, and reaction mixture was incubated at 30°C (200 rpm). After 48 h, cells were harvested by centrifugation at 7,000 ×g for 15 min. The effect of cell mass concentration was studied by incubating different concentrations of cells (25–250 mg/mL) with gold salt at 35°C (200 rpm) for 72 h. From time to time, the samples were taken from the reaction mixture, and nanoparticles were characterized according to the set procedures.

*2.5.2. Reaction Temperature.* Tetrachloroauric acid trihydrate ($AuCl_4{\cdot}3H_2O$) was added in 1 mM concentration to the resuspended whole cells, and the flasks were incubated at a range of temperature of 15 to 40°C (200 rpm). Samples were collected at a regular interval of time and characterized accordingly.

*2.5.3. Reaction pH.* To the cell suspension, tetrachloroauric acid trihydrate ($AuCl_4{\cdot}3H_2O$) was added to achieve a final concentration of 1 mM, and the flasks were incubated at 35°C (200 rpm). A range of pH (3–13) of the reaction mixture was chosen. Samples were collected at a regular interval of time and characterized accordingly.

*2.5.4. Metal Salts.* To the whole cell suspension, different concentrations (1 to 10 mM) of $AuCl_4{\cdot}3H_2O$ were added, and the flasks were incubated at 35°C (200 rpm). Samples were collected at a regular interval of time and characterized accordingly.

## 3. Results and Discussion

The shape, size, and size distribution of metal nanoparticles produced by the reduction of gold ion ($Au^{3+}$) by *Geotrichum candidum* in solution depend on various reaction conditions such as temperature, cell age, cell mass pH, and concentration of metal salt. After downstream processing, the nanoparticles were characterized by various techniques.

*3.1. Characterization of AuNPs*

*3.1.1. Visual Observation.* It has been reported [25] in the literature that the formation of AuNPs is detected by observing

FIGURE 1: Visual observations of AuNPs produced by whole cell *Geotrichum candidum*.

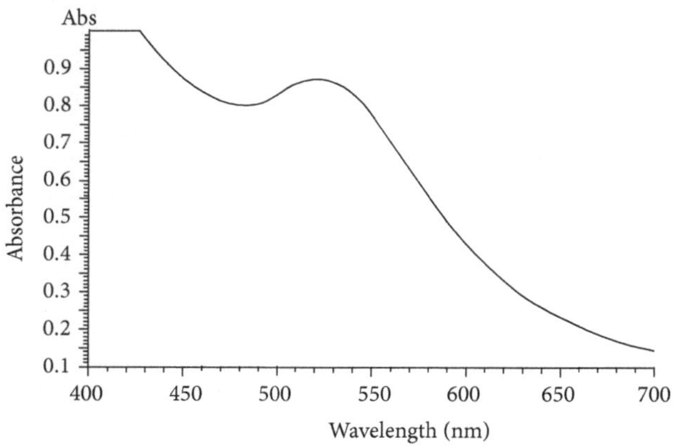


FIGURE 2: UV-Vis spectra of AuNPs produced by whole cell *Geotrichum candidum*.

the colour change of reaction mixture. In this case, also the formation of nanoparticle was preliminarily identified by observing the colour change of the reaction mixture (Figure 1). The cell mass of *G. candidum* turned light yellow (Figure 1(a)) to purple colour (Figure 1(b)) indicating the synthesis of AuNPs. The synthesis might be either intracellular or on the cell surface.

*3.1.2. UV-Vis Spectral Analysis.* As previously reported by other workers [26] on the colloidal solution of gold nanoparticle when subjected to UV-Vis spectrometry, a strong peak was absorbed at 520–530 nm (Figure 2).

*3.1.3. Control of Size and Shape of AuNPs.* To achieve the better size and shape of AuNPs synthesised by the cells of *Geotrichum candidum*, parameters such as cell age, pH, temperature, metal ion, and cell mass concentration were optimized. Metabolic activities of cell vary with the cell age. The cellmass harvested at 48 h of growth produced highest amount of AuNPs (Figure 3(a)). This may be due to expression of higher amount of reductant at that the time of harvest (48 h growth) resulting in higher reduction. With the increase in temperature, reaction rate increased up to a certain value and then it started decreasing. Temperature of 35°C was found to be optimum for nanoparticle synthesis, beyond which the absorption at 520 nm decreased (Figure 3(b)). This may be due to instability of reductive compounds (protein/peptide) at higher temperatures. Gericke and Pinches [27] also found similar effect for the Au-nanoparticle synthesis. Nanoparticle synthesis is a crystallisation phenomenon that depends upon solubility of salt in the media. pH affects the solubility and thus can affect the nanoparticle formation. The optimum pH for synthesis of gold nanoparticle was found to be 7 where the maximum absorption was observed (Figure 3(c)). Both above and below pH 7, aggregation took place. Higher concentration of auric chloride was found to be toxic to the microbial cell. Lower concentration of auric chloride (1 mM) was very efficiently reduced by the cells of *G. candidum* (Figure 3(d)). Cellmass concentration also showed prominent effect on the nanoparticle formation. Keeping all the other parameters constant, the cellmass concentration was varied from 25 to 250 mg/mL, and 50 mg/mL was found to give maximum yield of AuNPs (Figure 3(e)).

*3.1.4. Zetasizer.* The synthesized nanoparticles have average particle size of about 76.58 nm and PDI of 0.231 (Figure 4(a)) which indicates monodispersity, and Zeta potential of −17.5 mV (Figure 4(b)) shows that the synthesized AuNPs are capped by -vely charged groups and are moderately stable.

*3.1.5. Scanning Electron Microscopy (SEM) Analysis.* A representative SEM image of the synthesized AuNPs is shown in Figure 5. This SEM image shows that the synthesized nanoparticles lay on the surface of *Geotrichum candidum* (Figure 5(a)). The average nanoparticle size lies in the range of 65–70 nm. Figure 5(b) shows the harvested AuNPs isolated from the cellmass of *G. candidum*.

*3.1.6. Energy Dispersive Spectroscopy (EDS).* An area profile EDS spectra of AuNPs (Figure 6) synthesized by whole cells of *G. candidum* after purification showed the presence of strong signals of the gold atoms.

*3.1.7. FTIR Spectroscopy.* FTIR spectroscopy analysis was done to reveal the involvement of possible biomolecules affecting the reduction and subsequent capping of the metal nanoparticles [28] synthesized by *G. candidum*. The proteins can bind to nanoparticles through its various functional groups [29]. The FTIR spectra revealed the presence of different functional groups such as amide and -COOH linkages on amino acid residues in protein and synthesized AuNPs. The bands were observed at 1299 $cm^{-1}$, 3433.14 $cm^{-1}$, 2923.27 $cm^{-1}$, and 1735.66 $cm^{-1}$ (Figure 7), acknowledged as amide III band, –OH stretching, C–H stretching, and C–O stretching of carboxylic acid, respectively [30]. The band at 1647 $cm^{-1}$ represents C=O stretching of amide bond. The band at 1351.71 $cm^{-1}$ represents C–N stretching which is commonly present in proteins, indicating the presence of protein as ligand for AuNPs, which increases the stability

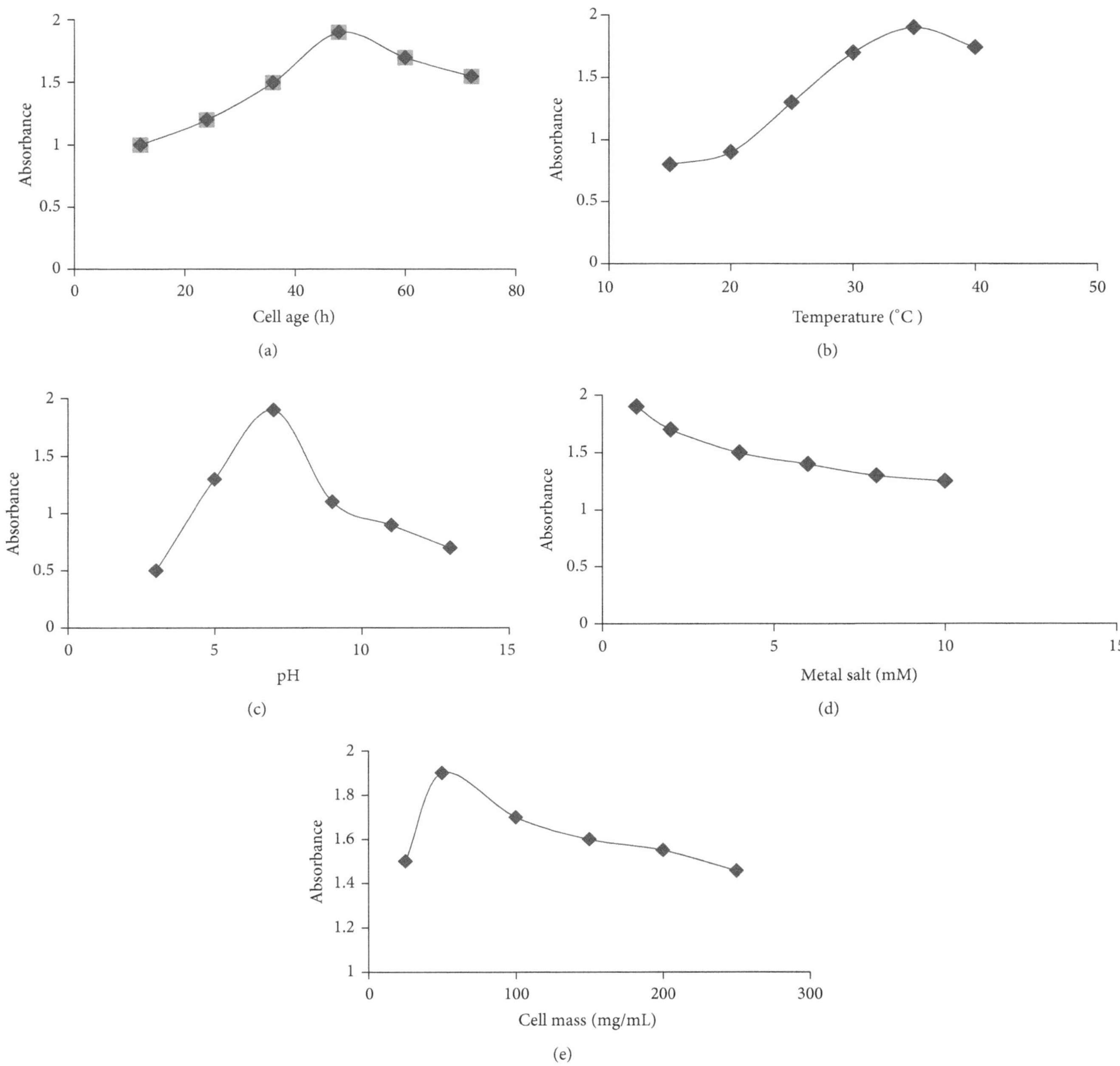


FIGURE 3: Optimization of various physicochemical parameters for AuNPs synthesis by whole cells of *Geotrichum candidum.*

of synthesized nanoparticles [31]. All peaks of FTIR spectra showed the presence of protein attachment to the AuNPs.

## 4. Conclusion

*Geotrichum candidum* was reported for the first time to synthesize AuNPs. Various reaction parameters like cell age, temperature, pH, cell mass, and metal ion concentration, were optimized to increase the yield and to improve the dispersity of nanoparticles. Proteins are responsible for the reduction of metal salts as well as its stabilization by capping. The gold nanoparticle throughout the cellmass suggests that $Au^{+3}$ ions entered the cells through a transport system. The presence of different metabolic enzymes inside the cells or cytoplasm probably reduced $Au^{+3}$ to $Au^{o}$ and capped it by protective peptide/proteins. The microorganism mediated synthesis of metal nanoparticles can be a viable alternative to physical and chemical methods and can be easily scaled up.

## Conflict of Interests

All the authors have no conflict of interests relevant to this paper.

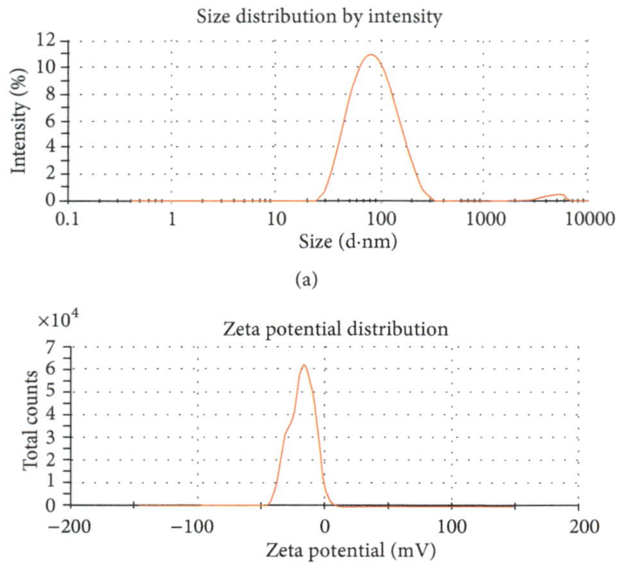


FIGURE 4: (a) Zetasizer distribution and (b) Zeta potential of AuNPs synthesized by whole cell of *Geotrichum candidum*.

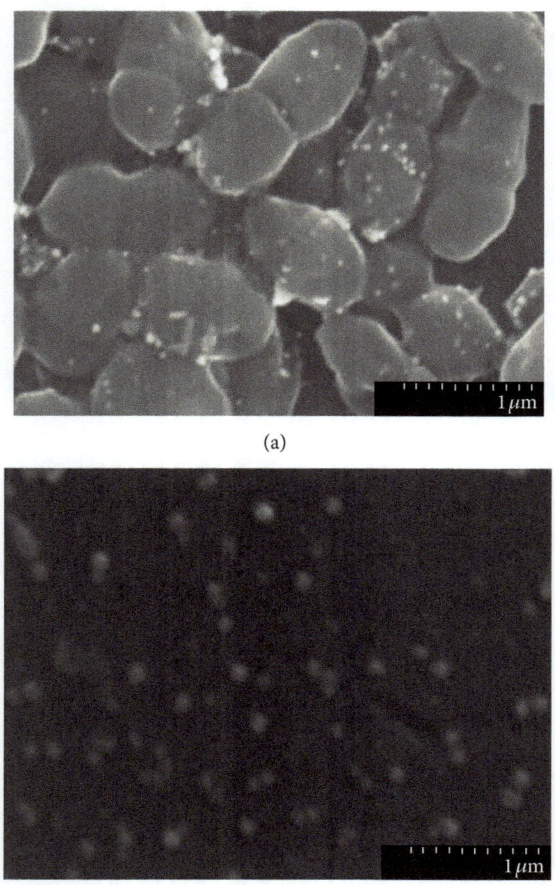


FIGURE 5: SEM images of AuNPs synthesized by *Geotrichum candidum* (a) in the whole cell and (b) after downstream processing.

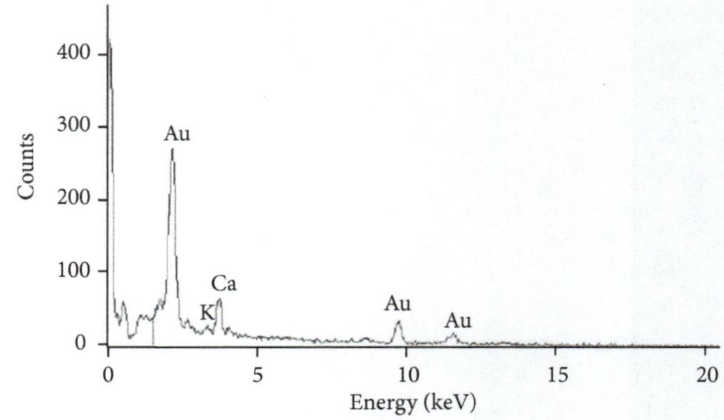


FIGURE 6: EDS spectra of AuNPs synthesized by whole cells *G. candidum*.

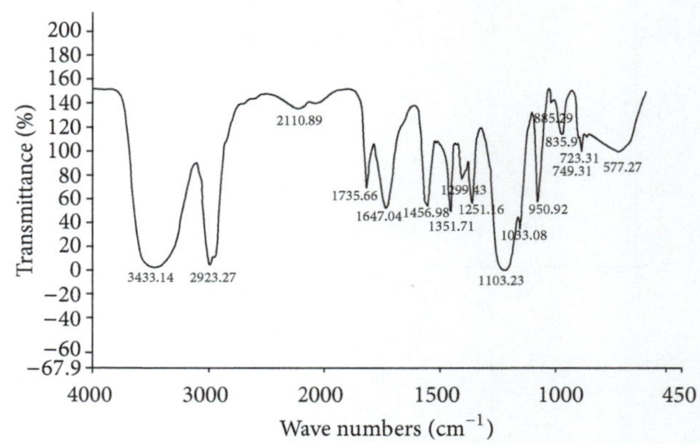


FIGURE 7: FTIR spectra of AuNPs synthesized by whole cell *Geotrichum candidum* NCIM 980 after harvesting.

## Acknowledgments

One of the authors (A. K. Mittal) gratefully acknowledges financial support from the CSIR, and both A. Kaler and A. V. Mulay gratefully acknowledge the financial support from DBT, New Delhi, India.

## References

[1] P. K. Jain, X. Huang, I. H. El-Sayed, and M. A. El-Sayed, "Noble metals on the nanoscale: optical and photothermal properties and some applications in imaging, sensing, biology, and medicine," *Accounts of Chemical Research*, vol. 41, no. 12, pp. 1578–1586, 2008.

[2] F. S. Rosarin and S. Mirunalini, "Nobel metallic nanoparticles with novel biomedical properties," *Journal of Bioanalysis & Biomedicine*, vol. 3, pp. 085–091, 2011.

[3] M. Solomon and G. G. M. D'Souza, "Recent progress in the therapeutic applications of nanotechnology," *Current Opinion in Pediatrics*, vol. 23, no. 2, pp. 215–220, 2011.

[4] E. Boisselier and D. Astruc, "Gold nanoparticles in nanomedicine: preparations, imaging, diagnostics, therapies and toxicity," *Chemical Society Reviews*, vol. 38, no. 6, pp. 1759–1782, 2009.

[5] J. A. Dahl, B. L. S. Maddux, and J. E. Hutchison, "Toward greener nanosynthesis," *Chemical Reviews*, vol. 107, no. 6, pp. 2228–2269, 2007.

[6] K. N. Thakkar, S. S. Mhatre, and R. Y. Parikh, "Biological synthesis of metallic nanoparticles," *Nanomedicine*, vol. 6, no. 2, pp. 257–262, 2010.

[7] R. Sanghi and P. Verma, "15 Microbes as green and eco-friendly nanofactories," in *Green Chemistry for Environmental Sustainability*, S. K. Sharma, Ed., p. 315, CRC Press, 2010.

[8] H. Korbekandi, S. Iravani, and S. Abbasi, "Production of nanoparticles using organisms Production of nanoparticles using organisms," *Critical Reviews in Biotechnology*, vol. 29, no. 4, pp. 279–306, 2009.

[9] A. Ahmad, P. Mukherjee, S. Senapati et al., "Extracellular biosynthesis of silver nanoparticles using the fungus *Fusarium oxysporum*," *Colloids and Surfaces B*, vol. 28, no. 4, pp. 313–318, 2003.

[10] A. K. Gade, P. Bonde, A. P. Ingle, P. D. Marcato, N. Durán, and M. K. Rai, "Exploitation of Aspergillus niger for synthesis of silver nanoparticles," *Journal of Biobased Materials and Bioenergy*, vol. 2, no. 3, pp. 243–247, 2008.

[11] L. Dykman and K. Nikolai, "AuNPs in biomedical applications: recent advances and perspectives," *Chemical Society Reviews*, vol. 41, no. 6, pp. 2256–2282, 2012.

[12] S. Marimuthu, A. A. Rahuman, G. Rajakumar et al., "Evaluation of green synthesized silver nanoparticles against parasites," *Parasitology Research*, vol. 108, no. 6, pp. 1541–1549, 2011.

[13] A. K. Mittal, Y. Chisti, and U. C. Banerjee, "Synthesis of metallic nanoparticles using plant extracts," *Biotechnology Advances*, vol. 31, no. 2, pp. 346–356, 2013.

[14] A. K. Mittal, A. Kaler, and U. C. Banerjee, "Free radical scavenging and antioxidant activity of silver nanoparticles synthesized from flower extract of Rhododendron dauricum," *Nano Biomedicine and Engineering*, vol. 4, no. 3, pp. 118–124, 2012.

[15] K. B. Narayanan and N. Sakthivel, "Biological synthesis of metal nanoparticles by microbes," *Advances in Colloid and Interface Science*, vol. 156, no. 1-2, pp. 1–13, 2010.

[16] Z. Sadowski, "Biosynthesis and application of silver and AuNPs," *Wroclaw University of Technology*, p. 257, 2009.

[17] A. Kaler, R. Nankar, M. S. Bhattacharyya, and U. C. Banerjee, "Extracellular biosynthesis of silver nanoparticles using aqueous extract of *Candida viswanathii*," *Journal of Bionanoscience*, vol. 5, no. 1, pp. 53–58, 2011.

[18] M. F. Lengke, M. E. Fleet, and G. Southam, "Biosynthesis of silver nanoparticles by filamentous cyanobacteria from a silver(I) nitrate complex," *Langmuir*, vol. 23, no. 5, pp. 2694–2699, 2007.

[19] Y. Konishi, T. Tsukiyama, K. Ohno, N. Saitoh, T. Nomura, and S. Nagamine, "Intracellular recovery of gold by microbial reduction of $AuCl^4$- ions using the anaerobic bacterium *Shewanella algae*," *Hydrometallurgy*, vol. 81, no. 1, pp. 24–29, 2006.

[20] Y. Konishi, K. Ohno, N. Saitoh et al., "Bioreductive deposition of platinum nanoparticles on the bacterium *Shewanella algae*," *Journal of Biotechnology*, vol. 128, no. 3, pp. 648–653, 2007.

[21] P. R. Selvakannan, S. Mandal, R. Pasricha, S. D. Adyanthaya, and M. Sastry, "One-step synthesis of hydrophobized gold nanoparticles of controllable size by the reduction of aqueous chloroaurate ions by hexadecylaniline at the liquid-liquid interface," *Chemical Communications*, no. 13, pp. 1334–1335, 2002.

[22] T. K. Sau and C. J. Murphy, "Room temperature, high-yield synthesis of multiple shapes of gold nanoparticles in aqueous solution," *Journal of the American Chemical Society*, vol. 126, no. 28, pp. 8648–8649, 2004.

[23] S. P. Chandran, M. Chaudhary, R. Pasricha, A. Ahmad, and M. Sastry, "Synthesis of gold nanotriangles and silver nanoparticles using Aloe vera plant extract," *Biotechnology Progress*, vol. 22, no. 2, pp. 577–583, 2006.

[24] B. Ankamwar, M. Chaudhary, and M. Sastry, "Gold nanotriangles biologically synthesized using tamarind leaf extract and potential application in vapor sensing," *Synthesis and Reactivity in Inorganic, Metal-Organic and Nano-Metal Chemistry*, vol. 35, no. 1, pp. 19–26, 2005.

[25] P. Mukherjee, S. Senapati, D. Mandal et al., "Extracellular synthesis of AuNPs by the fungus *Fusarium oxysporum*," *ChemBioChem*, vol. 3, no. 5, pp. 461–463, 2002.

[26] S. S. Shankar, A. Rai, A. Ahmad, and M. Sastry, "Rapid synthesis of Au, Ag, and bimetallic Au core-Ag shell nanoparticles using Neem (*Azadirachta indica*) leaf broth," *Journal of Colloid and Interface Science*, vol. 275, no. 2, pp. 496–502, 2004.

[27] M. Gericke and A. Pinches, "Biological synthesis of metal nanoparticles," *Hydrometallurgy*, vol. 83, no. 1–4, pp. 132–140, 2006.

[28] S. S. Shankar, A. Ahmad, and M. Sastry, "Geranium leaf assisted biosynthesis of silver nanoparticles," *Biotechnology Progress*, vol. 19, no. 6, pp. 1627–1631, 2003.

[29] C. M. Niemeyer, "Nanoparticles, proteins, and nucleic acids: biotechnology meets materials science," *Angewandte Chemie International Edition*, vol. 40, no. 22, pp. 4129–4158, 2001.

[30] B. D. Chithrani, A. A. Ghazani, and W. C. W. Chan, "Determining the size and shape dependence of gold nanoparticle uptake into mammalian cells," *Nano Letters*, vol. 6, no. 4, pp. 662–668, 2006.

[31] N. Vigneshwaran, N. M. Ashtaputre, P. V. Varadarajan, R. P. Nachane, K. M. Paralikar, and R. H. Balasubramanya, "Biological synthesis of silver nanoparticles using the fungus *Aspergillus flavus*," *Materials Letters*, vol. 61, no. 6, pp. 1413–1418, 2007.

# Rapid Green Synthetic Protocol for Novel Trimetallic Nanoparticles

**B. Karthikeyan and B. Loganathan**

*Department of Chemistry, Annamalai University, Annamalai Nagar, Tamil Nadu 608 002, India*

Correspondence should be addressed to B. Karthikeyan; bkarthi_au@yahoo.com

Academic Editor: Hamed Bahmanpour

Metallic nanocomposites and nanotubes can be rapidly prepared under microwave irradiation (MW) from an aqueous solution of metallic precursors in the presence of trisodium citrate as a reductant. For the nanotubes nanoparticles are stabilized by poly(N-vinyl-2-pyrrolidone) (PVP), a protecting agent. PVP is a typical capping and structure-directing agent used for the synthesis of various metallic nanostructures. In this work, we have demonstrated for the first time an MW irradiation approach for the synthesis of trimetallic nanocomposites and nanotubes. The resulting nanohybrids were characterized by UV-Vis spectroscopic analysis, high-resolution scanning electron microscopy (HR-SEM), energy dispersive X-ray spectroscopic analysis (EDX), and X-ray diffractometer (XRD) techniques.

## 1. Introduction

Over the past decade there has been an increased emphasis on the topic of green chemistry and chemical processes [1]. Colloidal metallic nanoparticles are emerging as an important family of multifunctional nanoscale materials that assist diverse applications in various fields [2–7]. For these essential applications, a variety of preparation routes have been reported for the syntheses of metallic nanosized particles. In order to increase the efficiency of Pt-based bimetallic and trimetallic nanoparticles catalysts have been prepared [8, 9]. Because, in usual metal catalysts, addition of other element can often improve the catalytic activity and selectivity, from the same viewpoint bimetallic and trimetallic nanoparticles are also often investigated. Trimetallic nanoparticles retain an ever greater degree of catalytic activity than the bimetallic and monometallic ones. However, there have been only few reports of trimetallic nanoparticles in the literature. The reported trimetallic nanoparticles were synthesized by only thermal heating method [10–14].

On the other hand, microwave (MW) dielectric heating is fast emerging as a widely accepted rapid processing technology for a variety of synthesis of nanoparticles. In the MW-assisted syntheses of nanoparticles, we found that MW heating is a rapid technology and it can control the size distribution of the nanoparticles than the conventional heating by thermal convection [15].

In the present work on the preparation of nanosized trimetallic nanocomposites and trimetallic nanotubes from an aqueous solution of corresponding metallic precursors, we have employed trisodium citrate as a reducing agent for the preparation of different combinations of trimetallic nanoparticles and poly(N-vinyl-2-pyrrolidone) (PVP) as a stabilizing agent. Using MW irradiation as a heating source, the rapid, homogeneous, and selective heating in the reaction system at the molecular level for rapid and size-controlled preparation of nanosized particles, which has never been achieved by other conventional methods is reported for the first time.

## 2. Materials and Methods

*2.1. Materials.* Hydrogen tetrachloroaurate (III) trihydrate ($HAuCl_4{\cdot}3H_2O$), hydrated hexachloroplatinic (IV) acid ($H_2PtCl_6{\cdot}6H_2O$), silver nitrate ($AgNO_3$), Palladium chloride ($PdCl_2$) trisodium citrate ($C_6H_5Na_3O_7{\cdot}2H_2O$), poly(N-vinyl-2-pyrrolidone) (PVP; $K = 30$, $M_W = 40000$), and

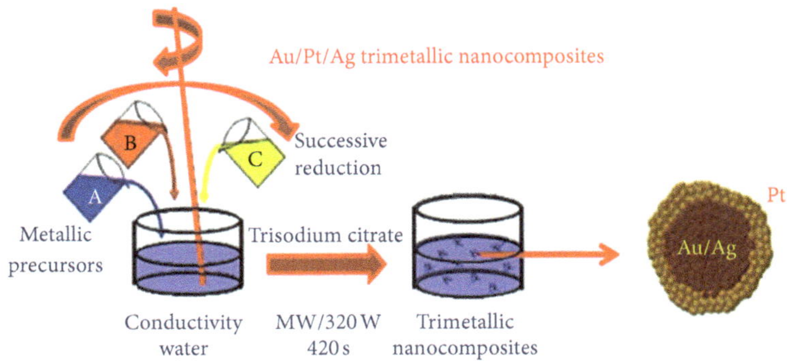


FIGURE 1: MW irradiation method for the preparation of Au-Pt-Ag trimetallic nanoparticles.

all other chemicals used were of analytical grade and the chemicals were purchased from HiMedia Laboratories Pvt. Ltd. (Mumbai, India). All chemicals were used as received without further purification. Conductivity water was used in all experiments.

*2.2. Synthesis of Trimetallic Nanoparticles.* The Au-Pt-Ag trimetallic nanocomposites and Au-Pd-Pt trimetallic nanotubes are prepared by MW irradiation method. This procedure was carried out in an MW oven (LG Grill, Intellowave, 160–800 W, consumption 800 W, output power 320 W, and frequency 2450 MHz) operating in a cyclic mode (ON 15 s, OFF 5 s) to prevent intense boiling of sol as well as aggregation. As synthesized trimetallic nanoparticles were sonicated in a Fast Clean-Ultrasonicator.

*2.2.1. Synthesis of Au-Pt-Ag Trimetallic Nanocomposites.* Citrate-stabilized Au nanoparticles were prepared according to the reported method by us [16]. Briefly, 10 mL of aqueous 0.1% $HAuCl_4 \cdot 3H_2O$ was heated to a boiling and 2 mL of 1% trisodium citrate was then added with stirring. The reaction mixture was heated for 4 min (240 sec) and cooled to room temperature. The solution was turned to vivid magenta from slight yellow, indicating the formation of Au nanoparticles. 10 mL of 0.1% $H_2PtCl_6 \cdot 6H_2O$ was then added to the Au nanoparticles followed by the addition of 2 mL 1% trisodium citrate with stirring. Finally 10 mL of 0.1% $AgNO_3$ was added into the Au/Pt nanoparticles. The overall reaction time for the preparation of trimetallic nanocomposites are 420 sec. Then the prepared sol was sonicated for 1 hr.

*2.2.2. Synthesis of Au-Pd-Pt Trimetallic Nanotubes.* In atypical synthesis of Pd containing trimetallic nanotubes, 45 mL of aqueous 1.20 mM $HAuCl_4 \cdot 3H_2O$ was heated to a boil and 2 mL of 1% trisodium citrate was then added with stirring. The reaction mixture was heated for 4 min (240 s) and cooled to room temperature. The solution was turned to vivid magenta from slight yellow, indicating the formation of Au nanoparticles. 3 mL of 1% PVP was then added to the solution and stirred at room temperature overnight. The prepared sol was sonicated for 40 min. Deaerated aqueous $PdCl_2$ (2.0 mL of 35 mM) was added to PVP-stabilized Au nanoparticles, with stirring under nitrogen atmosphere (3 min). After 15 min 2.0 mL of a 1% trisodium citrate solution was again added dropwise to the solution with constant stirring. The vivid magenta coloured solution turned brown indicating the formation of nano-Pd on the PVP-stabilized Au nanoparticles. The stirring was continued for an additional 3 hr. Then 2.0 mL of 20 mM $H_2PtCl_6 \cdot 6H_2O$ was added and the solution was stirred for another 2 hr at room temperature. Finally the prepared Au-Pd-Pt sol was sonicated for 1 hr to get the fine dispersion of nanoparticles.

*2.3. Characterization of Nanoparticles.* UV-vis (ultraviolet and visible light) absorbance spectra were measured over a range of 200–800 nm with a Shimadzu UV-1650PC recording spectrometer using a quartz cell with 10 mm of optical path length. High-resolution scanning electron microscopy (HR-SEM), elementary dispersive X-ray analysis (EDX) experiments were carried out on a FEI Quanta FEG 200 instrument with EDX analysis facility at 25°C. Wide-angle powder X-ray diffraction (XRD) pattern was recorded with an XPERT-PRO diffractometer equipped with monochromatic Cu K$\alpha$ radiation by using a step scan programme.

## 3. Results and Discussion

The MW irradiation method for the preparation of Au-Pt-Ag trimetallic nanocomposites and Au-Pd-Pt trimetallic nanotubes is shown in Figures 1 and 2. From the viewpoint of sequential electron transfer [17], we chose the combination of the trimetals. The ionization potentials of Au, Pt, Ag, and Pd are known to be 9.225, 8.620, 7.576, and 8.340 eV, respectively. Ionization energy of the corresponding bulk metals is also shown for the comparison in Figure 3 [18].

*3.1. UV-Vis Spectral Studies*

*3.1.1. Au-Pt-Ag Trimetallic Nanocomposites.* The absorbance spectrum of the sols containing various nanoparticles for the comparison has been given in Figure 4. The absorbance studies have been carried out after ensuring complete reduction of the metal ions based on the kinetics of reduction of individual metal ions. Absorbance spectra corresponding to the individual Au, Pt, and Ag nanoparticles are also given

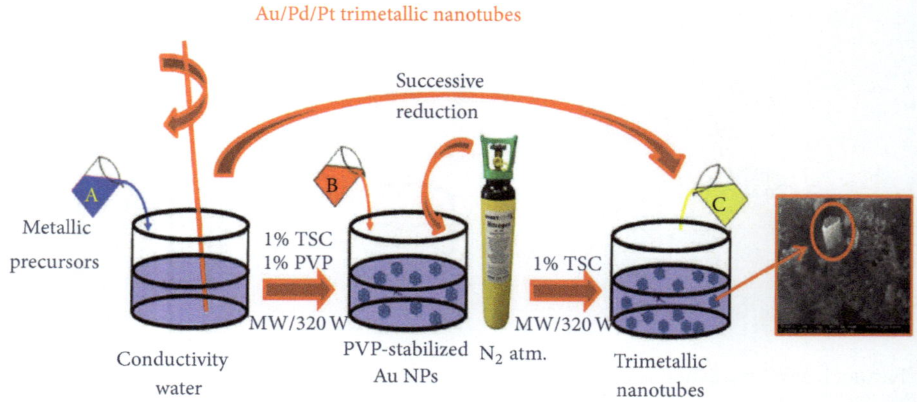


FIGURE 2: MW irradiation method for the preparation of Au-Pd-Pt trimetallic nanotubes.

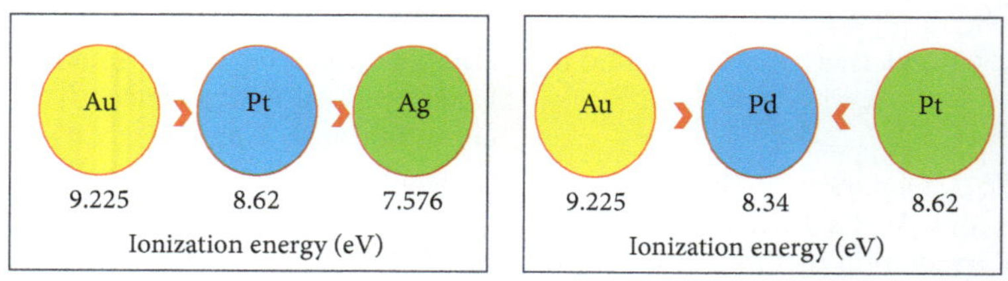


FIGURE 3: Ionization energy of the corresponding bulk metals.

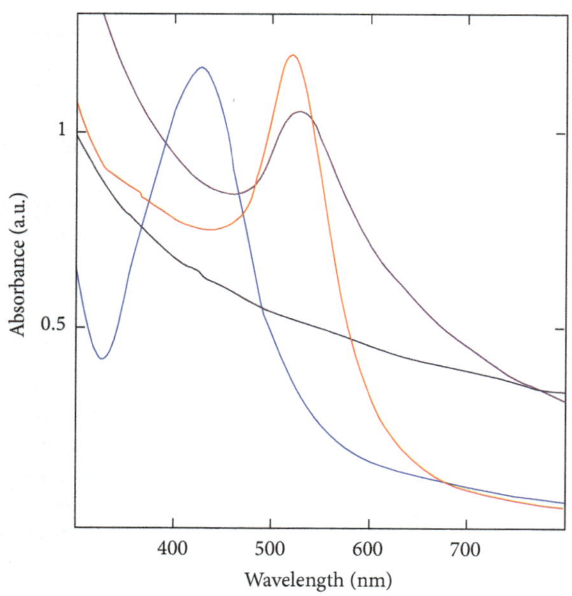


FIGURE 4: UV-vis spectrum of the colloidal dispersion of polymer protected Au nanoparticles.

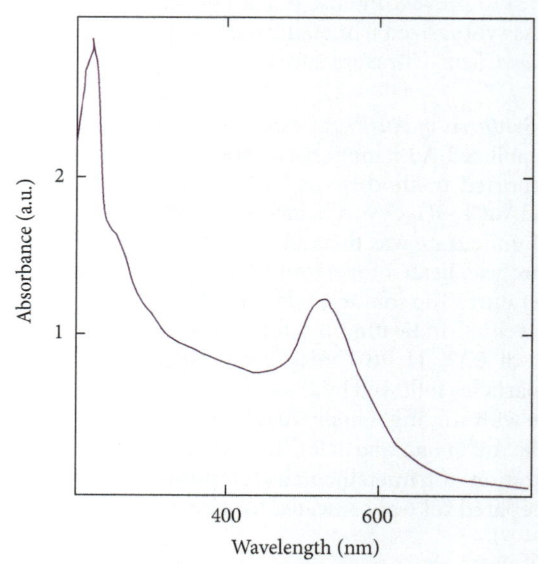


FIGURE 5: UV-vis spectrum of the colloidal dispersion of polymer protected Au nanoparticles.

for comparison. The absorption peak at 530 nm and 430 nm can be attributed to the surface plasmon absorption of Au and Ag nanoparticles (note that Pt colloids do not have an individual visible absorbance). The absence of peaks at 310 nm, characteristic of unreduced Au (III) ion, and peaks at 378, 460 nm indicates the complete reduction of Pt (IV) ion [19]. But, in the case of Au-Pt-Ag trimetallic nanocomposites, a clear single absorbance at 528 nm was attributed to the surface plasmon resonance of trimetallic nanoparticles. The absence of multiple bands in the spectrum also rules out the presence of individual Au, Pt, and Ag nanoparticles in the dispersion [16].

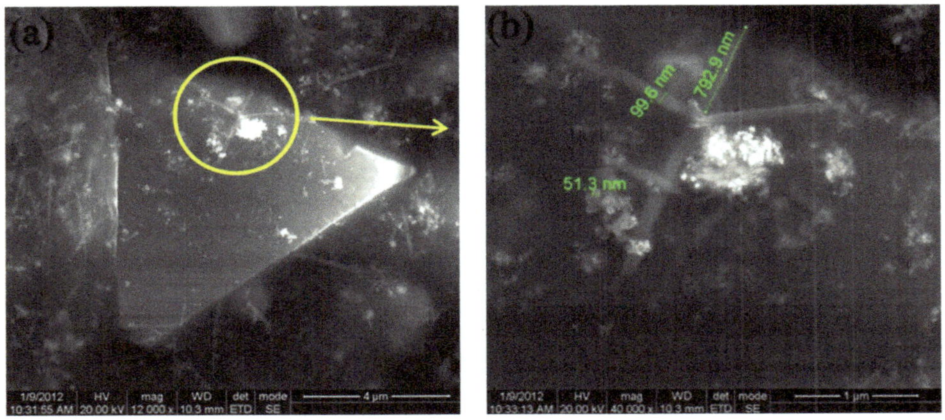


(a) (b)

FIGURE 6: A typical HR-SEM image of Au-Pt-Ag trimetallic nanocomposites. Magnifications at (a) 12000x (20 kV) and (b) 20000x (20 kV). Scale bars (a) 4 $\mu$m and (b) 1 $\mu$m.

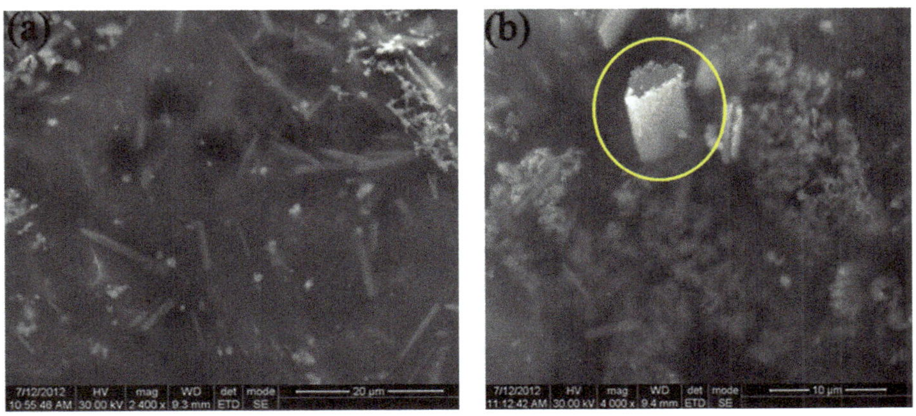


(a) (b)

FIGURE 7: A typical HR-SEM image of Au-Pd-Pt trimetallic nanotubes. Magnifications at (a) 2400x (30 kV) and (b) 7000x (30 kV). Scale bars (a) 20 $\mu$m and (b) 10 $\mu$m.

*3.1.2. Au-Pd-Pt Trimetallic Nanotubes.* Figure 5 shows the UV-Vis spectrum of the colloidal dispersion of polymer protected Au nanoparticles (note that Pd and Pt colloids do not have an individual visible absorbance). A clear surface plasmon absorbance peak which appeared at 528 nm is corresponding to the presence of Au nanoparticles and the vivid magenta in colour sol also indicating the presence of Au nanoparticles.

### *3.2. High-Resolution Scanning Electron Microscopic (HR-SEM) Studies*

*3.2.1. Au-Pt-Ag Trimetallic Nanocomposites.* HR-SEM images indicate that the particles are nearly nanowires (Au) and nanocomposite (Ag-Pt) structure, the length of the nanowire is 792 nm and the width is 99 and 51 nm (Figure 6(b)). From Figure 6(a), it is inferred that the nanocomposites are not well dispersed as individual nanoparticles and showed Au nanowires and Ag-Pt bimetallic nanoparticles. The length and width of the nanoparticles (Figure 6(b)) are indicated on the shape of the nanoparticles.

*3.2.2. Au-Pd-Pt Trimetallic Nanotubes.* Figure 7 shows HR-SEM images of as-synthesized Au-Pd-Pt trimetallic nanotubes. HR-SEM investigation clearly reveals that the nanoparticles were uniformly having nanotube shape. The length of the nanotube is 141 nm (Figure 7(b)) and width of that one is 152 nm.

*3.3. Energy Dispersive X-Ray Microanalysis (EDX).* One of the most revealing analytical methods for the composition of trimetallic nanoparticle is energy dispersive X-ray spectroscopy (EDX). Figures 8 and 9 show the EDX elemental analysis of as-synthesized Au-Pt-Ag trimetallic nanocomposites and Au-Pd-Pt trimetallic nanotubes. All the K and L emission peaks for trimetals are observed. The carbon and copper peaks are due to the HR-SEM holding grid. No other obvious peak attributable to impurity is detected.

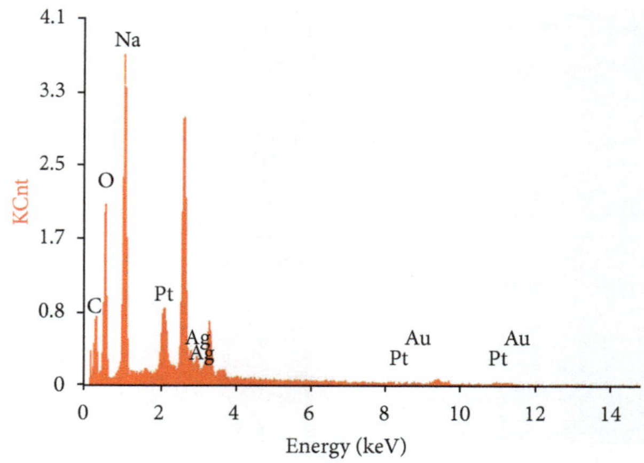


| Element | Wt% | At% |
|---|---|---|
| Au | 11.39 | 02.28 |
| Pt | 26.86 | 05.44 |
| Ag | 10.29 | 03.75 |

FIGURE 8: EDX elemental data of Au-Pt-Ag trimetallic nanocomposites.

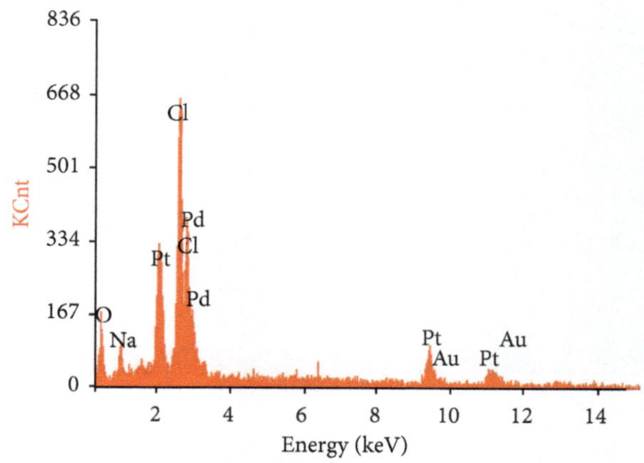


| Element | Wt% | At% |
|---|---|---|
| Au | 06.25 | 01.39 |
| Pd | 27.01 | 11.1 |
| Pt | 25.18 | 05.65 |

FIGURE 9: EDX elemental data of Au-Pd-Pt trimetallic nanotubes.

This result indicates that as-synthesized product is composed of high purity trimetallic nanoparticles. The carbon and oxygen signals were most likely due to the reducing agents on the surface of the prepared nanoparticles. EDX analysis of Au-Pt-Ag trimetallic nanocomposites gave weight of 11.39% Au, 26.86% Pt, and 10.23% Ag in the sample. It indicated the formation of trimetallic nanocomposites in the ratio of 1 : 3 : 1. For Au-Pd-Pt trimetallic nanotubes EDX analysis gave 6.25% Au, 25.18% Pd, and 27.01% Pt. It indicated the formation of trimetallic nanotubes in the ratio of 1 : 4 : 4. From the above results, the maximum percentage of Pt nanoparticles is present on the surface of Au-Pt-Ag trimetallic nanocomposites and Pd and Pt nanoparticles are present on the surface of as-synthesised Au-Pd-Pt trimetallic nanotubes.

*3.4. XRD Studies.* The diffractograms of the trimetallic combinations generally show broad bands while their monometallic counterparts exhibit fairly sharp bands [19]. The particle sizes of the trimetallic combinations determined from the XRD spectra correlate well with the sizes obtained from HR-SEM measurements. The particle size is calculated based on the Scherrer's equation:

$$D = \frac{K\lambda}{\beta s \cos\theta}, \quad (1)$$

where $D$ corresponds to the particle size, $K$ is the shape-dependent Scherrer's constant, $\lambda$ is the wavelength of radiation, $\beta s$ is the full peak width at half-maximum (FWHM) of the peak, and $\theta$ is the Bragg diffraction angle [20].

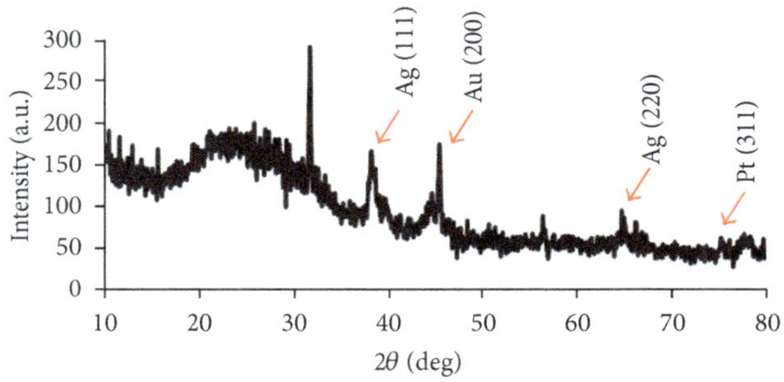


FIGURE 10: Wide-angle XRD pattern of Au-Pt-Ag trimetallic nanocomposites.

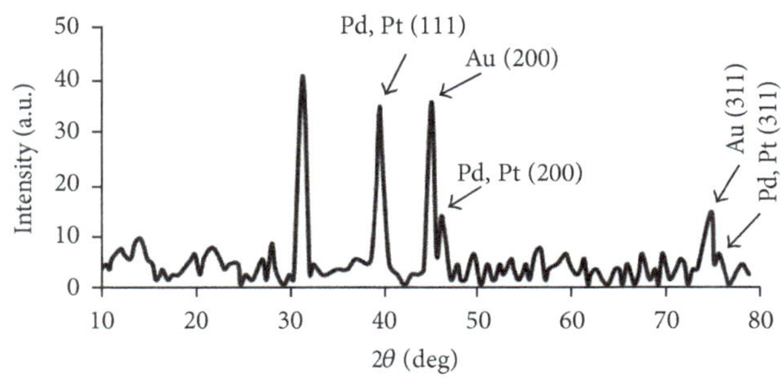


FIGURE 11: Wide-angle XRD pattern of Au-Pd-Pt trimetallic nanotubes.

*3.4.1. Au-Pt-Ag Trimetallic Nanocomposites.* Figure 10 shows the XRD of Au-Pt-Ag trimetallic nanocomposites. The diffraction patterns indicate a broad diffraction peak ($2\theta$ about 45.4°) corresponding to (2 0 0) reflection. This suggested that the prepared particles have a very small size and a face-centered cubic (fcc) structure. The XRD also indicated the reflection peak that appears at 38.12° ($2\theta$) corresponds to the (1 1 1) crystallographic plane and the peak at 64.83° ($2\theta$) corresponds to the (2 2 0) plane. The diffractograms show that (3 1 1) crystallographic surface is marked for Pt ($2\theta = 75.29°$). The above crystalline data are verified by JCPDS card number of Au, Pt, and Ag. Table 1 shows the crystalline sizes of nanoparticles and identification of nanoparticles present in trimetallic nanocomposites.

*3.4.2. Au-Pd-Pt Trimetallic Nanotubes.* Figure 11 shows the XRD pattern of Au-Pd-Pt trimetallic nanotubes. The wide-angle XRD pattern showed randomly oriented fcc crystals. A diffraction peak which appeared at $2\theta = 45.17°$ and 75.02° corresponds to Au (2 0 0) and Au (3 1 1) planes. The XRD also indicated that the reflection peaks that appeared at $2\theta =$ 39.62°, 46.68°, and 75.32° ascribed to Pd/Pt (1 1 1), Pd/Pt (2 0 0), and Pd/Pt (3 1 1) planes. Owing to the lattice mismatch factor [21], it was very difficult to resolve the peaks for Pd and Pt in the XRD pattern, whereas the peaks of Au and Pd/Pt could be readily distinguished by XRD. Table 2 shows the crystalline sizes of nanoparticles and the identification of nanoparticles present in trimetallic nanotubes. From all of the above above observation, Au in the prepared Au-Pd-Pt trimetallic nanotubes was confirmed as forming inner core.

TABLE 1: The crystalline sizes of nanoparticles and identification of nanoparticles present in Au-Pt-Ag trimetallic nanocomposites.

| Sl. no. | 2 Theta (°) | *d* (A) | FWHM (deg) | Crystalline size (nm) |
|---|---|---|---|---|
| 1 | 38.12 | 2.27293 | 0.49 | 3.66 |
| 2 | 45.40 | 2.00579 | 0.20 | 4.25 |
| 3 | 64.83 | 1.94330 | 0.79 | 4.79 |
| 4 | 75.29 | 1.26514 | 0.59 | 7.85 |

TABLE 2: The crystalline sizes of nanoparticles and identification of nanoparticles present in Au-Pd-Pt trimetallic nanotubes.

| Sl. no | 2 Theta (°) | *d* (A) | FWHM (deg) | Crystalline size (nm) |
|---|---|---|---|---|
| 1 | 39.62 | 2.27293 | 0.500 | 3.67 |
| 2 | 45.17 | 2.00579 | 0.304 | 4.80 |
| 3 | 46.68 | 1.94330 | 0.110 | 4.33 |
| 4 | 75.02 | 1.26514 | 0.310 | 7.47 |

*3.5. Surface Energy.* The surface energy of selected 4d and 5d metals is given in Table 3. In addition to that cohesive energy and atomic radius are also tabulated. From this viewpoint, Pt has more surface energy compared with Au, Ag, and Pd. So, it is one of the reasons for Pt is present on the surface.

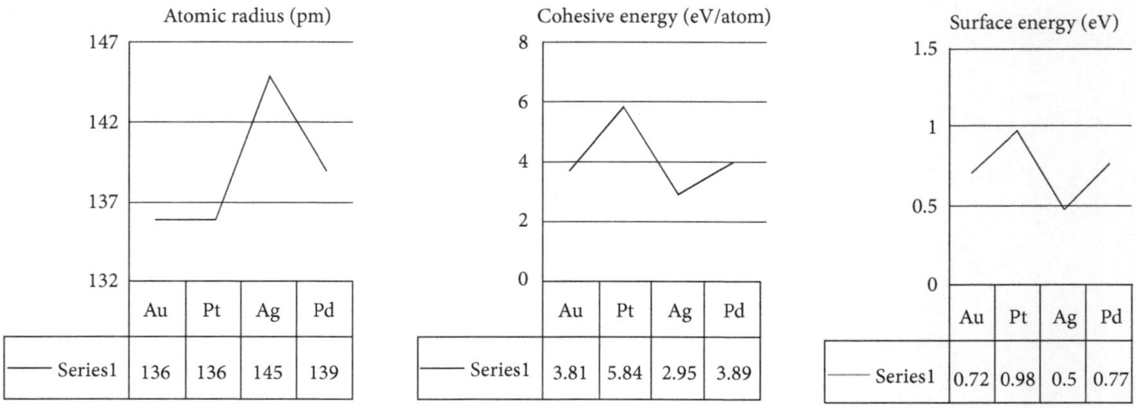


FIGURE 12: Surface energy, cohesive energy, and atomic radius are shown for comparison of Au, Pt, Ag, and Pd metals.

TABLE 3: Surface energy, cohesive energy and atomic radius of the Au, Pt, Ag and Pd metals.

| Property | Au | Pt | Ag | Pd |
|---|---|---|---|---|
| Surface energy | 0.72 eV | 0.98 eV | 0.50 eV | 0.77 eV |
| Cohesive energy | 3.81 eV/atom | 5.84 eV/atom | 2.95 eV/atom | 3.89 eV/atom |
| Atomic radius | 136 pm | 136 pm | 145 pm | 139 pm |

Surface energy, cohesive energy, and atomic radius are shown for comparison (Figure 12) of Au, Pt, Ag, and Pd metals.

## 4. Conclusions

In conclusion, for the first time, simple MW-irradiated syntheses of Au-Pt-Ag trimetallic nanocomposites and Au-Pd-Pt trimetallic nanotubes are reported. The UV-visible and HR-SEM confirmed the formation of nanocomposites and nanotube-shaped nanoparticles. EDX analysis confirmed that the percentage of Pt is maximum on the surface of trimetallic nanoparticles and nanotubes. XRD crystal data revealed that the particles are very small in size. The results reported here will inspire the further design of multilayered metallic nanoparticles.

## Acknowledgments

The author (B. Karthikeyan) is highly thankful for UGC, New Delhi, for granting a major research project. Electron microscopy was performed at the DSTUNS and SAIF-IIT Madras. XRD data were recorded at the Department of Physics, Alagappa University, Karaikudi. The coauthor (B. Loganathan) is thankful for UGC for the award of BSR-SAP fellowship.

## References

[1] P. Raveendran, J. Fu, and S. L. Wallen, "Completely "green" synthesis and stabilization of metal nanoparticles," *Journal of the American Chemical Society*, vol. 125, no. 46, pp. 13940–13941, 2003.

[2] M. R. Buck, J. F. Bondi, and R. E. Schaak, "A total-synthesis framework for the construction of high-order colloidal hybrid nanoparticles," *Nature Chemistry*, vol. 4, no. 1, pp. 37–44, 2012.

[3] C. Xu, J. Xie, D. Ho et al., "Au-$Fe_3O_4$ dumbbell nanoparticles as dual-functional probes," *Angewandte Chemie International Edition*, vol. 47, pp. 173–176, 2008.

[4] J. Choi, Y. Jun, S. Yeon, H. C. Kim, J. Shin, and J. Cheon, "Biocompatible heterostructured nanoparticles for multimodal biological detection," *Journal of the American Chemical Society*, vol. 128, no. 50, pp. 15982–15983, 2006.

[5] C. Xu, B. Wang, and S. Sun, "Dumbbell-like au-$fe_3O_4$ nanoparticles for target-specific platin delivery," *Journal of the American Chemical Society*, vol. 131, no. 12, pp. 4216–4217, 2009.

[6] H. Yin, C. Wang, H. Zhu, S. H. Overbury, S. Sun, and S. Dai, "Colloidal deposition synthesis of supported gold nanocatalysts based on Au-$Fe_3O_4$ dumbbell nanoparticles," *Chemical Communications*, no. 36, pp. 4357–4359, 2008.

[7] T. Mokari, E. Rothenberg, I. Popov, R. Costi, and U. Banin, "Selective growth of metal tips onto semiconductor quantum rods and tetrapods," *Science*, vol. 304, no. 5678, pp. 1787–1790, 2004.

[8] B. Karthikeyan and M. Murugavelu, "Nano bimetallic Ag/Pt system as efficient opto and electrochemical sensing platform towards adenine," *Sensors and Actuators B: Chemical*, vol. 163, no. 1, pp. 216–223, 2012.

[9] S. Patra and H. Yang, "Synthesis of trimetallic Au@Pb@Pt core-shell nanoparticles and their electrocatalytic activity toward formic acid and methanol," *Bulletin of the Korean Chemical Society*, vol. 30, no. 7, pp. 1485–1488, 2009.

[10] H. Zhang, M. Okumura, and N. Toshima, "Stable dispersions of PVP-protected Au/Pt/Ag trimetallic nanoparticles as highly active colloidal catalysts for aerobic glucose oxidation," *Journal of Physical Chemistry C*, vol. 115, no. 30, pp. 14883–14891, 2011.

[11] P. Venkatesan and J. Santhanalakshmi, "Core-shell bimetallic Au-Pd nanoparticles: synthesis, structure, optical and catalytic properties," *Journal of Nanoscience and Nanotechnology*, vol. 1, pp. 43–47, 2011.

[12] B. N. Wanjala, J. Luo, R. Loukrakpam et al., "Nanoscale alloying, phase-segregation, and core-shell evolution of gold-platinum nanoparticles and their electrocatalytic effect on oxygen reduction reaction," *Chemistry of Materials*, vol. 22, no. 14, pp. 4282–4294, 2010.

[13] L. Wang and Y. Yamauchi, "Strategic synthesis of trimetallic Au@Pd@Pt core-shell nanoparticles from poly(vinylpyrrolidone)-based aqueous solution toward highly active electrocatalysts," *Chemistry of Materials*, vol. 23, no. 9, pp. 2457–2465, 2011.

[14] P. Fang, S. Duan, X. Lin et al., "Tailoring Au-core Pd-shell Pt-cluster nanoparticles for enhanced electrocatalytic activity," *Chemical Science*, vol. 2, no. 3, pp. 531–539, 2011.

[15] H. Yin, T. Yamamoto, Y. Wada, and S. Yanagida, "Large-scale and size-controlled synthesis of silver nanoparticles under microwave irradiation," *Materials Chemistry and Physics*, vol. 83, no. 1, pp. 66–70, 2004.

[16] B. Karthikeyan and B. Loganathan, "Strategic green synthesis and characterization of Au/Pt/Ag trimetallic nanocomposites," *Materials Letters*, vol. 85, pp. 53–56, 2012.

[17] N. Toshima, *Macromolecular Complexes: Dynamic Interactions and Electronic Processes*, edited by E. Tsuchida, Japan Scientific Societies Press, Tokyo, Japan, 1991.

[18] N. Toshima, "Polymer-assisted composites of trimetallic nanoparticles with a three-layered core-shell structure for catalyses," *Nanohybridization of Organic-Inorganic Materials. Advances in Matetrials Research*, vol. 13, pp. 55–79, 2009.

[19] T. Yonezawa and N. Toshima, "Mechanistic consideration of formation of polymer-protected nanoscopic bimetallic clusters," *Journal of the Chemical Society, Faraday Transactions*, vol. 91, no. 22, pp. 4111–4119, 1995.

[20] D. S. Sheny, J. Mathew, and D. Philip, "Phytosynthesis of Au, Ag and Au-Ag bimetallic nanoparticles using aqueous extract and dried leaf of Anacardium occidentale," *Spectrochimica Acta A: Molecular and Biomolecular Spectroscopy*, vol. 79, no. 1, pp. 254–262, 2011.

[21] S. E. Habas, H. Lee, V. Radmilovic, G. A. Somorjai, and P. Yang, "Shaping binary metal nanocrystals through epitaxial seeded growth," *Nature Materials*, vol. 6, no. 9, pp. 692–697, 2007.

# 5

# *Adiantum philippense* L. Frond Assisted Rapid Green Synthesis of Gold and Silver Nanoparticles

**Duhita G. Sant,[1] Tejal R. Gujarathi,[1] Shrikant R. Harne,[1] Sougata Ghosh,[2] Rohini Kitture,[3] Sangeeta Kale,[3] Balu A. Chopade,[1] and Karishma R. Pardesi[1]**

[1] *Department of Microbiology, University of Pune, Pune, Maharashtra 411007, India*
[2] *Institute of Bioinformatics and Biotechnology, University of Pune, Pune, Maharashtra 411007, India*
[3] *Department of Applied Physics, Defence Institute of Advanced Technology, Girinagar, Pune 411025, India*

Correspondence should be addressed to Karishma R. Pardesi; karishma@unipune.ac.in

Academic Editor: Gunjan Agarwal

Development of an ecofriendly, reliable, and rapid process for synthesis of nanoparticles using biological system is an important bulge in nanotechnology. Antioxidant potential and medicinal value of *Adiantum philippense* L. fascinated us to utilize it for biosynthesis of gold and silver nanoparticles (AuNPs and AgNPs). The current paper reports utility of aqueous extract of *A. philippense* L. fronds for the green synthesis of AuNPs and AgNPs. Effect of various parameters on synthesis of nanoparticles was monitored by UV-Vis spectrometry. Optimum conditions for AuNPs synthesis were 1 : 1 proportion of original extract at pH 11 and 5 mM tetrachloroauric acid, whereas optimum conditions for AgNPs synthesis were 1 : 1 proportion of original extract at pH 12 and 9 mM silver nitrate. Characterization of nanoparticles was done by TEM, SAED, XRD, EDS, FTIR, and DLS analyses. The results revealed that AuNPs and AgNPs were anisotropic. Monocrystalline AuNPs and polycrystalline AgNPs measured 10 to 18 nm in size. EDS and XRD analyses confirmed the presence of elemental gold and silver. FTIR analysis revealed a possible binding of extract to AuNPs through $-NH_2$ group and to AgNPs through C=C group. These nanoparticles stabilized by a biological capping agent could further be utilized for biomedical applications.

## 1. Introduction

Nanotechnology is the study of materials that have at least one dimension in the range of 1–100 nm. Decrease in size is accompanied by elevated surface-area-to-volume ratio. Electronic and chemical properties of a material are dependent on its size. When the bulk material is reduced to nanoscale, its properties change. Such nanomaterials displaying novel properties have effective and wide utility in biological and biomedical applications. Noble metal nanoparticles such as Au, Ag, Pt, and Pd have been most effectively studied [1, 2].

Physical and chemical methods yield nanoparticles with well-defined shape and size, but these methods are expensive and potentially toxic to environment. This has created a need to develop clean, nontoxic, economical, and environmentally benign methods to synthesize nanoparticles. These concerns have led researchers to develop biological methods for synthesis of nanoparticles.

Castro et al. [3] have reported synthesis of AuNPs, by intracellular or extracellular reduction of tetrachloroauric acid using bacteria. Other biosynthetic routes include fungus, marine sponge, eggshell membrane, whole plant, leaf extract, flower, and tuber extract. Numerous methods available for biosynthesis of AgNPs include bacteria, mushroom, bark, peel extract, and enzyme as reducing agents [1, 4–8].

AuNPs are being considered in a wide range of diagnostic and therapeutic applications, such as cancer detection and cancer treatment, drug delivery, biological imaging, and biomedicine, as well as in electrochemistry. The prominent applications of silver nanomaterials include detection and diagnostics, antimicrobials and therapeutics, catalysis, and biolabeling [2].

Integration of phytochemicals in synthesis of nanomaterials is of extreme importance as it brings about integrity of plant science and nanotechnology, termed as green

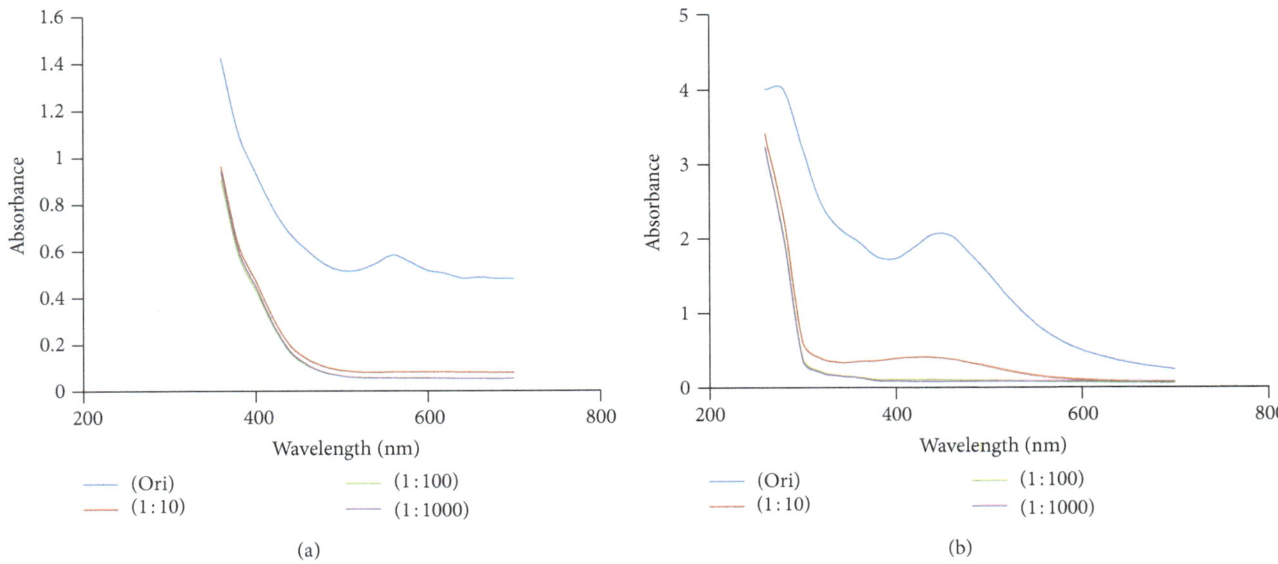


FIGURE 1: Surface plasmon peaks showing effect of varying extract concentrations (original extract, 1:10, 1:100, and 1:1000) on synthesis of AuNPs (a) and AgNPs (b).

technology. Biosynthesis of nanoparticles using dicotyledonous and monocotyledonous plants has been reported [1]. However, pteridophytes known to man for their use in folk medicine remain unexplored for this purpose.

The fern *Adiantum philippense* L. (family Adiantaceae) well known for its antioxidant potential [9] has been ethnomedicinally used in treatment of paralysis, blood diseases, epileptic fits, rabies, dysentery, elephantiasis, pimples, and wounds [10, 11]. It is also known for its antimicrobial activity [11]. Phytochemical analysis of the plant extracts belonging to genus *Adiantum* shows the presence of carbohydrates, phenols, flavonoids, and terpenoids [11–13]. Therefore, presence of these compounds may have a role in making *Adiantum* a potential candidate for bioreduction of gold and silver salts. In the present study, we have used *A. philippense* L. extract for the synthesis of AuNPs and AgNPs.

## 2. Material and Methods

*2.1. Preparation of the Aqueous Plant Extract.* Fresh fronds of *A. philippense* L. were obtained from Western Ghats (Sinhagad fort, Pune, India) region and identified from Botanical Survey of India (Western Circle), Pune, India. Fronds with mature sori were washed thoroughly and allowed to dry. Five grams of fronds were chopped into fine pieces and grounded into paste. This paste was extracted in 100 mL double distilled water. The extract was passed through muslin cloth and filtrate was centrifuged at 2500 g for 10 min at room temperature. Supernatant thus obtained was filtered through Whatman Filter Paper No. 1 and used for synthesis of nanoparticles.

*2.2. Synthesis of AuNPs and AgNPs.* Tetrachloroauric acid salt ($HAuCl_4{\cdot}4H_2O$, 49.9%) (SRL Chemicals, India) and silver nitrate salt ($AgNO_3$, 99.9%) (Qualigen Fine Chemicals, India) were procured and used without any further purification. Aqueous stock solutions of tetrachloroauric acid (10 mM) and silver nitrate (100 mM) were prepared in a stoppered volumetric flask and stored in amber color bottles.

Freshly prepared extract of *A. philippense* L. was mixed with aqueous solution of 10 mM tetrachloroauric acid in 1:1 proportion. The mixture was allowed to react at 30°C, with vigorous shaking. Synthesis of AuNPs was checked visually by determining the color change from yellow brown to dark blue. Similar procedure was followed for synthesis of AgNPs, indicated by a brick red color formation. Confirmation was done on the basis of a characteristic peak obtained in the absorbance range of 500–600 nm for AuNPs and 400–500 nm for AgNPs.

*2.3. Effect of Different Parameters on Synthesis of AuNPs and AgNPs.* Effect of different parameters such as concentration of extract (original, 1:10, 1:100, and 1:1000 in double distilled water), concentration of salt (1–10 mM), and pH (2–13) of original extract on synthesis of AuNPs and AgNPs was checked. Criterion for selecting optimum condition was observing for a narrow peak at characteristic wavelength giving maximum absorbance.

*2.4. Characterization of AuNPs and AgNPs.* Absorption spectrum of AuNPs and AgNPs was recorded on (JASCO V-630) UV-Vis spectrophotometric multiplate reader. Film X-ray diffraction (XRD) was performed using an X-ray diffractometer (Phillips PW1710, Holland) with CuK$\alpha$ radiation $\lambda = 1.5406$ Å over a wide range of Bragg angles ($20^\circ \le 2\theta \le 80^\circ$). Morphology and size of AuNPs were investigated by transmission electron microscopy (Tecnai $G^2$ 20U-TWIN (FEI, the Netherlands), TEM with $LaB_6$ electron source)

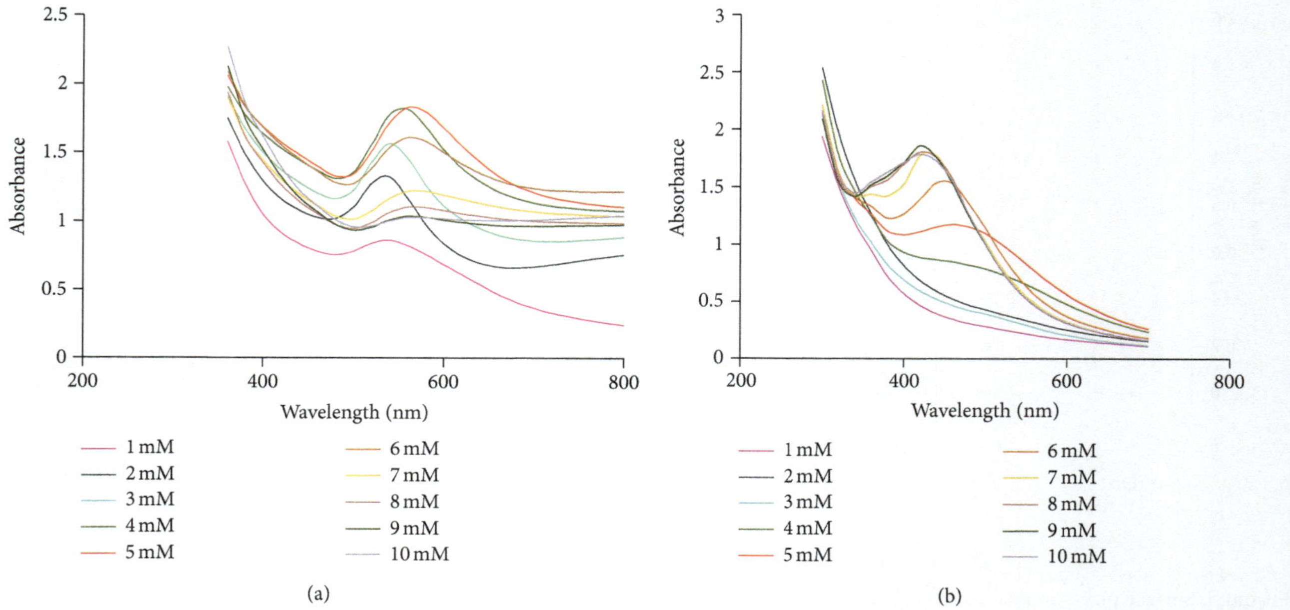


Figure 2: (a) Surface Plasmon Resonance of varying tetrachloroauric acid concentrations (1–10 mM) on synthesis of AuNPs. (b) Surface Plasmon Resonance of varying silver nitrate concentrations (1–10 mM) on synthesis of AgNPs.

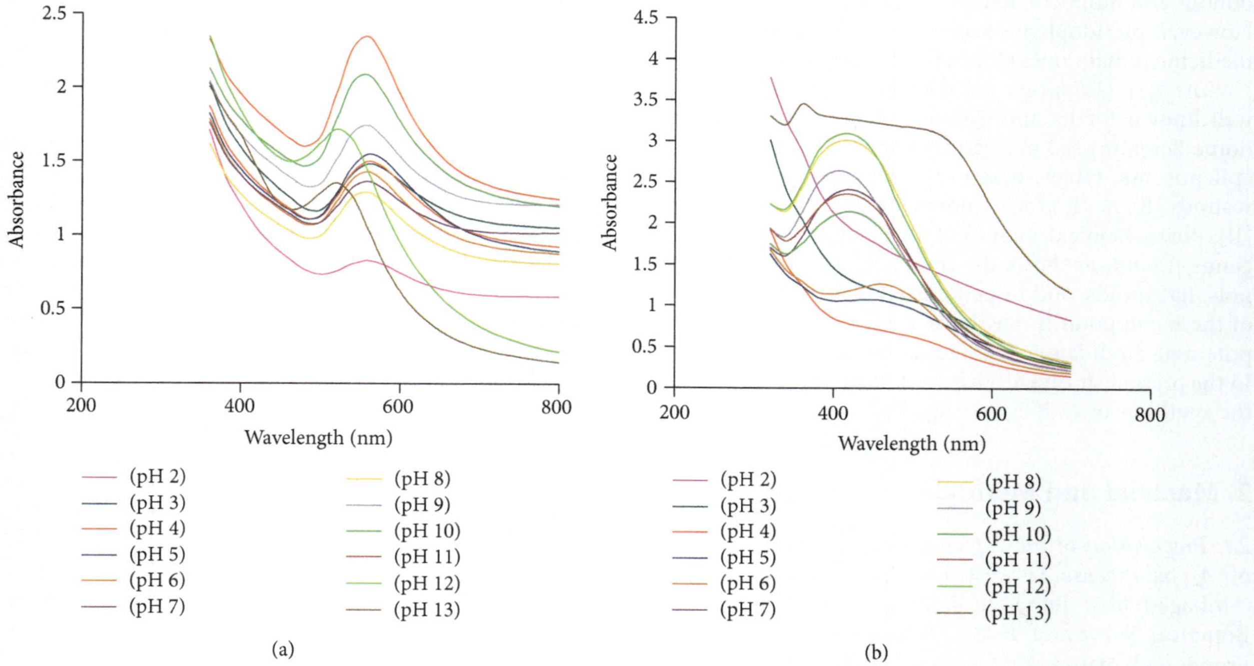


Figure 3: (a) Surface plasmon peaks of AuNPs with varying pH values (2–13). (b) Surface plasmon peaks of AgNPs with varying pH values (2–13).

with an accelerating voltage of 200 kV. An elemental analysis of the sample was examined by energy dispersive analyses of X-rays (EDAX) with 6360 (LA) instrument at an Acc voltage of 20 kV. Fourier transform infrared spectroscopic (FTIR) measurements were done using Perkin Elmer (16 PC-FT-IR) spectrophotometer. Particle size distribution was measured by a dynamic light scattering apparatus utilizing 663 nm red laser at 90° angles.

The samples for transmission electron microscopy (TEM) and energy dispersive scattering (EDS) and X-ray diffraction (XRD) were prepared by placing a drop of AuNPs and AgNPs onto a 1 cm ∗ 1 cm glass slide. Samples were then heat fixed and subjected for analysis. The AuNPs, AgNPs, and *A. philippense* L. extracts were pelletized separately in IR grade KBr (HiMedia, India) and were scanned over a range of 4000–400 $cm^{-1}$ in FTIR analysis.

## 3. Results and Discussion

### 3.1. Effects of Various Parameters on Synthesis of AuNPs and AgNPs

*3.1.1. Biosynthesis of AuNPs and AgNPs.* High antioxidant potential and ethnomedicinal importance of *A. philippense* L. intrigued us to explore it for biosynthesis of AuNPs and AgNPs. Microbial enzymes, polysaccharides, and plant phytochemicals with antioxidant or reducing properties are usually responsible for reduction of metal compounds into their respective nanoparticles [1, 5, 14–16].

A dark blue color was observed on addition of plant extract to 5 mM tetrachloroauric acid solution indicating AuNPs synthesis. UV-Vis spectroscopy of this solution confirmed the synthesis of AuNPs, as revealed by a characteristic absorption peak at 530 nm [17]. On the other hand, a characteristic brick red color was formed on addition of 9 mM silver nitrate solution to the plant extract, indicating synthesis of AgNPs, which was confirmed by an absorption peak at 420 nm [18]. The time for complete reduction of gold and silver ions in solution to nanoparticles was 10 min.

Shankar et al. [19] reported that reduction of the metal ions and stabilization of AuNP or AgNP are due to the presence of terpenoids or alkaloids present in the geranium extract. On the other hand, AuNP synthesis is caused by the reducing sugar present in the lemongrass extract [20]. Similarly, *A. philippense* L. is known to contain variety of such biologically active phytochemicals like carbohydrates, glycosides, alkaloids, tannins, flavonoids, terpenoids, and saponins. These phytochemicals are responsible for making the extract highly oxidant [12] and thus might play a role in bioreduction of gold and silver nanoparticles.

*3.1.2. Effect of Extract Concentration.* Various dilutions of plant extract, original 1 : 10, 1 : 100, and 1 : 1000 (diluted in double distilled water), were analyzed for their effect on synthesis of AuNPs and AgNPs. Original extract (5 g plant biomass in 100 mL distilled water) showed effective synthesis of nanoparticles with the characteristic color. However, further dilutions of the extract showed negligible synthesis of nanoparticles. UV-Vis spectra suggest that sharpness of absorption peak is dependent on concentration ratio of extract [21]. Accordingly, sharp peak was observed in original extract concentration (Figures 1(a) and 1(b)). There are similar reports showing effect of biomass quantity on synthesis of nanoparticles [22].

*3.1.3. Effect of Metal Salt Concentration.* Mixing plant extract with increasing concentration of tetrachloroauric acid/silver nitrate (1 to 10 mM) resulted in synthesis of nanoparticles as indicated by an increasing intensity of the characteristic color. In case of AuNPs, peak absorbance increased with tetrachloroauric acid concentration from 1 to 5 mM but decreased at higher metal ion concentrations (Figure 2(a)). Thus, the optimum salt concentration for AuNP synthesis was 5 mM, as revealed by a narrow peak at 530 nm. Similar results have been reported by Ghosh et al. [15] with 1 mM tetrachloroauric acid as optimum concentration for AuNP synthesis.

A steady rise in absorbance was also observed for AgNP synthesis using 1 to 10 mM silver nitrate solution. However, there was a marginal difference in absorbance using 9 mM and 10 mM salt concentration as indicated by overlapping peaks in UV-Vis spectra (Figure 2(b)). This indicates that 9 mM was the optimum concentration for AgNPs synthesis. Other researchers have also reported a significant effect of metal salt concentration on the synthesis of silver nanoparticles [5, 23].

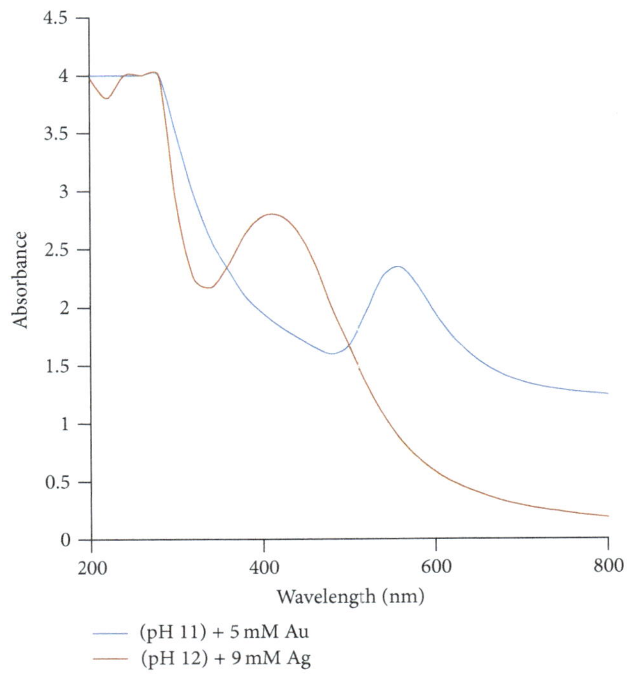


FIGURE 4: SPR at optimum conditions. For Au, original extract at pH 11 and 5 mM tetrachloroauric acid was kept for 10 min. For Ag, original extract at pH 12 and 9 mM silver nitrate was kept for 10 min.

*3.1.4. Effect of pH.* Formation of distinct peak depends on concentration of $H^+$ ions (pH). Reaction mixtures, containing extract at different pH units (2 to 13) and tetrachloroauric acid solution, showed blue color of varying intensities, indicating formation of AuNPs. UV-Vis spectroscopy further confirmed that a sharp peak obtained at pH 11 was optimum for synthesis of AuNPs (Figure 3(a)).

Similarly for silver, the alkaline pH range showed brick red color of varying intensities indicating the synthesis of AgNPs. UV-Vis spectrum confirmed a sharp peak at pH 12, to be optimum (Figure 3(b)). At lower pH, nanoparticles show lower and broader absorbance which can be due to larger size.

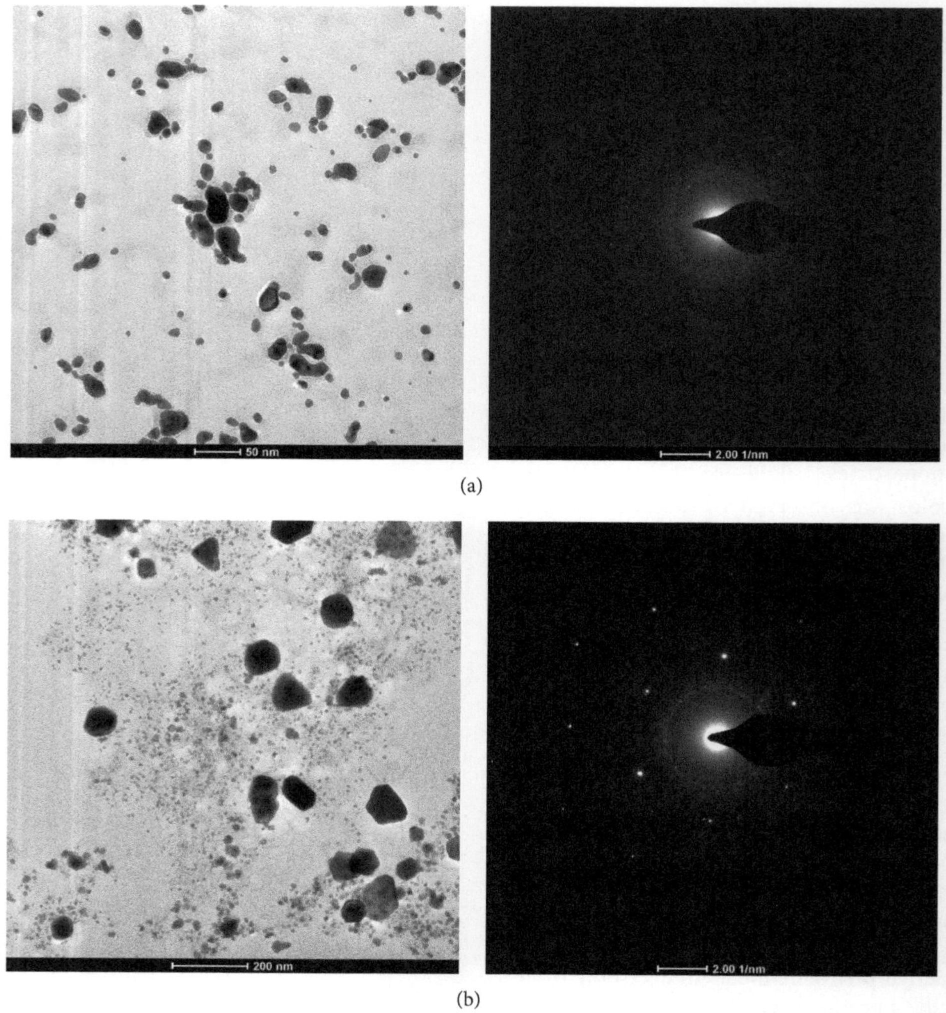


Figure 5: (a) TEM image (left) and SAED pattern (right) of AuNPs synthesized using extract of *A. philippense* L. (b) TEM image (left) and SAED pattern (right) of AgNPs synthesized using extract of *A. philippense* L.

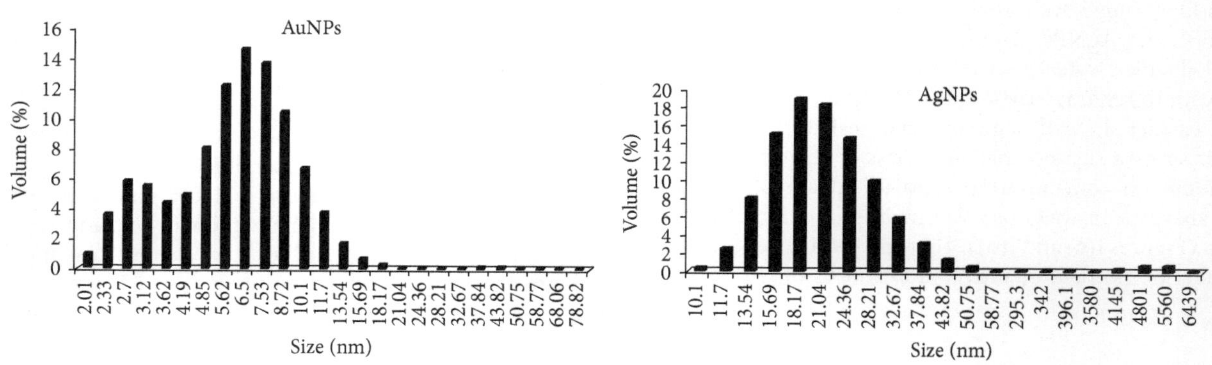


Figure 6: (a) Histogram of size distribution of AuNPs synthesized by *A. philippense* L. (b) Histogram of size distribution of AgNPs synthesized by *A. philippense* L.

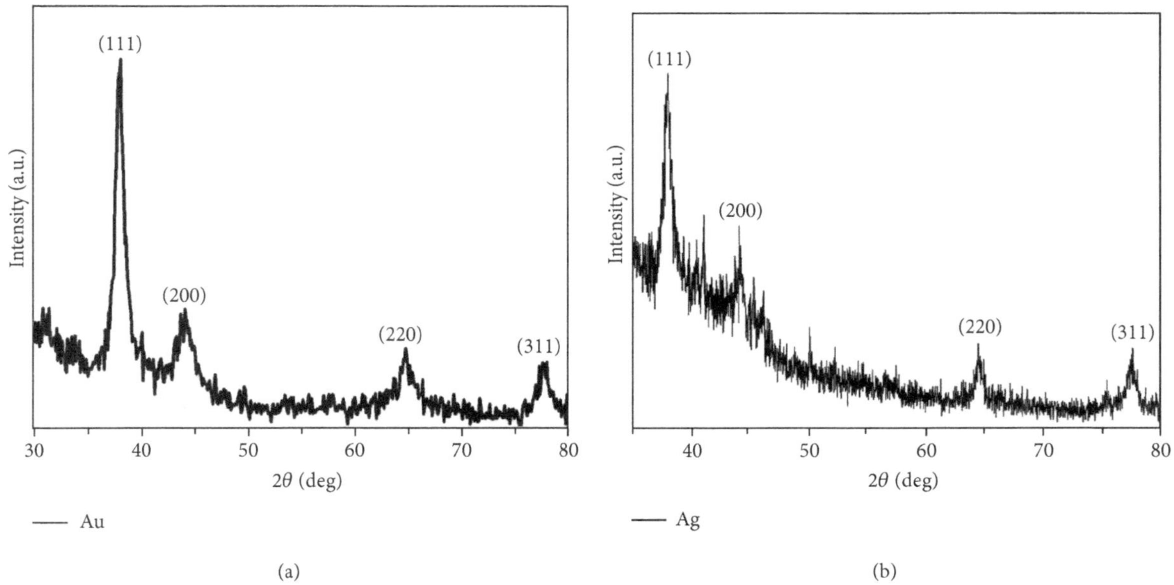


FIGURE 7: (a) XRD pattern of AuNPs. (b) XRD pattern of AgNPs.

Earlier reports also indicate that high absorbance and narrow peaks at alkaline pH favour synthesis of monodispersed and small sized nanoparticles [24]. Alkaline pH prevents agglomeration of synthesized nanoparticles; therefore, small sized particles are formed [4, 23].

Thus, optimum conditions for AuNPs synthesis were 1 : 1 proportion of original extract at pH 11 and 5 mM tetrachloroauric acid, whereas optimum conditions for AgNPs synthesis were 1 : 1 proportion of original extract at pH 12 and 9 mM silver nitrate. Time required for nanoparticle synthesis under optimum conditions was checked by reading absorbance in a UV-Vis spectrophotometer and was found to be 10 min (Figure 4).

*3.2. Characterization of AuNPs and AgNPs.* Confirmation of AuNPs and AgNPs formation was done by taking UV-Vis spectrum of the resultant solution, in the range of 200–1000 nm. The spectrum (Figure 4) showed a distinct peak, centered at 530 nm, specific for AuNPs [17]. Similarly, a distinct peak observed at 420 nm was specific for AgNPs (Figure 4) [18].

The result obtained from the transmission electron microscopy (TEM) and selected area electron diffraction (SAED) study gives clear indication about the shape, size, and size distribution of the nanoparticles. TEM and SAED pattern revealed that AuNPs were anisotropic and polycrystalline in nature (Figure 5(a)). However, TEM and SAED studies in case of AgNPs showed their anisotropic and monocrystalline nature (Figure 5(b)). Particle size analysis of TEM images was done using ImageJ software. Particle size calculated was also confirmed using Scherrer's formula that makes use of XRD data of the samples. Particle size of AuNPs and AgNPs ranged between 10 and 18 nm. This is in agreement with earlier studies reporting nanoparticles of 10 to 100 nm obtained using various plant extracts [1, 7, 23, 24]. Small size of nanoparticles obtained in the present study could be a result of alkaline pH used for their synthesis [7, 24].

Particle size distribution was measured by dynamic light scattering (DLS) (Figures 6(a) and 6(b)). Although the histograms show a significant proportion of nanoparticles lying in the size range as calculated by TEM and XRD analysis, other large agglomerates interfere with the accuracy of particle size determination. These agglomerates could be of biological origin present in the plant extract or could be aggregates of nanoparticles [25]. Das et al. [26] reported that the variation in particle size distribution in DLS (99.91 nm) as compared to that observed in HRTEM analysis (15.07 nm) could be due to water absorption on electrostatic stabilized gold nanoparticles.

The obtained XRD data was checked against the standard data for the respective samples, provided by the Joint Committee on Powder Diffraction Standards (JCPDS). Both the particles showed face centered cubic (FCC) structure. In case of AuNPs, the data matched with the JCPDS data card number 04-0784, with lattice constant $a = 4.078$ Å. The observed peaks correspond to the planes (111), (200), (220), and (311), respectively, for the peaks in ascending $2\theta$ order. The results are in agreement with previous reports [4, 8, 23, 26].

The XRD pattern of AgNPs matched with JCPDS data card number 04-0783. Standard lattice constant value for this structure is reported to be 4.086 Å. Lattice planes (111), (200), (220), and (311) for AgNPs were also observed in the XRD pattern. Broad peak in both the XRD patterns suggests that the particles size was very less. Crystallite size for both the particles was calculated using Scherrer's formula [27] as

$$D = \frac{0.9\lambda}{\beta \cos\theta}, \quad (1)$$

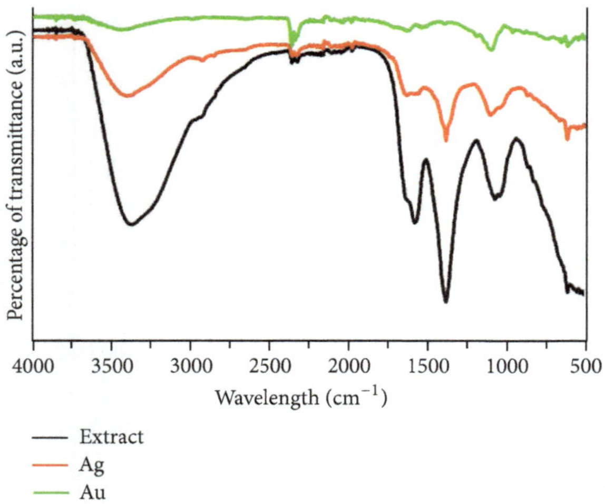


Figure 8: FTIR spectra for *A. philippense* L. extract, the AuNPs, and AgNPs.

where $D$ is particle size, $\lambda$ is wavelength of X-ray, $\beta$ is full width at half maxima for a peak, and $\theta$ is the corresponding Bragg angle. The particle size was estimated to be ~11 nm for AuNPs and ~13 nm for AgNPs (Figures 7(a) and 7(b)). This is in good agreement with the particle size obtained from TEM images.

FTIR spectroscopy was carried out to identify the biogroups that bound distinctively on the gold and silver surface and were involved in the synthesis of these nanoparticles. The major peaks observed in FTIR of the extract were 3369, 2360, 1585, 1384, 1076, and 514 $cm^{-1}$. Peak at 3369 $cm^{-1}$ is attributed to OH stretch in phenols. This suggests presence of phenolic compound in the extract [28]. Phytochemical evaluation of *A. philippense* L also reports the presence of phenols [11, 13]. Taraschewski et al. [29] report that the peak at 2360 $cm^{-1}$ results due to $CO_2$ vibration that may not be necessarily from the sample. Peaks at 1585, 1384, and 1076 $cm^{-1}$ are attributed to C=C stretching, $-NH_2$ symmetric stretch, and CO vibrations, respectively. The peak at 514 $cm^{-1}$ can be attributed to C–N–C bending [28, 30, 31].

As seen in the case of FTIR for AuNPs, the peak at 1384 $cm^{-1}$ had completely diminished, while the rest of the peaks from extract were present with reduced intensity. This suggests that the extract might be bound to the AuNPs through or by replacing $-NH_2$ group. In FTIR of AgNPs, the $-NH_2$ peak remained intact while the C=C stretching peak almost vanished. This may be due to conjugation of extract to the nanoparticle through C=C group (Figure 8). These functional groups of water soluble compounds such as flavonoids, terpenoids, and phenols present in the *A. philippense* L. extract appear to be responsible for synthesis of nanoparticles and their stabilization. Other studies have also reported that proteins or polysaccharides present in the biological material cap and stabilize nanoparticles [1, 17]. These interactions may occur between free amino groups, carboxylate group, or cysteine residues in the proteins [1, 3, 6, 32].

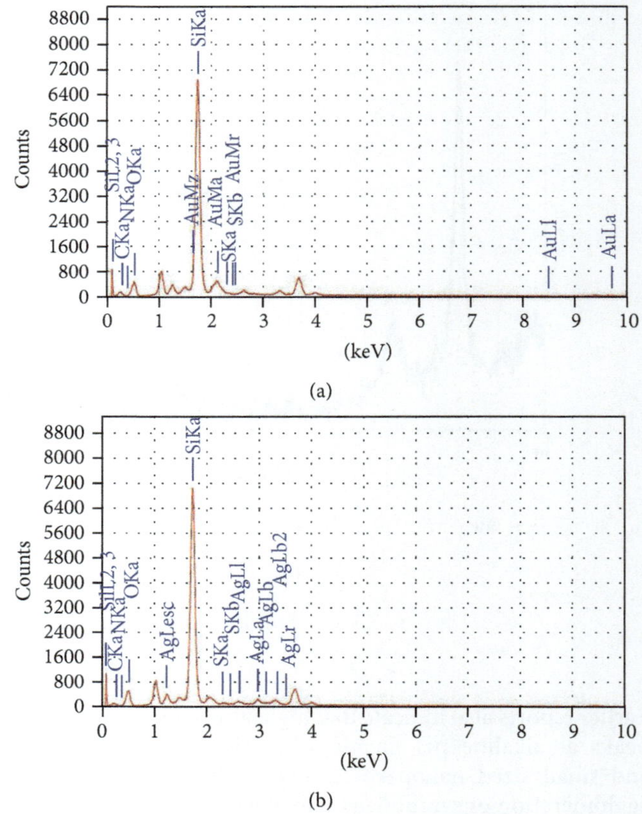


Figure 9: (a) EDS spectrum of AuNPs synthesized using extract of *A. philippense* L. (b) EDS spectrum of AgNPs synthesized using extract of *A. philippense* L.

Elemental composition of AuNPs and AgNPs and their surrounding was carried out by energy dispersive scattering (EDS). EDS studies for AuNPs and AgNPs showed presence of elemental gold (8.95%) and silver (5.04%) (Figures 9(a) and 9(b)). Detection of carbon and oxygen suggested the presence of carbohydrate compounds containing sulphur that were either capped on AuNPs and AgNPs or were present in the background. Further characterization of the capping agent and its probable involvement in biomedical application needs to be investigated.

## 4. Conclusion

We have developed a method for synthesis of gold and silver nanoparticles using fern *A. philippense* L., at room temperature, within 10 min. The TEM and EDS analysis results both demonstrate that nanoparticles are successfully synthesized using *A. philippense* L. extract in this study. To the best of our knowledge, use of pteridophytes in nanoparticles synthesis has not yet been reported. Medicinal importance of ferns makes them a suitable candidate for biosynthesis of nanoparticles and application in disease treatment. Use of such medicinally important plants has a promising potential in nanoscience for drug delivery and biomedical application. Exploitation of synthesized capped nanoparticles would enable us to know the nature of capping agent and utilize them for medicinal and biomedical applications.

## Conflict of Interests

The authors declare no conflict of interests.

## Acknowledgments

The authors acknowledge Dr. S. V. Ravat from the Botanical Survey of India, Western Circle, Pune, India, for his help in plant specimen identification, Mr. R. Sridharkrishna from the Department of Physics, University of Pune, and Dr. M. Jayakannan from IISER, Pune, for technical help in material characterization. S. Ghosh thanks Council of Scientific and Industrial Research (CSIR, Government of India) for Senior Research Fellowship (09/137(0516)/2012-EMR-I). The authors also thank DST-PURSE for funding this research.

## References

[1] T. C. Prathna, L. Mathew, N. Chandrasekaran, A. M. Raichur, and A. Mukherjee, *Biomimetic Synthesis of Nanoparticles: Science, Technology & Applicability, Biomimetics Learning from Nature*, edited by A. Mukherjee, 2010.

[2] R. R. Arvizo, S. Bhattacharyya, R. Kudgus, K. Giri, R. Bhattacharya, and P. Mukherjee, "Intrinsic therapeutic applications of noble metal nanoparticles: past, present and future," *Chemical Society Reviews*, vol. 41, no. 7, pp. 2943–2970, 2012.

[3] L. Castro, M. L. Blázquez, F. González, J. A. Muñoz, and A. Ballester, "Extracellular biosynthesis of gold nanoparticles using sugar beet pulp," *Chemical Engineering Journal*, vol. 164, pp. 92–97, 2010.

[4] A. D. Dwivedi and K. Gopal, "Biosynthesis of silver and gold nanoparticles using *Chenopodium album* leaf extract," *Colloids and Surfaces A*, vol. 369, no. 1–3, pp. 27–33, 2010.

[5] S. Ghosh, S. Patil, M. Ahire et al., "Synthesis of silver nanoparticles using *Dioscorea bulbifera* tuber extract and evaluation of its synergistic potential in combination with antimicrobial agents," *International Journal of Nanomedicine*, vol. 7, pp. 483–496, 2012.

[6] D. Philip, "Biosynthesis of Au, Ag and Au-Ag nanoparticles using edible mushroom extract," *Spectrochimica Acta Part A*, vol. 73, pp. 374–381, 2009.

[7] M. Sathishkumar, K. Sneha, S. W. Won, C. W. Cho, S. Kim, and Y. S. Yun, "*Cinnamon zeylanicum* bark extract and powder mediated green synthesis of nano-crystalline silver particles and its bactericidal activity," *Colloids and Surfaces B*, vol. 73, no. 2, pp. 332–338, 2009.

[8] S. Kaviya, J. Santhanalakshmi, B. Viswanathan, J. Muthumary, and K. Srinivasan, "Biosynthesis of silver nanoparticles using citrus sinensis peel extract and its antibacterial activity," *Spectrochimica Acta Part A*, vol. 79, no. 3, pp. 594–598, 2011.

[9] O. Sawant, V. J. Kadam, and R. Ghosh, "*In vitro* free radical scavenging and antioxidant activity of *Adiantum lunulatum*," *Journal of Herbal Medicine and Toxicology*, vol. 3, no. 2, pp. 39–44, 2009.

[10] V. Karthik, K. Raju, M. Ayyanar, K. Gowrishankar, and T. Sekar, "Ethno medicinal uses of pteridophytes in Kolli Hills, Eastern Ghats of Tamil Nadu, India," *Journal of Natural Product and Plant Resources*, vol. 1, no. 2, pp. 50–55, 2011.

[11] C. Pan, Y. G. Chen, X. Y. Ma, J. H. Jiang, F. He, and Y. Zhang, "Phytochemical constituents and pharmacological activities of plants from the Genus *Adiantum*: a review," *Tropical Journal of Pharmaceutical Research*, vol. 10, no. 5, pp. 681–692, 2011.

[12] M. J. Mithraja, J. M. Antonisamy, M. Mahesh, Z. M. Paul, and S. Jeeva, "Inter-specific variation studies on the phytoconstituents of Christella and Adiantum using phytochemical methods," *Asian Pacific Journal of Tropical Biomedicine*, vol. 2, no. 1, pp. 40–45, 2012.

[13] M. S. Ali, M. R. Amin, C. M. I. Kamal, and M. A. Hossain, "In vitro antioxidant, cytotoxic, thrombolytic activities and phytochemical evaluation of methanol extract of the *A. philippense* L. leaves," *Asian Pacific Journal of Tropical Biomedicine*, vol. 3, no. 6, pp. 464–469, 2013.

[14] S. Ghosh, S. Patil, M. Ahire et al., "Gnidia glauca flower extract mediated synthesis of gold nanoparticles and evaluation of its chemocatalytic potential," *Journal of Nanobiotechnology*, vol. 10, no. 17, 2012.

[15] S. Ghosh, S. Patil, M. Ahire et al., "Synthesis of gold nanoanisotrops using *Dioscorea bulbifera* tuber extract," *Journal of Nanomaterials*, vol. 2011, Article ID 354793, 8 pages, 2011.

[16] Y. Park, Y. N. Hong, A. Weyers, Y. S. Kim, and R. J. Linhardt, "Polysaccharides and phytochemicals: a natural reservoir for the green synthesis of gold and silver nanoparticles," *IET Nanobiotechnology*, vol. 5, no. 3, pp. 69–78, 2011.

[17] K. Govindaraju, S. K. Basha, V. G. Kumar, and G. Singaravelu, "Silver, gold and bimetallic nanoparticles production using single-cell protein (*Spirulina platensis*) Geitler," *Journal of Materials Science*, vol. 43, no. 15, pp. 5115–5122, 2008.

[18] R. Vaidyanathan, K. Kalishwaralal, S. Gopalram, and S. Gurunathan, "Nanosilver—the burgeoning therapeutic molecule and its green synthesis," *Biotechnology Advances*, vol. 27, no. 6, pp. 924–937, 2009.

[19] S. S. Shankar, A. Rai, B. Ankamwar, A. Singh, A. Ahmad, and M. Sastry, "Biological synthesis of triangular gold nanoprisms," *Nature Materials*, vol. 3, no. 7, pp. 482–488, 2004.

[20] S. S. Shankar, A. Rai, A. Ahmad, and M. Sastry, "Controling the optical properties of lemongrass extract synthesized gold nanotriangles and potebtial application in infrared absorbing optical coatings," *Applied Nanoscience*, vol. 1, pp. 69–77, 2005.

[21] D. S. Sheny, J. Mathew, and D. Philip, "Phytosynthesis of Au, Ag and Au-Ag bimetallic nanoparticles using aqueous extract and dried leaf of *Anacardium occidentale*," *Spectrochimica Acta Part A*, vol. 79, no. 1, pp. 254–262, 2011.

[22] K. D. Arunachalam, S. K. Annamalai, and S. Hari, "One-step green synthesis and characterization of leaf extract-mediated biocompatible silver and gold nanoparticles from *Memecylon umbellatum*," *International Journal of Nanomedicine*, vol. 8, pp. 307–1315, 2013.

[23] S. P. Dubey, M. Lahtinen, and M. Sillanpää, "Tansy fruit mediated greener synthesis of silver and gold nanoparticles," *Process Biochemistry*, vol. 45, no. 7, pp. 1065–1071, 2010.

[24] T. P. Amaladhas, S. Sivagami, T. A. Devi, N. Ananthi, and S. P. Velammal, "Biogenic synthesis of silver nanoparticles by leaf extract of *Cassia angustifolia*," *Advances in Natural Sciences*, vol. 3, pp. 1–8, 2012.

[25] A. Bootz, V. Vogel, D. Schubert, and J. Kreuter, "Comparison of scanning electron microscopy, dynamic light scattering and analytical ultracentrifugation for the sizing of poly(butyl cyanoacrylate) nanoparticles," *European Journal of Pharmaceutics and Biopharmaceutics*, vol. 57, no. 2, pp. 369–375, 2004.

[26] S. Das, P. Roy, S. Mondal, T. Bera, and A. Mukherjee, "One pot synthesis of gold nanoparticles and application in chemotherapy of wild and resistant type visceral leishmaniasis," *Colloids and Surfaces B*, vol. 107, pp. 27–34, 2013.

[27] R. Parmar, M. H. Mangrola, B. H. Parmar, and V. G. Joshi, "A software to calculate crystalline size by Debey-Scherrer Formula using VB.NET," *Multi Disciplinary Edu Global Quest*, vol. 1, no. 1, 2012.

[28] R. Silverstein and F. Webster, *Spectrometric Identification of Organic Compounds*, Wiley India, 6th edition, 2006.

[29] M. Taraschewski, H. K. Cammenga, R. Tuckermann, and S. Bauerecker, "FTIR Study of $CO_2$ and $H_2O/CO_2$ nanoparticles and their temporal evolution at 80 K," *Journal of Physical Chemistry A*, vol. 109, no. 15, pp. 3337–3343, 2005.

[30] C. Y. Kim, T. Sekino, and K. Niihara, "Synthesis of bismuth sodium titanate nanosized powders by solution/sol-gel process," *Journal of the American Ceramic Society*, vol. 86, no. 9, pp. 1464–1467, 2003.

[31] Y. F. Zhang, J. X. Zhang, Q. M. Lu, and Q. Y. Zhang, "Synthesis and characterization of $Ca_3Co_4O_9$ nanoparticles by citrate sol-gel method," *Materials Letters*, vol. 60, no. 20, pp. 2443–2446, 2006.

[32] S. P. Chandran, M. Chaudhary, R. Pasricha, A. Ahmad, and M. Sastry, "Synthesis of gold nanotriangles and silver nanoparticles using Aloe vera plant extract," *Biotechnology Progress*, vol. 22, no. 2, pp. 577–583, 2006.

# 6

# Shape Control of CdTe Nanocrystals: Influence of the Solvent Composition and Ligand Effects

**Xiaoping Jin, Jürgen Parisi, and Joanna Kolny-Olesiak**

*Energy and Semiconductor Research Laboratory, Department of Physics, Carl von Ossietzky University of Oldenburg, Carl von Ossietzky Straße 9–11, 26111 Oldenburg, Germany*

Correspondence should be addressed to Joanna Kolny-Olesiak; joanna.kolny@uni-oldenburg.de

Academic Editor: Vijaya Rangari

CdTe nanocrystals were synthesized by the hot-injection method with a mixture of oleylamine and octadecene as a solvent. The influence of the composition of the solvent and of the injection solution on the shape of CdTe nanoparticles was investigated. Various shapes of CdTe nanocrystals, such as nanodots, nanorods, multipods, and nanowires, could be obtained by changing the reaction conditions. Tuning the reactivity of both the cadmium and the tellurium precursors at the same time was found to be the main reason for the shape control of CdTe nanocrystals in this reaction system. The reactivity of the Cd precursor was controlled by the composition of the solvent, while the activity of the Te precursor could be influenced by using trioctylphosphine and tributylphosphine in the injection solution.

## 1. Introduction

CdTe is a direct semiconductor with a bandgap in the bulk of 1.475 eV and a Bohr exciton radius of 7.3 nm. CdTe nanocrystals attract interest due to their pronounced quantum size effect and optical activity both in the visible and near-infrared spectral regimes. This makes them interesting candidates for a variety of applications, for example, in solar energy conversion [1, 2] or biolaballing [3, 4]. Thus, size and shape control of CdTe nanocrystals have been a pursued activity for the past few years [5–11]. Various shapes of CdTe have been obtained by different kinds of synthetic procedures. The hot-injection technique especially allows for synthesizing particularly uniform particles and offers a means to control not only the size of the particles, but also their shape. Uniform quasispherical nanocrystals, nanorods, and tetrapods as well as nanowires were grown by this method [10, 12–17].

In the hot-injection method, monomer activity during the nucleation and growth processes plays a key role for shape control of nanocrystals. Several parameters influence the monomer activity, like precursor concentration, precursor activation, and ligands effect. Ligands as an important part in the synthesis of nanocrystals influence the monomer activity in several ways. Yu et al. [12] demonstrated how ligands affect the shape of CdTe nanocrystals in a noncoordinating solvent. In the present study we investigated the influence of ligands and solvents onto the shape control of CdTe nanocrystals growing in a coordinating solvent (oleylamine, (OLAM)). In our previous works we showed that the growth process of CdTe nanocrystals in OLAM and with oleic acid (OA) as stabilizer can be strongly influenced by even small changes in the reaction conditions, which allows for shape control while keeping the principal reagents fixed [15]. Here we investigate further the possibility to control the morphology of CdTe nanocrystals synthesized in OLAM with OA as ligand. A variety of shapes (dots, rods, multipods, and nanowires) have been realized by the hot-injection method by changing the compositions of the solvent and of the injection solution. The resulting materials were characterized by UV-Vis absorption spectroscopy, powder X-ray diffraction (XRD) measurements, and transmission electron microscopy (TEM).

## 2. Experimental

*2.1. Chemicals.* All chemicals were used as received without further purification: cadmium oxide (CdO, 99%,

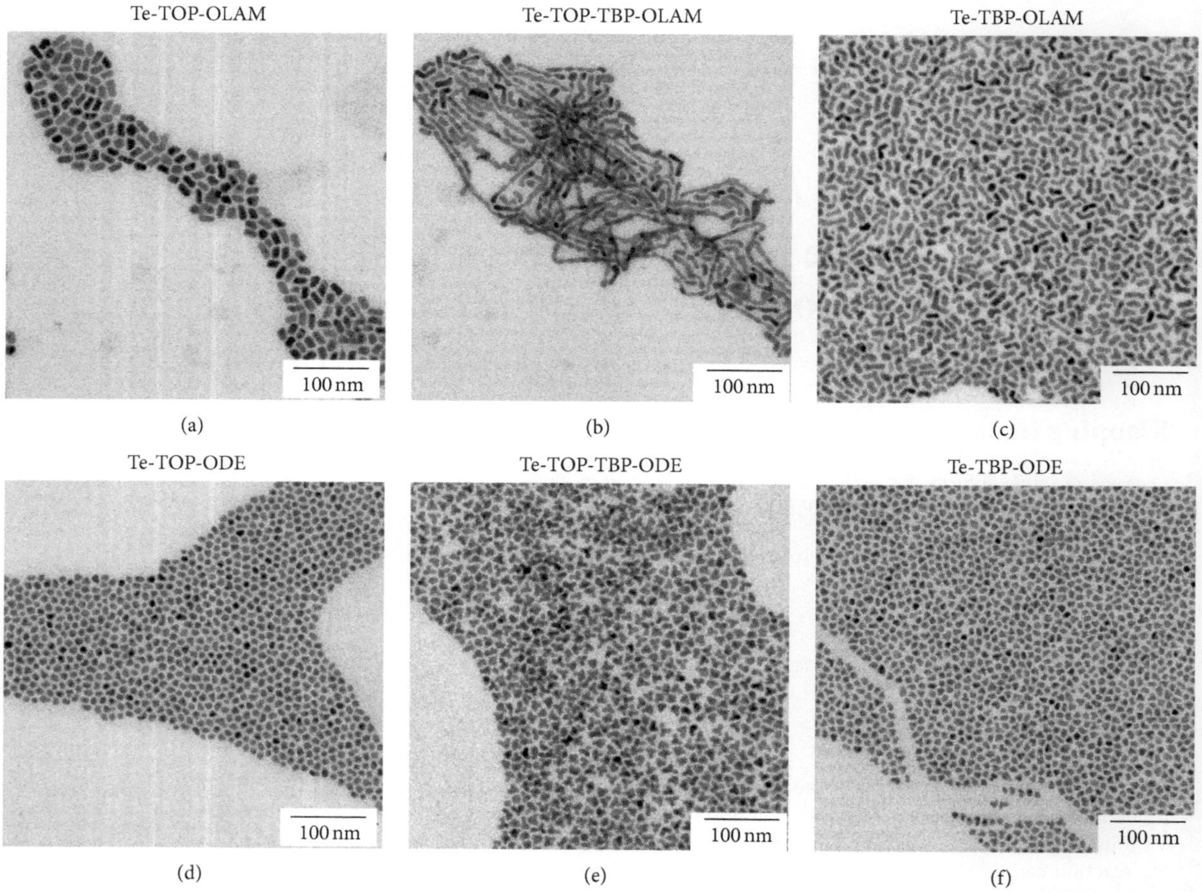


FIGURE 1: TEM images of CdTe NCs synthesized from different tellurium precursor solutions. (a) Te-TOP-OLAM; (b) Te-TOP-TBP-OLAM; (c) Te-TBP-OLAM; (d) Te-TOP-ODE; (e) Te-TOP-TBP-ODE; (f) Te-TBP-ODE.

Fluka), tri-n-octylphosphine (TOP, 90%, Aldrich), tri-n-butylphosphine (TBP, 90%, Aldrich), oleic acid (OA, 90%, Aldrich), 1-octadecene (ODE, Merck), tellurium powder (Te, purum p., Fluka), and oleylamine (OLAM, approximately C18-content 80–90%, Acros Organics).

*2.2. Te Precursor Preparation.* The general procedure for preparing Te precursor solutions is as follows. Te was added to TOP, TBP, or the mixture of them in the glove box. The mixture was further heated to 200°C under nitrogen flow till the dissolution of Te. The Te injection solution was obtained by diluting this Te solution with OLAM or ODE (Table 1).

*2.3. Synthesis of Colloidal CdTe Nanocrystals with Various Injection Solutions.* A mixture of CdO (0.1 mmol), OA (0.16 mL), and OLAM (10 mL) was degassed at 100°C for 1 hour in a 50 mL three-necked flask connected to a condenser. The mixture was heated slowly under nitrogen until most of the CdO decomposed at ~290°C, then another 0.2 mL OA was added to form a clear and colourless solution. Te precursor solution (including 0.05 mmol tellurium) was injected quickly at 300°C. The temperature dropped to 260°C and maintained this value throughout the synthesis. All syntheses were stopped after 5 minutes by removing the heating mantle and by rapidly cooling the flask. Aliquots were taken out at different reaction times to monitor the reaction process by measuring the UV-Vis absorbance and TEM images.

*2.4. Synthesis of Colloidal CdTe Nanocrystals with Various OLAM : OA Ratios.* All manipulations were performed using standard air-free techniques. A mixture of CdO (0.1 mmol), OA (0.16 mL), and OLAM + ODE (10 mL) was degassed at 110°C for 40 minutes in a 50 mL three-necked flask connected to a condenser. The mixture was heated slowly under nitrogen until most of the CdO decomposed at ~290°C, then another 0.2 mL OA was added to form a clear and colourless solution. A Te-TOP-TBP-ODE (0.124 mL, Cd/Te = 10 : 7) precursor solution was injected quickly at 300°C. The temperature dropped to 260°C and was kept at this value throughout the synthesis. Samples were taken out at different reaction times between 1 and 5 minutes.

*2.5. Characterization.* UV/Vis spectra were recorded using a Carry 100 absorption spectrophotometer. The samples were dispersed in 1 cm path length quartz cells filled with hexane.

TABLE 1: The composition of the Te precursor solutions.

| Injection solutions | Te (g) | TOP (g) | TBP (g) | Te solution (g) | OLAM mL | ODE mL |
|---|---|---|---|---|---|---|
| Te-TOP-OLAM | 0.5 | 4.5 | 0 | 0.0638 | 2 | 0 |
| Te-TOP-ODE | 0.5 | 4.5 | 0 | 0.0638 | 0 | 2 |
| Te-TBP-OLAM | 0.5 | 0 | 4.5 | 0.0638 | 2 | 0 |
| Te-TBP-ODE | 0.5 | 0 | 4.5 | 0.0638 | 0 | 2 |
| Te-TOP-TBP-OLAM | 0.7018 | 6.5428 | 2.1054 | 0.1190 | 2 | 0 |
| Te-TOP-TBP-ODE | 0.7018 | 6.5428 | 2.1054 | 0.1190 | 0 | 2 |

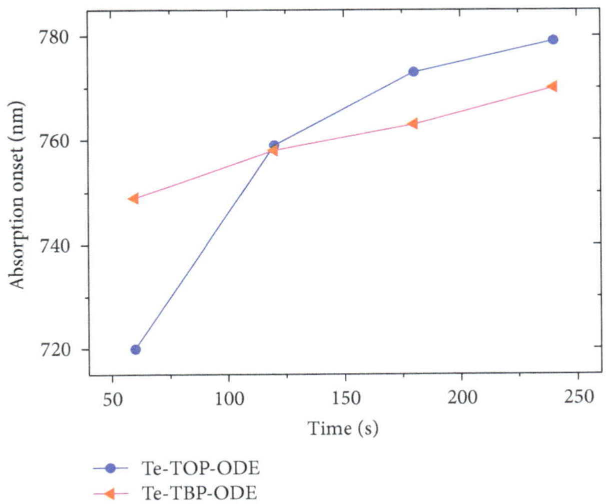


FIGURE 2: Absorption onsets of CdTe NCs synthesized with different tellurium precursor solutions: Te-TOP-ODE and Te-TBP-ODE.

The absorbance values were used to calculate the molar particle extinction coefficients, the NCs diameters, and their concentration.

*2.5.1. Transmission Electron Microscopy (TEM).* The size and morphology of the nanocrystals were studied with a Zeiss EM 902 A transmission electron microscope with an accelerating voltage of 80 kV. TEM specimens were prepared by putting a 10 $\mu$L aliquot drop of a very dilute hexane solution onto a carbon-coated copper grid, which was dried in air at room temperature.

*2.5.2. Powder X-Ray Diffraction (XRD).* Powder X-ray diffraction (XRD) was measured with a PANalytical X'Pert PRO MPD diffractometer operating with Cu K$\alpha$ radiation, Bragg-Brentano $\theta$-2$\theta$ geometry, a goniometer radius of 240 mm and variable slits. The samples were measured on low background silicon sample holders and prepared for the XRD measurements as follows. Nanocrystals in hexane solution were dropped on the silicon substrate at room temperature. The XRD data was analyzed using the X'Pert HighScore Plus software in conjunction with the ICDD database, version 2.0902.

## 3. Results and Discussion

In this work, we investigated the influence of the reaction conditions on the shape of CdTe nanoparticles, which were synthesized by a hot-injection method in reaction of Cd-oleate with trialkylphoshine telluride in OLAM as solvent. OLAM is a common organic compound which can be surfactant and solvent at the same time. OLAM reacts with OA to form an amide in the presence of metal ions [15, 17], which leads to a destabilization of the nanocrystals and, for example, facilitates the formation of nanowires by oriented attachment [17]. Moreover, OLAM was found to activate the metal precursors; thus, it influences the growth of nanoparticles [15, 17–19]; diluting OLAM with a noncoordinating solvent opens up the possibility to tune the activity of the metal monomers. This turned out to be an efficient way to control the shape of CdTe nanocrystals [15]. In the following, we evaluate the possibility to control the morphology of CdTe nanoparticles by additionally changing the reactivity of the Te-monomers.

*3.1. Effect of Ligands and Solvents in the Tellurium Precursor Solution on the Shape of CdTe Nanocrystals.* Various shapes of CdTe nanocrystals, such as nanodots, nanorods, multipods, and nanowires, can be synthesized in pure OLAM solution, only by changing the composition of the tellurium injection solution (Figure 1). Dissolving tellurium in TOP, TBP, or the mixture of them and diluting the resulting tri-alkylphosphine telluride with OLAM or ODE represent a suitable way to prepare tellurium injection solutions. TEM images of CdTe nanocrystals obtained with different Te precursor solutions are showed in Figure 1. In these six reactions we varied the ligands (TOP, TBP, or the mixture of them) and the solvent (OLAM and ODE), but we kept the Te concentration constant.

Comparison between the upper and the lower panel in Figure 1 shows that even a relatively small amount of solvent in the injection solution has a big impact on the morphology of CdTe nanocrystals (Figure 1). This can be rationalized when taking into account that amine compounds are known to activate the metal carboxylate precursors [15, 17–19]. Therefore, the ODE solvent in tellurium precursor solution decreases the activity of the Cd monomers, because it lowers the overall concentration of the amine and, thus, reduces the activation of the Cd monomers. All reactions with injection solutions containing OLAM (upper panel in

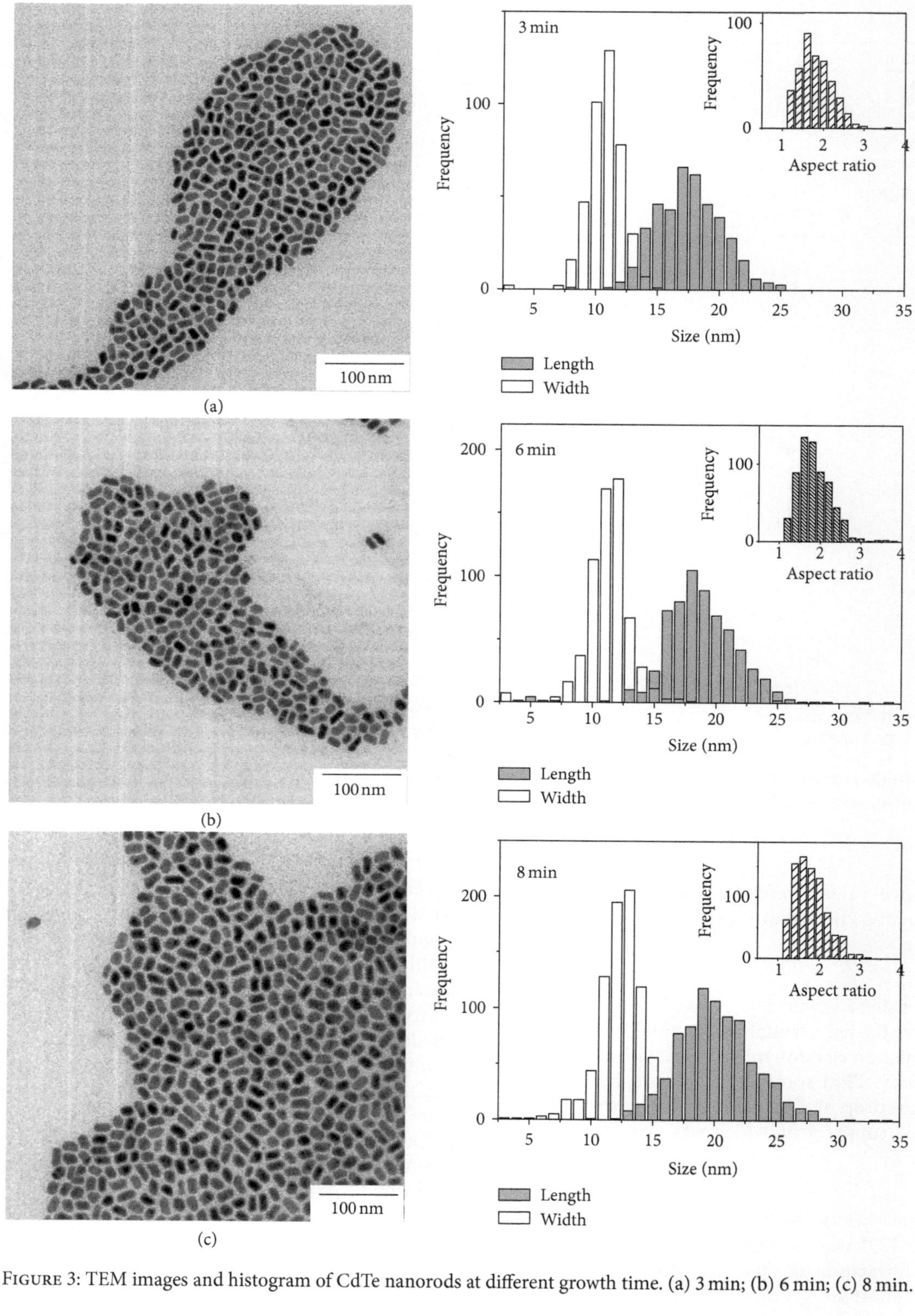


FIGURE 3: TEM images and histogram of CdTe nanorods at different growth time. (a) 3 min; (b) 6 min; (c) 8 min.

Figure 1) yield particles with elongated shapes. This preferential one-dimensional growth is an indication for a higher monomer activity, compared with reactions with injection solutions containing ODE.

The use of TOP and TBP as ligands in tellurium precursor solutions is a good way to control the activity of the Te monomers. TOP and TBP are similar organic compounds, with C8 chains in TOP and C4 chains in TBP. Ligands with

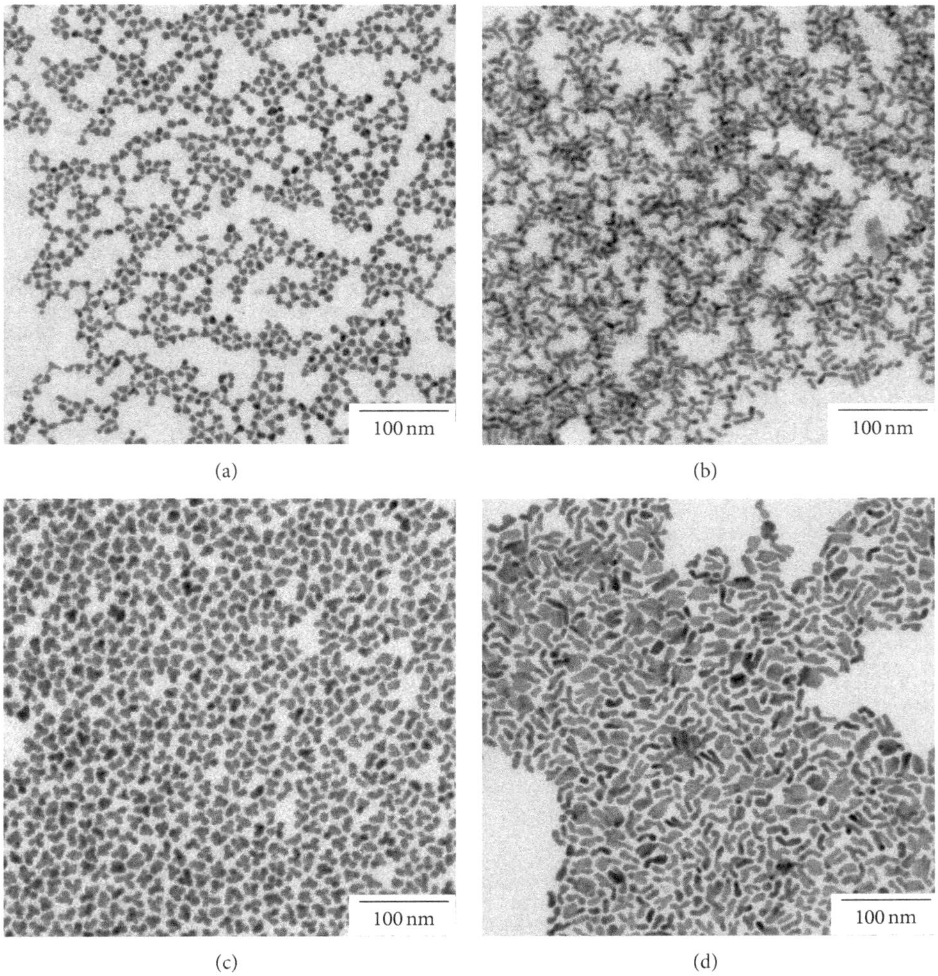


FIGURE 4: TEM images of CdTe NCs synthesized with different OLAM : OA ratio. (a) No OLAM; (b) OLAM : OA = 1; (c) OLAM : OA = 4; (d) OLAM : OA = 20.

shorter carbon chains have a higher diffusion coefficient, which results in a higher monomer activity, compared to compounds with longer carbon chains. Furthermore, TOP-Te and TBP-Te differ in the strength of their Te-P bonds; TOP-Te being more stable. Thus, TBP-Te is more reactive than TOP-Te. In spite of that, the results of reaction with TBP-Te and TOP-Te show some similarities; in OLAM both Te-precursors lead to the formation of nanorods, while the growth of quasispherical particles can be observed in ODE. However, a closer look at the development of the reaction shows that the underlying mechanisms are different. We compared the absorption onsets of aliquots taken during reactions with Te-TOP-ODE and Te-TBP-ODE (Figure 2). In the reaction with TBP, the particles forming during the first minute are larger than in the reaction with TOP. This is due to the higher reactivity of the Te precursor containing TBP. However, because of the formation of larger particles, a smaller amount of monomers is left for further growth. Thus, higher monomer reactivity leads to a fast monomer depletion of the solution, which slows down the growth rate of the particles at later growth stages. In the reaction with TOP-Te, the Te precursor is less reactive, which results in the formation of smaller seeds. There are more monomers left for further growth, which is, however, slower, compared with the case of TBP-Te, because of the lower monomer activity. In the reactions of both Te precursors with more reactive Cd-precursor (in pure OLAM) nanorods are formed, due to the relatively high chemical potential of the solution. A slightly lower Cd precursor reactivity (with ODE in the injection solution) leads to the formation of nanodots.

In reactions with Te-TOP-TBP, two Te precursors with different reactivities are present in one reaction. The more reactive Te-TBP precursor should react first and be preferentially involved in the seed formation process. The amount of monomers used up during this growth stage will be smaller, than in the case of pure Te-TBP precursor, because of the lower concentration of Te-TBP. Thus, the chemical potential of the solution remains relatively high, and also enough monomers are left for further one-dimensional growth of the nanocrystals. In the reactions with a more reactive Cd precursor (in pure OLAM), this leads to the formation of nanowires, while a slightly lower Cd-precursor reactivity

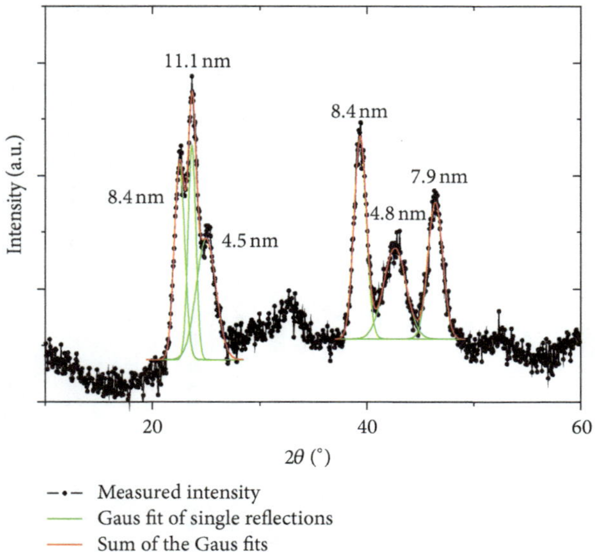


FIGURE 5: XRD pattern of CdTe NCs synthesized from OLAM : OA = 4 (TEM image in Figure 4(c)). Each reflection was fitted with a single Gaussian (green lines), and the size of the crystallites was calculated from the broadening of the reflections with the Debye-Scherrer formula.

(with ODE in the injection solution) result, in the formation of tetrapods.

With TOP as ligand and OLAM as solvent in tellurium precursor solution, uniform CdTe nanorods were synthesized. Their growth was studied in more detail by TEM (Figure 3). Their mean aspect ratio is around 1.6, and it is almost not changing during the reaction (Figures 3(a), 3(b), and 3(c)). The elongated shape forms in the beginning of the reaction and does not change during further growth. Also the size distribution of the nanorods remains the same. This shows that the monomer activity does not decrease significantly during the first 8 minutes of the reaction. In this case the shape of the nanocrystals would change from elongated to quasi-spherical; however, the monomer activity is not sufficient for the further one-dimensional growth of the nanorods.

*3.2. Morphology Control by Changing OLAM : OA Ratio.* In this section, we investigated the possibility to influence further the morphology of CdTe tetrapods. We used a Te-TOP-TBP-ODE injection solution, which turned out to be suitable to synthesize uniform tetrapods (see Figure 1(e)) and changed the ratio between OLAM and OA. In these experiments, we took the same amount of OA and kept the same sum volume of OLAM and ODE constant (10 mL); thus, the overall CdTe concentration was fixed. In this way we could exclude that the effects we observe are solely due to changes in the concentration of the reactants.

Figure 4 shows the different morphologies of CdTe nanocrystals obtained by changing OLAM : OA ratio from 0 to 20. In our experiment, the use of OA as the only surfactant leads to the formation of triangular (in the two-dimensional projection) nanoparticles (Figure 4(a)). Without OLAM the reactivity of the Cd-monomers is low (no amine activation); thus, there is no one-dimensional growth. The presence of the same volume of OLAM and OA in the reaction system causes the formation of tripods and tetrapods (Figure 4(b)). With larger amount of OLAM, the thickness of the arms of the tetrapods increases, while their length decreases (Figure 4(c)). With even higher OLAM concentrations, formation of more irregular shapes of branched CdTe nanocrystals, such as Y-shaped multipods, can be observed. Also the fraction of flat, two-dimensional CdTe nanoparticles increases with higher OLAM concentrations (Figure 4(d)). Thus, the presence of OLAM influences the growth direction of the CdTe nanocrystals.

Figure 5 shows the XRD patterns of CdTe nanocrystals shown in Figure 4(c). Each reflection was fitted with a single Gaussian function (green curves in Figure 5); the sum of the fitting curves, which corresponds well to the measured diffraction pattern, is shown as red lines in Figure 5. The diffraction peaks can be assigned to wurtzite CdTe (ICDD number 01-073-2871). Results of the calculation of the crystallite dimension in the corresponding directions using the Scherrer equation with $K = 1$ for undefined geometry are included in Figure 5. Because of the anisotropic growth of the nanocrystals, we obtain different values for different reflections. The relatively sharp (002) peak in the diffraction pattern is consistent with the preferential growth of the CdTe NCs in the [001] direction. The crystallite size in the [001] direction (11.1 nm) corresponds well to the length of the arms of the tetrapods measured from the TEM images.

## 4. Conclusions

In conclusion, we demonstrated that various shapes of CdTe nanocrystals could be synthesized by controlling the composition of the solvent and of the tellurium precursor solution. Tuning the reactivity of both the cadmium and the tellurium precursors is the main reason for the shape control of CdTe nanocrystals in such a simple system. The activity of the Cd precursor can be controlled by the concentration of OLAM in the solvent, while the use of TOP, TBP, or a mixture of both phosphines in the tellurium injection solution turned out to be a suitable way to control the reactivity of the Te monomers. Changing the reactivity of the Cd and Te precursors at the same time opens up the possibility to influence the seed formation and the growth process of the CdTe nanocrystals and, thus, to control their shape.

## Acknowledgments

The authors would like to thank Erhard Rhiel and Heike Oetting (University of Oldenburg) for assistance in obtaining TEM images and Marta Kruszynska (University of Oldenburg) for taking the XRD data. They gratefully acknowledge funding of the EWE Research Group "Thin Film Photovoltaics" by the EWE AG, Oldenburg.

## References


[1] A. I. Hochbaum and P. Yang, "Semiconductor nanowires for energy conversion," *Chemical Reviews*, vol. 110, no. 1, pp. 527–546, 2010.

[2] P. V. Kamat, "Quantum dot solar cells. Semiconductor nanocrystals as light harvesters," *The Journal of Physical Chemistry C*, vol. 112, pp. 18737–18753, 2008.

[3] Q. Ma and X. Su, "Near-infrared quantum dots: synthesis, functionalization and analytical applications," *The Analyst*, vol. 135, no. 8, pp. 1867–1877, 2010.

[4] P. Alivisatos, "The use of nanocrystals in biological detection," *Nature Biotechnology*, vol. 22, no. 1, pp. 47–52, 2004.

[5] Z. Tang, N. A. Kotov, and M. Giersig, "Spontaneous organization of single CdTe nanoparticles into luminescent nanowires," *Science*, vol. 297, no. 5579, pp. 237–240, 2002.

[6] Z. Tang, Z. Zhang, Y. Wang, S. C. Glotzer, and N. A. Kotov, "Self-assembly of CdTe nanocrystals into free-floating sheets," *Science*, vol. 314, no. 5797, pp. 274–278, 2006.

[7] H. Qian, C. Dong, J. Weng, and J. Ren, "Facile one-pot synthesis of luminescent, water-soluble, and biocompatible glutathione-coated CdTe nanocrystals," *Small*, vol. 2, no. 6, pp. 747–751, 2006.

[8] Y. A. Yang, H. Wu, K. R. Williams, and Y. C. Cao, "Synthesis of CdSe and CdTe nanocrystals without precursor injection," *Angewandte Chemie—International Edition*, vol. 44, no. 41, pp. 6712–6715, 2005.

[9] H. Zhang, D. Wang, and H. Möhwald, "Ligand-selective aqueous synthesis of one-dimensional CdTe nanostructures," *Angewandte Chemie—International Edition*, vol. 45, no. 5, pp. 748–751, 2006.

[10] J. Y. Zhang and W. W. Yu, "Formation of CdTe nanostructures with dot, rod, and tetrapod shapes," *Applied Physics Letters*, vol. 89, no. 12, Article ID 123108, 2006.

[11] R. Zheng, S. Guo, and S. Dong, "Synthesis of CdTe nanocrystals using Te nanorods as the Te source and the formation of microtubes with red fluorescence," *Inorganic Chemistry*, vol. 46, no. 17, pp. 6920–6923, 2007.

[12] W. W. Yu, Y. A. Wang, and X. Peng, "Formation and stability of size-, shape-, and structure-controlled CdTe nanocrystals: ligand effects on monomers and nanocrystals," *Chemistry of Materials*, vol. 15, no. 22, pp. 4300–4308, 2003.

[13] J. Kolny-Olesiak, V. Kloper, R. Osovsky, A. Sashchiuk, and E. Lifshitz, "Synthesis and characterization of brightly photoluminescent CdTe nanocrystals," *Surface Science*, vol. 601, no. 13, pp. 2667–2670, 2007.

[14] V. Kloper, R. Osovsky, J. Kolny-Olesiak, A. Sashchiuk, and E. Lifshitz, "The growth of colloidal cadmium telluride nanocrystal quantum dots in the presence of Cd0 nanoparticles," *Journal of Physical Chemistry C*, vol. 111, no. 28, pp. 10336–10341, 2007.

[15] X. Jin, J. Li, J. Parisi, and J. Kolny-Olesiak, "Shape control of CdTe nanocrystals synthesized in presence of in situ formed CdO particles," *Journal of Nanoparticle Research*, vol. 13, no. 12, pp. 6963–6970, 2011.

[16] J. W. Cho, H. S. Kim, Y. J. Kim et al., "Phase-tuned tetrapod-shaped CdTe nanocrystals by ligand effect," *Chemistry of Materials*, vol. 20, no. 17, pp. 5600–5609, 2008.

[17] X. Jin, M. Kruszynska, J. Parisi, and J. Kolny-Olesiak, "Catalyst-free synthesis and shape control of CdTe nanowires," *Nano Research*, vol. 4, no. 9, pp. 824–835, 2011.

[18] L. S. Li, N. Pradhan, Y. Wang, and X. Peng, "High quality ZnSe and ZnS nanocrystals formed by activating zinc carboxylate precursors," *Nano Letters*, vol. 4, no. 11, pp. 2261–2264, 2004.

[19] N. Pradhan and X. Peng, "Efficient and color-tunable Mn-doped ZnSe nanocrystal emitters: control of optical performance via greener synthetic chemistry," *Journal of the American Chemical Society*, vol. 129, no. 11, pp. 3339–3347, 2007.

# 7

# Metal Oxide Sol-Gels ($ZrO_2$, AlO(OH), and $SiO_2$) to Improve the Mechanical Performance of Wood Substrates

**Véronic Landry, Pierre Blanchet, and Gabrielle Boivin**

*FPInnovations, 319 Franquet, Quebec, QC, Canada G1P 4R4*

Correspondence should be addressed to Pierre Blanchet; pierre.blanchet@fpinnovations.ca

Academic Editor: Joseph Koo

Wood is a renewable material widely used in many applications due to its unique properties and distinctive look. However, as wood is organically constituted, it is slowly destroyed by the long-term impact of oxygen, UV radiations, water, and biological attacks (Mahltig et al., 2008). Therefore, protective treatments are necessary to improve the mechanical, thermal, and chemical properties of wood. In order to improve the mechanical properties of sugar maple (*Acer saccharum* Marsh.), as this species is widely used in the wood products industry, samples of sugar maple were impregnated with sols of metal oxides (AlO(OH), $SiO_2$, and $ZrO_2$). The weight gain and two different techniques of microscopy were used to evaluate the efficiency of the impregnation in the wood samples. The mechanical properties were evaluated using hardness test, scratch test, and impact test. It was shown that the maple samples impregnated with $ZrO_2$ led to the greatest improvement of the mechanical properties.

## 1. Introduction

Wood is a renewable material widely used in many applications, such as outdoor (sidings, shingles, etc.) and indoor products (hardwood floors, kitchen cabinets, furniture, etc.), due to its excellent properties and distinctive look. However, as wood is organically constituted, it is slowly destroyed by the long-term impact of oxygen, UV radiations, water, and biological attacks [1]. Therefore, protective treatments have to be used to increase its life in service. In addition the growing environmental concerns push the industry to develop and use nontoxic wood preservatives [2, 3].

Recently, it has been observed that one promising method to improve and provide new properties for wood materials is to modify the wood surface by the deposition of hybrid inorganic-organic materials with the sol-gel process. Hybrid inorganic-organic materials offer the opportunity to combine the desirable properties (i.e., toughness, elasticity, etc.) of organic polymers with those of inorganic solids (i.e., hardness, chemical resistance, etc.) [4]. The sol-gel process is simple, low cost, and it allows room-temperature deposition of hybrid inorganic-organic thin films on a wide range of substrates, including wood [2, 3]. This technique is appropriate for commodity production such as wood product industry. Depending on the sol-gel synthesized, various wood properties can be improved, such as color retention, mechanical properties, dimensional stability, and more.

Over the last two decades, the sol-gel process has been the subject of many studies. However, most of the studies on sol-gel process were carried on different substrates such as glass, metal, or textile. Using this process on substrates like wood brings out many special features, such as thermal treatments. Usually, coating metal oxide sols on substrates requires a really high temperature that cannot be used for wood substrates as the organic components would degrade. The application of sols to wood can be done in many ways; unlike the other substrates, the sol can be impregnated into the wood, and not just applied to the wood surface. Depending on the type of application or impregnation techniques used, the surface protection is different: using a brush or nozzle protects only on the surface, while a vacuum or vacuum/pressure impregnation gives full protection [1].

Nanoparticles such as silica ($SiO_2$) and aluminum oxide (AlO(OH)) are widely used in the coatings industry as they improve the scratch and abrasion resistance and the hardness of the coatings [5]. Table 1 compares the hardness (Mohs

TABLE 1: Hardness (Mohs scale) and refractive index of frequently used particles in the coating industry at the micrometric scale.

| Properties | $Al_2O_3$ | $SiO_2$ | $ZrO_2$ | $TiO_2$ | ZnO |
|---|---|---|---|---|---|
| Hardness (Mohs scale) | 8.0-9.0 | 7.0 | 8.0 | 5.5-6.5 | 5.0 |
| Refractive index | 1.7-1.8 | 1.55 | 2.15 | 2.6 | 2.02 |

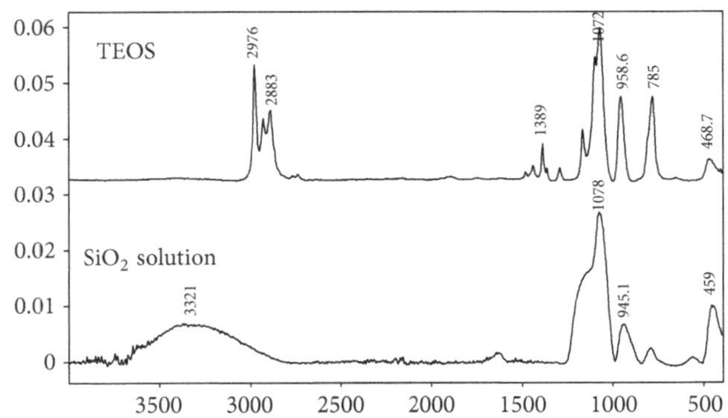


FIGURE 1: Infrared spectra of TEOS, the organic precursor, and the $SiO_2$ product found after the drying.

scale) and the refractive index of frequently used particles in the coating industry. Based on high values of the Mohs scale and refractive index that should not alter the perceived wood color, sols containing AlO(OH), $SiO_2$, and $ZrO_2$ were selected for this study.

The objective of this work was to improve the mechanical properties of sugar maple (*Acer saccharum* Marsh.) by sol-gel treatments with three different nanoparticles (AlO(OH), $ZrO_2$, and $SiO_2$). First, the penetration of the sol was evaluated by two different techniques of microscopy and by weight gain. Then, the hardness, impact resistance, and scratch resistance of the samples treated were evaluated.

## 2. Materials

*2.1. Wood Samples.* The wood species selected for this project was the sugar maple (*Acer saccharum* Marsh.), as it is widely used in the north American wood product industry. Sugar maple is a diffuse porous wood, with a specific density of 620 to 680 kg/m$^3$ [6]. Sugar maple is a hardwood, but as a commercial species, it faces competition from tropical wood due to the latter's high density and hardness. Samples used in this study were 5 cm × 5 cm × 1.27 cm, longitudinal, tangential, and radial wood directions, respectively, and were sanded at 150 grit with a wide belt sander (SCM, Sandya 20, Italy) and conditioned at 20°C and 50% relative humidity (RH) until constant mass before the treatment.

*2.2. Sol Preparation.* Aluminum isopropoxide (AIP), tetraethylorthosilicate (TEOS), 70% w/w solution of zirconium n-propoxide, n-propanol, and acetyl acetone (acac) were purchased from Sigma-Aldrich. Ethanol anhydrous

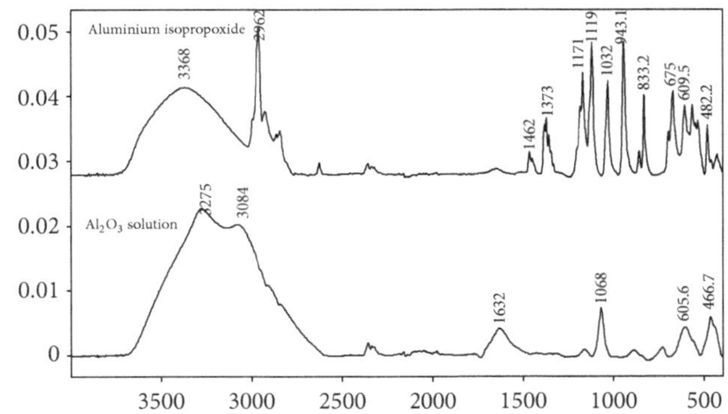


FIGURE 2: Infrared spectra of the aluminium precursor (aluminium isopropoxide) and the AlO(OH) solution.

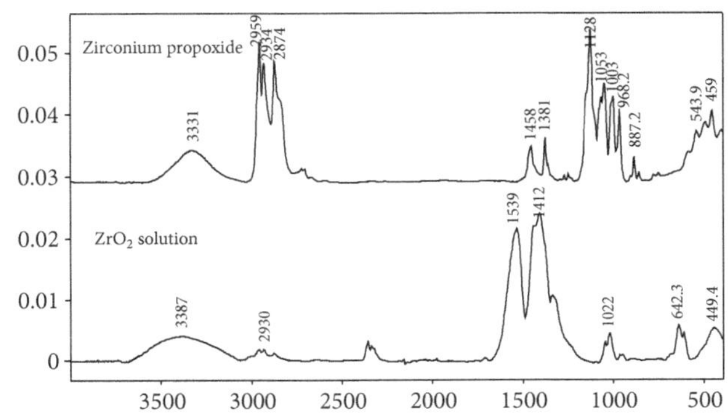


FIGURE 3: Infrared spectra of the zirconium precursor (zirconium propoxide) and the $ZrO_2$ solution.

was purchased from SAQ (Quebec, Canada) and p-toluene sulfonic acid was purchased from Laboratoire MAT (Quebec, Canada).

*2.2.1. Synthesis of AlO(OH) Sol.* AIP ($Al(O(CH(CH_3)_2)_3)$) was mixed in deionized water (200 mole of water for 1 mole of alkoxide) under vigorous stirring for two hours at 85°C. Then, 0.2 mole of hydrochloric acid (1 M) for each mole of alkoxide was added to the solution as a catalyst and mixed for one hour at 85°C.

*2.2.2. Synthesis of $SiO_2$ Sol.* TEOS ($Si(OC_2H_5)_4$), anhydrous ethanol ($CH_3CH_2OH$), and distilled water were mixed together in a molar ratio of 1 : 4 : 16, respectively. A pH value of 3 was reached by adding hydrochloric acid (1 M) drop by drop. The sol gel was then heated for 1.5 hours at 60°C.

*2.2.3. Synthesis of $ZrO_2$ Sol.* A 70% w/w solution of zirconium n-propoxide ($Zr(OCH_2CH_2CH_3)_4$ in n-propanol was mixed with n-propanol ($CH_3CH_2CH_2OH$) and acac ($CH_3C(O)CH_2C(O)CH_3$). A solution of p-toluene sulfonic acid ($CH_3C_6H_4SO_3H$) in distilled water (4 M) was added to this mixture. The resulting solution was heated for two

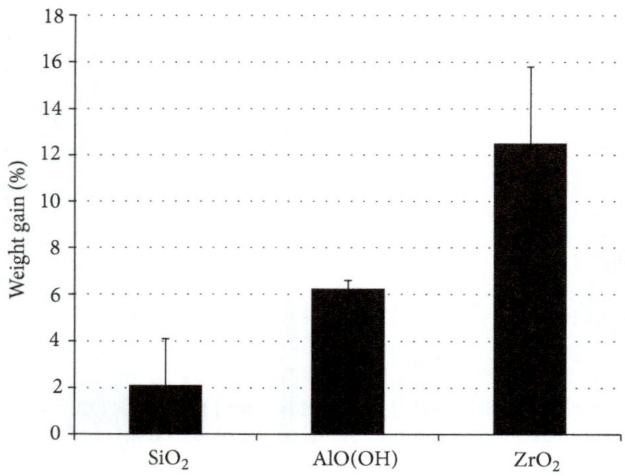


FIGURE 4: Weight gains of maple samples treated with the $SiO_2$, AlO(OH), and $ZrO_2$ sol-gels.

hours at 60°C. The final zirconium concentration in the sol-gel was 0.8 M and the molar ratio of acac and distilled water to zirconium was kept to 1 : 10 : 1.

*2.3. Wood Samples Treatment.* Thirty samples of sugar maple were used for each treatment. First, the wood samples were weighed and then soaked in the solution. A vacuum at 81.2 MPa was generated for one hour and then the samples were returned to ambient pressure for two hours. After the treatment, the wood samples were slightly wiped with a paper towel and they were dried at room temperature for 24 hours. The samples were then placed in an oven for 24 hours at 103°C for water and solvent evaporation and to promote the condensation reaction. After curing, the wood samples were weighed again to determine the gel retention (weight gain).

## 3. Methods

Infrared spectroscopy was used to follow the reaction of the different sols. The apparatus used in this project is a Tensor 37 Fourier transform infrared spectrophotometer equipped with a Platinum ATR (diamond crystal) from Bruker, using a resolution of 4 $cm^{-1}$.

Weight gain of the samples was obtained by measuring the initial and final weight (after sol-gel impregnation and curing) of the samples. Sample weighing was done at the same temperature and humidity conditions before and after the sol-gel treatment (20°C, 50% RH).

Two different microscopes were used to assess the penetration of the sol-gels in the wood.

Transmission electron microscopy (TEM) was used for the observation of gel penetration in the wood substrate. Each wood sample was cut to obtain 200 nm thick wood slices, using a Reichert-Jung ultramicrotome, model ultracut E (Vienna, Austria). The transmission electron microscope used was a JEM-1230 (JEOL) and the images were recorded at an acceleration voltage of 80 kV.

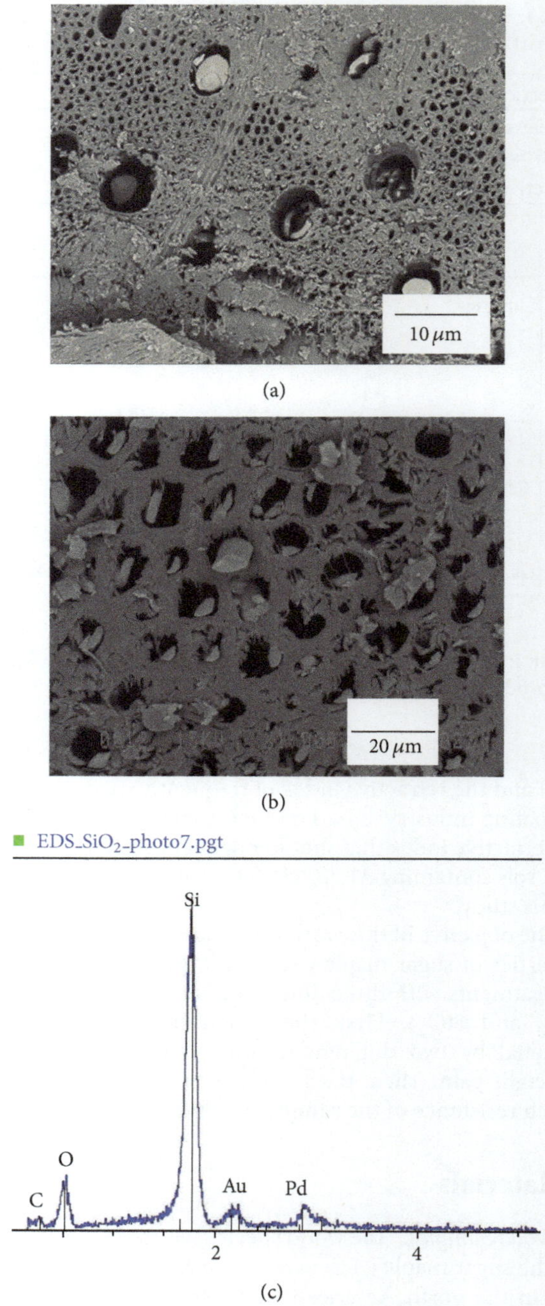


FIGURE 5: Scanning electron microscopy images of the maple samples treated with the $SiO_2$ sol ((a) and (b)) elemental analysis spectra recorded for the particles found in the wood cells (c).

Scanning electron microscopy with energy-dispersive X-ray spectrometry (SEM-EDS) was performed using a JSM-840A (JEOL) microscope equipped with a NORAN EDS detector. Prior to SEM investigation, the wood samples were microtomed and then gold palladium was coated in a vacuum sputter. In addition to the images obtained, this technique allows the precise identification of the different chemical elements at a specific location.

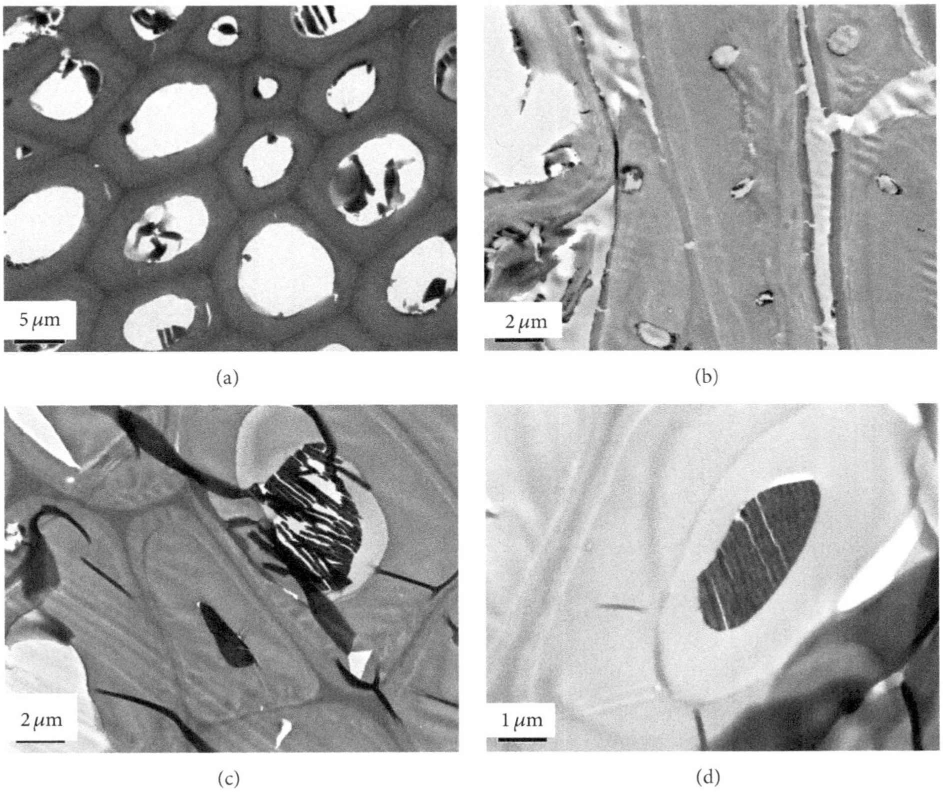


FIGURE 6: Transmission electron microscopy images of the wood cells of maple samples impregnated with the $SiO_2$ sol.

*3.1. Mechanical Properties.* Brinell hardness (HB) can be described by the following equation:

$$HB = \frac{2 * F}{g * \pi * D * \left(D - \sqrt{D^2 - d^2}\right)}, \quad (1)$$

where $F$ being the applied force, $g$ the gravity acceleration, $D$ the ball diameter, and $d$ the print diameter. The method used here is based on the British Standard Brinell EN 1534-2000 method; a steel ball is pushed into the material at a constant speed until a penetration depth of 1 mm. An Alliance Rt/50 system from MTS Systems (MN, USA) was used. The test bench with a charge cell measures the applied force needed to achieve these results.

The scratch test was divided into two steps; first the scratches were made with a Taber Multi-Finger Scratch/Mar Tester using a weight of 15 N and a tungsten carbide 0.7 mm diameter tip. The test was performed back and forth on a pneumatic table at a speed of 10 cm/s. The depth and width of the scratches were measured with a 3D profilometer, the ContourGT-K1 from Bruker.

The impact resistance of the samples was measured using an impact tester; a ball weighing 0.91 kg was dropped on the sample from a 25 cm height. The impact print was manually measured using a caliper for the diameter (two measures were taken for each print) and a vertical tip mounted micrometer to measure the depth. This test is based on the standard test method ASTM D 2794-93.

## 4. Results and Discussion

FT-IR spectra of the sol-gel precursors and the final product obtained for each sol-gel synthesis were taken.

*4.1. $SiO_2$ Sol.* The infrared spectra of TEOS, the organic precursor, and of the $SiO_2$ product obtained after drying the solution are presented in Figure 1. The infrared spectrum of $SiO_2$ is available in the literature. The strong band at 3321 $cm^{-1}$ represents the OH stretching at the surface of the silica particles, the peak at 1078 $cm^{-1}$ represents the Si-O-Si stretching vibration, the peak at 935 $cm^{-1}$ represents the Si-OH stretching band, and finally the band at 459 $cm^{-1}$ represents the Si-O out of plane deformation. The presence of these peaks confirms the synthesis of a $SiO_2$ gel.

Figure 2 presents the organic precursor, the aluminium isopropoxide, and the AlO(OH) product obtained after drying. The surface of the AlO(OH) is covered by OH groups (boehmite) and the band due to these species is presented above 3000 $cm^{-1}$. This band can be associated with the OH stretching of the absorbed water, with the bridged hydroxyl group with molecular water or with other OH groups, and

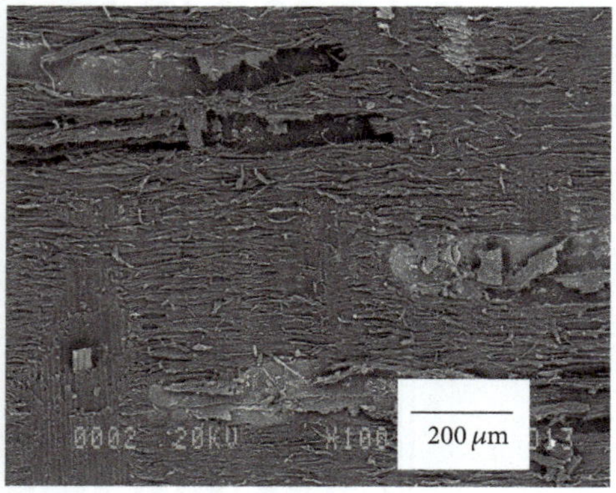


(a)

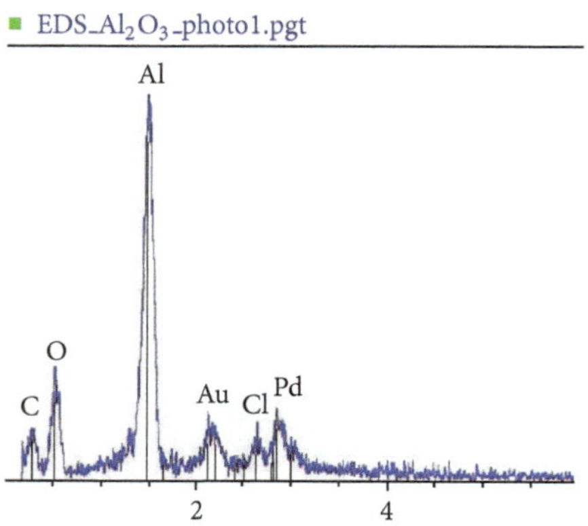


(b)

FIGURE 7: Scanning electron microscopy of the maple samples impregnated with the boehmite sol (a) and elemental analysis of the particles found in the maple samples impregnated with the boehmite sol.

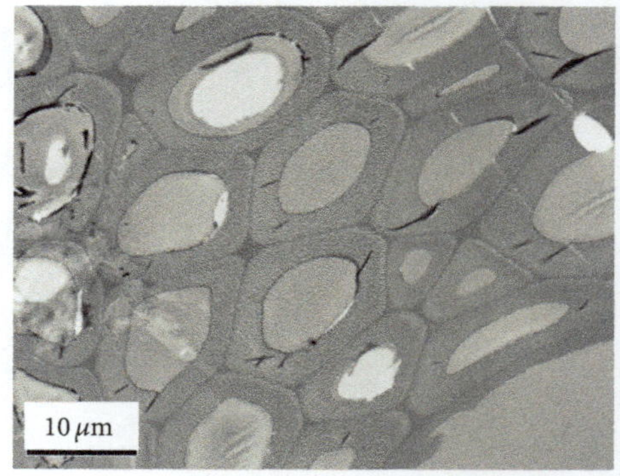


(a)

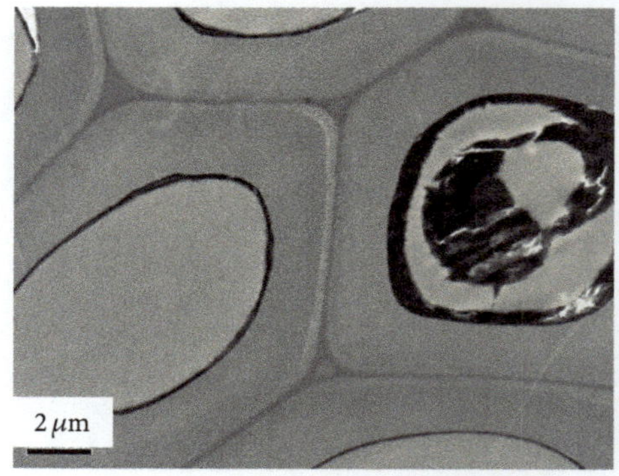


(b)

FIGURE 8: Transmission electron microscopy images of the wood cells of maple samples impregnated with the boehmite sol.

with isolated OH groups. The disappearance of most of the aluminium isopropoxide peak also revealed that a reaction occurred.

*4.2. $ZrO_2$ Sol.* Figure 3 presents the spectra of the zirconium propoxide and of the $ZrO_2$ gel after drying. As $ZrO_2$ spectrum is not reported clearly in the literature, it is difficult to analyze the spectra obtained for the $ZrO_2$ gel. However, the disappearance of the group of peaks around 2900 $cm^{-1}$ and 1000 $cm^{-1}$ is a good sign of the sol-gel reaction.

### 4.3. Gel Retention

*4.3.1. Weight Gain.* Weight gains obtained for the samples treated with the three different sols are presented in Figure 4. Significant variations in weight gain can be observed from one sample to another. Samples treated with $SiO_2$ have a lower weight gain percentage than those treated with $ZrO_2$, which have the highest. Factors like molecular weight of the precursors could explain this difference. However, as the molecular weight of the principal precursor is smaller in the following order: AIP < TEOS < zirconium n-propoxide, there is reason to think that other factors affect the penetration of the sols, such as the hydrolysis time of alkoxides and the viscosity of the solution.

### 4.4. Microscopy

*4.4.1. $SiO_2$ Gel.* Figure 5 shows scanning electron microscopy images taken from the maple samples treated with the $SiO_2$ sol. On Figures 5(a) and 5(b), white aggregates can be observed in the wood cells. These aggregates represent $SiO_2$ gel particles as identified by the elemental analysis (Figure 6(c)). In fact, it was found that the particles present in the wood cells are mainly constituted of silicon and oxygen, thus the $SiO_2$ gel product. It is also possible to see that the particles did not bond to the wood cell surface and that they did not occupy the whole cavity of the wood cells. Transmission electron microscopy images (Figures 6(a) and 6(c)) also present this phenomenon. Figure 6(b) shows $SiO_2$ particles in the wood cells but also in the punctuations of the tracheids. The presence of $SiO_2$ particles in the punctuations could prevent the penetration of the sol further into the wood.

*4.4.2. AlO(OH) Gel.* Scanning electron microscopy images of a maple sample treated with AlO(OH) sol is showed in Figure 7(a). The white aggregates in the wood vessels represent the AlO(OH) gel as verified by the elemental analysis presented on Figure 7(b). Aluminum and oxygen, the two main components of the gel, were dominantly found,

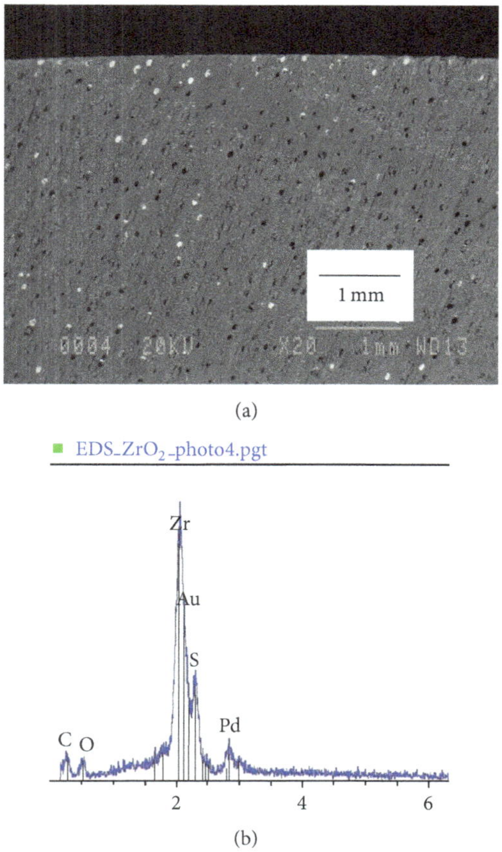


FIGURE 9: (a) Scanning electron microscopy image of the $ZrO_2$ sol-gel treated maple samples and (b) elemental analysis of the white spots observed.

then some carbon which probably comes from impurities or unreacted reactive, and finally chlorine which comes from the hydrochloric acid used during the synthesis to control the pH. Figure 7(a) also revealed that boehmite particles completely occupy some of the wood vessels and that they are bonded to the cell walls, which was not the case for the $SiO_2$ particles. However, transmission electron microscopy of the boehmite treated maple samples (Figure 8) shows that the gel is predominantly located on the cell walls only and do not occupy the whole cavity. This suggests that a condensation reaction has occurred between the cell walls and the gel. This reaction is favorable as it would create a composite effect and reinforce the wood samples.

*4.4.3. $ZrO_2$ Gel.* On the scanning electron microscopy images (Figure 9(a)), white aggregates can be observed not only at the surface but also deeper in the wood. These white aggregates were identified as the $ZrO_2$ gel by the elemental analysis (Figure 9(b)). Moreover, it can be observed that the $ZrO_2$ particles present in the wood cells seem to be firmly bonded to the wood and to occupy the whole wood cells cavity. Transmission electron microscopy images (Figure 10) present a totally different sol-gel compared to the one seen for $SiO_2$ and AlO(OH). In fact, Figures 10(b), 10(c), and 10(d) clearly show nanoparticles located not only on the cell walls but also on a 3D network.

*4.5. Mechanical Properties.* The hardness, scratch resistance, and impact resistance of the different treated maple samples were evaluated and compared with the results obtained for the untreated sugar maple samples. As the species are hardwood, achieving improvement of the mechanical properties is a challenge.

*4.6. Hardness.* The results obtained with the Brinell modified hardness test are presented in Figure 11. These results show that the Brinell modified hardness of maple samples was improved by the $ZrO_2$ treatment. In fact, sol-gel treatment with $ZrO_2$ nanoparticles increases the hardness by 20% more than untreated samples. The other two treatments ($SiO_2$ and AlO(OH)) did not increase the hardness significantly. When referring to the Mohs scale hardness, AlO(OH) is 9.0, $SiO_2$ is 7.0, and $ZrO_2$ is 8.0. Sol-gel treatments with AlO(OH) would be expected to have the greatest hardness improvement, followed by $ZrO_2$, and then $SiO_2$. However, other factors such as the nature of the interactions between the inorganic materials and the wood cells (condensation on the cell walls or no), as well as gel retention, are also expected to play an important role in the mechanical properties. The microscopy images showed a better chemical grafting for $ZrO_2$ followed by AlO(OH) and $SiO_2$. It is indeed possible to see particles bonded on the cell walls for $ZrO_2$ and AlO(OH), but not for $SiO_2$. This proves that chemical grafting (condensation) to the cell walls is important in achieving an improvement of the mechanical properties, regardless of the Mohs hardness of the material used. It is also possible that the $SiO_2$ treatment leads to the leaching of some wood components which could reduce the hardness.

*4.7. Scratch Resistance.* Scratch resistance was evaluated by measuring the depth and width of the scratches prepared by the multi-finger scratch/mar tester. Figures 12 and 13 present the depth and the width of the scratches, respectively. For both parameters, the $ZrO_2$ treated maple samples led to the best results. In fact, with the $ZrO_2$ sol-gel process, the depth of the scratches was decreased by 50%, and the width by 20%. AlO(OH) treated maple samples also present a good scratch resistance compared with the untreated maple samples. Since AlO(OH) is harder than $ZrO_2$, the better performance of the $ZrO_2$ samples could be explained by the higher sol-gel retention. A more significant retention would lead to a denser material and this could help in improving the surface mechanical properties. The $SiO_2$ treatment did not show significant improvements. The lack of bonding between the sol-gel material and the wood (no condensation with the hydroxyl groups on the cell walls), as well as the low material retention, could explain these poor results.

*4.8. Impact Resistance.* The impact resistance of the different samples, treated and untreated, is presented in Figure 14. The

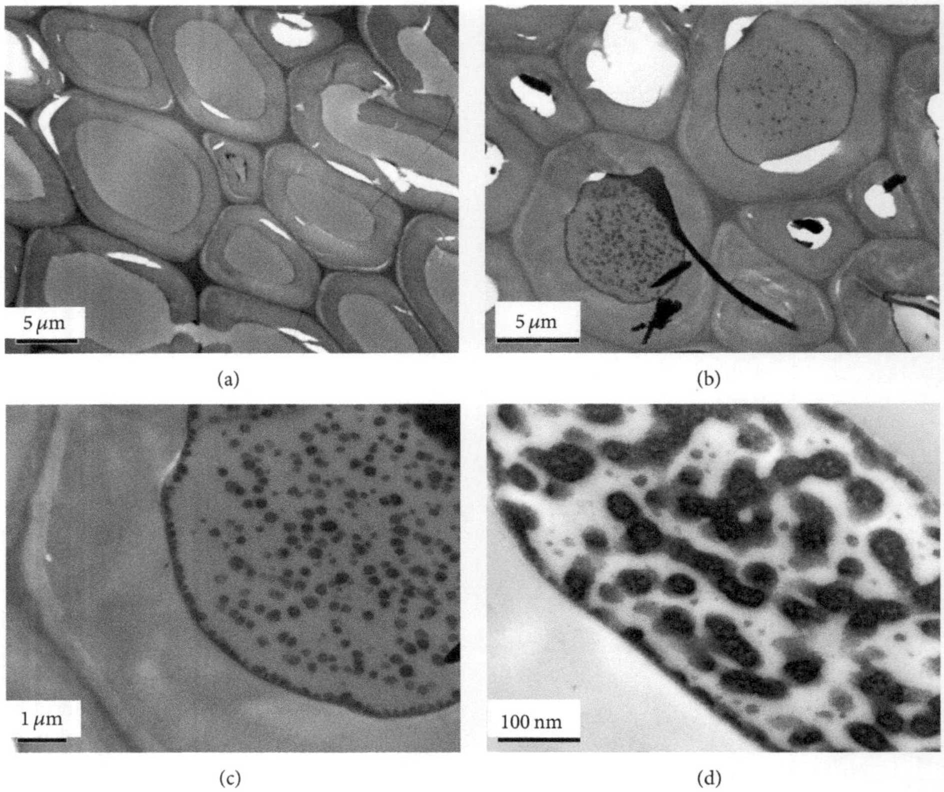


FIGURE 10: Transmission electron microscopy images of the wood cells of the maple samples impregnated with the $ZrO_2$ sol ((a) and (b)), $ZrO_2$ nanoparticles distribution in the wood cells.

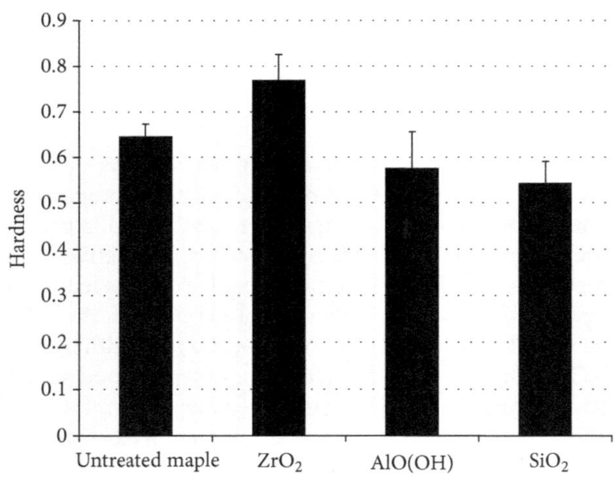


FIGURE 11: Brinell modified hardness of the untreated and sol-gel treated maple samples.

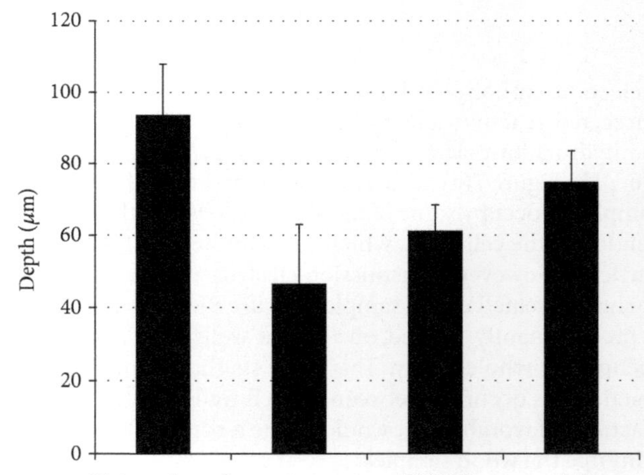


FIGURE 12: Depth of the scratches prepared on the untreated, $ZrO_2$, AlO(OH), and $SiO_2$ sol-gel treated maple samples.

$ZrO_2$ and AlO(OH) treated samples led to the best results with an impact defect decrease of 30%. The $SiO_2$ treatment did not seem to improve nor worsen the impact resistance. As previously stated, the gel retention and the good chemical grafting for $ZrO_2$ and AlO(OH) could explain the good results obtained in this section.

## 5. Conclusions

Sugar maple samples were treated with sols containing three different nanoparticles (AlO(OH), $ZrO_2$, and $SiO_2$), in order to achieve better mechanical properties. The different nanoparticles were selected for their good mechanical

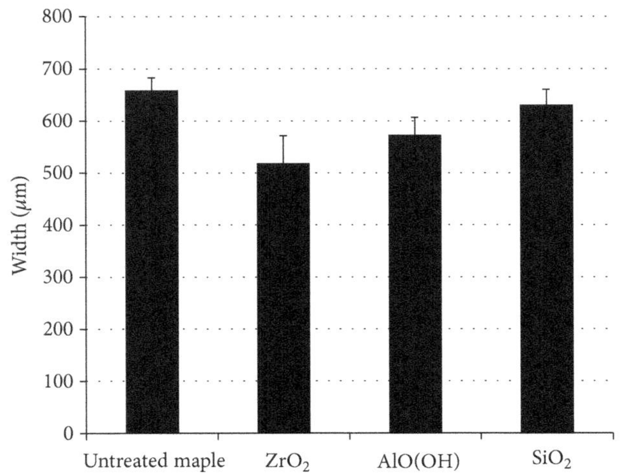


FIGURE 13: Width of the scratches prepared on the untreated, $ZrO_2$, AlO(OH), and $SiO_2$ sol-gel treated maple samples.

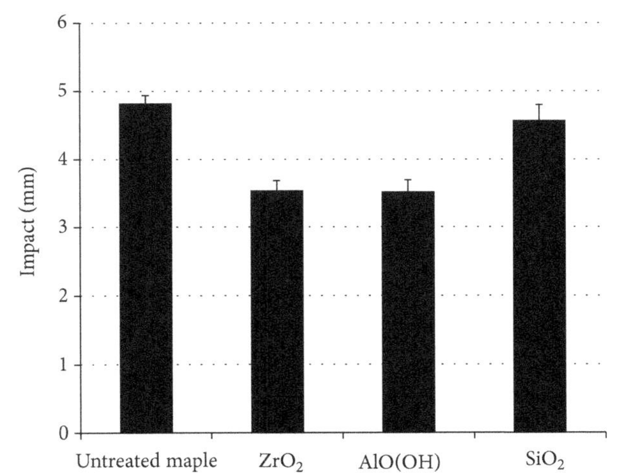


FIGURE 14: Impact resistance of the untreated and the sol-gel treated samples.

properties. The penetration of the sols was evaluated by two different kinds of microscopy and by weight gain.

The mechanical properties, that is, the hardness, scratch resistance, and impact resistance, were measured for the untreated and the sol-gel treated maple samples. The $ZrO_2$ treatment was found to be the most efficient treatment and led to the best mechanical performance, followed by the AlO(OH) treatment and the $SiO_2$ treatment. In fact, with the $ZrO_2$ sol-gel process, the hardness was increased by 20%, the depth of the scratches was decreased by 50%, and the width of the scratches by 20%, while the impact defect was decreased by 30%. These good results can be explained in part by the sol-gel retention, which was more efficient for the $ZrO_2$ than for the other two. Moreover, as reported in the literature, chemical bonding is expected to lead to good properties. In the case of $ZrO_2$ and AlO(OH), the sol-gel was attached to the cell walls, but that was not the case with $SiO_2$; the sol-gel material seemed to have only been deposited there. The improvement of the wood mechanical properties through this technique will help increase wood's service life and make it even more competitive in the market.

## Acknowledgments

The author would like to thank the Natural Resources Canada (NRCan) for their financial support through the Transformative Technologies program.

## References

[1] B. Mahltig, C. Swaboda, A. Roessler, and H. Böttcher, "Functionalising wood by nanosol application," *Journal of Materials Chemistry*, vol. 18, no. 27, pp. 3180–3192, 2008.

[2] M. A. Tshabalala, "Sol-gel deposition of inorganic alkoxides on wood surfaces to enhance their durability under exposure to sunlight and moisture," in *Surface Life Prediction: Challenging the Status Quo*, pp. 385–391, Bluebell, 2005.

[3] M. A. Tshabalala and L. P. Sung, "Wood surface modification by in-situ sol-gel deposition of hybrid inorganic-organic thin films," *Journal of Coatings Technology Research*, vol. 4, no. 4, pp. 483–490, 2007.

[4] S. Wang, R. Mahlberg, S. Jämsä et al., "Surface properties and moisture behaviour of pine and heat-treated spruce modified with alkoxysilanes by sol-gel process," *Progress in Organic Coatings*, vol. 71, no. 3, pp. 274–282, 2011.

[5] C. Sow, B. Riedl, and P. Blanchet, "UV-waterborne polyurethane-acrylate nanocomposite coatings containing alumina and silica nanoparticles for wood: mechanical, optical, and thermal properties assessment," *Journal of Coatings Technology Research*, vol. 8, no. 2, pp. 211–221, 2011.

[6] E. Kallio and C. H. Thubbs, *Sugar Maple*, U.S. Department of Agriculture. Forest Service AM Wood services, 1980.

8

# Synthesis of ZnO Nanostructures by Microwave Irradiation Using Albumen as a Template

**T. Prakash,[1] R. Jayaprakash,[1] G. Neri,[2] and Sanjay Kumar[3]**

[1] *Nanotechnology Laboratory, Department of Physics, Sri Ramakrishna Mission Vidyalaya College of Arts and Science, Coimbatore, Tamil Nadu 641020, India*
[2] *Department of Electronic Engineering, Chemistry and Industrial Engineering, University of Messina, 98166 Messina, Italy*
[3] *Department of Physics, B. R. Ambedkar Bihar University, Muzaffarpur, Bihar 842001, India*

Correspondence should be addressed to R. Jayaprakash; jprakash jpr@rediffmail.com

Academic Editor: Amir Kajbafvala

ZnO nanostructures have been successfully prepared by a microwave irradiation method. The role of albumen as a template in addressing the size and morphology of ZnO has been investigated by X-ray diffraction (XRD), Fourier transform infrared spectroscopy (FT-IR), thermogravimetric analysis (TG-DTA), scanning electron microscopy (SEM), and transmission electron microscopy (TEM) techniques. A heterogeneous mixture of $Zn(OH)_2$ and ZnO was obtained in absence of albumen. Pure ZnO nanostructures, consisting of rod- and whisker-like nanoparticles embedded in a sheet matrix, were obtained in the presence of albumen. Optical and photoluminescence (PL) properties of the synthesized samples were also compared. Results obtained indicate that the microwave-assisted method is a promising low temperature, cheap, and fast method for the production of ZnO nanostructures.

## 1. Introduction

Recently, metal oxides which exhibit morphologies at the nanoscale (<100 nm) have become increasingly important. One of them is ZnO, a wide band gap n-type semiconductor (3.37 eV) with large exciton binding energy of (60 meV) at room temperature, which has attracted much attention due to its unique properties. ZnO nanoparticles have found numerous applications, such as in gas sensors [1], transparent electrodes [2], pH sensors [3], biosensors [4], acoustic wave devices [5], and UV photodiodes [6].

Up to now, a number of investigations have focused on the synthesis of nano- and microarchitectures through different methods like hydrothermal [7], sol-gel [8], electrochemical deposition [9], and vapor-phase process [10]. Previous studies on the microwave radiation synthesis of ZnO with a variety of morphologies, such as nanowires [11], nanorods [12], nanoneedles [13], hollow structures, and self-assembled architectures [14], were widely reported.

In this work, we synthesized ZnO nanostructures via a microwave irradiation method in the presence of albumen as a biotemplate. Microwave irradiation as a heating method has found a number of applications in chemistry. The utilization of microwave irradiation in the preparation of nanoparticles has been reported in recent years [15]. Compared to the conventional methods, the microwave synthesis has the advantages of producing small particle size metal oxides with high purity owing to short reaction time [15–19]. Due to the "*in situ*" mode of energy conversion, the microwave heating process is fundamentally different from conventional heating processes. Heat will be generated internally within the material, instead of originating from external sources. By means of this method, many functional materials and compounds with novel structures and properties have been obtained [20]. In recent studies, it has been shown that nanoparticles, such as $Co_3O_4$ [21], $LiMn_{2-x}Cr_xO_4$ [22], $SnO_2$ [23] $Ce_{1-x}Gd_xO_{2-1/2x}$ [24], MgO [25], ZnO [26], and $TiO_2$ [27], can be prepared by the microwave synthesis process. Though the application

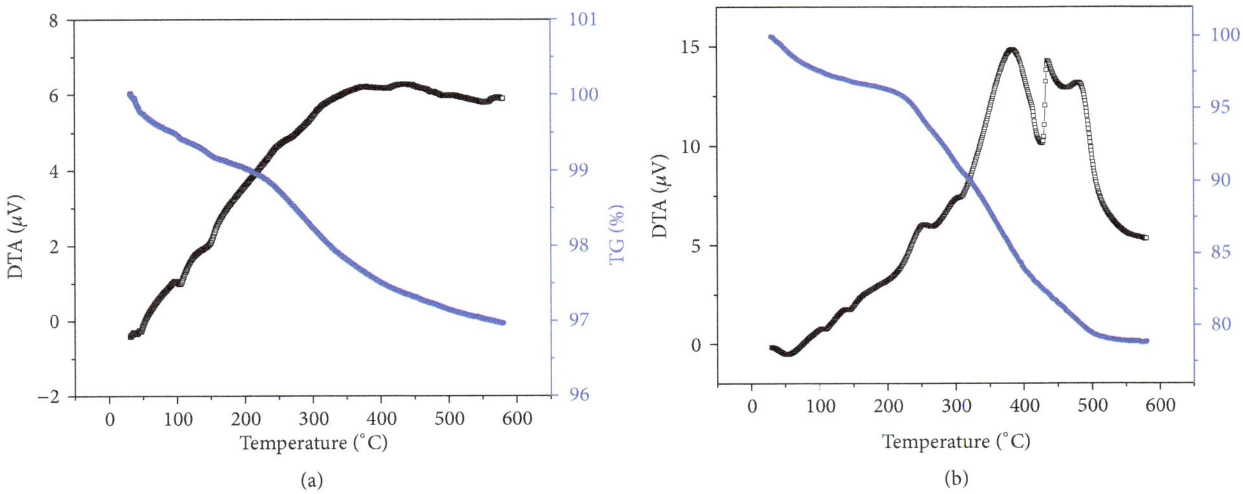


FIGURE 1: TG-DTA analysis of microwave synthesized samples: (a) sample A, (b) sample B.

of microwave irradiation in synthetic chemistry started in 1980s, only in the recent years, its use has provided a new, efficient, and environmentally benign methodology for the synthesis of various metal oxide nanoparticles of diverse morphologies and sizes. These radiations have unique properties like homogenous volumetric heating, which causes heating directly inside the sample, high reaction rate, selectivity, and increased product yield. Moreover, it is an energy saving process [14, 28, 29].

The process here proposed to synthesize ZnO nanostructures by a microwave-assisted process is simple, cheap, and fast and is characterized by a low synthesis temperature (130°C). As compared to conventional heating, microwaves cause the uniform distribution of temperature between the surface and the bulk material thereby leading to the fast formation of ZnO nanoparticles [30, 31]. Another advantage is that no postannealing process is necessary to obtain ZnO crystalline nanometer-sized structures compared to other microwave irradiation methods reported in the literature [32, 33]. To investigate the role of template in the synthesis of ZnO nanostructure by this method, a comparative study has been carried characterizing the products obtained in the presence or not of albumen as a biotemplate. The ZnO nanomaterials obtained were thoroughly characterized through X-ray diffraction (XRD), Fourier transform infrared spectroscopy (FT-IR), thermogravimetric analysis (TG-DTA), scanning electron microscopy (SEM), and transmission electron microscopy (TEM) techniques. Furthermore, their optical properties were assessed by UV-Vis DRS spectra and photoluminescence (PL) measurements.

## 2. Experimental Procedure

*2.1. Materials.* Zinc nitrate $Zn(NO_3)_2$ and ammonia solution ($NH_4OH$) were supplied from (Merk, 98%) Mumbai, India. All the chemicals were of analytical grade and used as received without further purification. The albumen, extracted from white part of egg, was used for the synthesis. Double distilled water was used through the experiments.

*2.2. Synthesis.* The synthesis of ZnO nanostructures by the microwave irradiation method was carried out as follows. First, a 0.1 M of zinc hydroxyl solution was prepared by dissolving zinc nitrate ($Zn(NO_3)_2$) in deionized water. Then pH of the solution was maintained at 8 by adding liquid ammonia solution dropwise. The resulting product was filtered and washed with double distilled water and ethanol until it became free from impurities. The precipitate was irradiated for 5 minutes in household microwave (radiation frequency 2.45 GHZ, Power up to 1 KW) with convection mode, giving a white product. Finally the sample was dried at 130°C (sample A) for 5 hours.

The same experimental procedure was followed for the synthesis of ZnO nanostructures carried out in the presence of albumen (sample B), by adding drop by drop 5 mL of freshly extracted albumen, mixed with 25 mL of deionized water and stirred, to 0.1 M zinc hydroxyl solution.

*2.3. Characterization.* The sample microstructure was analyzed by X-ray diffraction (XRD) using a Bruker AXS D8 Advance instrument and the monochromatic $CuK_{\alpha 1}$ wavelength of 1.5406 Å. The average crystalline size of the crystallites was evaluated using Scherrer's formula, $d = K\lambda/\beta\cos\theta$, where $d$ is the mean crystalline size, $K$ is a grain shape dependent constant (0.9), $\lambda$ is the wavelength of the incident beam, $\theta$ is a Bragg reflection angle, and $\beta$ is the full width at half maximum (FWHM) of the main diffraction peak. The sample morphology was observed by scanning electron microscopy (SEM), using a JEOL 5600LV microscope at an accelerating voltage of 10 kV. High resolution transmission electron microscopy (HRTEM) and selected-area electron diffraction (SAED) were recorded on a Tecnai G20-stwin

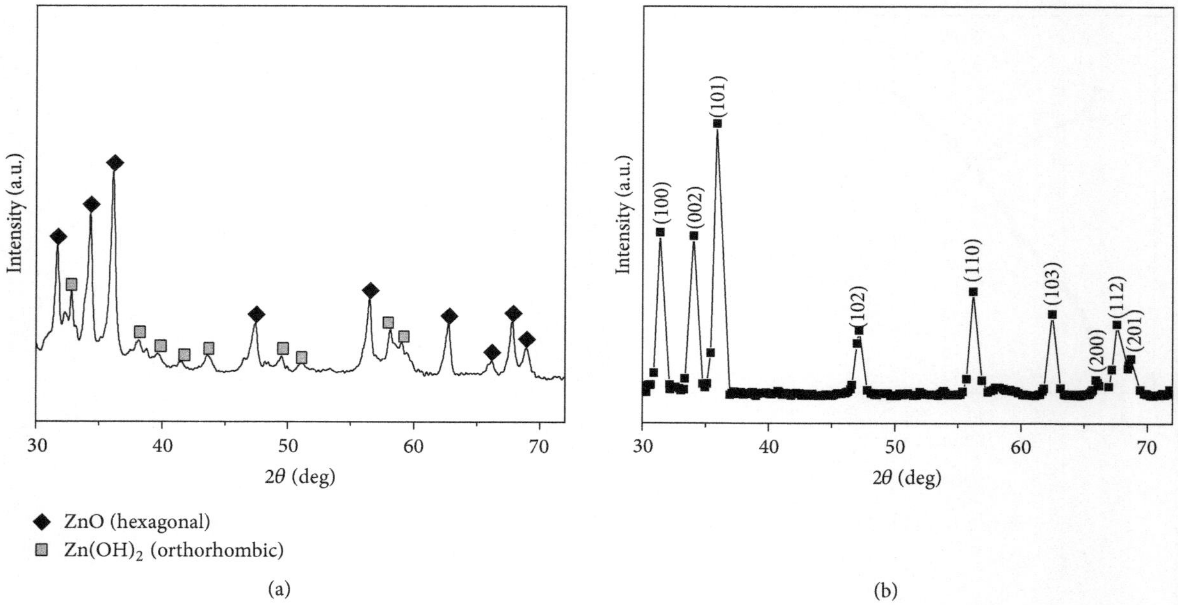


FIGURE 2: XRD analysis of microwave synthesized samples: (a) sample A, (b) sample B.

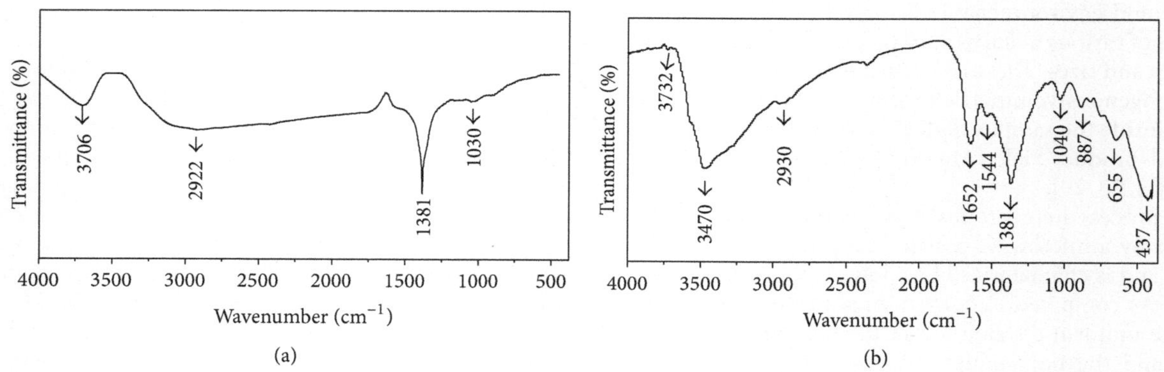


FIGURE 3: FT-IR spectra of microwave synthesized samples: (a) sample A, (b) sample B.

using an accelerating voltage of 200 kV. The Fourier transform infrared spectra (FT-IR) of the samples were recorded by using a Nicolet 5DX FTIR spectrometer. Thermal analysis was carried out by using a thermogravimetric and differential scanning calorimeter apparatus (TG-DSC Netzsch—Model STA 409PC). The analyses were carried out with a heating rate of 10°C/min in static air up to 800°C. The ultraviolet (UV) spectrum of the ZnO samples was recorded on a Perkin Elmer UV-visible DRS spectrophotometer. The room-temperature PL spectrum was performed on a spectrofluorometer instrument (JY Fluorolog-FL3-11).

## 3. Results and Discussion

### *3.1. Morphological and Microstructural Analysis*

*3.1.1. Thermogravimetric Analysis.* The thermal behavior of ZnO samples synthesized has been first investigated by TG-DTA. Figure 1(a) shows the TG-DTA curves of sample A. The weight loss monitored in the range 25–600°C is low and has been calculated to be less than 3.3% in total. This weight loss is ascribed to the removal of absorbed water molecules and ethanol and the loss of hydroxyl ions. The final step at higher temperature is attributed to the burnout of organic species still remained in the dried powder. By focusing the attention on sample B, TGA indicated that the weight loss is higher (about 20% in total). The first weight loss in the temperature range between 25°C and 150°C is ascribed to the removal of adsorbed water molecules. The second weight loss in the range between 230°C and 520°C comes from the decomposition of the organic template, which is also confirmed by the exothermic peak at about 384°C in the DTA curve.

*3.1.2. X-Ray Diffraction.* The phase identification of the synthesized ZnO nanostructures was determined by X-ray diffraction (XRD). Figures 2(a) and 2(b) show the XRD

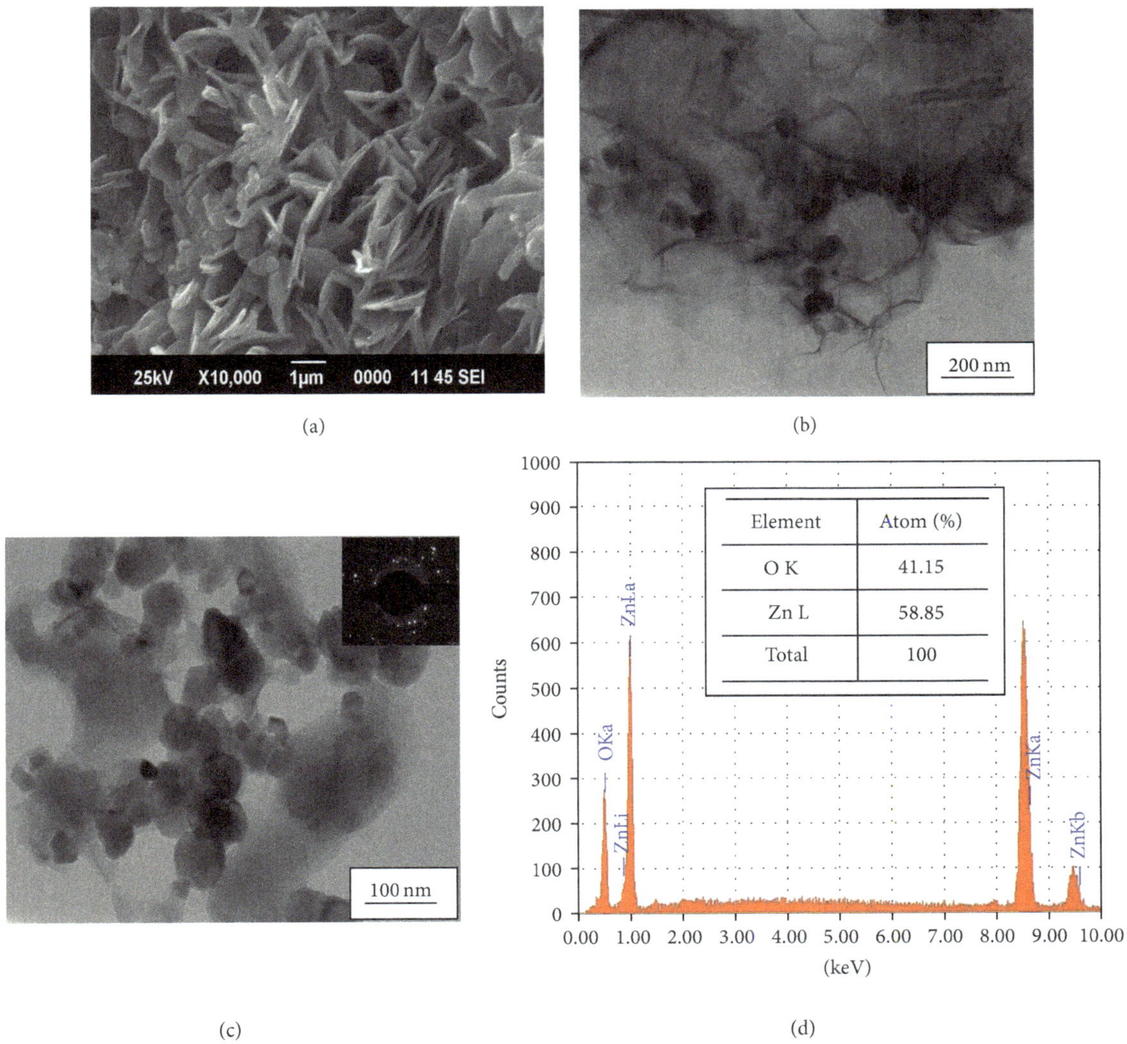


FIGURE 4: SEM and TEM micrographs of sample A. (a) SEM image, (b) and (c) TEM images of the sheet-like and spherical-like ZnO nanostructure, and (d) EDX analysis. Inset in Figure (c) shows the SAED pattern.

patterns of both samples. In Figure 2(a), XRD of sample A indicated that two phases, like ZnO hexagonal wurtzite phase structure (JCPDS number 897102) and $Zn(OH)_2$ orthorhombic phase structure (JCPDS number 890138), are present. On the basis of diffraction peaks intensity, ZnO is the predominant phase. According to Scherrer's equation, the average particle size of ZnO nanoparticles is about 29 nm.

Figure 2(b) shows the XRD pattern of sample B. The diffraction planes (100), (002), (101), (102), (110), (103), (200), (112), and (201) in the patterns can be perfectly indexed to the hexagonal wurtzite phase structure (JCPDS number 897102). Also no diffraction peaks corresponding to Zn, $Zn(OH)_2$, and other impurities are observed. The value of lattice parameters ($a$ = 3.2896 Å and $c$ = 5.2624 Å), as calculated by the XRD data, is in good agreement with those reported by other authors [33]. The average particle size of ZnO is estimated to be around 14 nm.

XRD analysis indicates that the pure ZnO phase is obtained only in the presence of albumen. Furthermore, the average particle size of ZnO phase is markedly lower. It can be supposed that, in the formation processes in absence of albumen, the zinc hydroxide species was first produced step by step by the reaction between the Zn species and the hydroxyl ions coming from dissociation of hydrated ammonia molecules, and then in a consecutive reaction ZnO nanoparticles are formed. In the presence of template molecules, it appears that this second reaction is strongly promoted; so all initial zinc hydroxide species are transformed into ZnO.

*3.1.3. Fourier Transform Infrared Spectroscopy.* Figure 3(a) shows the typical FT-IR spectrum of sample A. The broad peak at 3706 $cm^{-1}$ corresponds to the vibrational mode of O–H bond, indicating the presence of zinc hydroxide. The

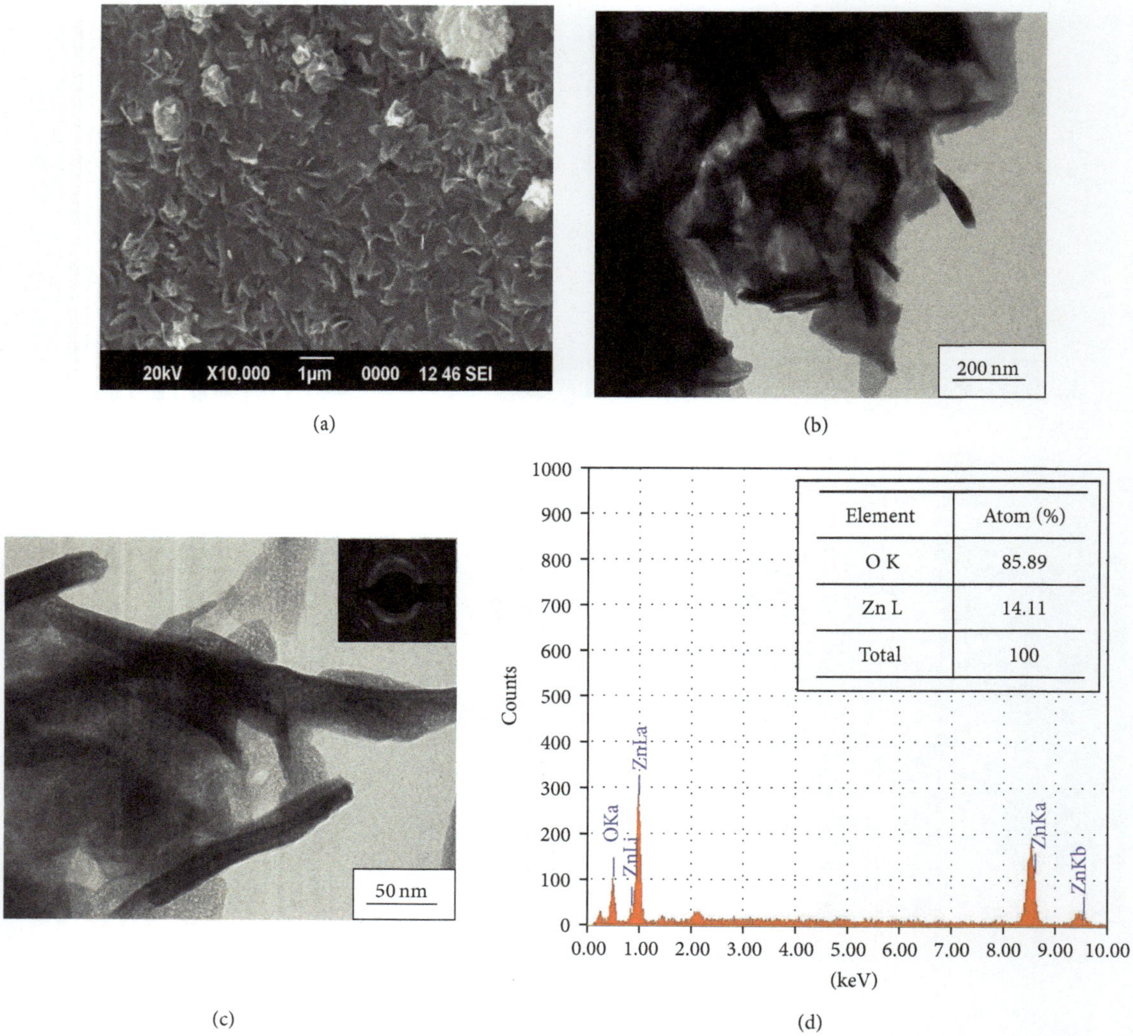


FIGURE 5: SEM and TEM micrographs of sample B. (a) SEM image, (b) and (c) TEM images of the whisker-like and rod-like ZnO nanostructure, and (d) EDX analysis. Inset in Figure (c) shows the SAED pattern.

stretching mode of vibrations in asymmetric and symmetric C=O bonds that is observed at 1381 $cm^{-1}$ is associated with residual organic impurity in the dried powder as suggested by TG-DTA analysis.

In the spectrum of sample B (Figure 3(b)), the appearance of a sharp band at 437 $cm^{-1}$ in the FT-IR spectrum, characteristic for the Zn–O stretching vibration [34], confirms the presence of well-crystallized ZnO. Additionally broad absorption peaks centered at around 3470 $cm^{-1}$ and 1652 $cm^{-1}$ are caused by the O–H stretching of the absorbed water reabsorption during the storage of the sample in ambient air [35]. The absence of features at 3706 $cm^{-1}$ is an indication of the absence of $Zn(OH)_2$ phase on this sample. In this sample are also seen the signatures of the organic template. The broad peak at 3470 $cm^{-1}$ is in fact due to superimposition of N–H stretching of amide group at 3449 $cm^{-1}$ of the biotemplate and O–H stretching mode of water molecules. Also some of the absorption bands in the range 1000–1600 $cm^{-1}$ are due to C=O and C=C stretching vibrational modes of albumen, as well the weak peaks located at 2930 $cm^{-1}$ are due to symmetric and asymmetric C–H bonds, respectively [36]. A very small band originated at 887 $cm^{-1}$ is probably due to the carbonate moieties which are generally observed when FT-IR samples are measured in air [37, 38].

*3.1.4. Scanning and Transmission Electron Microscopy.* SEM and TEM images of sample A are reported in Figure 4. SEM analysis shows extensive sheet-like nanostructure where spherical-shaped particles are also present. TEM analysis evidences clearly this composite structure and the large diameter distribution of the spherical-shaped nanoparticles. No attempt has been made to assign the two phases present, ZnO and $Zn(OH)_2$, to each of these nanostructures. The thickness of the sheets and spherical particles is in the range of about 13–50 nm. The SAED pattern taken from one of

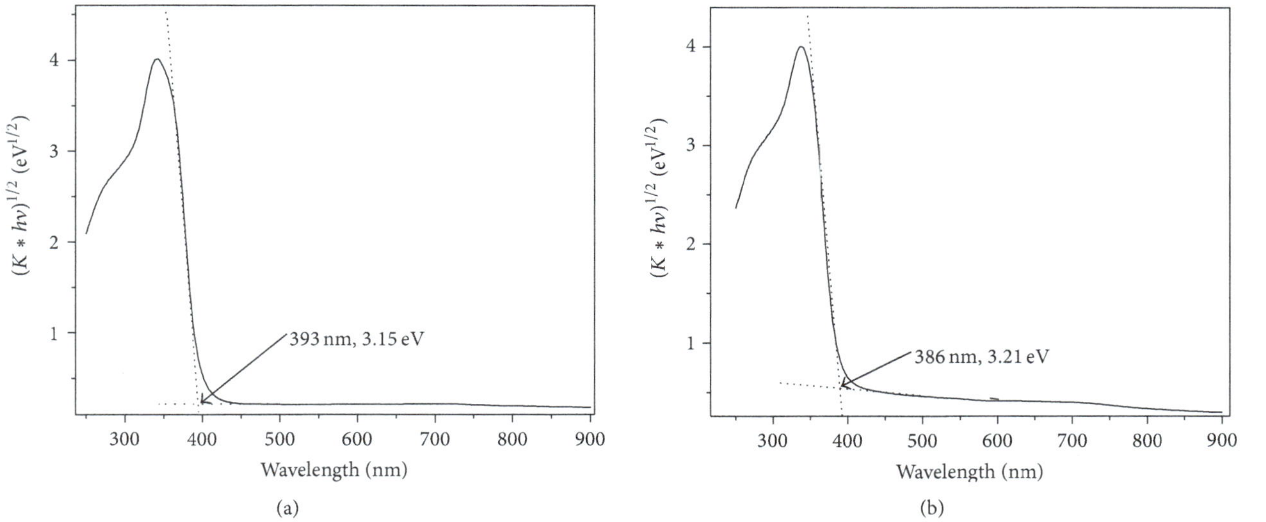


FIGURE 6: UV-Vis DRS of the typical ZnO nanoparticles: (a) sample A, (b) sample B. Extrapolation of the linear portion to the photon energy axis to obtain the $E_g$ value is shown.

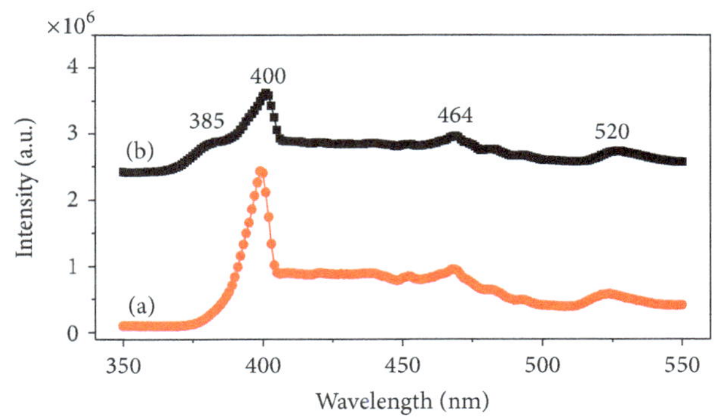


FIGURE 7: PL spectra of (curve a) sample A and (curve b) sample B.

the particles is shown in the inset, and it further confirms the crystalline nature of the sample. In addition, the EDS spectrum of the sample confirms the presence of Zn and O in sample without any other element contamination. The atomic percent ratio of O and Zn in dried sample is 41.15 : 58.85, which might indicate that there is a small content of oxygen vacancy defects on the surface of as-obtained product.

Figure 5 shows SEM and TEM images of sample B. SEM analysis shows that this sample is characterized by a more uniform surface sheet-like nanostructure. TEM analysis revealed the presence of nanowhiskers and nanorods of different length and diameter embedded within the sheet matrix. The thickness of the nanorods and nanowhiskers is in the range of about 10–57 nm. The SAED pattern of ZnO nanostructures further confirms the crystalline nature of the sample. EDS analysis reports that the atomic percent ratio of O and Zn in the dried sample is 85.59 : 14.11, which might indicate that there is a larger content of oxygen on the surface with respect to sample A.

*3.2. Optical Properties.* In order to determine the precise value of optical band gap UV-visible DRS of the ZnO nanoparticles in samples A and B, we used the optical absorption method on reflectance spectrum. The reflectance values were converted to absorbance by application of the Kubelka-Munk function [39–42]. The Kubelka-Munk theory is generally used for analyzing the diffuse reflectance spectra obtained from weakly absorbing samples. The Kubelka-Munk formula can be expressed by the following relation [43]:

$$K = \frac{(1-R)^2}{2R}, \tag{1}$$

where $K$ is the reflectance transformed according to Kubelka-Munk and $R$ is the reflectance (%).

The relationship $(k * h\vartheta)^{1/2} = f(h*\vartheta)$ is shown in Figures 6(a) and 6(b). The $E_g$ value can be obtained by extrapolating the linear portion to the photon energy axis. The reflectance shoulder onset peaks are located at 393 nm and 386 nm corresponding to the band gaps of 3.15 eV and 3.21 eV, respectively. The $E_g$ values are smaller than that of 3.3 eV reported for single crystalline ZnO samples. On comparison with ZnO nanoparticles prepared by other methods, ZnO nanoparticles are obtained by this microwave irradiation method showing larger band gap energy. It is well known that the difference in the band gap is related to the presence of vacancies and dopants [44]. The red shift of the band gap energy also relates to structural morphologies, particle size, and surface microstructures [45]. Then, the higher value of band gap of the sample B synthesized in the presence of albumen can be attributed to a minor amount of defects, due essentially to its more ordered crystalline structure.

The study of the photoluminescence properties is interesting because it can provide valuable information on the quality, purity, and the structural properties of the material. Moreover, recording PL spectrum is of paramount importance for evaluating their optical characteristics for various applications. The PL spectra of samples A and B are depicted in Figures 7(a) and 7(b). On sample A, three emission peaks, a blue light at 400 nm, a blue-green around 464 nm, and a green light at around 520 nm, were recorded. The main strong UV emission at 400 nm corresponds to the near-band-edge emission, while the blue-green and green emission peaks at 464 nm and 520 nm are instead possibly associated with oxygen vacancies [46, 47]. Specifically, the blue-green transition at 464 nm is caused due to transition from the level of ionized oxygen vacancy to the valence band, while the green emission, called deep-level emission, occurs by the recombination of the photogenerated holes with singly ionized oxygen vacancies in ZnO. On sample B, an additional UV emission peak at 385 nm, due to the recombination of photogenerated electrons and holes, appears as a shoulder of the main peak at 400 nm. The higher intensity of near-band-edge emission peaks compared to green emission peaks is an indication of the good quality of ZnO nanostructures produced.

## 4. Conclusions

The morphology, structure, and optical properties of ZnO nanostructures synthesized by microwave irradiation have been investigated in detail. The composition and morphology of the nanostructures are affected by the presence of albumen as a template. The optical band gap of the polycrystalline ZnO nanoparticles increases from 3.22 to 3.25 eV for the sample synthesized in the presence of albumen because of the defects' decrease. The good quality of ZnO nanostructures produced was also demonstrated by PL studies which evidenced a higher intensity of near-band-edge emission peaks compared to green emission peaks. In summary, results obtained indicate that the microwave-assisted method is a promising low temperature, cheap, and fast method for the production of ZnO nanostructures.

## References


[1] B. B. Rao, "Zinc oxide ceramic semi-conductor gas sensor for ethanol vapour," *Materials Chemistry and Physics*, vol. 64, no. 1, pp. 62–65, 2000.

[2] W. H. G. Horsthuis, "ZnO processing for integrated optic sensors," *Thin Solid Films*, vol. 137, no. 2, pp. 185–192, 1986.

[3] Asif, Nur, Willander, and Danielsson, "Selective calcium ion detection with functionalized ZnO nanorods-extended gate MOSFET," *Biosensors and Bioelectronics*, vol. 24, no. 11, pp. 3379–3382, 2009.

[4] C. Xia, N. Wang, L. Lidong, and G. Lin, "Synthesis and characterization of waxberry-like microstructures ZnO for biosensors," *Sensors and Actuators B*, vol. 129, no. 1, pp. 268–273, 2008.

[5] Krishnamoorthy and Iliadis, "Development of high frequency ZnO/SiO2/Si Love mode surface acoustic wave devices," *Solid-State Electronics*, vol. 50, no. 6, pp. 1113–1118, 2006.

[6] Liu, Zhang, Lu et al., "Fabrication and characterization of ZnO film based UV photodetector," *Journal of Materials Science*, vol. 20, no. 3, pp. 197–201, 2009.

[7] Y. Wang and M. Li, "Hydrothermal synthesis of single-crystalline hexagonal prism ZnO nanorods," *Materials Letters*, vol. 60, no. 2, pp. 266–269, 2006.

[8] Lee, Easteal, Pal, and Bhattacharyya, "Evolution of ZnO nanostructures in sol-gel synthesis," *Current Applied Physics*, vol. 9, no. 4, pp. 792–796, 2009.

[9] Wang, Liu, Chang, and Tang, "Synthesis of sulfur-doped ZnO nanowires by electrochemical deposition," *Materials Science in Semiconductor Processing*, vol. 10, no. 6, pp. 241–245, 2007.

[10] J. Park, Y. Choi, and J. Park, "Synthesis of ZnO nanowires and nanosheets by an O2-assisted carbothermal reduction process," *Journal of Crystal Growth*, vol. 280, no. 1-2, pp. 161–167, 2005.

[11] H. Cheng, J. Cheng, Y. Zhang, and Q. Wang, "Large-scale fabrication of ZnO micro-and nano-structures by microwave thermal evaporation deposition," *Journal of Crystal Growth*, vol. 299, no. 1, pp. 34–40, 2007.

[12] J. Liu, J. Cao, Z. Li, G. Ji, and M. Zheng, "A simple microwave-assisted decomposing route for synthesis of ZnO nanorods in the presence of PEG400," *Materials Letters*, vol. 61, no. 22, pp. 4409–4411, 2007.

[13] S. Cho, D. Shim, S. Jung, E. Oh, B. R. Lee, and K. Lee, "Fabrication of ZnO nanoneedle arrays by direct microwave irradiation," *Materials Letters*, vol. 63, no. 9-10, pp. 739–741, 2009.

[14] Krishnakumar, Jayaprakash, N. Pinna, Singh, Mehta, and Phani, "Microwave-assisted synthesis and characterization of flower shaped zinc oxide nanostructures," *Materials Letters*, vol. 63, no. 2, pp. 242–245, 2009.

[15] I. Y. Y. Bu, "Rapid synthesis of ZnO nanostructures through microwave heating process," *Ceramics International*, vol. 39, pp. 1189–1194, 2013.

[16] S. T. Aruna and A. S. Mukasyan, "Combustion synthesis and nanomaterials," *Current Opinion in Solid State and Materials Science*, vol. 12, no. 3-4, pp. 44–50, 2008.

[17] Mohebbi, Ebadzadeh, and Hesari, "Synthesis of nano-crystalline (Ni/NiO)-YSZ by microwave-assisted combustion synthesis method: the influence of pH of precursor solution," *Journal of Power Sources*, vol. 178, no. 1, pp. 64–68, 2008.

[18] M. A. Bhosale, K. D. Bhatte, and B. M. Bhanage, "A rapid, one pot microwave assisted synthesis of nanosize cuprous oxide," *Powder Technology*, vol. 235, pp. 516–519, 2013.

[19] P. Dinka and A. S. Mukasyan, "In situ preparation of oxide-based supported catalysts by solution combustion synthesis," *Journal of Physical Chemistry B*, vol. 109, no. 46, pp. 21627–21633, 2005.

[20] Kasapoğlu, Baykal, Köseoğlu, and Toprak, "Microwave-assisted combustion synthesis of CoFe2O4 with urea, and its magnetic characterization," *Scripta Materialia*, vol. 57, no. 5, pp. 441–444, 2007.

[21] L. Ai and Jiang, "Rapid synthesis of nanocrystalline Co3O4 by a microwave-assisted combustion method," *Powder Technology*, vol. 195, no. 1, pp. 11–14, 2009.

[22] Y. Fu, Y. Su, and C. Lin, "Comparison of microwave-induced combustion and solid-state reaction for synthesis of LiMn2-xCrxO4 powders and their electrochemical properties," *Solid State Ionics*, vol. 166, no. 1-2, pp. 137–146, 2004.

[23] A. Srivastava, Lakshmikumar, Srivastava, Rashmi, and K. Jain, "Gas sensing properties of nanocrystalline SnO2 prepared in solvent media using a microwave assisted technique," *Sensors and Actuators B*, vol. 126, no. 2, pp. 583–587, 2007.

[24] Y. Fu, Y. Chang, and S. Wen, "Microwave-induced combustion synthesis and electrical conductivity of Ce1-xGdxO2-1/2x ceramics," *Materials Research Bulletin*, vol. 41, no. 12, pp. 2260–2267, 2006.

[25] E. Esmaeili, A. Khodadadi, and Y. Mortazavi, "Microwave-induced combustion process variables for MgO nanoparticle synthesis using polyethylene glycol and sorbitol," *Journal of the European Ceramic Society*, vol. 29, no. 6, pp. 1061–1068, 2009.

[26] C. Hwang and T. Wu, "Combustion synthesis of nanocrystalline ZnO powders using zinc nitrate and glycine as reactants - Influence of reactant composition," *Journal of Materials Science*, vol. 39, no. 19, pp. 6111–6115, 2004.

[27] M. Selvakumar and D. K. Bhat, "Microwave synthesized nanostructured $TiO_2$-activated carbon composite electrodes for supercapacitor Selvakumar," *Applied Surface Science*, vol. 263, pp. 236–241, 2012.

[28] I. Bilecka, P. Elser, and M. Niederberger, "Kinetic and thermodynamic aspects in the microwave-assisted synthesis of ZnO nanoparticles in benzyl alcohol," *ACS Nano*, vol. 3, no. 2, pp. 467–477, 2009.

[29] P. Zhu, J. Zhang, Z. Wu, and Z. Zhang, "Microwave-assisted synthesis of various ZnO hierarchical nanostructures: effects of heating parameters of microwave oven," *Crystal Growth and Design*, vol. 8, no. 9, pp. 3148–3153, 2008.

[30] N. F. Hamedani, A. R. Mahjoub, A. A. Khodadadi, and Y. Mortazavi, "Microwave assisted fast synthesis of various ZnO morphologies for selective detection of CO, CH4 and ethanol," *Sensors and Actuators B*, vol. 156, no. 2, pp. 737–742, 2011.

[31] J. ZHU, J. ZHANG, H. ZHOU, W. QIN, L. CHAI, and Y. HU, "Microwave-assisted synthesis and characterization of ZnO-nanorod arrays," *Transactions of Nonferrous Metals Society of China (English Edition)*, vol. 19, no. 6, pp. 1578–1582, 2009.

[32] Savary, Marinel, Colder, Harnois, Lefevre, and Retoux, "Microwave sintering of nano-sized ZnO synthesized by a liquid route," *Powder Technology*, vol. 208, no. 2, pp. 521–525, 2011.

[33] L. C. Nehru, V. Swaminathan, and C. Sanjeeviraja, "Rapid synthesis of nanocrystalline ZnO by a microwave-assisted combustion method," *Powder Technology*, vol. 226, pp. 29–33, 2012.

[34] Al-Hajry, A. Umar, Hahn, and Kim, "Growth, properties and dye-sensitized solar cells-applications of ZnO nanorods grown by low-temperature solution process," *Superlattices and Microstructures*, vol. 45, no. 6, pp. 529–534, 2009.

[35] T. Prakash, R. Jayaprakash, D. Sathya Raj et al., "Sensing properties of ZnO nanoparticles synthesized by using albumen as a biotemplate for acetic acid monitoring in aqueous mixture," *Sensors and Actuators B*, vol. 176, pp. 560–568, 2013.

[36] Vafaee and M. S. Ghamsari, "Preparation and characterization of ZnO nanoparticles by a novel sol-gel route," *Materials Letters*, vol. 61, no. 14-15, pp. 3265–3268, 2007.

[37] A. Umar, Rahman, M. Vaseem, and Y. Hahn, "Ultra-sensitive cholesterol biosensor based on low-temperature grown ZnO nanoparticles," *Electrochemistry Communications*, vol. 11, no. 1, pp. 118–121, 2009.

[38] A. Umar, Rahman, Al-Hajry, and Hahn, "Highly-sensitive cholesterol biosensor based on well-crystallized flower-shaped ZnO nanostructures," *Talanta*, vol. 78, no. 1, pp. 284–289, 2009.

[39] A. E. Morales, E. S. Mora, and U. Pal, "Use of diffuse reflectance spectroscopy for optical characterization of un-supported nanostructures," *Revista Mexicana De Fisica S*, vol. 53, pp. 18–22, 2007.

[40] Senthilkumar, Vickraman, and Ravikumar, "Synthesis of fluorine doped tin oxide nanoparticles by sol-gel technique and their characterization," *Journal of Sol-Gel Science and Technology*, vol. 53, no. 2, pp. 316–321, 2010.

[41] F. Yakuphanoglu, "Electrical characterization and device characterization of ZnO microring shaped films by sol-gel method," *Journal of Alloys and Compounds*, vol. 507, no. 1, pp. 184–189, 2010.

[42] Yakuphanoglu, Mehrotra, Gupta, and M. Oz, "Nanofiber organic semiconductors: the effects of nanosize on the electrical charge transport and optical properties of bulk polyanilines," *Journal of Applied Polymer Science*, vol. 114, no. 2, pp. 794–799, 2009.

[43] J. Tauc, R. Grigorovici, and A. Vancu, "Optical properties and electronic structure of amorphous germanium," *Physica Status Solidi B*, vol. 15, pp. 627–637, 1966.

[44] Wang, Chen, and Yang, "Microstructure and optical properties of polycrystalline ZnO films sputtered under different oxygen flow rates," *Journal of Alloys and Compounds*, vol. 488, no. 1, pp. 232–237, 2009.

[45] A. K. Zak, M. E. Abrishami, W. H. Abd Majid, R. Yousefi, and S. M. Hosseini, "Effects of annealing temperature on some structural and optical properties of ZnO nanoparticles prepared by a modified sol-gel combustion method," *Ceramic International*, vol. 37, pp. 393–398, 2011.

[46] P. Yang, H. Yan, S. Mao et al., "Controlled growth of ZnO nanowires and their optical properties," *Journal of Advanced Functional Materials*, vol. 12, pp. 323–331, 2002.

[47] P. Jiang, J. Zhou, H. Fang, C. Wang, Z. L. Wang, and S. Xie, "Hierarchical shelled ZnO structures made of bunched nanowire arrays," *Advanced Functional Materials*, vol. 17, no. 8, pp. 1303–1310, 2007.

# Photocatalytic Properties of Microwave-Synthesized $TiO_2$ and ZnO Nanoparticles Using Malachite Green Dye

**A. K. Singh[1,2] and Umesh T. Nakate[1]**

[1] *Nanomaterials and Sensors Laboratory, Defence Institute of Advanced Technology, Girinagar, Pune, Maharashtra 411025, India*
[2] *DRDO Centre for Piezoceramics and Devices, ARDE, Pashan, Pune, Maharashtra 411021, India*

Correspondence should be addressed to A. K. Singh; aksingh@diat.ac.in

Academic Editor: Amir Kajbafvala

$TiO_2$ and ZnO nanoparticles (NPs) were synthesized using microwave-assisted method. Synthesized NPs were characterized for their structure, morphology, and elemental composition using X-ray diffraction (XRD), scanning electron microscopy (SEM), and energy dispersive spectroscopy (EDS). The crystallite size of synthesized NPs of $TiO_2$ and ZnO was about 12.3 and 18.7 nm as obtained from the Scherrer formula from the most intense XRD peak. The synthesized NPs have been found to be in stoichiometric ratio having anatase and hexagonal wurtzite structure for $TiO_2$ and ZnO, respectively, and are spherical in shape. Surface area of $TiO_2$ and ZnO NPs was found to be about 43.52 $m^2/g$ and 7.7 $m^2/g$. Photocatalytic (PC) properties of synthesized NPs were studied for malachite green (MG) dye under UV light. $TiO_2$ NPs were found to be highly photocatalytically active among the two, having efficiency and apparent photodegradation rate of 49.35% and $1.32 \times 10^{-2}$, respectively.

## 1. Introduction

Nowadays, organic pollutants like dyes are becoming a serious health problem, which are being discharged by many industries and houses through wastewater in the environment. The discharged wastes containing dyes are toxic to microorganisms and aquatic life [1]. Malachite green (MG) dye is an extensively used biocide in the global aquaculture industry. It is also used as a food colouring agent, food additive, medical disinfectant, and anthelmintic as well as in silk, wool, jute, leather, cotton, paper, and acrylic industries. However, MG has now become a highly controversial compound due to the risks it poses to the consumers of treated fish, including its effects on the immune system and reproductive system and its genotoxic and carcinogenic properties [2]. In the light of the basic and applied researches reviewed, PC oxidation method appears to be a promising route for the treatment of wastewater contaminated with phenols and dyes. In recent years, there has been a wide research interest in developing semiconductor photocatalysts having high PC activities for environmental applications such as water disinfection, hazardous waste remediation, and water purification [3–5]. It is reported that nanosize semiconductors act as an efficient PC agent resulting from either an extremely large-surface-area-to-volume ratio or quantum confinement effects of charge carriers for the degradation of organic pollutants in water under UV irradiation [6–10]. PC function is similar to the chlorophyll in the photosynthesis. In a PC system, photo-induced molecular transformation or reaction takes place at the surface of the catalyst. A basic mechanism of PC reaction on the generation of electron-hole pair and its destination is as follows: when a photocatalyst is illuminated by the light stronger than its bandgap energy, electron-hole pairs diffuse out to the surface of the PCs and participate in the chemical reaction with the electron donor and acceptor. Those free electrons and holes transform the surrounding oxygen or water molecules into OH free radicals with super strong oxidization [11]. One of the major problems in the PC process is the electron-hole recombination. To avoid this problem, UV irradiation source stands up among other sources [12] because the energy of the UV irradiation is large compared to the bandgap energy of the catalysts. The problem of electron-hole recombination is not fully solved but largely avoided with the use of UV source.

Many photocatalysts like $TiO_2$, ZnO, $ZrO_2$, $CeO_2$, $Fe_2O_3$, CdS, and ZnS have been attempted for the PC degradation of a wide variety of environmental contaminants. $TiO_2$ [7] is one of the most widely used semiconductors for PC degradation of waste due to its biological and chemical inertness, high photoreactivity, nontoxicity, and photostability. It is proven that among the three crystalline polymorphs of $TiO_2$, that is, rutile, anatase, and brookite, the anatase phase or the mixed anatase/rutile phase exhibits better PC properties. The anatase form of $TiO_2$ has the desirable properties of being chemically stable, readily available, and active as a catalyst for oxidation processes. Bandgap of about 3.2 eV matches the output of a wide variety of readily available lamps. Recently, ZnO [13–17] has also been reported for PC activity having bandgap close to $TiO_2$. However, only a handful of studies have been attempted which compare the efficiency of different catalysts for a particular organic compound under identical experimental conditions. Serpone and Pelizzetti [18] have reported that $TiO_2$ and ZnO are the two most active catalysts.

During the recent past, ZnO and $TiO_2$ were the most widely investigated [19, 20] semiconductors. There are several reported methods of NPs synthesis such as sol-gel [21], organometallic [22], hydrothermal [23, 24], and microwave methods [25–27]. Due to the intense friction and the collision of molecules created by microwave irradiation, microwave irradiation not only provides the energy for heating but also greatly accelerates the nucleation. With microwave irradiation on the reactant solution, temperature and concentration gradients can be avoided leading to uniform nucleation. Microwave-based synthesis method is one of the easiest, energy-saving, green, and quick methods for large-scale production of nanomaterials. Microwave synthesis is the novel route of synthesis of metal oxide semiconductor NPs which is a clean, cost-effective, energy-efficient, eco-friendly, rapid, and convenient method of heating and results in higher yields in shorter reaction times [26, 27].

In the present work, $TiO_2$ and ZnO NPs were synthesized using the microwave method. These NPs were characterized using XRD, SEM, and EDS. The PC activity of these NPs was studied for MG dye under UV light.

## 2. Experimental

*2.1. Synthesis.* For the synthesis of $TiO_2$, 0.5 M titanium butoxide solution was prepared in 100 mL butanol and stirred for 15 min; further 30 mL deionized water was added dropwise in the above solution to allow hydrolysis. This solution was stirred for 30 min which gave rise to white precipitation. The obtained white precipitate was microwave irradiated for 5 min at 700 W power using Raga's microwave system. The microwave used for this experiment was having a power range of 140 W to 700 W. This obtained solution was left 24 hr for aging at room temperature and then centrifuged at 2000 rpm for 15 min. Obtained precipitate was dried at 80°C for 12 hrs. After complete drying, powder was crushed and calcinated in air at 500°C for 2 hr to remove hydroxide impurities and recrystallization.

For the synthesis of ZnO, NaOH (0.4 M), and zinc acetate (0.2 M), solutions were mixed slowly with molar ratio of 2 : 1, respectively. The above solution was stirred for 10 min. After that, a small quantity of tea was added, and stirring continued for another 10 min. This solution was put for microwave irradiation at 700 W in two steps, that is, 40°C for 20 min and 60°C for 30 min. Resulting precipitate was washed with DI water 2-3 times before drying at 70°C for 4 hr, then crushed using mortar pestle, and calcinated in air at 500°C for 1 hr.

*2.2. Characterization.* The prepared samples were characterized for their structure, morphology, and elemental composition using X-ray diffraction (XRD) analysis, scanning electron micrograph (SEM), and energy dispersive spectrometry (EDS), respectively. XRD was performed using Bruker AXS, Germany (Model D8 Advanced), diffractometer in the scanning range of 20–70° ($2\theta$) using Cu Ka radiations of wavelength 1.5406 Å. A JEOL ASM 6360A SEM was used to study the morphology and elemental analysis of the NPs. Measurement of BET surface area was carried out for nitrogen adsorption using a Micromeritrics Ins., USA.

*2.3. Photocatalytic Activity.* Synthesized NPs were used for the photodegradation of MG dye, chemical formula: $[C_6H_5C(C_6H_4N(CH_3)_2)_2]Cl$-4, IUPAC name: [(4-dimethylaminophenyl)phenyl-methyl]-*N*,*N*-dimethylaniline, under UV light. In each experiment, 40 mL of MG solution of 0.025 g $mL^{-1}$ concentration and 0.005 gr of catalyst were used. This solution was stirred for 10 minutes in the dark for equilibrium of adsorption and desorption process of MG with NPs. After stirring, the solution was irradiated by UV lamp (DYMAX Corporation) under medium intensity spot lamp. The UV lamp was having radiation wavelength range of 200 nm–600 nm with nominal intensity 1000–2000 mW/$cm^2$. The intensity of absorption peaks, and peak absorbance of NPs was examined by an Ocean Optics UV-Vis high-resolution spectrophotometer (Ocean Optics HR 4000). From which concentrations of MG solution initially and after every 4 minutes of irradiation of UV light were measured from the Lambert–Beer Law [28] that is the absorbance ($A$) of MG solution is proportional to its concentration ($c$), which generally followed the following equation:

$$A = \varepsilon bc, \quad (1)$$

where $\varepsilon$ is the molar absorption coefficient, and $b$ is the thickness of the absorption cell. In our experiment, all the testing parameters were kept constant, so that $\varepsilon$ and $b$ could be considered constant. Therefore, changes in the concentration of MB aqueous solution can be determined by a UV-Vis spectrophotometer.

## 3. Results and Discussion

*3.1. X-Ray Diffraction (XRD).* XRD spectra of calcinated samples of $TiO_2$ and ZnO are shown in Figures 1(a) and 1(b). The obtained diffraction pattern has been compared with JCPDS datasheet JCPDS-080-0074 and JCPDS-84-1284, respectively. From Figures 1(a) and 1(b), it is confirmed from the XRD spectra that the given phase of $TiO_2$ and ZnO is anatase and hexagonal wurtzite, respectively. The crystallite

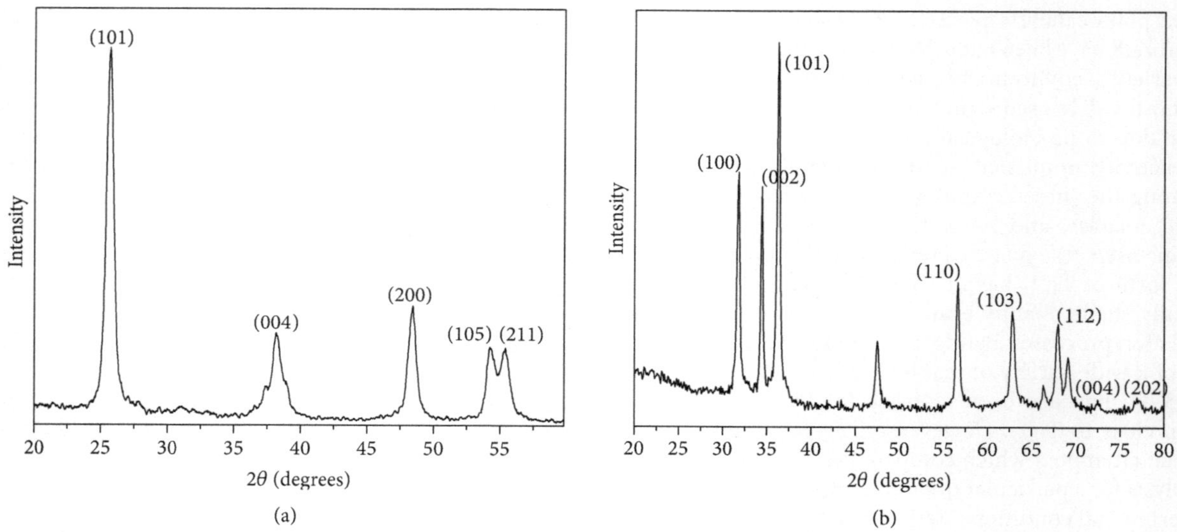


FIGURE 1: XRD of microwave-synthesized NPs: (a) $TiO_2$ and (b) ZnO.

TABLE 1: PC degradation of MG dye performance parameters of all the samples.

| Sample | Crystallite size (nm) | $K_{app}$ ($\times 10^{-2}$) | $X_{(32\ min)}$ (%) |
|---|---|---|---|
| $TiO_2$ | 12.3 | 1.32 | 49.35 |
| ZnO | 18.7 | 0.84 | 23.31 |

size, $D$, of NPs has been calculated by XRD line broadening of the most intense peak using the Scherrer formula [29]

$$D = \frac{K\lambda}{\beta \cos\theta}, \quad (2)$$

where $\lambda$ is the wavelength of X-ray, $\beta$ is the full width and half maxima, $\theta$ is Bragg's angle, and $K$ is the shape factor. The dimensionless shape factor has a typical value of about 0.9 but varies with the actual shape of the crystallite. Here, $K$ is taken as 0.9. Crystallite sizes for these NPs are about 12.3 nm and 18.7 nm for $TiO_2$ and ZnO, respectively, calculated from the most intense peak and shown in Table 1. The sharpness of peaks shows that $TiO_2$ NPs are highly crystalline in nature.

*3.2. Morphology and Elemental Analysis.* Morphology of synthesized NPs was studied using scanning electron microscope (SEM) and is shown in Figures 2(a), 2(b), 2(c), and 2(d) at two different magnifications for $TiO_2$ and ZnO, respectively. Synthesized NPs of $TiO_2$ and ZnO were found to be spherical in shape, and ZnO NPs are of lower size than $TiO_2$ with size uniformity. Some agglomeration was also observed in the SEM image of $TiO_2$, which may be due to calcination NPs. The measured BET surface area of $TiO_2$ and ZnO NPs has been found to be, respectively, 43.52 $m^2/g$ and 7.7 $m^2/g$. Elemental analysis of NPs was done by using energy dispersive spectrometer (EDS); the plot of spectrum is shown in Figures 3(a) and 3(b), for $TiO_2$ and ZnO, respectively. For $TiO_2$ and ZnO, respectively, the at% of elements detected from EDX is Ti, and O is 54.29 and 45.71; Zn and O are 43.09 and 56.91, respectively, which shows that synthesized NPs are in stoichiometric ratio. The peaks of the plot show the presence of titanium and oxygen elements, zinc and oxygen elements in the corresponding images, that is, Figures 3(a) and 3(b), respectively.

*3.3. Photocatalytic Activity.* The UV-visible absorbance spectra for photodegradation of MG dye using $TiO_2$ and ZnO are shown in Figures 4(a) and 4(b), respectively, and corresponding UV-visible spectra of $TiO_2$ and ZnO are shown in Figure 5. From the plot, it is observed that as irradiation time increases, the concentration of MG dye decreases, which is shown by the decrease in UV-absorbance spectra. The absorption of photon leads to charge separation due to promotion of an electron ($e^-$) from the valence band of the semiconductor catalyst to the conduction band, thus generating a hole ($h^+$) in the valence band. Light source provides photon energy required to excite the semiconductor electron from the valence band (VB) region to the conduction band (CB) region. With the increase in irradiation having sufficient intensity, photodegradation efficiency increases because electron-hole formation is predominantly increasing, and electron-hole recombination is negligible at sufficiently higher intensity, whereas at lower light intensity, electron-hole pair separation competes with recombination which in turn decreases the formation of free radicals, thereby, causing less effect on the percentage degradation of the dyes.

Mechanism for photodegradation of MG dye can be explained as follows. On illumination of catalyst surface with enough energy (equals or higher than the bandgap energy, $E_{bg}$, of the catalyst), it leads to the formation of a hole ($h^+$) in the valence band and an electron ($e^-$) in the conduction band. The hole oxidizes either pollutant directly or water to produce $OH^-$ radicals, whereas the electron in the conduction band reduces the oxygen adsorbed on the catalyst. The activation

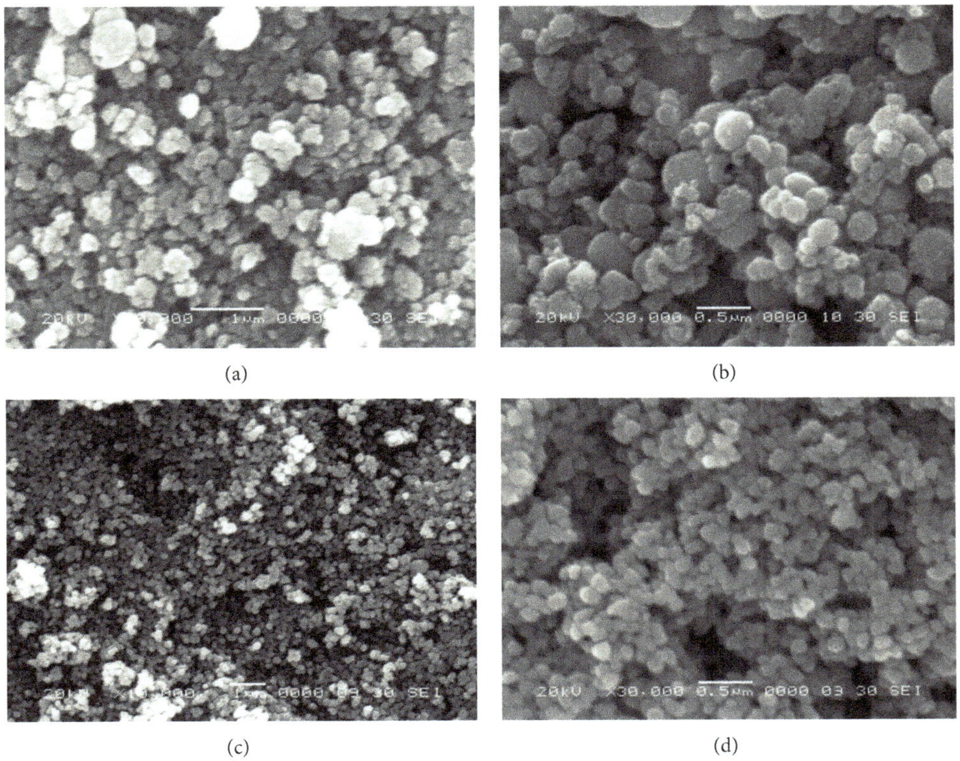


FIGURE 2: SEM image of microwave-synthesized NPs at different magnifications: (a)-(b) for $TiO_2$ and (c)-(d) for ZnO NPs.

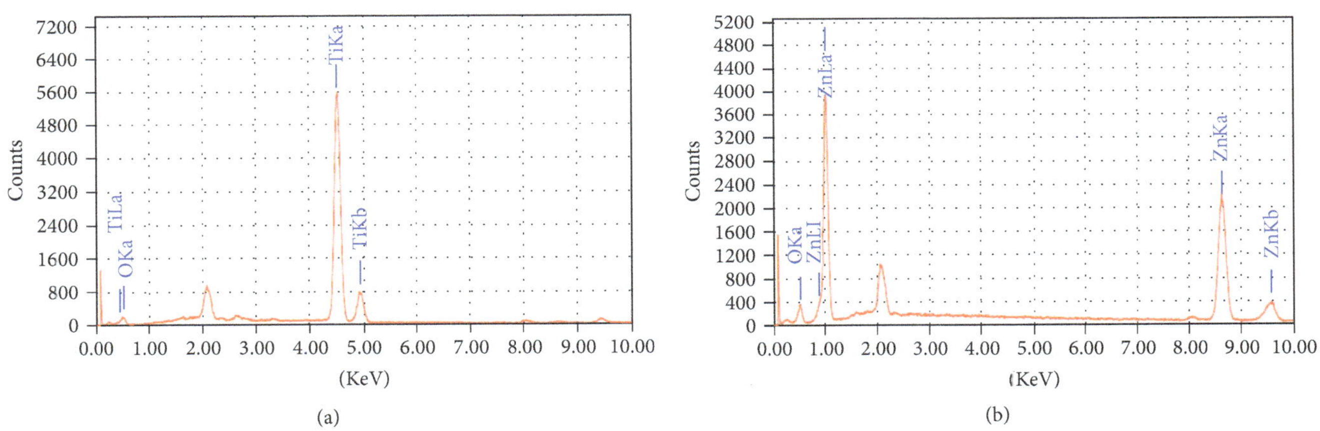


FIGURE 3: EDS spectrum of microwave-synthesized samples: (a) $TiO_2$ and (b) ZnO NPs.

of catalyst ($TiO_2$ or ZnO) by UV light can be represented by the following steps:

$$\text{Catalyst} + h\nu \longrightarrow e^- + h^+ \quad (3)$$

$$e^- + O_2 \longrightarrow O_2^- \quad (4)$$

In this reaction, $h^+$ and $e^-$ are powerful oxidizing and reductive agents, respectively. The oxidative and reductive reaction steps are expressed as follows:

$$h^+ + MG \longrightarrow \text{degradation compounds} \quad (5)$$

$$h^+ + H_2O \longrightarrow OH^- + H^+ \quad (6)$$

$$OH^- + MG \longrightarrow \text{degradation compounds} \quad (7)$$

One of the major problems in the photocatalytic process is the electron-hole recombination. To avoid this problem, UV irradiation source stands up among other sources [12]. The energy of UV irradiation is large compared to the bandgap energy of the catalysts. Hence, the problem of electron-hole recombination is not fully but largely avoided with UV

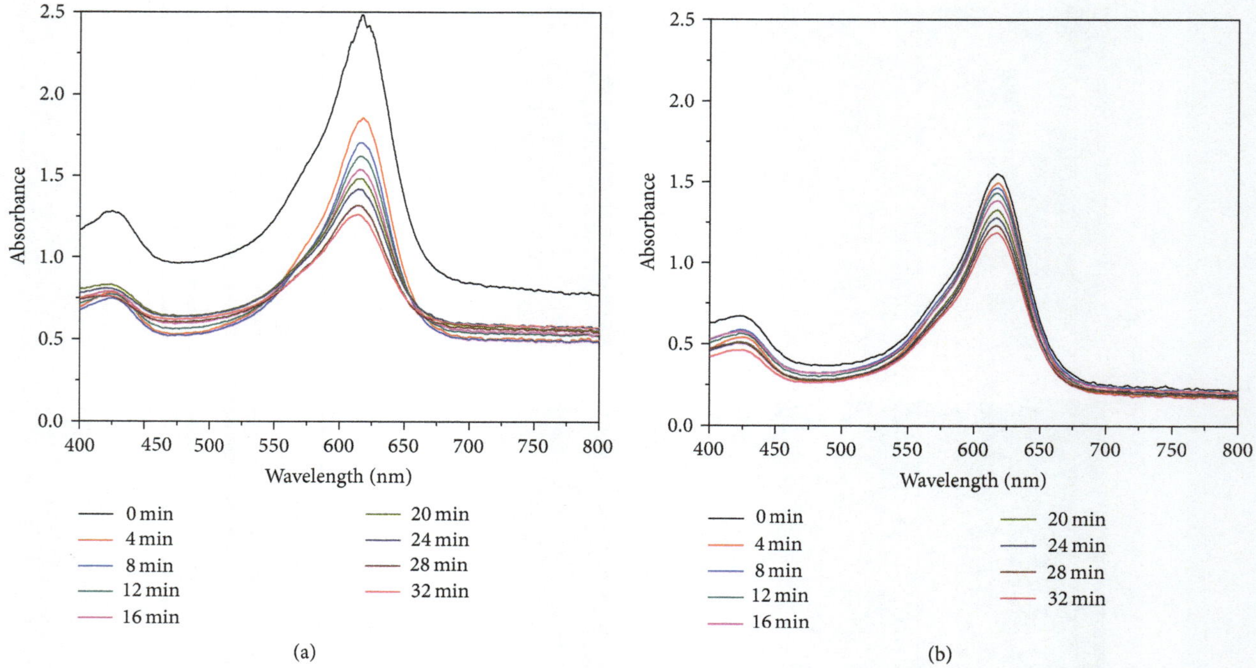


FIGURE 4: UV absorbance spectra of catalyst during photodegradation of MG dye for (a) $TiO_2$ NPs and (b) ZnO NPs.

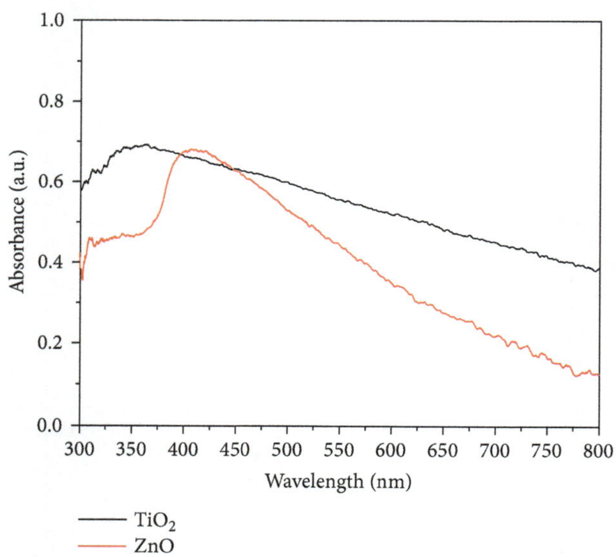


FIGURE 5: UV absorbance spectra of synthesized $TiO_2$ NPs and ZnO NPs.

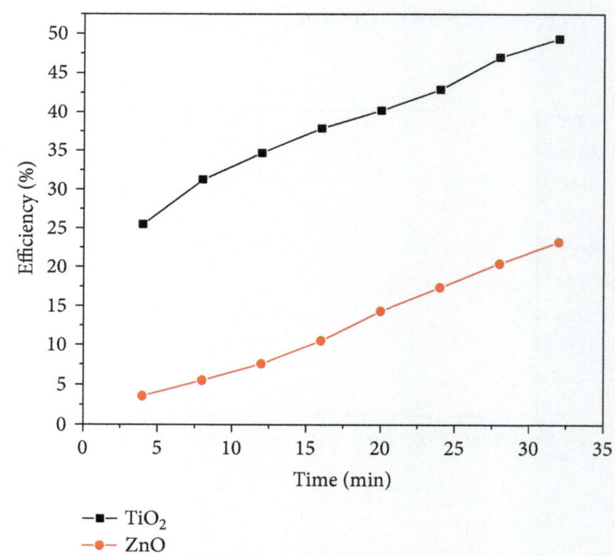


FIGURE 6: Photodegradation percentage efficiency for $TiO_2$ and ZnO catalysts.

source. The percentage efficiency of photodegradation was determined using the following equation [30]:

$$X = \frac{C_o - C}{C_o} \times 100, \quad (8)$$

where $C_0$ is the initial solution concentration of MG and $C$ is the solution concentration of MG after degradation. The graph of percentage efficiency with irradiation time is shown in Figure 6. It is clear from the figure that as irradiation time increases the efficiency of nanoparticles to degrade MG dye also increases. Here, $TiO_2$ NPs are found to be more efficient catalysts than ZnO NPs. The apparent rate constant for degradation of MG was determined using the following equation [30]:

$$\ln\left(\frac{C_0}{C}\right) = K_{app}t. \quad (9)$$

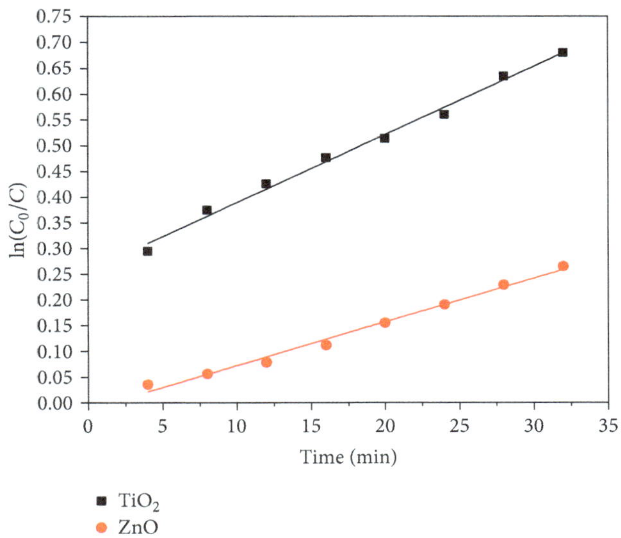


FIGURE 7: Apparent rate constant for degradation of MG dye by synthesized $TiO_2$ and ZnO NPs catalysts.

$K_{app}$ ($min^{-1}$) is the observed first-order rate constant. In the early stage of the reaction (i.e., up to 30 min), MG degradation is found to follow first-order kinetics according to the equation. This empirical equation has been used from common first-order kinetics and proposed for the simple description of the decay process [31]. The plot of ln ($C_0/C$) with irradiation time is shown in Figure 7 where the slopes represent the apparent rate constant for degradation of MG dye. The apparent rate constants of $TiO_2$ and ZnO catalysts are given in Table 1. Here, we found that $TiO_2$ NPs have higher apparent rate constant than other ZnO NPs; this is due to $TiO_2$ having wide bandgap and highly sensitivity to UV radiations, as can be seen in Figure 5. The bandgap value of anatase $TiO_2$ is around 3.2 eV, which enables UV light of wavelengths smaller than 400 nm to activate the catalyst [32].

## 4. Conclusions

In this study, synthesis of uniform size $TiO_2$ and ZnO nanoparticles was achieved using microwave synthesis which is considered to be a green, efficient, and cost-effective method having potential for large-scale synthesis. Photocatalytic properties of synthesized $TiO_2$ and ZnO nanoparticles were studied which shows that $TiO_2$ and ZnO nanoparticles have 49.35% and 23.31% photodegradation efficiency, respectively, for MG dye under UV light. $TiO_2$ nanoparticles show more than two-order photodegradation property for MG dye as compared to ZnO.

## Acknowledgments

The authors are thankful to the director of ARDE, Pashan, Pune, for granting permission to publish this work. Thanks are also due to the Department of Physics, University of Pune, for SEM and XRD of samples.

## References

[1] P. Borker and A. V. Salker, "Photocatalytic degradation of textile azo dye over $Ce_{1-x}Sn_xO_2$ series," *Materials Science and Engineering B*, vol. 133, pp. 55–60, 2006.

[2] C. C. Chen, C. S. Lu, Y. C. Chung, and J. L. Jan, "UV light induced photodegradation of malachite green on $TiO_2$ nanoparticles," *Journal of Hazardous Materials*, vol. 141, no. 3, pp. 520–528, 2007.

[3] A. Fujishima, T. N. Rao, and D. A. Tryk, "Titanium dioxide photocatalysis," *Journal of Photochemistry and Photobiology C*, vol. 1, pp. 1–21, 2000.

[4] C. Kormann, D. W. Bahnemann, and M. R. Hoffmann, "Photolysis of chloroform and other organic molecules in aqueous $TiO_2$ suspensions," *Environmental Science and Technology*, vol. 25, no. 3, pp. 494–500, 1991.

[5] J. Peral and D. F. Ollis, "Heterogeneous photocatalytic oxidation of gas-phase organics for air purification: acetone, 1-butanol, butyraldehyde, formaldehyde, and m-xylene oxidation," *Journal of Catalysis*, vol. 136, no. 2, pp. 554–565, 1992.

[6] S. Chakrabarti and B. K. Dutta, "Photocatalytic degradation of model textile dyes in waste water using ZnO as semiconductor catalyst," *Journal of Hazardous Materials B*, vol. 112, pp. 269–278, 2004.

[7] G. P. Fotou and S. E. Pratsinis, "Photocatalytic destruction of phenol and salicylic acid with aerosol-made and commercial titania powders," *Chemical Engineering Communications*, vol. 151, pp. 251–269, 1996.

[8] M. L. Curridal, R. Comparelli, P. D. Cozzli, G. Mascolo, and A. Agostiano, "Colloidal oxide nanoparticles for the photocatalytic degradation of organic dye," *Materials Science and Engineering C*, vol. 23, no. 1-2, pp. 285–289, 2003.

[9] S. B. Park and Y. C. Kang, "Photocatalytic activity of nanometer size ZnO particles prepared by spray pyrolysis," *Journal of Aerosol Science*, vol. 28, pp. S473–S474, 1997.

[10] R. Y. Hong, T. T. Pan, J. Z. Qian, and H. Z. Li, "Synthesis and surface modification of ZnO nanoparticles," *Chemical Engineering Journal*, vol. 119, no. 2-3, pp. 71–81, 2006.

[11] Y. J. Jang, C. Simer, and T. Ohm, "Comparison of zinc oxide nanoparticles and its nano-crystalline particles on the photocatalytic degradation of methylene blue," *Materials Research Bulletin*, vol. 41, no. 1, pp. 67–77, 2006.

[12] B. Neppolian, H. C. Choi, S. Sakthivel, B. Arabindoo, and V. Murugesan, "Solar/UV-induced photocatalytic degradation of three commercial textile dyes," *Journal of Hazardous Materials B*, vol. 89, no. 2-3, pp. 303–317, 2002.

[13] Y. Zhang, F. Zhu, J. Zhang, and L. Xia, "Converting layered zinc acetate nanobelts to one-dimensional structured ZnO nanoparticle aggregates and their photocatalytic activity," *Nanoscale Research Letters*, vol. 3, no. 6, pp. 201–204, 2008.

[14] M. Zhang, G. Sheng, J. Fu, T. An, X. Wang, and X. Hu, "Novel preparation of nanosized ZnO–$SnO_2$ with high photocatalytic activity by homogeneous co-precipitation method," *Materials Letters*, vol. 59, no. 28, pp. 3641–3644, 2005.

[15] C. Wang, X. Wang, B. Q. Xu et al., "Enhanced photocatalytic performance of nanosized coupled ZnO/$SnO_2$ photocatalysts for methyl orange degradation," *Journal of Photochemistry and Photobiology A*, vol. 168, no. 1-2, pp. 47–52, 2004.

[16] W. Cun, Z. Jincai, W. Xinming et al., "Preparation, characterization and photocatalytic activity of nano-sized ZnO/$SnO_2$ coupled photocatalysts," *Applied Catalysis B*, vol. 39, no. 3, pp. 269–279, 2002.

[17] A. Kajbafvala, H. Ghorbani, A. Paravar, J. P. Samberg, E. Kajbafvala, and S. K. Sadrnezhaad, "Effects of morphology on photocatalytic performance of Zinc oxide nanostructures synthesized by rapid microwave irradiation methods," *Superlattices and Microstructures*, vol. 51, no. 4, pp. 512–522, 2012.

[18] N. Serpone and E. Pelizzetti, Eds., *Photocatalysis: Fundamentals and Applications*, Wiley, New York, NY, USA, 1989.

[19] A. K. Singh, "Synthesis, characterization, electrical and sensing properties of ZnO nanoparticles," *Advanced Powder Technology*, vol. 21, pp. 609–613, 2010.

[20] J. Wang, J. Polleux, J. Lim, and B. Dunn, "Pseudocapacitive contributions to electrochemical energy storage in $TiO_2$ (anatase) nanoparticles," *Journal of Physical Chemistry C*, vol. 111, no. 40, pp. 14925–14931, 2007.

[21] T. Sugimoto, X. Zhou, and A. Muramatsu, "Synthesis of uniform anatase $TiO_2$ nanoparticles by gel-sol method: 3. Formation process and size control," *Journal of Colloid and Interface Science*, vol. 259, no. 1, pp. 43–52, 2003.

[22] J. Tang, F. Redl, Y. Zhu, T. Siegrist, L. E. Brus, and M. L. Steigerwald, "An organometallic synthesis of $TiO_2$ nanoparticles," *Nano Letters*, vol. 5, no. 3, pp. 543–548, 2005.

[23] S. Jeon and P. V. Braun, "Hydrothermal synthesis of Er-doped luminescent $TiO_2$ nanoparticles," *Chemistry of Materials*, vol. 15, no. 6, pp. 1256–1263, 2003.

[24] S. K. Sahoo, M. Mohapatra, A. K. Singh, and S. Anand, "Hydrothermal synthesis of single crystalline nano $CeO_2$ and its structural, optical, and electronic characterization," *Materials and Manufacturing Processes*, vol. 25, no. 9, pp. 982–989, 2010.

[25] A. K. Singh, "Microwave synthesis, optical, structural and magnetic properties of ZnO/Mn doped ZnO nanostructures," *Journal of Optoelectronics and Advanced Materials*, vol. 12, pp. 2255–2260, 2010.

[26] P. Periyat, N. Leyland, D. E. McCormack, J. Colreavy, D. Corr, and S. C. Pillai, "Rapid microwave synthesis of mesoporous $TiO_2$ for electrochromic displays," *Journal of Materials Chemistry*, vol. 20, no. 18, pp. 3650–3655, 2010.

[27] A. Kajbafvala, J. P. Samberg, H. Ghorbani, E. Kajbafvala, and S. K. Sadrnezhaad, "Effects of initial precursor and microwave irradiation on step-by-step synthesis of zinc oxide nanoarchitectures," *Materials Letters*, vol. 67, no. 1, pp. 342–345, 2012.

[28] H. Kumazawa, M. Inoue, and T. Kasuya, "Photocatalytic degradation of volatile and nonvolatile organic compounds on titanium dioxide particles using fluidized beds," *Industrial and Engineering Chemistry Research*, vol. 42, no. 14, pp. 3237–3244, 2003.

[29] P. B. Bagdare, S. B. Patil, and A. K. Singh, "Phase evolution and PEC performance of $Zn_xCd_{(1-x)}S$ nanocrystalline thin films deposited by CBD," *Journal of Alloys and Compounds*, vol. 506, no. 1, pp. 120–124, 2010.

[30] S. Ghasemi, S. Rahimnejad, S. Rahman Setayesh, S. Rohani, and M. R. Gholami, "Transition metal ions effect on the properties and photocatalytic activity of nanocrystalline $TiO_2$ prepared in an ionic liquid," *Journal of Hazardous Materials*, vol. 172, no. 2-3, pp. 1573–1578, 2009.

[31] G. Mascolo, R. Comparelli, M. L. Curri, G. Lovecchio, A. Lopez, and A. Agostiano, "Photocatalytic degradation of methyl red by $TiO_2$: comparison of the efficiency of immobilized nanoparticles versus conventional suspended catalyst," *Journal of Hazardous Materials*, vol. 142, no. 1-2, pp. 130–137, 2007.

[32] A. H. C. Chan, C. K. Chan, J. P. Barford, and J. F. Porter, "Solar photocatalytic thin film cascade reactor for treatment of benzoic acid containing wastewater," *Water Research*, vol. 37, no. 5, pp. 1125–1135, 2003.

# 10

# The Growth of Bismuth Sulfide Nanorods from Spherical-Shaped Amorphous Precursor Particles under Hydrothermal Condition

**Pravas Kumar Panigrahi and Amita Pathak**

*Department of Chemistry, Indian Institute of Technology Kharagpur, Kharagpur, West Bengal 721302, India*

Correspondence should be addressed to Amita Pathak; ami@chem.iitkgp.ernet.in

Academic Editor: Fabien Grasset

A surfactant/solid-template-free hydrothermal process has been developed for the synthesis of single-crystalline nanorods of bismuth sulfide ($Bi_2S_3$) using triethanolamine as a complexing agent for the $Bi^{3+}$ ions and elemental sulfur, solubilized in monoethanolamine, as the sulfur source. X-ray diffraction and morphological studies of a series of samples synthesized at different reaction conditions suggest that the growth of nanorods occurred at the expense of the low-crystalline spherical precursor particles of aminium compounds of bismuth sulfide or bismuth sulfate formed at room temperature. In the process, the reaction condition is optimized for obtaining crystalline nanorods of pure $Bi_2S_3$ with high aspect ratio. From the XRD, XPS, and HRTEM analysis of the samples, the growth of nanorods was assessed to be due to the cooperative effects of solid-solution-solid transformation and controlled oriented attachment. The hydrothermal process parameters and the presence of water in the reaction system have been found to play a crucial role in the formation of high aspect ratio nanorods. The optical band gap of the synthesized sample at optimized conditions is found to be 1.46 eV as calculated from its diffused reflectance spectrum at room temperature.

## 1. Introduction

Bismuth sulfide ($Bi_2S_3$) is a direct band gap semiconductor with large absorption coefficient, high band gap energy (1.3–1.7 eV) [1], and having energy conversion efficiency close to theoretical maximum attainable value. These qualities make $Bi_2S_3$ potentially suitable for the fabrication of optoelectronic and thermoelectric devices as well as applications in photovoltaic thermoelectric transport, photoconductivity, electrical photoresponse, and field emission [2–7]. Nanostructures of $Bi_2S_3$, especially the one-dimensional (1D) nanostructures (such as wires, rods, tubes, and ribbons), are in fact considered to be the best contenders for these applications due to quantum confinement effect [8, 9]. Due to quantum confinement of the charge carriers in 1D nanostructure, the thermoelectric efficiency of a material is enhanced through the increase in its figure-of-merit value brought about by the increase in the Seebeck coefficient and the decrease in the coefficient of thermal conductivity through boundary scattering of the heat carriers [10]. Similarly, the electron transport in crystalline nanowires is expected to be several folds faster than through a polycrystalline structure. It is because of these reasons that the optoelectronic properties of single-crystalline 1D nanostructures of $Bi_2S_3$ have been widely studied for the design and fabrication of fast, miniaturized optoelectronic devices (such as optical switches and photodiode arrays) [11, 12] or for augmenting the performance of existing ones.

Due to the immense technological importance of 1D nanostructured $Bi_2S_3$ in nanoscale devices fabrication, there has been a surge of interest in their synthesis. There has been a variety of synthetic approaches reported so far for their synthesis, which include evaporation-condensation route [1], molecular precursor decomposition route [13], solventless synthesis route [14], electrochemical deposition route [15], solvothermal/hydrothermal route [16, 17], and ionic liquid-assisted route [18]. Among these processes, solvothermal/hydrothermal method is one of the most studied solution-based synthesis route for the fabrication of 1D nanostructures of $Bi_2S_3$ because this technique allows the opportunities to control the morphology of the final products through the variation of different reaction parameters, which include reaction temperature, reaction time, reaction medium (or solvent), bismuth and sulfur sources, complexing agent, and surfactant [19–21]. Besides, in the context of the possibilities to modify the growth behaviour of $Bi_2S_3$ crystal to anisotropic one, the use of soft templates (such as surfactant, complexing agent, and block copolymer) as growth-directing agent has received significant amount of research interest in the recent past. Among various growth-directing agents, the use of a complexing agent to change the growth behaviour (i.e., both nucleation and growth rate) and to confine it in a desired direction remained an interesting aspect. The complexing agents are generally polyfunctional ligands that coordinate to the surface of either the ions or the nanoparticles formed, modify the particle surface energy, and thereby enforce the unidirectional growth of the crystal. These bi- or multidentate ligands in some cases also act as a soft template in the growth process.

There are a number of reports on the synthesis of 1D nanostructured $Bi_2S_3$ by using organic ligands, such as long chain amines, alcohols, and acids as complexing agents. Yu et al., for example, reported the use of EDTA as a coordinating agent for the preparation of $Bi_2S_3$ nanorods [22]. EDTA plays multiple roles as a coordinating agent for $Bi^{3+}$ as well as a soft template for the growth of nanorods under solvothermal condition [23]. On the other hand, Ota and Srivastava [24] utilized a multifunctional organic molecule (i.e., tartaric acid) as a capping agent for producing nanorods of $Bi_2S_3$. Similarly, Chen et al. [25] synthesized highly crystalline $Bi_2S_3$ nanorods using bismuth citrate and thiourea as precursor materials in the presence of cetyltrimethylammonium bromide. They claimed that the growth of nanorods is assisted by linear bismuth citrate polymer template. In addition, a number of polyols, for example, ethylene glycol [26], diethylene glycol (DEG) [27], and so forth, have been reported to assist in the direction-arrested one-dimensional growth of $Bi_2S_3$. In this process, the multiple –OH groups cap the formed particles of $Bi_2S_3$ and make them stable. These polyols could also exist in long chains due to the effective hydrogen bonding between them that serves as a template for growth of $Bi_2S_3$ nuclei to form nanorods [27]. Triethanolamine (TEA) is another important tetradentate chelating ligand containing three –OH and one amine group, which, like diethylene glycol, exists as a long chain polymeric framework in the presence of water due to the effective hydrogen bonding [28]. Moreover, TEA is miscible with water in all proportions and can be used in hydrothermal process successfully. Hence, it is expected that the use of this organic ligand can be an alternative route for the synthesis of $Bi_2S_3$ nanorods. However, the use of TEA for the synthesis of 1D nanostructure of $Bi_2S_3$ powders is limited only to a few reports [29].

Therefore, the present work aims at synthesizing nanorods of $Bi_2S_3$ by using TEA as a complexing agent under hydrothermal condition. In the developed process, a novel and inexpensive sulfur source, that is, elemental sulfur solubilized in monoethanolamine (MEA) has been used. The essence of this work lies in the formation of amorphous spherical precursors at room temperature and subsequently their transformation into nanorods under hydrothermal condition. A growth mechanism is also proposed.

## 2. Experimental Details

*2.1. Materials.* Bismuth nitrate ($Bi(NO_3)_3{\cdot}5H_2O$), triethanolamine ($(HOC_2H_4)_3N$), monoethanolamine ($HOC_2H_4NH_2$), and elemental sulfur were purchased from Merck, India, and used as received without further purification.

*2.2. Synthesis.* In a typical preparation, 1 g of bismuth nitrate was solubilized in a mixture of 20 mL of TEA and 120 mL of distilled water by stirring continuously for 1 h. The clear solution was obtained possibly through the formation of $[Bi(TEA)_x{}^{3+}]$ complex [29, 30]. Elemental sulfur (0.132 g) was separately dissolved in 10 mL of MEA, and it was then poured into the solution of $[Bi(TEA)_x]^{3+}$ complex. A brown precipitate was obtained immediately, which turned gray within few seconds. The entire solution mixture containing the gray colored solid precursor (i.e., the precursor sol) was then transferred to a Teflon-lined stainless steel autoclave and subjected to hydrothermal reaction at 160°C. After cooling the autoclave naturally to room temperature, the precipitates were collected, washed with water for several times and finally with ethanol, and dried. For the investigation of the reaction mechanism, the precursor sol formed at room temperature (prior to hydrothermal treatment) was also collected separately and characterized. This sample was referred to as "precursor sample" in the later part of the paper.

*2.3. Characterization.* The phase and structure analysis of the synthesized samples were carried out on Philips PW 1729 X-ray diffractometer using Co $K_\alpha$ radiation ($\lambda$ = 1.79 Å). X-ray photoelectron spectroscopy (XPS) studies of the samples were carried out on an XPS system made by Omicron NanoTechnology using Mg $K_\alpha$ (1253.6 eV) as the excitation source. The morphologies of the samples were studied by using JEOL JSM-5800 model and Zeiss EVO-60 model scanning electron microscope. The chemical compositions of the samples were confirmed through energy dispersive analysis by X-ray (EDAX) using Oxford INCA electron microprobe attached to scanning electron microscope. JEOL JEM-2100 model high-resolution transmission electron microscope (HRTEM) was used to examine the microstructural details of the prepared samples. Raman spectroscopy was performed using Renishaw RM-1000 model instrument equipped with an Ar ion laser of wavelength 514.4 nm. The diffuse reflectance spectra of the as-synthesized powders were recorded at room temperature using Varian Cary 5000 UV-Vis-NIR spectrophotometer.

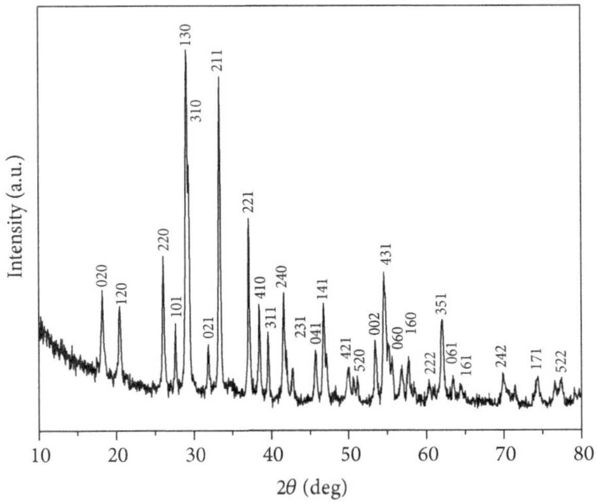


FIGURE 1: XRD pattern of the $Bi_2S_3$ sample obtained after 18 h of hydrothermal treatment at a temperature of 160°C.

## 3. Results and Discussion

The XRD analysis of the product, obtained after 18 h of hydrothermal treatment at 160°C, is displayed in Figure 1. The positions and relative intensities of all the peaks are in good agreement with those of the orthorhombic crystal structure of $Bi_2S_3$ and are indexed according to JCPDS file number 17-0320. The calculated cell parameters ($a = 11.133$ Å, $b = 11.304$ Å, and $c = 3.977$ Å) are also found to be well matched with the standard values (i.e., $a = 11.14$ Å, $b = 11.30$ Å, and $c = 3.98$ Å). Within the detection limit of the XRD method, no unidentified peaks in the diffractogram are noticed, suggesting that the final product is free from impurities. Energy dispersive analysis by X-ray (EDAX) of the sample reveals S-to-Bi atomic ratio as 1.45 against the ideal value of 1.5 for pure $Bi_2S_3$.

The composition of this sample was also verified by X-ray photoelectron spectroscopy (XPS) (Figure 2). The survey spectrum (Figure 2(a)) reveals the presence of elements, such as Bi, S, O, and C, which indicates the high purity of the resulting product. The peak for O arises, possibly, due to the adsorbed gases and/or oxides of C on the surfaces of the samples [29]. This observation is common in case of ultrafine powder samples when exposed to atmosphere. The high resolution XPS spectra in Bi region (Figure 2(b)) shows two peaks at 157.7 and 163 eV, which can be assignable, respectively, to the binding energies of Bi $4f_{7/2}$ and Bi $4f_{5/2}$ in $Bi_2S_3$. The broad peak at around 224.9 eV (Figure 2(c)) corresponds to the binding energy of S 2s. The observed values are found to be in close agreement with the data reported by Grigas et al. [31]. Raman spectroscopy studies were carried out to further validate the composition of the samples. Room-temperature Raman spectrum (recorded between 200 and 1400 $cm^{-1}$) for the sample obtained after 18 h of hydrothermal treatment at 160°C is displayed in Figure 3. The spectrum depicts five distinct vibrational peaks at about 260, 306, 422, 606, and 965 $cm^{-1}$, which corresponds to the characteristic Raman bands of crystalline $Bi_2S_3$ samples [24]. These are also in good agreement with the nanoparticles of $Bi_2S_3$ reported by Rabin et al. [7, 32].

The particle morphologies of the synthesized $Bi_2S_3$ powders were studied by electron microscopy. Representative SEM and bright field TEM micrographs of the sample obtained after 18 h of hydrothermal treatment at 160°C are shown in Figures 4(a) and 4(b), respectively. The micrographs reveal the formation of nanorods of $Bi_2S_3$ with average diameters and lengths ranging between 70 and 170 nm and 0.550–3 $\mu$m, respectively. TEM image of a typical nanorod having diameter of 90 nm and length of 2.5 $\mu$m and its detailed lattice fringe pattern are illustrated in Figures 4(c) and 4(d), respectively. The calculated fringe width of 0.56 nm (shown in Figure 4(d)) matches with the d-spacing of the (200) planes of the orthorhombic $Bi_2S_3$ crystal [12], suggesting that the growth of the nanorods of $Bi_2S_3$ occurred along the preferential [001] ($c$-axis) direction. Furthermore, the spotted pattern of the selected area electron diffraction (SAED) of the selected nanorod (shown as an inset in Figure 4(d)) infers single-crystalline nature of the synthesized sample.

In order to investigate the possible growth mechanism and the chemistry involved in the synthesis of $Bi_2S_3$ nanorods, a series of samples were synthesized at different reaction conditions and were analyzed by XRD and electron microscopy. Figure 5 represents the XRD patterns of the room-temperature precipitate that was separated prior to hydrothermal treatment and the samples obtained after 1, 3, and 12 h of hydrothermal treatment at a fixed temperature of 160°C. XRD pattern in Figure 5(a) showed broad and diffused peaks centered at around $2\theta = 30.88°$, 35.23°, and 51.73° indicating the precursor sample to be predominantly amorphous. The positions of these diffused peaks are slightly shifted from the standard values of major peaks reported for the orthorhombic $Bi_2S_3$ phase (JCPDS file no. 17-0320) but could not be assigned to any compound reported in the available literature. These unidentified broad peaks were however observed to diminish in intensity and eventually disappeared to give way to diffraction lines characteristic of the pure $Bi_2S_3$ phase. This could be achieved through hydrothermal treatment of the precursors at temperatures ≥160°C and incubation periods ≥6 h (see Figure S1 in Supplementary Material available online at http://dx.doi.org/10.1155/2013/367812). It was interesting to note that though the orthorhombic phase of $Bi_2S_3$ was realized through hydrothermal processing of the precursors at temperatures as low as 120°C and incubation period of just 3 h, but it was accompanied by the highest intense peak (at $2\theta \approx 30.28°$) of the unidentified phase. It was also realized that this peak persisted even though it prolonged the incubation period to 24 h at a temperature of 120°C. The peak however diminished in intensity with an increase in hydrothermal temperatures beyond 120°C, and in effect, the pure orthorhombic phase of $Bi_2S_3$ was accomplished on hydrothermal processing of the precursors at 160°C for 12 h (Figure 5(d)), although the products after 1 and 3 h of heating periods contain some amount of unidentified phase (as indicated by * in Figures 5(b) and 5(c)). On the basis of optimization studies, however, the pure and crystalline phases of $Bi_2S_3$ with relatively high aspect ratio were obtained

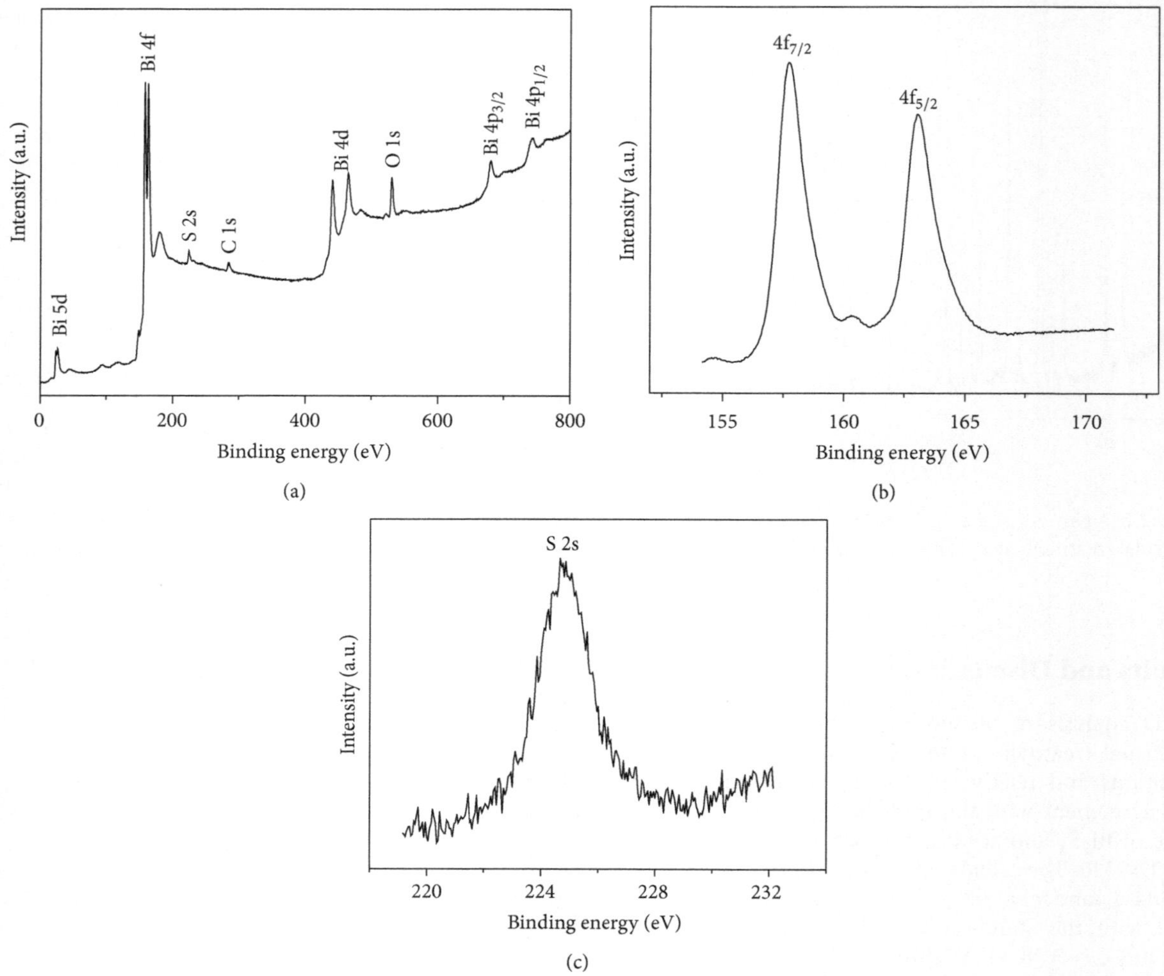


FIGURE 2: XPS analysis of $Bi_2S_3$ synthesized after 18 h of hydrothermal treatment at a temperature of 160°C; (a) survey spectra, (b) Bi region, and (c) S region.

on hydrothermal processing of the precursors at 160°C for 18 h, as depicted in Figure 1.

It may thus be inferred that the broad peaks of the precursors (at $2\theta$ = 30.88°, 35.23°, and 51.73°) correspond to an intermediate compound that are poorly crystalline and decompose to $Bi_2S_3$ at hydrothermal processing temperatures ≥160°C and incubation period ≥6 h. Liu et al. [16] reported the formation of similar intermediates, which decomposed to generate $Bi_2S_3$ nanoribbons. They assigned the broad peaks located at $2\theta \approx$ 30.57°, 29.62°, and 52.75° to cubic $NaBiS_2$, which got formed through hydrothermal reaction between $Bi^{3+}$-glycerol complexes and elemental sulfur/$Na_2S_2O_3$ in aqueous solution of NaOH. Ota and Srivastava [33] also predicted the formation of $Bi_2S_3$ nanotubes via decomposition of $NH_4BiS_2$ intermediates that were claimed to be produced from the reaction of $[Bi(TEA)_x]^{3+}$ complex with $CS_2$ and $NH_3$ in presence of Triton X-100 under hydrothermal condition. It was noted that the observed $2\theta$ values of our precursors were slightly shifted from those reported for the cubic $NaBiS_2$ phase. Therefore, based on these observations and drawing analogies with the reported literature in conjunction with the knowledge of the reagents involved in the present preparation, we can rationally predict the intermediates (precursors) to be aminium compounds of bismuth sulfide $[(BiS_2)^-]$ or thiosulfate $[[Bi(S_2O_3)]^{3-}]$. Hydrothermal processing of these precursors probably generates nanorods of $Bi_2S_3$. It however needs to be mentioned that though $Na_3Bi(S_2O_3)_3{\cdot}nH_2O$ and $NaBiS_2$ are reported in the literature [16, 34], the existence of aminium/ammonium intermediates is difficult to establish through XRD studies because of their poor crystalline nature and nonavailability of the standard literature claiming their formation in the crystalline form.

In order to substantiate the prediction from XRD analysis, the composition of the precursor sample (collected prior to hydrothermal treatment) was further studied by XPS. The survey spectrum, shown in Figure 6(a), reveals the presence of Bi, S, O, N, and C. Here, C 1s with standard binding energy (BE) value of 284.6 eV is taken as a reference. As C 1s peak in Figure 6(a) lies at 285.5 eV, instead of its ideal

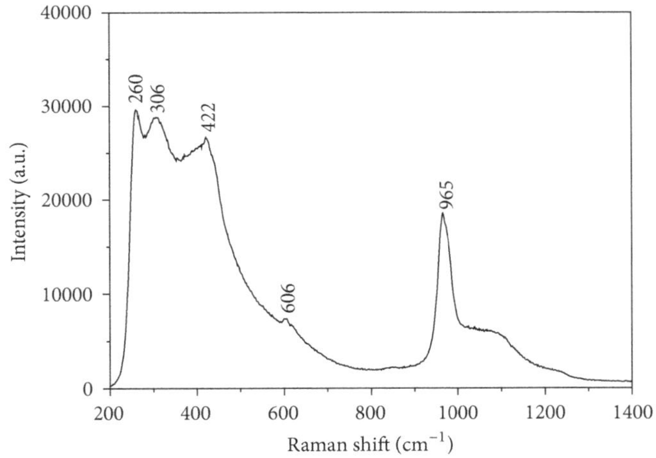


FIGURE 3: Raman spectra of the $Bi_2S_3$ sample synthesized through hydrothermal process at 160°C and incubation time period of 18 h.

FIGURE 4: $Bi_2S_3$ sample obtained through hydrothermal process at 160°C after 18 h of incubation period; (a) SEM image, (b) corresponding TEM image, (c) TEM image of a selected nanorod, and (d) HRTEM image of the selected nanorod with the corresponding SAED pattern (inset).

value of 284.6 eV, all the observed BEs in the XPS analysis are corrected accordingly [35, 36]. The small peak around 400 eV in survey spectrum may be attributed to N 1s of amine analogue [37, 38] present in the precursor sample, that is, aminium compounds of $(BiS_2)^-$ and $[Bi(S_2O_3)_3]^{3-}$. The standard 2s binding energy for sulfur in their elemental state is reported to be 229 eV [39], which is lowered by 3 eV for $S^{2-}$ state in $Bi_2S_3$ [31]. So, the peak at 228 eV in Figure 6(b) may be assigned to 2s BE for $S^{2-}$ state in $(BiS_2)^-$ species. On the other hand, the peak at 231.5 eV, which is higher than 2s BE for $S^0$ but lower than $S^{6+}$ state in sulfate ion [38], may be assigned to the presence of $S_2O_3{}^{2-}$ species ($S^{2+}$ state) in the precursor sample. In addition, the strong peak for O 1s in the survey spectrum (Figure 6(a)) could be deconvoluted into two peaks (531.1 and 533.6 eV). These two deconvoluted peaks could respectively be assigned to $S_2O_3{}^{2-}$ species and

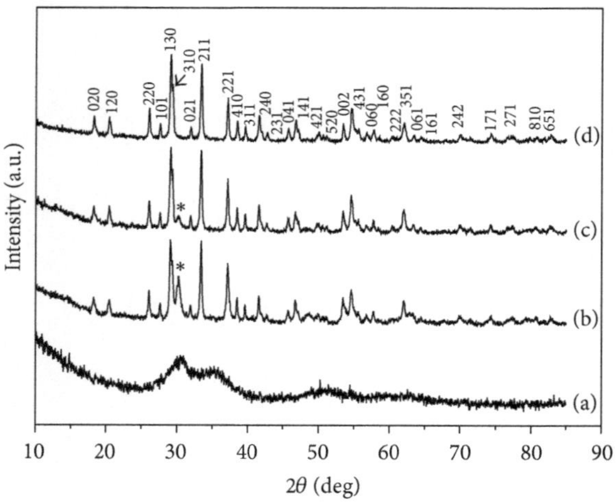


FIGURE 5: XRD patterns of the samples synthesized at different conditions; (a) at room temperature, after hydrothermal treatment at 160°C for (b) 1 h, (c) 3 h, and (d) 12 h.

to adsorbed $CO_2$ in the sample [38]. The peak pair at 159.5 and 164.5 eV (in Figure 6(c)) is close to binding energies of Bi $4f_{7/2}$ and Bi $4f_{5/2}$ in $Bi_2S_3$ [13] and can be assigned to $Bi^{3+}$ in $(BiS_2)^-$ or $[Bi(S_2O_3)_3]^{3-}$ species. This is because $Bi^{3+}$ is not very sensitive to chemical environment; therefore, $(BiS_2)^-$ or $[Bi(S_2O_3)_3]^{3-}$ species are expected to show ionization potential values comparable to $Bi_2S_3$.

Based on the above experimental results, the possible chemistry involved in the preparation of $Bi_2S_3$ can be proposed. As it is already discussed in our previous paper [40] that the disproportionation reaction of solution of elemental sulfur in MEA generates a number of ions including thiosulfate ($S_2O_3{}^{2-}$) and $HS^-$ ions in the presence of $OH^-$ ions in the reaction mixture. The TEA chelated complexes of bismuth ions react then with the $S^{2-}$ and $S_2O_3{}^{2-}$ ions at room temperature to produce aminium compounds of bismuth sulfide and thiosulfate (i.e., $(BiS_2)^-$ and $[Bi(S_2O_3)]^{3-}$) as per reactions (1) and (2):

$$Bi^{3+} + 2S^{2-} \rightleftharpoons [BiS_2]^- \quad (1)$$

$$Bi^{3+} + 3S_2O_3{}^{2-} \rightleftharpoons [Bi(S_2O_3)_3]^{3-} \quad (2)$$

Under hydrothermal processing, the aminium compounds of $[Bi(S_2O_3)]^{3-}$ are first converted into the corresponding $(BiS_2)^-$, which subsequently undergoes decomposition to generate $Bi_2S_3$ nanocrystals. These nanocrystals probably grow into 1D nanostructures of $Bi_2S_3$ with increasing the incubation period. The whole process thus can be summarized through the following set of chemical reactions:

$$[Bi(S_2O_3)_3]^{3-} \xrightleftharpoons{\text{Hydrothermal processing}} [BiS_2]^- \quad (3)$$

$$2[BiS_2]^- \xrightleftharpoons{\text{Hydrothermal processing}} Bi_2S_3\,(\text{nanocrystals}) + S^{2-} \quad (4)$$

$$n\,(Bi_2S_3) \xrightarrow{\text{hydrothermal}} (Bi_2S_3)_n \quad (5)$$

To further study the possible growth mechanism of the formation of one-dimensional structures of $Bi_2S_3$, the effect of reaction temperature and reaction time on the morphologies of the samples were studied exhaustively by electron microscopy. Figure 7(a) depicts TEM image of the precursor sample obtained prior to hydrothermal treatment, which suggests the presence of mostly aggregation of nearly spherical nanoparticles of diameters in the range of 20–40 nm. These aggregates are amorphous in nature as indicated from its diffused SAED patterns (inset of Figure 7(a)). This finding is also supported by XRD analysis (Figure 5(a)). Figure 7(b) reveals that the nanorod-shaped morphologies get formed even after 1 h of hydrothermal processing at 160°C, though some agglomerated spherical particles are still present. With the increase of the incubation to 12 h, the number of rod-shaped particles of $Bi_2S_3$ increased, while the spherical aggregates reduced drastically (Figure 7(c)). Pure crystalline nanorods of $Bi_2S_3$ were eventually realized on increasing the incubation period to 18 h at the hydrothermal reaction temperature of 160°C (as shown in Figure 4(a)). It is interesting to note that on increasing the heating period to 24 h, a few nanowires of $Bi_2S_3$ (indicated by arrows in Figure 7(d)) are also formed amidst a pool of nanorods.

The effect of reaction temperature on the morphology of the synthesized $Bi_2S_3$ sample was also investigated by varying the reaction temperature in the range of 120–180°C for a fixed incubation period (i.e., 6 h). The SEM images of the synthesized samples at hydrothermal temperatures 120, 140, 160, and 180°C (Figure S2, Supporting Information) suggest that rod-like morphologies with diameters in nanometer range are formed after 6 h of incubation period irrespective of the hydrothermal reaction temperature. However, some aggregated masses were visible, when the reaction was carried out at a lower temperature, which gradually disappeared as the temperature was increased. It can thus be inferred that the growth of nanorods occurs probably at the expense of amorphous spherical particles.

Based on the morphological findings and other experimental observations, it is thus possible to propose a growth mechanism for the preparation of $Bi_2S_3$. As it was already mentioned earlier, a large numbers of $Bi_2S_3$ nuclei are generated as a result of the decomposition of the precursors at the initial stage of the hydrothermal reaction. These freshly formed nanocrystals or nuclei in the solution are unstable due to the presence of a large number of dangling bonds, defects, or traps on the nuclei surfaces. Hence, under hydrothermal

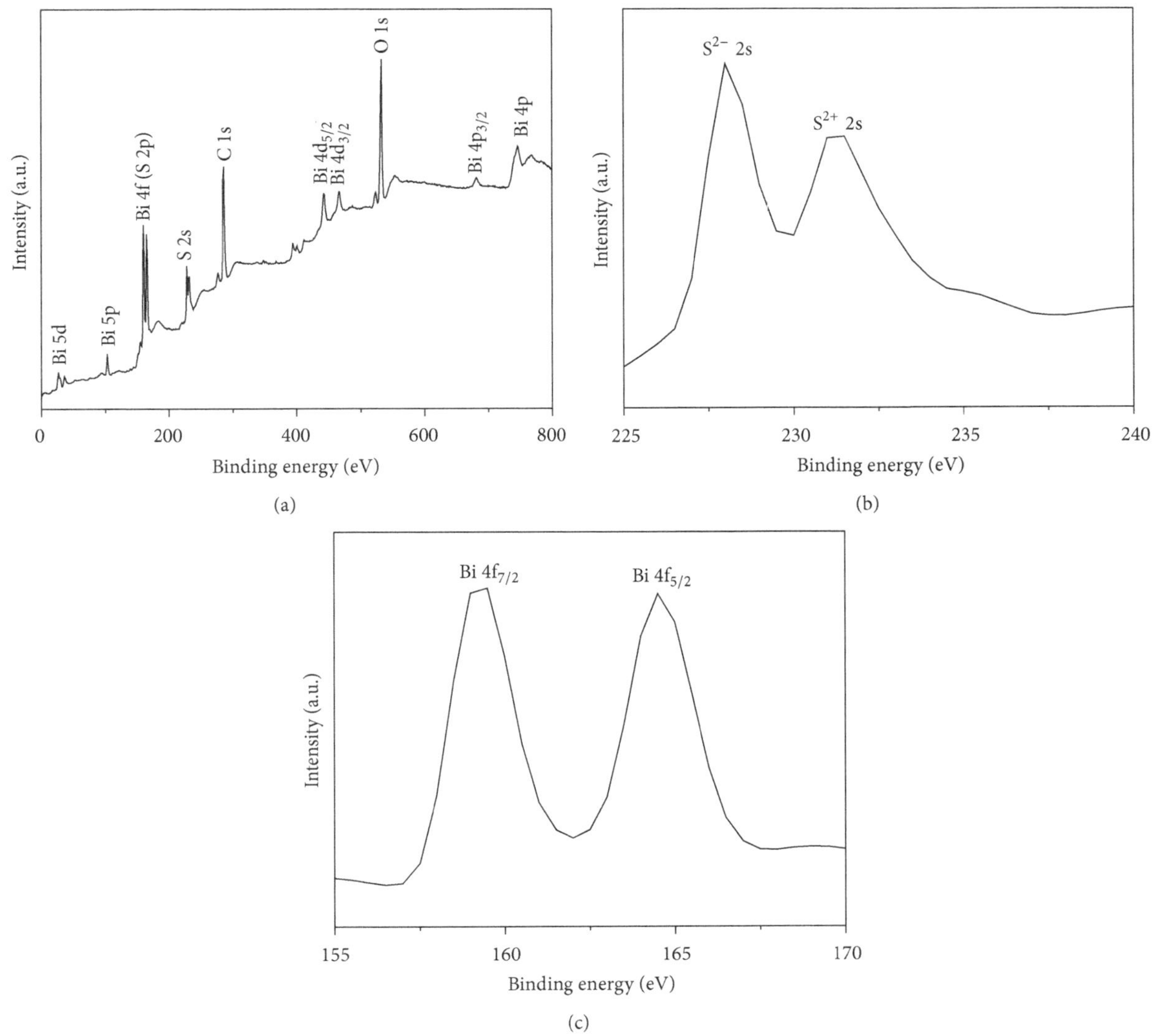


FIGURE 6: XPS analysis of the precursor sample; (a) survey spectra, (b) S region, and (c) Bi region.

condition, the smaller $Bi_2S_3$ nanocrystals precipitate onto comparatively larger particles, leading to growth of the crystals along the energetically favorable direction (i.e., *c*-direction) to form the nanorods of $Bi_2S_3$. The proposed growth mechanism agrees with the morphological findings, where it is evident that with the increase in the incubation period, the $Bi_2S_3$ particles grew more rapidly along the length compared to the diameters. This oriented crystal growth can be attributed to the typical anisotropic layered structure of $Bi_2S_3$. The process is also thermodynamically driven by the decrease in the total interfacial energy of the particles as a result of merging. An analogous mechanism was suggested by Zhu et al. [41] for the growth of nanorods from the initially formed $Bi_2S_3$ particles through a dynamic equilibrium: $2Bi^{3+} + 3S^{2-} \leftrightarrow Bi_2S_3$. They proposed that the $Bi^{3+}$ and $S^{2-}$ ions, which were in equilibrium in solution, precipitated onto larger particles, thereby decreasing the total interfacial energy between the particles and the solution. Liu et al. [16] recognized a similar growth sequence for the fabrication of $Bi_2S_3$ nanoribbons from $NaBiS_2$ precursors through solvothermal process, but they choose to call the mechanism as "solid-solution-solid (SSS) transformation". Gates et al. [42] also proposed the synthesis of single-crystal nanowires of selenium from spherical colloidal particles of amorphous selenium in aqueous solution through similar mechanism and called it to be SSS transformation. Based on the discussion, the formation and the overall growth process of the $Bi_2S_3$ nanorods through hydrothermal process can be schematically represented by a sketch as shown in Figure 8.

On the other hand, the formation of longer rods or wires of $Bi_2S_3$ (as shown in Figure 7(d)) probably occurred at the expense of small rod-shaped particles, which merged to reduce the net interfacial energy of the particles. This prediction is supported by bright field TEM images obtained for the sample synthesized at a hydrothermal temperature of 160°C for a period of 18 h (Figure 9). The contrast variation

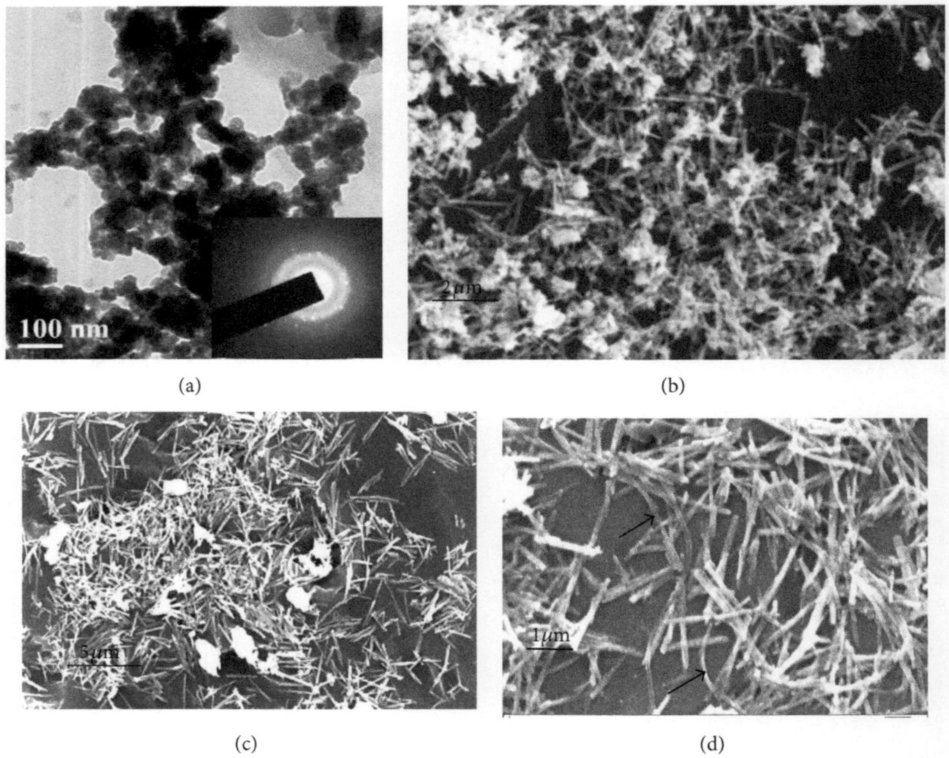


FIGURE 7: (a) TEM image of precursor sample and the corresponding SAED pattern (inset) and SEM images of the sample obtained through hydrothermal treatment at 160°C, (b) after 1 h, (c) after 12 h, and (d) after 24 h.

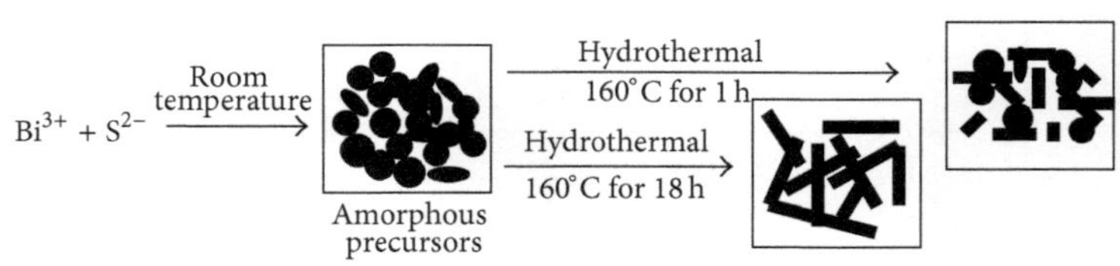


FIGURE 8: Schematic representation of the growth of $Bi_2S_3$ nanorods from spherical nanoparticles of precursors through hydrothermal process.

of TEM images (shown in Figures 9(a) and 9(b)) clearly depicts the joining of a pair of $Bi_2S_3$ nanorods. It can be seen from the images that rods are continuous across the connected part barring at some portions. For further insight into the microstructure at the junction, the encircled portion of the TEM image was magnified (Figure 9(c)), which shows the lattice planes at the joint. Lattice spacing, calculated from the fringe pattern of the HRTEM image, is found to be 0.476 nm corresponding to (120) plane of orthorhombic phase of $Bi_2S_3$. It can be seen from images that fringes are almost continuous across the joint with disrupted continuity only at selected area. The disrupted portion is indicated by an arrow in Figure 9(c).

It can thus be predicted that under hydrothermal conditions, fusion of the particles at the interface takes place only when joining particles are suitably positioned so as to share a common crystallographic orientation. A similar morphological sequence has also been observed by Yu et al. [43] for the synthesis of nanowires of $Bi_2S_3$ from spherical crystals. They suggested that the formation of nanowires of $Bi_2S_3$ occurred via merging of short nanorods under hydrothermal reaction at 180°C (for 3 days) of the solution containing spherical crystals of $Bi_2S_3$. Deng et al. [23] explained the fabrication of sheet rods of $Bi_2Te_3$ on the basis of EDTA-assisted controlling oriented attachment of individual sheets. They too proposed that the joining process was thermodynamically driven by the reduction in the net surface energy of the particles, and the joining of particles was possible, only when they had common crystallographic orientation. On the basis of the present discussion, it can thus be inferred that both controlled oriented attachment and SSS mechanism cooperate in the formation of nanorods/wires of $Bi_2S_3$ from smaller rods/spherical particles in the hydrothermal process.

In the developed process, TEA, which serves as a stabilizing agent for retaining the bismuth ions in solution through complex formation as well as an efficient capping

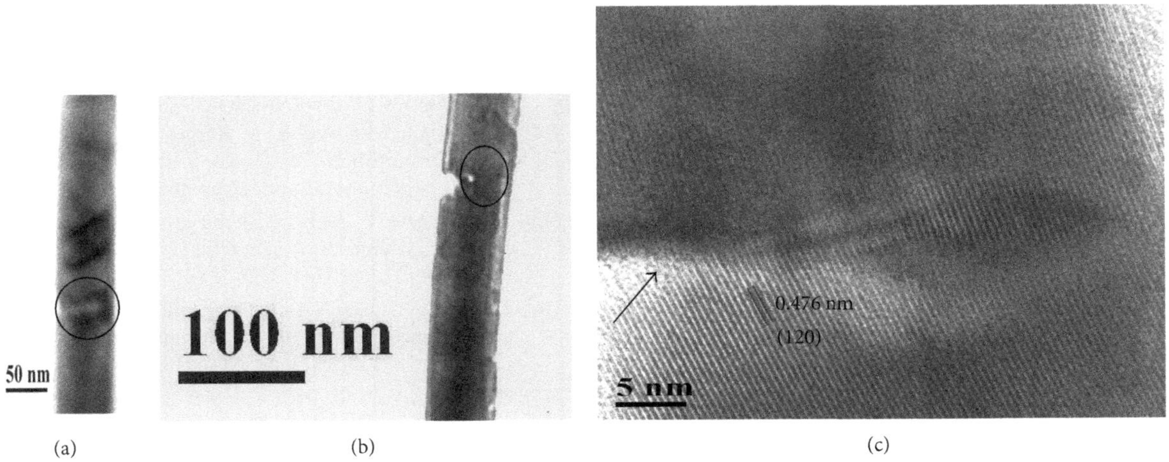


FIGURE 9: TEM images showing the joining of a pair of $Bi_2S_3$ rods (marked by circles); (a) at low resolution, (b) high resolution of image of the encircled portion in (b), and (c) high resolution of image showing the fringe patterns at the joint.

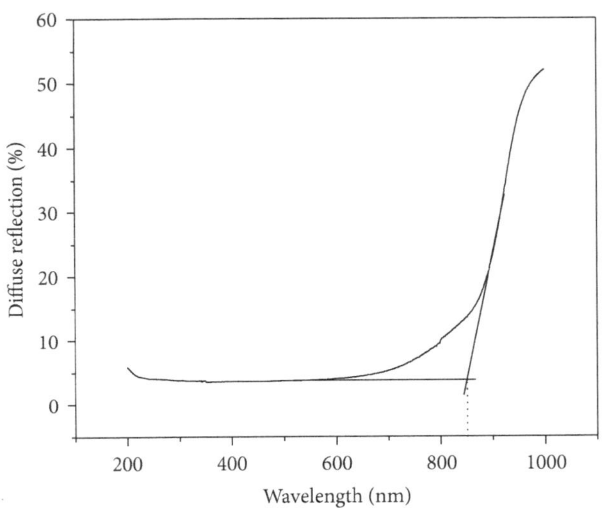


FIGURE 10: Diffuse reflectance spectra of the $Bi_2S_3$ samples, synthesized through hydrothermal process at a temperature of 160°C and incubation period of 18 h.

agent [44, 45], also plays another important role in the formation of shape-controlled 1D structures of $Bi_2S_3$. TEA is a tetradentate ligand with three –OH groups and one amine group that effectively coordinates to the ions or particles, and thereby modifies the surface energy states of the particles, making them to grow in one dimension. In the aqueous reaction mixture, the unchelated, free TEA probably forms long chain polymeric framework [28] due to hydrogen bonding. These structures may help in assembling the initially formed nearly spherical particles to grow in one dimension under hydrothermal condition leading to the generation of uniformly sized pure and crystalline 1D structures of $Bi_2S_3$. To investigate the role of water in this mechanism, further experiments were carried out by heating a reaction mixture containing only TEA as a solvent (i.e., without water) in an autoclave at 160°C for 6 h and studying the morphology of the synthesized sample. In this case, $Bi_2S_3$ nanorods with diameters and lengths in the range of 77–575 nm and 1200–3700 nm, respectively (Figure S3, Supporting Information), were obtained in contrast to the nanorods (diameters and lengths in the range of 44–140 nm and 945–2150 nm, resp.) produced under similar reaction conditions with TEA and water as reaction medium. The results validated the utility of water in the reaction medium for the generation of uniformly sized nanorods with high aspect ratio.

$Bi_2S_3$ has moderately high band gap energy ($E_g$) compared to the other sulfides of the same group, making them important materials for various optoelectronic applications. So, the diffuse reflectance spectrum of the $Bi_2S_3$ sample prepared at optimized hydrothermal reaction conditions (i.e., at 160°C for 18 h) has been carried out at room temperature, and the result is depicted in Figure 10. The optical band gap was calculated from the point of intersection of the tangent of the absorption edge and the extrapolated line of the diffuse reflectance at lower wavelength, and the value was found to be 1.46 eV. This calculated value lies in the band gap range reported for bulk $Bi_2S_3$, which may be attributed to their larger average particle diameters compared to the exciton Bohr radius of $Bi_2S_3$ (29.8 nm) [46].

## 4. Conclusions

Uniformly sized single-crystalline $Bi_2S_3$ nanorods with average diameters and lengths in the range of 70–170 nm and 0.55–3 $\mu$m, respectively, have been synthesized through hydrothermal reaction at a temperature as low as 160°C. These nanorods have been found to be of pure orthorhombic phase, as evident from XRD, XPS, and Raman spectroscopy. Nanorods of $Bi_2S_3$ have been generated at the expense of the amorphous precursor precipitates, formed at room temperature, which has been found to consist of aminium compounds of $[Bi(S_2O_3)]^{3-}$ and $(BiS_2)^-$. The precursors undergo decomposition and subsequent growth into nanorods/wires through SSS as well as oriented-aggregation-growth process under hydrothermal conditions. It is found that $Bi_2S_3$ nanorods with relatively good aspect ratio and smooth surfaces are produced, only when all the reaction conditions, such as hydrothermal temperature, incubation period, and solvents, are tuned.

## Acknowledgments

The authors acknowledge their sincere thanks to Professor B. Viswanathan and Dr. K. Thirunavukkarasu of the Indian Institute of Technology Madras for their help in characterizing the samples by XPS. They also express their gratitude to Professor M. K. Panigrahi of the Indian Institute of Technology Kharagpur for obtaining Raman spectra.

## References

[1] C. Ye, G. Meng, Z. Jiang, Y. Wang, G. Wang, and L. Zhang, "Rational growth of $Bi_2S_3$ nanotubes from quasi-two-dimensional precursors," *Journal of the American Chemical Society*, vol. 124, no. 51, pp. 15180–15181, 2002.

[2] S. K. Batabyal, C. Basu, A. R. Das, and G. S. Sanyal, "Nanostructures of bismuth sulphide: synthesis and electrical properties," *Journal of Nanoscience and Nanotechnology*, vol. 7, no. 2, pp. 565–569, 2007.

[3] J. D. Desai and C. D. Lokhande, "Chemical deposition of $Bi_2S_3$ thin films from thioacetamide bath," *Materials Chemistry and Physics*, vol. 41, no. 2, pp. 98–103, 1995.

[4] M. T. S. Nair and P. K. Nair, "Photoconductive bismuth sulphide thin films by chemical deposition," *Semiconductor Science and Technology*, vol. 5, no. 12, pp. 1225–1230, 1990.

[5] G. Konstantatos, L. Levina, J. Tang, and E. H. Sergeant, "Sensitive solution-processed $Bi_2S_3$ nanocrystalline photodetectors," *Nano Letters*, vol. 8, no. 11, pp. 4002–4006, 2008.

[6] S. C. Liufu, L. D. Chen, Q. Yao, and C. F. Wang, "Assembly of one-dimensional nanorods into $Bi_2S_3$ films with enhanced thermoelectric transport properties," *Applied Physics Letters*, vol. 90, Article ID 112106, 3 pages, 2007.

[7] X. Yu and C. Cao, "Photoresponse and field-emission properties of bismuth sulfide nanoflowers," *Crystal Growth & Design*, vol. 8, pp. 3951–3955, 2008.

[8] R. S. Mane, B. R. Sankapal, and C. D. Lokhande, "Photoelectrochemical cells based on chemically deposited nanocrystalline $Bi_2S_3$ thin films," *Materials Chemistry and Physics*, vol. 60, no. 2, pp. 196–203, 1999.

[9] R. Suarez, P. K. Nair, and P. V. Kamat, "Photoelectrochemical behavior of $Bi_2S_3$ nanoclusters and nanostructured thin films," *Langmuir*, vol. 14, no. 12, pp. 3236–3241, 1998.

[10] M. S. Dresselhaus, G. Chen, M. Y. Tang et al., "New directions for low-dimensional thermoelectric materials," *Advanced Materials*, vol. 19, no. 8, pp. 1043–1053, 2007.

[11] H. Bao, X. Cui, C. M. Li, G. Ye, J. Zhang, and J. Guo, "Photoswitchable semiconductor bismuth sulfide ($Bi_2S_3$) nanowires and their self-supported nanowire arrays," *The Journal of Physical Chemistry C*, vol. 111, no. 33, pp. 12279–12283, 2007.

[12] H. Bao, C. M. Li, X. Cui, Q. Song, H. Yang, and J. Guo, "Single-crystalline $Bi_2S_3$ nanowire network film and its optical switches," *Nanotechnology*, vol. 19, no. 33, Article ID 335302, 2008.

[13] Y. W. Koh, C. S. Lai, A. Y. Du, E. R. T. Tiekink, and K. P. Loh, "Growth of bismuth sulfide nanowire using bismuth trisxanthate single source precursors," *Chemistry of Materials*, vol. 15, no. 24, pp. 4544–4554, 2003.

[14] M. B. Sigman Jr. and B. A. Korgel, "Solventless synthesis of $Bi_2S_3$ (bismuthinite) nanorods, nanowires, and nanofabric," *Chemistry of Materials*, vol. 17, pp. 1655–1660, 2005.

[15] X. S. Peng, G. W. Meng, J. Zhang et al., "Electrochemical fabrication of ordered $Bi_2S_3$ nanowire arrays," *Journal of Physics D*, vol. 34, no. 22, pp. 3224–3228, 2001.

[16] Z. P. Liu, J. B. Liang, S. Li, S. Peng, and Y. T. Qian, "Synthesis and growth mechanism of $Bi_2S_3$ nanoribbons," *European Journal of Chemistry*, vol. 10, no. 3, pp. 634–640, 2004.

[17] H. Zhang, Y. Ji, X. Y. Ma, J. Xu, and D. Yang, "Long $Bi_2S_3$ nanowires prepared by a simple hydrothermal method," *Nanotechnology*, vol. 14, no. 9, pp. 974–977, 2003.

[18] J. Jiang, S. H. Yu, W. T. Yao, H. Ge, and G. Z. Zhang, "Morphogenesis and crystallization of $Bi_2S_3$ nanostructures by an ionic liquid-assisted templating route: synthesis, formation mechanism, and properties," *Chemistry of Materials*, vol. 17, no. 24, pp. 6094–6100, 2005.

[19] T. Thongtem, A. Phuruangrat, S. Wannapop, and S. Thongtem, "Characterization of $Bi_2S_3$ with different morphologies synthesized using microwave radiation," *Materials Letters*, vol. 64, no. 2, pp. 122–124, 2010.

[20] L. Tian, H. Y. Tan, and J. J. Vittal, "Morphology-controlled synthesis of $Bi_2S_3$ nanomaterials via single- and multiple-source approaches," *Crystal Growth & Design*, vol. 8, no. 2, pp. 734–738, 2008.

[21] J. Wu, F. Qin, G. Cheng et al., "Large-scale synthesis of bismuth sulfide nanorods by microwave irradiation," *Journal of Alloys and Compounds*, vol. 509, no. 5, pp. 2116–2126, 2011.

[22] H. S. Yu, J. Yang, Y. S. Wu, Z. H. Han, Y. Xie, and Y. T. Qian, "Hydrothermal preparation and characterization of rod-like ultrafine powders of bismuth sulfide," *Materials Research Bulletin*, vol. 33, no. 11, pp. 1661–1666, 1998.

[23] Y. Deng, C. W. Nan, and L. Guo, "A novel approach to $Bi_2Te_3$ nanorods by controlling oriented attachment," *Chemical Physics Letters*, vol. 383, no. 5-6, pp. 572–576, 2004.

[24] J. R. Ota and S. K. Srivastava, "Polypyrrole coating of Tartaric acid-assisted synthesized $Bi_2S_3$ nanorods," *The Journal of Physical Chemistry C*, vol. 111, no. 33, pp. 12260–12264, 2007.

[25] R. Chen, M. H. So, C. M. Che, and H. Sun, "Controlled synthesis of high crystalline bismuth sulfide nanorods: using bismuth citrate as a precursor," *Journal of Materials Chemistry*, vol. 15, no. 42, pp. 4540–4545, 2005.

[26] G. Z. Shen, D. Chen, K. B. Tang, F. Q. Li, and Y. T. Qian, "Large-scale synthesis of uniform urchin-like patterns of $Bi_2S_3$ nanorods through a rapid polyol process," *Chemical Physics Letters*, vol. 370, no. 3-4, pp. 334–337, 2003.

[27] H. Zhang and L. Wang, "Synthesis and characterization of $Bi_2S_3$ nanorods by solvothermal method in polyol media," *Materials Letters*, vol. 61, no. 8-9, pp. 1667–1670, 2007.

[28] A. Pathak, S. Mohapatra, S. Mohapatra et al., "Preparation of nanosized mixed-oxide powders," *American Ceramic Society Bulletin*, vol. 83, no. 8, pp. 9301–9306, 2004.

[29] H. Wang, J. J. Zhu, J. M. Zhu, and H. Y. Chen, "Sonochemical method for the preparation of bismuth sulfide nanorods," *The Journal of Physical Chemistry B*, vol. 106, no. 15, pp. 3848–3854, 2002.

[30] A. Jana, C. Bhattacharya, S. Sinha, and J. Datta, "Study of the optimal condition for electroplating of $Bi_2S_3$ thin films and their photoelectrochemical characteristics," *The Journal of Solid State Electrochemistry*, vol. 13, no. 9, pp. 1339–1350, 2009.

[31] J. Grigas, E. Talik, and V. Lazauskas, "X-ray photoemission spectra and electronic structure of $Bi_2S_3$ crystals," *Physica Status Solidi B*, vol. 232, pp. 220–230, 2002.

[32] O. Rabin, J. M. Perez, J. Grimm, G. Wojtkiewicz, and R. Weissleder, "An X-ray computed tomography imaging agent based on long-circulating bismuth sulphide nanoparticles," *Nature Materials*, vol. 5, pp. 118–122, 2006.

[33] J. R. Ota and S. K. Srivastava, "Low temperature micelle-template assisted growth of $Bi_2S_3$ nanotubes," *Nanotechnology*, vol. 16, pp. 2415–2419, 2005.

[34] M. S. Bhadraver, "Physico-chemical studies on the composition of thiosulfates of metals. I. Thermometric studies of bismuth thiosulfate complexes," *Bulletin of the Chemical Society of Japan*, vol. 35, no. 11, pp. 1768–1770, 1962.

[35] X. H. Liao, H. Wang, J. J. Zhu, and H. Y. Chen, "Preparation of $Bi_2S_3$ nanorods by microwave irradiation," *Materials Research Bulletin*, vol. 36, no. 13-14, pp. 2339–2346, 2001.

[36] W. B. Zhao, J. J. Zhu, Y. Zhao, and H. Y. Chen, "Photochemical synthesis and characterization of $Bi_2S_3$ nanofibers," *Materials Science and Engineering B*, vol. 110, no. 3, pp. 307–313, 2004.

[37] W. F. Egelhoff Jr., "$N_2$ on Ni(100): angular dependence of the N1s XPS peaks," *Surface Science*, vol. 141, no. 2-3, pp. L324–L328, 1984.

[38] R. Malakooti, L. Cademartiri, Y. Akcakir, S. Petrov, A. Migliori, and G. A. Ozin, "Shape-Controlled $Bi_2S_3$ anocrystals and their plasma polymerization into flexible films," *Advanced Materials*, vol. 18, no. 16, pp. 2189–2194, 2006.

[39] C. D. Wagner, W. M. Riggs, L. E. Davis, and J. F. Moulder, *Handbook of X-Ray Photoelectron Spectroscopy*, Edited by G. E. Mulenberg, Perkin-Elmer, Minneapolis, Minn, USA, 1979.

[40] P. K. Panigrahi and A. Pathak, "Microwave-assisted synthesis of $WS_2$ nanowires through tetrathiotungstate precursors," *Science and Technology of Advanced Materials*, vol. 9, no. 4, Article ID 045008, 2008.

[41] G. Q. Zhu, P. Liu, J. P. Zhou et al., "Effect of mixed solvent on the morphologies of nanostructured $Bi_2S_3$ prepared by solvothermal synthesis," *Materials Letters*, vol. 62, no. 15, pp. 2335–2338, 2008.

[42] B. Gates, Y. Yin, and Y. Xia, "A solution-phase approach to the synthesis of uniform nanowires of crystalline selenium with lateral dimensions in the range of 10–30 nm," *Journal of the American Chemical Society*, vol. 122, no. 50, pp. 12582–12583, 2000.

[43] Y. Yu, C. H. Jin, R. H. Wang, Q. Chen, and L. M. Peng, "High-quality ultralong $Bi_2S_3$ nanowires: structure, growth, and properties," *The Journal of Physical Chemistry B*, vol. 109, no. 40, pp. 18772–18776, 2005.

[44] A. K. Singh, V. Viswanath, and V. C. Janu, "Synthesis, effect of capping agents, structural, optical and photoluminescence properties of ZnO nanoparticles," *Journal of Luminescence*, vol. 129, no. 8, pp. 874–878, 2009.

[45] M. S. Mohajerani, M. Mazloumi, A. Lak, A. Kajbafvala, S. Zanganeh, and S. K. Sadrnezhaad, "Self-assembled zinc oxide nanostructures via a rapid microwave-assisted route," *Journal of Crystal Growth*, vol. 310, no. 15, pp. 3621–3625, 2008.

[46] B. Pejova and I. Grozdanov, "Structural and optical properties of chemically deposited thin films of quantum-sized bismuth(III) sulfide," *Materials Chemistry and Physics*, vol. 99, no. 1, pp. 39–49, 2006.

# 11

# Synthesis and Evaluation of Magnetic Nanoparticles for Biomedical Applications


**Nydeia W. Bolden,[1] Vijaya K. Rangari,[1] Shaik Jeelani,[1] Seyhan Boyoglu,[2] and Shree R. Singh[3]**

[1] *Department of Materials Science and Engineering, Tuskegee University, Tuskegee, AL 36088, USA*
[2] *Department of Pediatrics, Emory-Children's Center, Emory University, Atlanta, GA 30322, USA*
[3] *Center for NanoBiotechnology Research, Alabama State University, Montgomery, AL 36104, USA*

Correspondence should be addressed to Vijaya K. Rangari; rangariv@mytu.tuskegee.edu



Academic Editor: Fabien Grasset



In this study, iron oxide (IO) nanoparticles from various precursors have been synthesized using sonochemical method and characterized for their structural variability and toxicity. The iron oxide (IO) precursor solutions were prepared from iron acetate (IA), iron pentacarbonyl (IP), decalin, PEG (poly(ethylene glycol)), EG (ethylene glycol), PVA (poly(vinyl alcohol)), $\beta$-cyclodextrin (CD), and distilled water. These precursor solutions were irradiated with high power ultrasound for 3 hours and heat treated as needed. These as-prepared iron oxide nanoparticles were characterized using X-ray diffraction (XRD), Mössbauer spectroscopy, transmission electron microscopy (TEM), and magnetization measurements. XRD results show that all the particles are highly crystalline in nature and the particles sizes measured from TEM are approximately 5–20 nm. The maximum magnetization was observed for IO-IP at approximately 60.17 emu/g and the minimum was approximately 30.56 emu/g for IO-IA. These results confirm that the particles are superparamagnetic (SPM) in nature. Mössbauer spectroscopy verified the magnetic nanoparticles are purely $Fe_3O_4$ and particles sizes varied by the nature of the precursor and coatings.


## 1. Introduction

Iron based nanomaterials have found applications in many areas of science and technology due to their unique magnetic properties [1–5]. Recently, they are also used in biomedical applications such as MRI contrast agents for imaging [6], magnetic hyperthermia [7], and targeted drug delivery [8] because of their size compatibility [9] to cells, genes, and viruses. A size reduction of these materials from bulk to the nanoscale permits them to display various size related properties. Among other changes, a significant decrease in size can alter reactivity, increase surface area, and change the magnetic properties [10–12]. Iron oxide, for example, exhibits ferromagnetic behavior in bulk where it can reach full magnetization and the effect persists even after removal of the magnetic field. These same particles below a certain size (<20 nm) may exhibit a behavior similar to paramagnetism even when at temperatures below the Curie point. This is a nanoscale phenomenon, where the energy required to change the direction of the magnetic moment of a particle is comparable to the ambient thermal energy. The energy required to change the direction of magnetization of a crystallite is called the crystalline anisotropy energy and depends both on the material properties and the crystallite size. As the crystallite size decreases, so does the crystalline anisotropy energy, resulting in a decrease in the temperature at which the material becomes superparamagnetic. These materials require the presence of a field to align their magnetic moments and magnetize. They lose most or all of their magnetization once the field is removed.

SPM materials are potential candidates for targeted drug delivery applications because of their size compatibility and ability to become highly magnetized only in the presence of a magnetic field. Magnetic particles performing as drug carriers [13] involve systems that can be directed through the body to a specific location by an external magnet. The significance of these drug systems is perhaps most imperative in the area of chemotherapy where severe side effects such as morphological alterations and functional impairment of the heart can occur [14]. Cancer drugs perform by killing cancer cells, but are not restricted and consequently affect healthy cells as well. The side effects of these drugs may decrease while the effectiveness is increased if a majority of the drug is able to reach its targeted site. With the aid of an external magnet placed near the desired location, more of these drugs should be able to reach its destination and harm fewer healthy cells.

Due to the nature of these applications, which involves interactions with a magnetic field outside the body, it is necessary to use materials that can reach the highest possible magnetization. Currently, iron oxide, particularly magnetite ($Fe_3O_4$), is commonly studied for these applications over other iron oxide structures such as maghemite ($\gamma$-$Fe_2O_3$) due to a higher magnetic saturation [15]. Similar synthesis routes can be used to produce either structure and require detailed analysis such as XRD and Mössbauer spectroscopy to determine the specific phase. Peaks for both $Fe_3O_4$ and $\gamma$-$Fe_2O_3$ occur at nearly the same $2\theta$ in XRD, making it difficult to distinguish. $Fe_3O_4$ can be identified by Mössbauer spectra taken at room temperature by the display of hyperfine magnetic splitting that is not seen in $Fe_2O_3$ [16, 17].

Iron oxide nanoparticles for drug delivery applications can be synthesized through various methods including solution precipitation, decomposition, ultrasound chemistry, and microwave heating [18–20]. Sonochemical techniques, the method used in this study proceeds by the creation, formation, and bursting of cavitations created from ultrasound waves. When the bubble collapses, it generates pressures and temperatures high enough to cause bond breakage. Because the cooling rates are also high, the particles' size remains low, and only nanosized materials are produced. This method has been proven by Suslick [21] and others [22–25] as a reliable, reproducible, and relatively easy way to synthesize both uncoated and in situ coated materials. All particles created for biorelated applications must be biocompatible, and therefore may require additional coatings. Coatings can also provide water dispersion stability and protection from degradation as well as increasing circulation time [26]. Coatings may be added to the surface of the oxides through chemical or physical bonding [27] using techniques such as solvent evaporation and coupling agents [28] and can take place either in situ or after-synthesis. Commonly used coatings include poly(lactic-co-glycolic acid) (PLGA) [29], Polyethylene glycol (PEG) [30], Poly vinyl alcohol (PVA), [31] and pluronic-127 [32]. The efficiency of these coatings can be determined through cell viability studies at various concentrations of nanoparticles [33]. In this study a uniform superparamagnetic $Fe_3O_4$ nanoparticles were synthesized and characterized for their morphology, structural, magnetic and biocompatibility.

## 2. Materials and Methods

*2.1. Materials.* All materials including iron pentacarbonyl (99.99%), iron(II)acetate (99.99%), decalin (decahydronaphthalene, mix cis & trans, reagent grade 98%), PEG (avg $20M_w$), EG, $\beta$-Cyclodextrin hydrate, and PVA (99+ % hydrolyzed, avg $M_w$ 85,000–146,000) were purchased from Sigma Aldrich.

*2.2. Synthesis of $Fe_3O_4$ (IO-IP).* IO-IP magnetic nanoparticles were synthesized using ultrasonic irradiation of iron pentacarbonyl (2 mL), decalin (40 mL), and PEG (20 mL) for 3 hrs at 50% power using a high intensity ultrasonic horn (Ti-horn, 20 kHz, 100 W/cm$^2$). The reaction was carried out in open air within a stainless steel vessel at 30°C maintained by circulating water using a thermostat. The solution was diluted in ethanol and centrifuged (Allegra 64R centrifuge) at 15,000 rpm at 5°C. The precipitate was washed several times with ethanol and dried overnight under vacuum. The as-prepared sample was analyzed using XRD, TEM, Mossbauer spectroscopy, and magnetic measurements.

*2.3. Synthesis of PVA-Coated $Fe_3O_4$ (PVA-IO).* Following the same procedure as shown above, except the reaction was carried out in the presence of PVA (0.5 g); the as-prepared sample was heated at 300°C for 3 hrs to produce crystalline particles.

*2.4. Synthesis of $Fe_3O_4$ (IO-IA).* Additional $Fe_3O_4$ nanoparticles were synthesized through ultrasonic irradiation of iron acetate (1.0 g), ethylene glycol (40 mL), and distilled $H_2O$ solutions (100 mL). The reaction was carried out in open air in a glass beaker placed in a water bath at room temperature to prevent overheating and followed the same procedure as described earlier to obtain final product for characterization.

*2.5. Synthesis of $\beta$-Cyclodextrin-Coated $Fe_3O_4$ (CD-IA).* $\beta$-cyclodextrin-coated $Fe_3O_4$ was produced from iron acetate (1 g), ethylene glycol (40 mL), distilled $H_2O$ (100 mL), and $\beta$-cyclodextrin (500 mg), using the IO-IA procedure.

*2.6. Testing Method.* XRD analysis was carried out on a Rigaku Diffractometer (D-MAX2000) from 3–80° of $2\theta$ at a scan speed of 0.200°/min. Samples were prepared by spreading the powdered specimen on a glass holder using ethanol; the powder dried before testing. TEM studies were conducted using a JEOL-2010 transmission electron microscope. Samples were prepared by dropping a colloidal solution of ethanol nanoparticles onto a carbon-coated (200 mesh) copper grid. Magnetization curves were obtained at room temperature in an induced magnetic field of 2–20 kOe using a Princeton Measurement Micromag. These samples were prepared by placing a 200 mg of solid sample at the end of the flexible fiber and mounted perpendicular to magnetic field. The mass of the sample was calculated excluding the coating of PVA or CD.

The Mössbauer measurements were carried out using a conventional constant acceleration spectrometer operated in multichannel scaling mode. The gamma ray source used for the sample determinations and for iron-57 velocity calibrations consisted of 12 mCi of $Co^{57}$ in a rhodium metal matrix that was maintained at ambient temperature throughout spectral determinations. Using the latter source allowed for realization of reasonably high S/N velocity spectra in collection times approximately 24 to 48 hours. The spectrometer was calibrated using a 6-micron thick natural abundance iron foil. The center of the magnetic hyperfine pattern of the latter foil (at ambient temperature) is taken as zero velocity. The line widths of the inner-most pair of $\Delta M_I = \pm 1$ transitions of the latter Zeeman pattern were reproducibly determined to be 0.214 mm/s. The as-prepared samples were placed in nylon sample holders of 1 $cm^2$ cross-sectional area for transmission Mössbauer spectroscopy experiments.

*2.7. Cell Cytotoxicity of Nanoparticles.* HEp-2 cells were purchased from the American Type Culture collection (ATCC, Manassas, VA; CCL-23) and were propagated by standard methods using minimum essential medium (MEM) supplemented with 10% fetal bovine serum (FBS), 2 mM L-Glutamine, 75 U/mL penicillin, 100 $\mu$g/mL kanamycin, and 75 $\mu$g/mL streptomycin. The effect of nanoparticles on cytotoxicity was measured by MTS (3-(4,5-dimethylthiazol-2-yl)-5-(3-carboxymethoxyphenyl)-2-(4-sulfophenyl)-2H-tetrazolium) dye reduction. HEp-2 cells were seeded in a 96-well plate at a density of 20,000 cells per well in MEM containing 5% FBS and grown overnight. After 24, 48, and 72 h after-incubation of nanoparticles complexes, media were removed and cells were washed twice with sterile PBS. 20 $\mu$L of MTS (5 mg/mL) in sterile filtered PBS was added to each well and then incubated for 2 h to allow formation of formazan crystals at 37°C. After 2 h, the absorbance of formazan products was measured at 490 nm using a microplate reader.

## 3. Results and Discussion

XRD analysis was carried out to determine the amorphous/crystalline nature and impurities in the as-prepared nanoparticles. Figure 1 shows the XRD patterns of (a) as-prepared IO-IP, (b) PVA-IP heated at 300°C, (c) IO-IP heated at 300°C, (d) as-prepared CD-IA, and (e) as-prepared IO-IA. The powder X-ray pattern of Figure 1(a) clearly shows the as-prepared uncoated, unheated sample synthesized from iron pentacarbonyl is highly amorphous and no crystalline peaks were observed. To further analyze the amorphous material, XRD was carried out for IO-IP (Figure 1(c)) heated to 300°C in an argon atmosphere. Figure 1(c) shows that the particles are highly crystalline and all the diffraction peaks match very well with magnetite JCPDS file number 19-0629. No impurities were observed in this sample. Figure 1(b) shows PVA-IO heated at 300°C in argon atmosphere is also highly crystalline in nature and all peaks match with magnetite JCPDS file number 19-0629.

The XRD diffraction patterns of as-prepared particles synthesized from iron acetate (IO-IA) are shown in Figure 1(e). These diffraction peaks illustrate the highly crystalline nature of the particles. No XRD detectable impurities were observed in this sample as well. Unlike the iron oxide particles produced from the iron pentacarbonyl precursor, these iron oxides were highly crystalline and did not need any additional heat treatments. The peaks were also matched

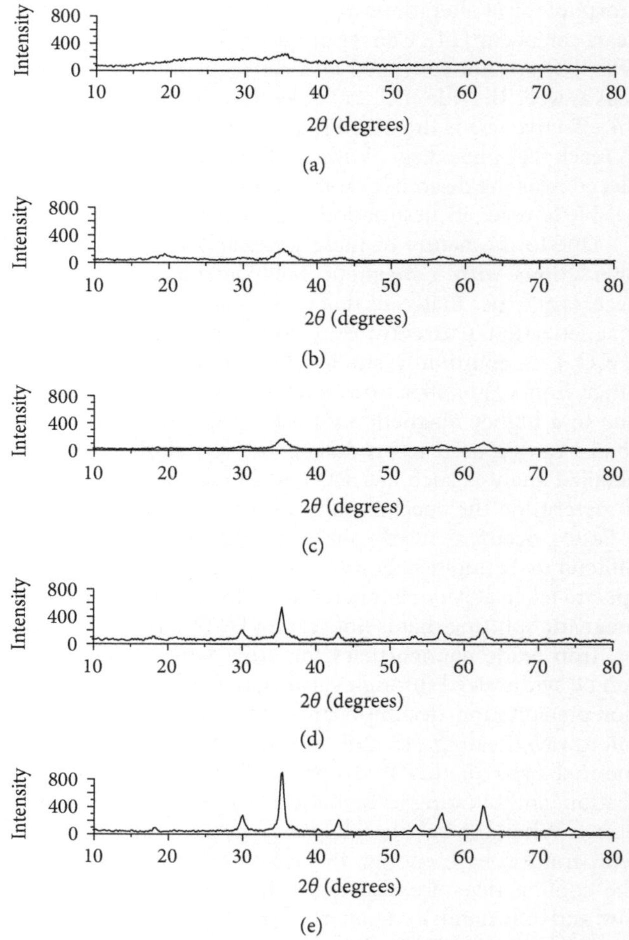


FIGURE 1: XRD patterns of (a) as-prepared IO-IP, (b) PVA-IP heated at 300°C, (c) IO-IP heated at 300°C, (d) as-prepared CD-IA, and (e) as-prepared IO-IA.

TABLE 1: Magnetization and coercive field values of $Fe_3O_4$ nanoparticles synthesized from various precursors and solvents.

| Sample | $H_c$ (Oe) | $M_s$ (emu/g) |
|---|---|---|
| IO-IP | 4.2 | 60.17 |
| PVA-IO | 9.9 | 31.32 |
| IO-IA | 16.9 | 30.56 |
| CD-IA | 25.4 | 36.92 |

with the JCPDS file number 19-0629 for magnetite. Figure 1(d) depicts the XRD of as-prepared CD-IA; the diffraction peaks clearly suggest that the as-prepared particles are also highly crystalline and all peaks match very well with the JCPDS number 19-0629 for magnetite. These XRD results visibly suggest that all the particles synthesized from various precursors and solvents produced only magnetite $Fe_3O_4$ nanoparticles with various coating.

TEM analysis was conducted to understand the morphology, surface coatings and crystal nature of the particles. Figure 2 represents the TEM micrograph of (a) IO-IP heated

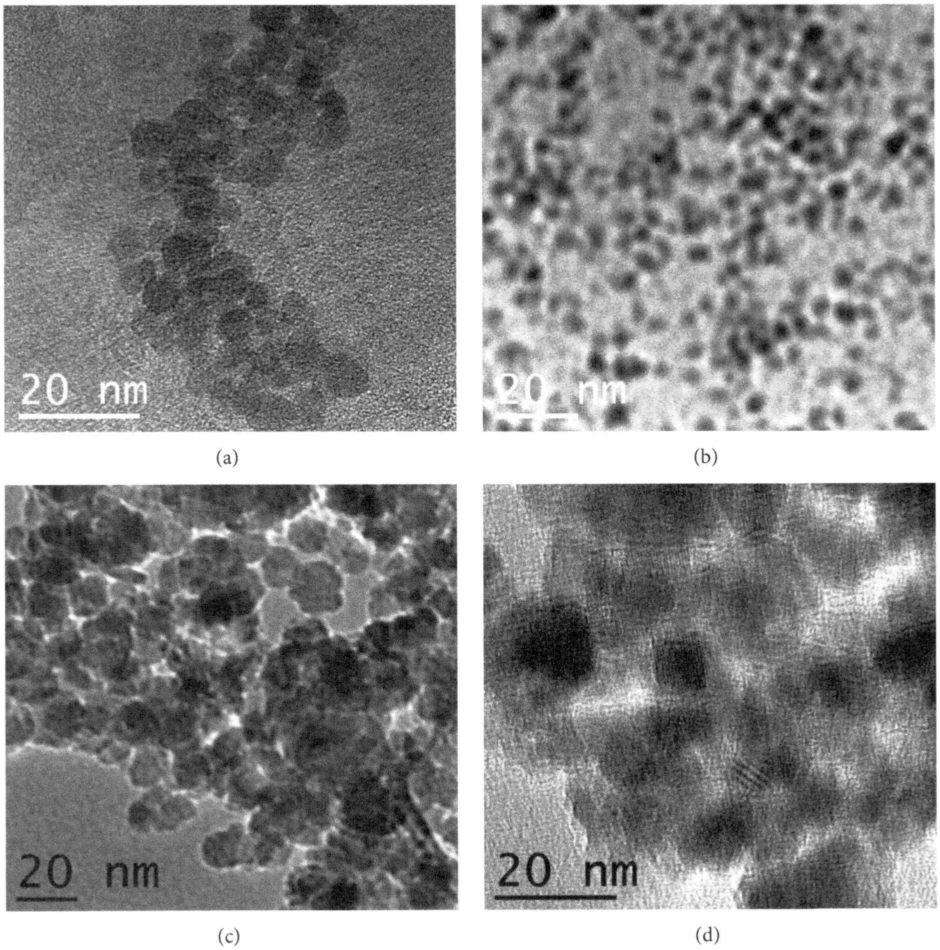


FIGURE 2: TEM micrograph of (a) IO-IP heated at 300°C, (b) PVA-IO heated at 300°C, (c) as-prepared IO-IA, and (d) as-prepared CD-IA.

at 300°C, (b) PVA-IO heated at 300°C, (c) as-prepared IO-IA, and (d) as-prepared CD-IA. Figure 2(a) shows that the iron oxide nanoparticles prepared from iron pentacarbonyl precursor and heated at 300°C are highly crystalline in nature and spherical in shape. The particles sizes measured from the micrograph are about approximately 5–10 nm. The nanoparticles prepared from PVA and iron pentacarbonyl are presented in Figure 2(b). The TEM micrograph shows that the particles are spherical in shape and are relatively less agglomerated when compared to the IO-IP heated sample with similar sizes. Figure 2(c) depicts the TEM micrograph of as-prepared iron oxide nanoparticles from IA. These particles are also spherical in shape and sizes measured are approximately 15–20 nm with a little agglomeration. The iron oxide nanoparticles prepared from IA in the presence of CD are presented in Figure 2(d). These particles are also spherical in shape, and uniformly dispersed, and the sizes measured are approximately 10 nm with minimum agglomeration. This micrograph also clearly shows the lattice structure of iron oxide nanoparticles. It is noteworthy to mention that all the iron oxide nanoparticles prepared are mostly spherical. Most importantly, the particle sizes observed in all of the samples fall within the critical limits for drug delivery applications and are small enough to allow for size increase due to additional coatings or drug loading without reaching above the limits. The PVA-coated particles seem to be completely embedded in the PVA, thereby providing an environment for drug loading, protection from potential oxidation encountered *in vivo*, and yielding a much more biocompatible material.

Magnetic properties were studied to determine the magnetic saturation and coercive field of (a) IO-IP heated at 300°C, (b) PVA-IO heated at 300°C, (c) as-prepared IO-IA, and (d) as-prepared CD-IA nanoparticles at room temperature are shown in Figure 3 and the results were summarized in Table 1. The magnetization/coercive field measured from Figure 3 is 60.17 emu/g/4.23 Oe, 31.32 emu/g/9.92 Oe, 30.32 emu/g/16.93 Oe, and 36.92 emu/g/25.38 Oe, respectively, for (a) IO-IP heated at 300°C, (b) PVA-IO heated at 300°C, (c) as-prepared IO-IA, and (d) as-prepared CD-IA nanoparticles.

The uncoated oxides (IO-IP) heated to 300°C were found to reach the highest magnetic saturation of approximately 60 emu/g as expected. With the addition of PVA, there is a significant drop in $M_s$, which is attributed to the polymer.

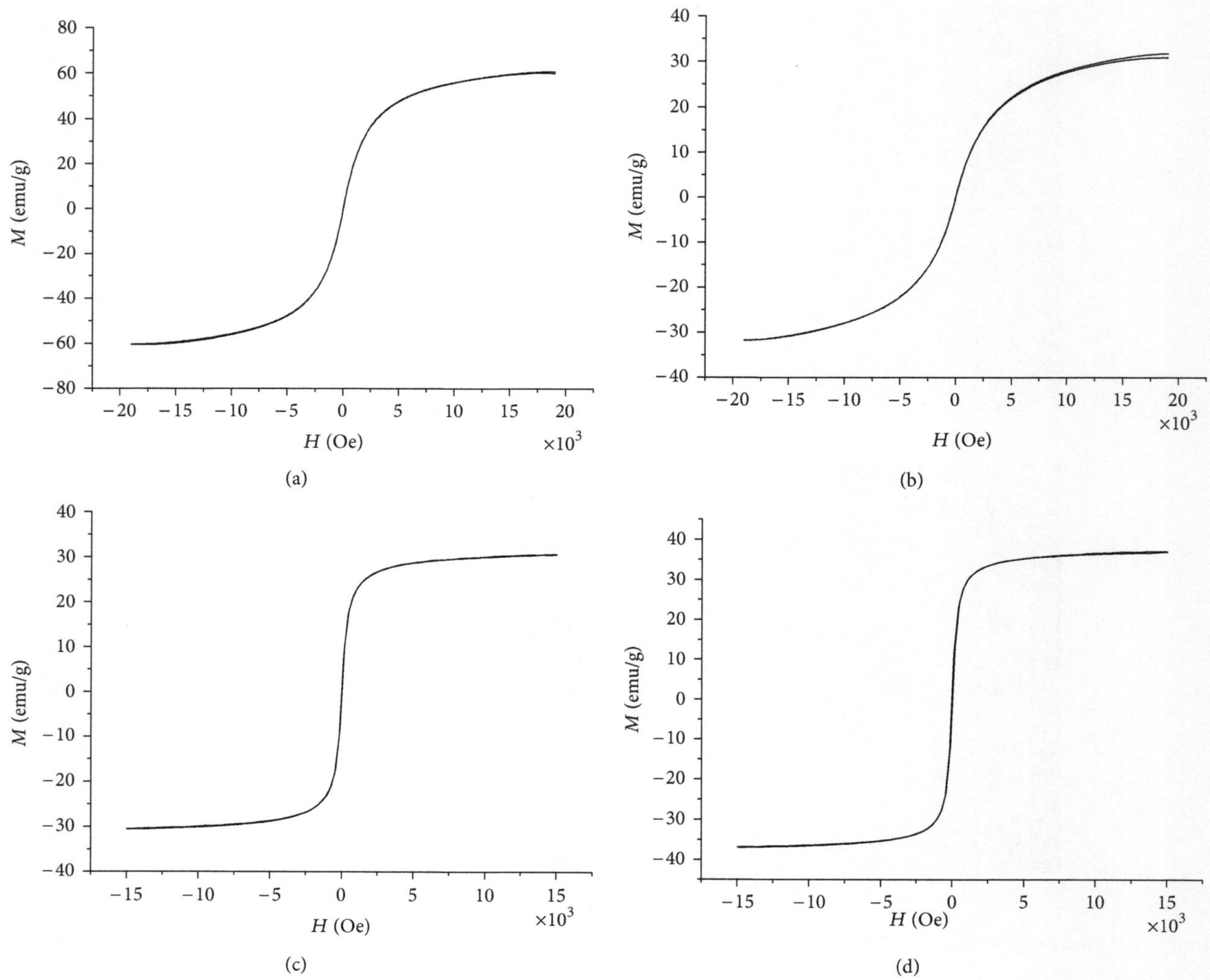


FIGURE 3: Magnetic measurements of (a) IO-IP heated at 300°C, (b) PVA-IO heated at 300°C, (c) as-prepared IO-IA, and (d) as-prepared CD-IA nanoparticles.

It was anticipated that a polymer coating would decrease the magnetization; here; it is decreased by half. The iron oxides synthesized from iron acetate were found to have magnetic saturations much less than those uncoated and produced by iron pentacarbonyl precursors. The CD-IA oxides unexpectedly reached a higher degree of magnetization of approximately 36.92 emu/g, compared to that of the uncoated oxides (IO-IA) which reached approximately 31.32 emu/g. The increase in magnetization of CD-IA is most likely due to the presence of the high crystalline nature of particles, which are ultimately more magnetic.

For comparison, we have noted that the saturation magnetization and coercive field for commercial [34] $Fe_3O_4$ are 96.3 emu/g and 293 Oe, respectively. However, in our case, the magnetite particles are much smaller than the commercial magnetite; the magnetization of these magnetite nanoparticles at 1.6 T does not fully reach saturation. This huge difference in magnetization and coercive field is a clear indication of the nanostructure of the samples. The total magnetization of a specimen consists of small particles and decreases with decreasing particle size due to the increased dispersion in the exchange integral [35, 36] and finally reaches superparamagnetic (SPM) state when each particle acts as a single domain dipole moment with suppressed exchange interaction between particles.

Mössbauer spectroscopy was used to determine the fine structure of iron oxide nanoparticles and their relative sizes. The ambient temperature Mössbauer spectra of (a) as-prepared IO-IP, (b) PVA-IO heated at 300°C, (c) as-prepared IO-IA, and (d) as-prepared CD-IA are shown in Figure 4 wherein one can see a progression from rapidly relaxing (super) paramagnetism (a) to slow paramagnetic relaxation (d) but not a fully resolved magnetic hyperfine splitting. Stoichiometric magnetite can be written as $[Fe^{3+}]_{tet}[Fe^{2+}Fe^{3+}]_{oct}O_4$ [16] and displays two sets of superimposed hyperfine resonance lines representing the iron at the octahedral and tetrahedral sites where only one set of lines is present in $Fe_2O_3$. Also, magnetite displays hyperfine splitting at room temperature, that

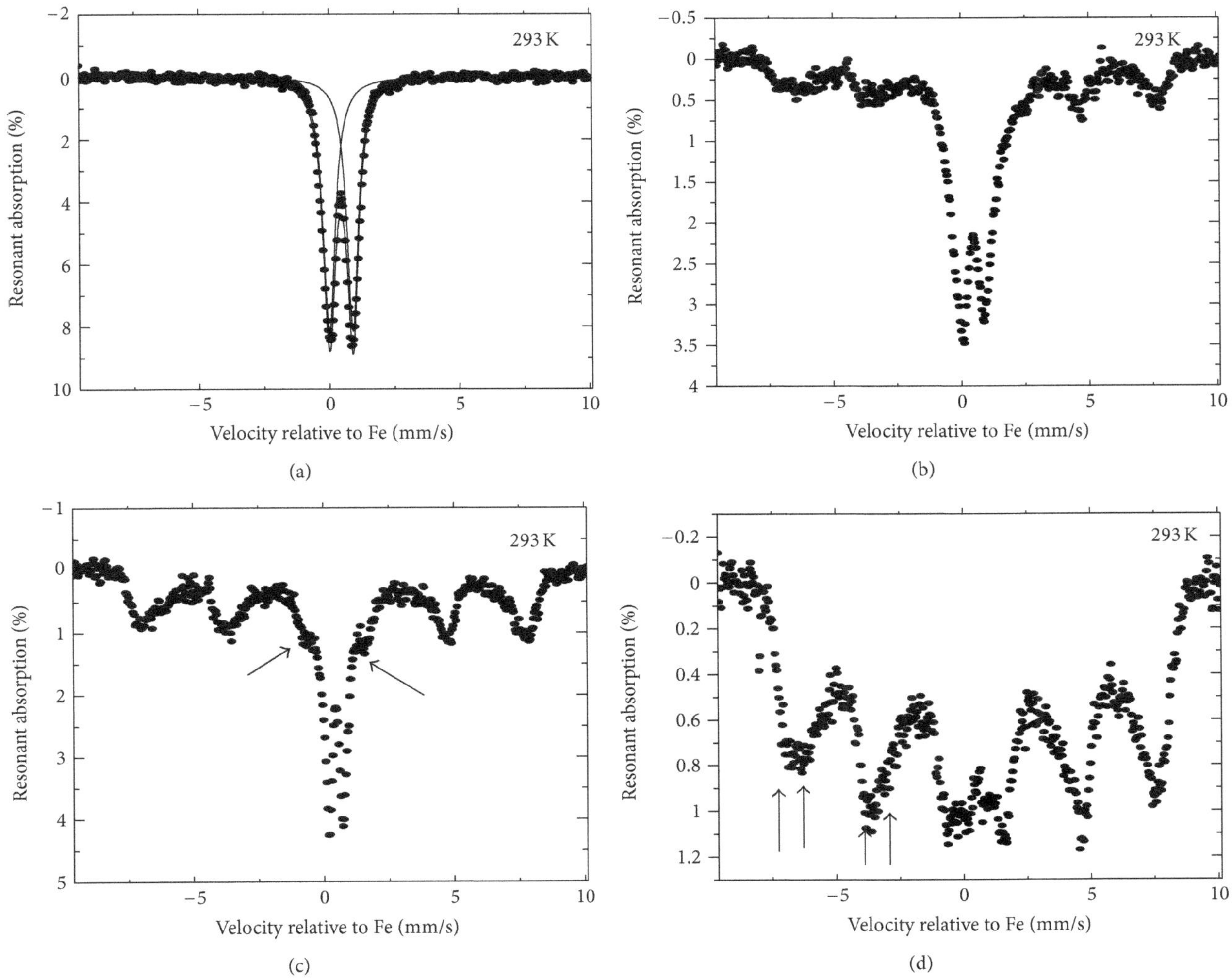


FIGURE 4: Mössbauer spectra of (a) as-prepared IO-IP, (b) PVA-IO heated at 300°C, (c) as-prepared IO-IA, and (d) as-prepared CD-IA.

is absent in maghemite. Spectra of IO-IP (Figure 4(a)) show evidence of the characteristic magnetite splitting.

Figure 4(b) which is of the PVA-coated particles illustrate widening peaks and a developing sextet, indicating an increase in particle size which is expected due to the additional polymer coating. The sextet becomes increasingly evident in the oxides synthesized from iron acetate, signifying a greater increase in particle size. As expected, the CD-IA has the most defined sextet, indicative of the largest size particles. This is a typical, systematic trend with increasing particle (actually magnetic domain) sizes for a fixed temperature and is also observed by other researchers [37, 38] for a specific case of hematite ($Fe_2O_3$) for a number of specific particle sizes. That is, with increasing particle size, the blocking temperature for the transition from rapidly relaxing superparamagnetism (for which the long time average molecular field is zero) to long range ordered bulk magnetic behavior increases. Thus the hierarchy for the particle size of the present samples implied by the ambient temperature results is particle size (a) < (b) < (c) < (d). On the other hand, for a fixed (small) particle size, one sees effects similar to those of increasing particle size (at fixed temperature) by decreasing the temperature [39]. The spectrum of the largest particle size is for (d), of Figure 4 begins to exhibit the low velocity doublets (indicated by vertical arrows in the figure) as seen in the larger particle size bulk $Fe_3O_4$. This confirms that the syntheses procedures used for the preparation of samples (a) through (d) maintain the intrinsic "$Fe_3O_4$" identity of these materials.

Toxicity studies were carried out to determine the effect of the as-prepared coated and uncoated iron oxide nanoparticles on human cell viability at increasing concentrations and exposure time. The cytotoxicity of the nanoparticles was evaluated in HEp-2 cell line by the MTS assay. The nanoparticle concentrations ranged from 5 to 400 $\mu$g/mL which is well above the conventional drug dosage concentration of 2 $\mu$g/mL. The cell viability was observed at 24, 48, and 72 hours. 5 $\mu$g/mL of IO-IP and PVA-IO were added to the cells, 90% of the cell viability were observed. Cell viability decreased to 80% when IO-IP and PVA-IO concentrations

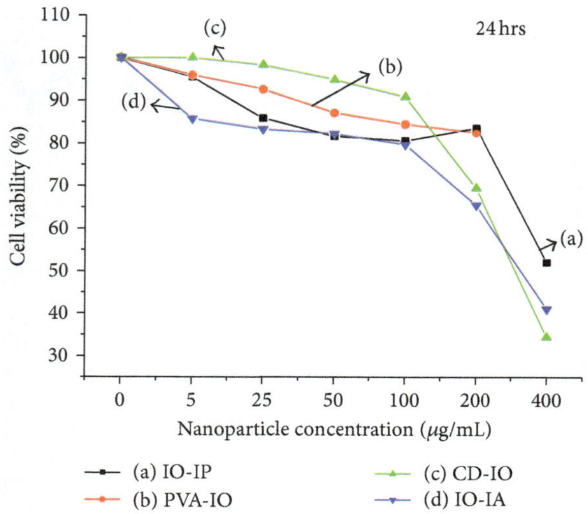


(a)

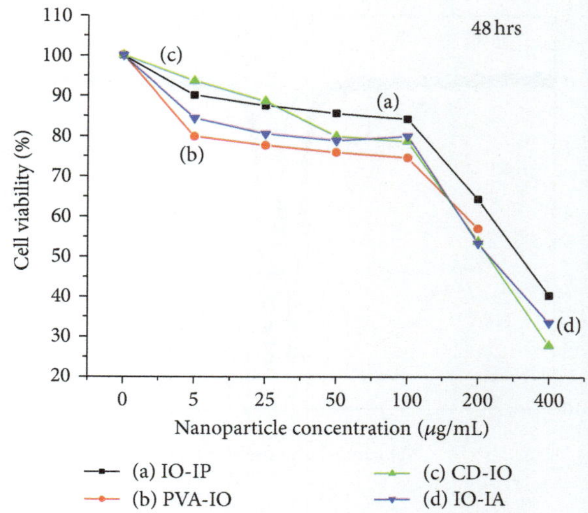


(b)

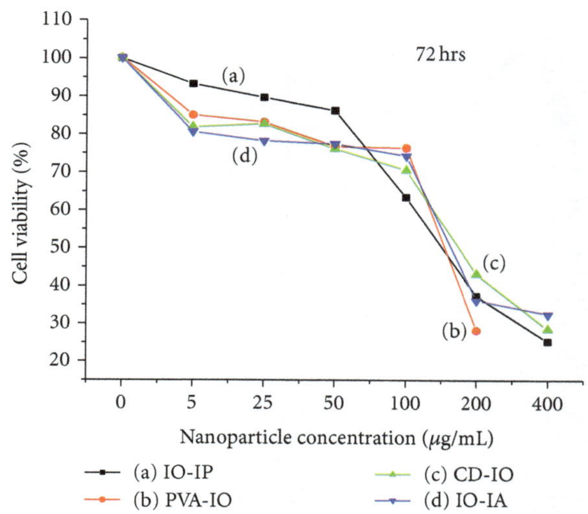


(c)

FIGURE 5: Toxicity studies of as-prepared oxide nanoparticles evaluated in HEp-2 cell lines by MTS assay at (a) 24 h, (b) 48 h, and (c) 72 h.

were increased to 200 μg/mL after 24 h after infection. There was also a 80% of the cell viability when 100 μg/mL of CD-IA and IO-IA were added to the cells. Cell viability continued decreasing about 60% when CD-IA and IO-IA were increased to 200 μg/mL after 24 h after infection (Figure 5(a)). At the exposure time (48 h) and concentration (100 μg/mL), the cells sustained at least 85% cell viability (Figure 5(b)). But, cell death was more prominent when cells were analyzed after 72 h as shown in Figure 5(c). Overall, all the particles were determined to have minimal toxicity. As the exposure time increased, the coatings on the nanoparticles became less effective.

## 4. Conclusions

Uncoated and coated $Fe_3O_4$ nanoparticles were prepared using simple one-step synthesis followed by heat treatments as needed. The use of ultrasonic chemistry provides an easy method to produce both coated and uncoated iron oxide nanoparticles that are of similar size, shape, and crystallinity by either the addition or omission of a polymer. In some instances, the polymer presence does decrease the potential magnetic susceptibility; however it provides an environment for drug loading. From toxicity studies, both the coated and uncoated particles were determined to have low toxicity.

## Acknowledgments

The authors would like to thank the NSF-IGERT, NSF-RISE, and NSF-EPSCoR for their financial support. They also would like to thank Dr. J. W. Harrell's from the University of Alabama, MINT Center, for allowing them to use their research facilities.

## References


[1] D. L. J. Thorek, A. K. Chen, J. Czupryna, and A. Tsourkas, "Superparamagnetic iron oxide nanoparticle probes for molecular imaging," *Annals of Biomedical Engineering*, vol. 34, no. 1, pp. 23–38, 2006.

[2] E. Amstad, T. Gillich, I. Bilecka, M. Textor, and E. Reimhult, "Ultrastable iron oxide nanoparticle colloidal suspensions using dispersants with catechol-derived anchor groups," *Nano Letters*, vol. 9, no. 12, pp. 4042–4048, 2009.

[3] R. Qiao, C. Yang, and M. Gao, "Superparamagnetic iron oxide nanoparticles: from preparations to in vivo MRI applications," *Journal of Materials Chemistry*, vol. 19, no. 35, pp. 6274–6293, 2009.

[4] S. T. Amyn and K. Pei-Yoong, "Synthesis, properties, and applications of magnetic iron oxide nanoparticles," *Progress in Crystal Growth and Characterization of Materials*, vol. 55, no. 1-2, pp. 22–45, 2009.

[5] L. Wang, J. Luo, M. M. Maye et al., "Iron oxide-gold core-shell nanoparticles and thin film assembly," *Journal of Materials Chemistry*, vol. 15, no. 18, pp. 1821–1832, 2005.

[6] K. Tanaka, N. Kitamura, M. Morita, T. Inubushi, and Y. Chujo, "Assembly system of direct modified superparamagnetic iron oxide nanoparticles for target-specific MRI contrast agents," *Bioorganic and Medicinal Chemistry Letters*, vol. 18, no. 20, pp. 5463–5465, 2008.

[7] M. Gonzales-Weimuller, M. Zeisberger, and K. M. Krishnan, "Size-dependant heating rates of iron oxide nanoparticles for magnetic fluid hyperthermia," *Journal of Magnetism and Magnetic Materials*, vol. 321, no. 13, pp. 1947–1950, 2009.

[8] B. Gaihre, M. S. Khil, D. R. Lee, and H. Y. Kim, "Gelatin-coated magnetic iron oxide nanoparticles as carrier system: drug loading and in vitro drug release study," *International Journal of Pharmaceutics*, vol. 365, no. 1-2, pp. 180–189, 2009.

[9] D. L. Leslie-Pelecky, V. D. Labhasetwar, and R. H. Kraus, *Advanced Magnetic Nanostructures*, Kluwer, New York, NY, USA, 2007.

[10] D. K. Kim, Y. Zhang, W. Voit, K. V. Rao, and M. Muhammed, "Synthesis and characterization of surfactant-coated superparamagnetic monodispersed iron oxide nanoparticles," *Journal of Magnetism and Magnetic Materials*, vol. 225, no. 1-2, pp. 30–36, 2001.

[11] T. Neuberger, B. Schöpf, H. Hofmann, M. Hofmann, and B. Von Rechenberg, "Superparamagnetic nanoparticles for biomedical applications: possibilities and limitations of a new drug delivery system," *Journal of Magnetism and Magnetic Materials*, vol. 293, no. 1, pp. 483–496, 2005.

[12] R. Asmatulu, M. A. Zalich, R. O. Claus, and J. S. Riffle, "Synthesis, characterization and targeting of biodegradable magnetic nanocomposite particles by external magnetic fields," *Journal of Magnetism and Magnetic Materials*, vol. 292, pp. 108–119, 2005.

[13] V. P. Torchilin, "Drug targeting," *European Journal of Pharmaceutical Sciences*, vol. 11, no. 2, pp. S81–S91, 2000.

[14] F. C. Maluf and D. Spriggs, "Anthracyclines in the treatment of gynecologic malignancies," *Gynecologic Oncology*, vol. 85, no. 1, pp. 18–31, 2002.

[15] M. A. Morales, T. K. Jain, V. Labhasetwar, and D. L. Leslie-Pelecky, "Magnetic studies of iron oxide nanoparticles coated with oleic acid and Pluronic block copolymer," *Journal of Applied Physics*, vol. 97, no. 10, Article ID 10Q905, 3 pages, 2005.

[16] I. Banerjee, Y. B. Khollam, C. Balasubramanian et al., "Preparation of $\gamma$-$Fe_2O_3$ nanoparticles using DC thermal arc-plasma route, their characterization and magnetic properties," *Scripta Materialia*, vol. 54, no. 7, pp. 1235–1240, 2006.

[17] Z. Surowiec, M. Wiertel, M. Budzyński, J. Sarzyński, and J. Goworek, "Magnetite nanowires in MCM-41 type mesoporous silica templates," *Journal of Non-Crystalline Solids*, vol. 354, no. 35–39, pp. 4271–4274, 2008.

[18] P. Tartaj, M. Del Puerto Morales, S. Veintemillas-Verdaguer, T. González-Carreño, and C. J. Serna, "The preparation of magnetic nanoparticles for applications in biomedicine," *Journal of Physics D*, vol. 36, no. 13, pp. R182–R197, 2003.

[19] J. Huang, H. Pen, Z. Xu, and C. Yi, "Magnetic $Fe_3O_4$/poly (styrene-co-acrylamide) composite nanoparticles prepared by microwave-assisted emulsion polymerization," *Reactive and Functional Polymers*, vol. 68, no. 1, pp. 332–339, 2008.

[20] J. Pinkas, V. Reichlova, R. Zboril, Z. Moravec, P. Bezdicka, and J. Matejkova, "Sonochemical synthesis of amorphous nanoscopic iron(III) oxide from $Fe(acac)_3$," *Ultrasonics Sonochemistry*, vol. 15, no. 3, pp. 257–264, 2008.

[21] K. S. Suslick, S. B. Choe, A. A. Cichowlas, and M. W. Grinstaff, "Sonochemical synthesis of amorphous iron," *Nature*, vol. 353, no. 6343, pp. 414–416, 1991.

[22] R. Vijayakumar, Y. Koltypin, I. Felner, and A. Gedanken, "Sonochemical synthesis and characterization of pure nanometer-sized $Fe_3O_4$ particles," *Materials Science and Engineering A*, vol. 286, no. 1, pp. 101–105, 2000.

[23] A. Gedanken, "Using sonochemistry for the fabrication of nanomaterials," *Ultrasonics Sonochemistry*, vol. 11, no. 2, pp. 47–55, 2004.

[24] G. Q. Zhang, H. P. Wu, M. Y. Ge, Q. K. Jiang, L. Y. Chen, and J. M. Yao, "Ultrasonic-assisted preparation of monodisperse iron oxide nanoparticles," *Materials Letters*, vol. 61, no. 11-12, pp. 2204–2207, 2007.

[25] A. Gedanken, *Ultrasonic Processing to Produce Nanoparticles*, Encyclopedia of Materials, Science and Technology Elsevier Science, Oxford, UK, 2008.

[26] P. Dallas, A. B. Bourlinos, D. Niarchos, and D. Petridis, "Synthesis of tunable sized capped magnetic iron oxide nanoparticles highly soluble in organic solvents," *Journal of Materials Science*, vol. 42, no. 13, pp. 4996–5002, 2007.

[27] R. V. Ramanujan and Y. Y. Yeow, "Synthesis and characterisation of polymer-coated metallic magnetic materials," *Materials Science and Engineering C*, vol. 25, no. 1, pp. 39–41, 2005.

[28] J. L. Arias, V. Gallardo, S. A. Gómez-Lopera, R. C. Plaza, and A. V. Delgado, "Synthesis and characterization of poly(ethyl-2-cyanoacrylate) nanoparticles with a magnetic core," *Journal of Controlled Release*, vol. 77, no. 3, pp. 309–321, 2001.

[29] L. Seung-Jun, J. Jong-Ryul, S. Sung-Chul et al., "Magnetic enhancement of iron oxide nanoparticles encapsulated with poly(D,L-latide-co-glycolide)," *Colloids and Surfaces A*, vol. 255, no. 1–3, pp. 19–25, 2005.

[30] H. Yaci Acar, R. S. Garaas, F. Syud, P. Bonitatebus, and A. M. Kulkarni, "Superparamagnetic nanoparticles stabilized by polymerized PEGylated coatings," *Journal of Magnetism and Magnetic Materials*, vol. 293, no. 1, pp. 1–7, 2005.

[31] P. J. Reséndiz-Hernández, O. S. Rodríguez-Fernández, and L. A. García-Cerda, "Synthesis of poly(vinyl alcohol)-magnetite ferrogel obtained by freezing-thawing technique," *Journal of Magnetism and Magnetic Materials*, vol. 320, pp. e373–e376, 2008.

[32] T. K. Jain, M. A. Morales, S. K. Sahoo, D. L. Leslie-Pelecky, and V. Labhasetwar, "Iron oxide nanoparticles for sustained delivery of anticancer agents," *Molecular Pharmaceutics*, vol. 2, no. 3, pp. 194–205, 2005.

[33] S. Hanessian, J. A. Grzyb, F. Cengelli, and L. Juillerat-Jeanneret, "Synthesis of chemically functionalized superparamagnetic nanoparticles as delivery vectors for chemotherapeutic drugs," *Bioorganic and Medicinal Chemistry*, vol. 16, no. 6, pp. 2921–2931, 2008.

[34] X. Cao, "Preparation and characterization of amorphous nanometre sized $Fe_3O_4$ powder," *Journal of Materials Chemistry*, vol. 7, no. 6, pp. 1007–1009, 1997.

[35] S. R. Elliott, *Physics of Amorphous Materials*, Longman, London, UK, 1984.

[36] C. C. Daniela, M. Gabriela, P. Luminiţa, and D. Lucian, "Synthesis and characterization of maltol modified magnetite nanoparticles," *Revue Roumaine de Chimie*, vol. 55, no. 2, pp. 131–135, 2010.

[37] N. N. Greenwood and T. C. Gibb, *Mössbauer Spectroscopy*, Chapman and Hall, London, UK, 1971.

[38] K. R. Vijaya, Y. Koltypin, X. N. Xu, Y. Yeshurun, A. Gedanken, and I. Felner, "Fabrication of magnetite nanorods by ultrasound irradiation," *Journal of Applied Physics*, vol. 89, no. 11, p. 6324, 2001.

[39] W. Kündig, H. Bömmel, G. Constabaris, and R. H. Lindquist, "Some properties of supported small $\alpha$-$Fe_2O_3$ particles determined with the Mössbauer effect," *Physical Review*, vol. 42, no. 2, pp. 327–333, 1966.

# 12

# Direct Fabrication of Cobalt Oxide Nanoparticles Employing Sucrose as a Combustion Fuel

**M. Th. Makhlouf, B. M. Abu-Zied, and T. H. Mansoure**

*Department of Chemistry, Faculty of Science, Assiut University, Assiut 71516, Egypt*

Correspondence should be addressed to B. M. Abu-Zied; babuzied@yahoo.com

Academic Editor: Raphael Schneider

Combustion method has been used as a fast and facile method to prepare nanocrystalline $Co_3O_4$ spinel employing sucrose as a combustion fuel. The products were characterized by thermal analyses (TGA and DTA), X-ray diffraction technique (XRD), Fourier transform infrared spectroscopy (FTIR), scanning electron microscopy (SEM), and transmission electron microscopy (TEM) techniques. Experimental results revealed that the molar ratio of fuel/oxidizer (F/O) plays an important role in controlling the crystallite size of $Co_3O_4$ nanoparticles. Transmission electron microscopy indicated that the crystallite size of $Co_3O_4$ nanocrystals was in the range of 13–32 nm. X-ray diffraction confirmed the formation of CoO phase with spinel $Co_3O_4$. The effect of calcination temperature on crystallite size and morphology has been, also, discussed.

## 1. Introduction

Spinel cobalt oxide ($Co_3O_4$), an important antiferromagnetic *p*-type semiconductor, is a technologically important and functional material, owing to its unique structure, intriguing properties, and potential practical applications in several important technological fields such as heterogeneous catalysis [1], solid state sensors [2], electrochromic sensors [3], anode materials in Li ion rechargeable batteries [4], energy storage [5], pigments and [6]. It is well known that the morphology and size of $Co_3O_4$ have a great influence on its properties, which are thus a key factor to their ultimate performance and applications. In this regard, it is desirable to tailor-synthesize nanoparticles with predesigned morphology and size distributions. $Co_3O_4$ with nanosized high surface area is expected to lead to even more attractive applications in conjunction of their traditional arena and nanotechnology [7]. Therefore, it is important to prepare $Co_3O_4$ with defined morphologies and a narrow range of size distribution.

Much effort has been made to synthesize nanocrystalline $Co_3O_4$, with various particle sizes, from economical and practical aspects point of view including thermal decomposition of cobalt oxalate (60–200 nm) [8], one-pot hydrothermal reaction (average size 30 nm) [9], thermal decomposition of sol-gel derived oxalates (15–20 nm) [10], and solution combustion method (23–90 nm) [11]. Focusing our attention to the combustion route, it involves a self-sustained reaction between an oxidizer (e.g., metal nitrate) and a fuel (e.g., urea, sucrose, glycine, and hydrazides). This process not only yields nanosize oxide materials, but also allows uniform (homogeneous) doping of trace amounts of rare-earth impurity ions in a single step. In addition, there are some reports involving the synthesis of $Co_3O_4$ nanocrystallites with various morphologies using combustion method. For instance, $Co_3O_4$ nanocrystallites with different morphologies such as cubic (10–50 nm) [7] and foamy porous (average size 67 nm) [12] morphology were obtained by solution combustion method. Moreover, Toniolo et al. [11] have prepared nanocrystalline $Co_3O_4$ with crystallites sizes from 23 to 90 nm employing urea and glycine as combustion fuels. Jiu et al. [13] have prepared nanocrystalline $Co_3O_4$ with crystallites sizes of 33 nm employing polyvinyl alcohol (PVA) as combustion fuel using polymer combustion method.

Metal ions can be uniformly distributed in molecular scale through polyfunctional groups (e.g., –COOH, $–NH_2$, –OH, etc.) from chelating ligands such as citric acid, glycine [14], ethylenediaminetetraacetic acid (EDTA), and polyacrylate [15] to form organic matrices. Thermal decomposition of the matrices provides fine particles at reasonably lower temperature. Sucrose is a type of water-soluble and inexpensive agricultural product. In the acidic condition, –COOH and –OH groups can be generated, which hence can form stable binding with metal ions in homogeneous solution. Therefore, sucrose can also act as an effective chelating agent like citric acid to produce fine particles [16]. In recent years, attempts have been made by several researchers to prepare nanopowders using sucrose as a chelating agent and fuel [17–19]. Das et al. [17] synthesized single phase $\alpha$-alumina nanocrystallites with high surface area at a low temperature of 600°C. Lazarraga et al. [18] obtained nanosized $LiNi_yMn_{2-y}O_4$ spinels for rechargeable lithium batteries.

To our knowledge, there is a lack of information concerning the synthesis of $Co_3O_4$ nanoparticles using sucrose as a combustion fuel. Accordingly, in this work, we report a novel chemical route for the preparation of $Co_3O_4$ nanoparticles via combustion synthesis using sucrose as a combustion fuel and cobalt nitrate as an oxidizer. The process involves dehydration of a Co ion-sucrose complex solution, followed by the decomposition of the obtained product. Complete dehydration of the complex solution to dryness produces a black precursor mass. Heat treatment of the dried precursor mass in a furnace air environment results in a voluminous, crushable, highly fragile phase-pure nanocrystalline $Co_3O_4$. Decomposition of sucrose generates excess heat and huge amounts of gases that help to produce the porosity in the final product. These characteristics, together with the water-soluble nature of sucrose, have motivated the initiation of the present study. The effect of the fuel-to-oxidizer molar ratio as well as calcination temperature on controlling particle size of the product was investigated. The textural and structure properties of the prepared $Co_3O_4$ nanoparticles were characterized by means of thermal analyses (TGA and DTA), X-ray powder diffraction (XRD), Fourier transform infrared spectroscopy (FT-IR), and electron microscopy (SEM and TEM).

## 2. Experimental

*2.1. Preparation Procedure.* Cobalt nitrate, $Co(NO_3)_2 \cdot 6H_2O$, and sucrose, $(C_{12}H_{22}O_{11})$, were of analytical grade reagents and were used without further purification. Distilled water was used in all of preparations. Two series of samples were prepared. In the first one (series I), we have investigated the effect of changing the sucrose/cobalt nitrate molar ratios (F/O), 0.5, 1, 2, 5, 8, and 16, on the morphology and crystallite size of $Co_3O_4$. In a typical procedure, the required amounts of cobalt nitrate (3.76 g) and sucrose (2.21, 4.42, 8.84, 22.1, 35.36, 70.72 g, and resp.) were weighed to the nearest milligram. Cobalt nitrate and sucrose were dissolved in distilled water (total solutions volume was 100 mL) to form a pink homogeneous solution; the solution was then heated on a hotplate at about 100°C to evaporate the excess water. After that the obtained viscous gel was calcined in muffle furnace at 400°C for 3 h in a static air atmosphere. In series II, we have investigated the effect of changing the calcination temperature, 350–1000°C, employing the same procedure at sucrose/cobalt nitrate molar ratio of 0.5.

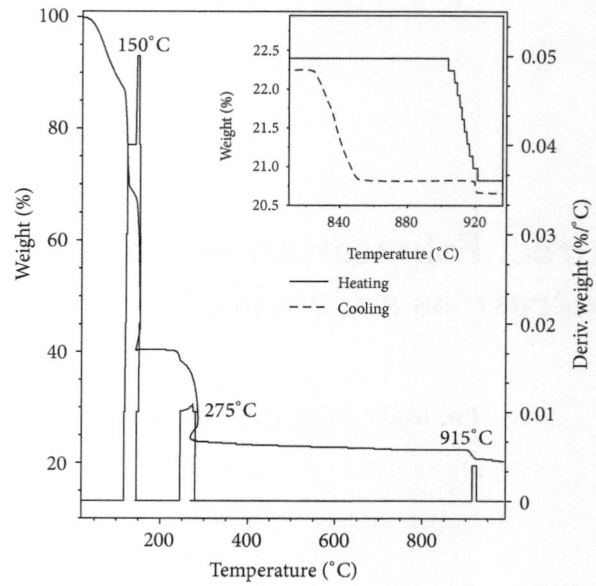


FIGURE 1: TGA and DTG curves obtained by heating the dried precursor having F/O equals to 0.5 in air atmosphere as a carrier gas.

*2.2. Characterization.* Simultaneous TGA and DTA curves were recorded with a Shimadzu DTG-60 instrument apparatus using a heating rate of 10°C $min^{-1}$ in air atmosphere (flow rate at 40 mL/min). Powder X-ray diffraction (XRD) patterns were recorded using Philips diffractometer (type PW 103/00) with CuK$\alpha$ radiation ($\lambda$ = 1.5405 Å) at 35 kV and 20 mA with a scanning rate in $2\theta$ of 0.06° $min^{-1}$. FTIR spectra were performed employing the KBr disc technique in the wavelength range of 4000–400 $cm^{-1}$, using Thermo-Nicolet-6700 FTIR spectrophotometer. Scanning electron micrographs were obtained using a JEOL scanning microscope (model JSM-5400 LV). Transmission electron pictures were taken using JEOL transmission microscope (model JEMTH-100 II).

## 3. Results and Discussion

*3.1. Thermal Analysis.* Figure 1 shows the weight loss (TG) and the associated derivative thermogram (DTG) of the precursor synthesized at F/O equal to 0.5 in air atmosphere as a carrier gas. This figure manifests that the precursor underwent decomposition with increasing the temperature from ambient degree 1000°C in several steps. The first step starts from room temperature to around 115°C which brings a weight loss (WL) of 15%. This step could be attributed to the dehydration of the parent mixture. TGA thermogram demonstrates the presence of two consecutive WL steps at the 115–165°C temperature range. Such steps could be ascribed to the decomposition of the precursor constituents as well as the combustion of the produced gases. Another step is located at 200–280°C temperature range which brings a weight loss of 16%. Such step could be related to the completion

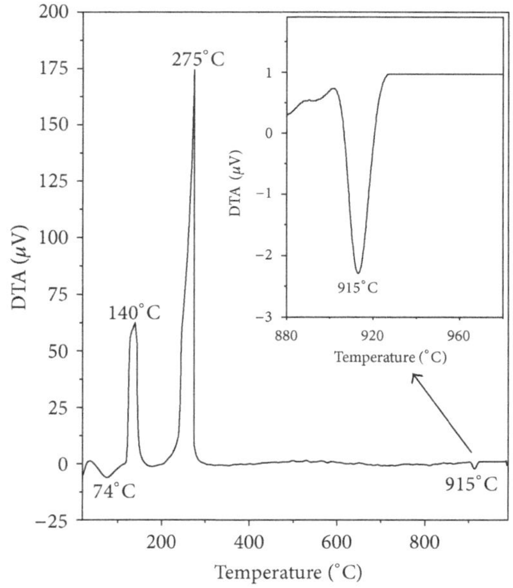


FIGURE 2: DTA thermogram obtained by heating the precursor (F/O equals to 0.5) in air atmosphere.

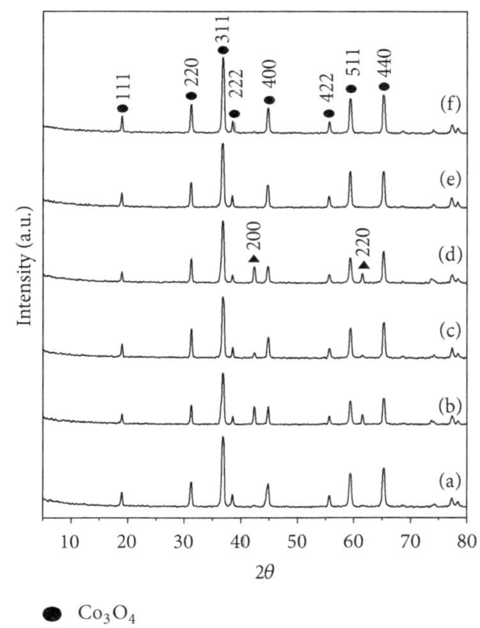


FIGURE 3: XRD patterns of nanocrystalline $Co_3O_4$ samples calcined at 400°C being prepared at different F/O molar ratios: (a) 0.5, (b) 1, (c) 2, (d) 5, (e) 8, and (f) 16.

of the precursor decomposition and the formation of the nanocrystalline $Co_3O_4$. On further heating the precursor up to 1000°C, one can observe a fifth step being maximized at 915°C. This step is accompanied by weight loss of 1.6% due to the decomposition of $Co_3O_4$ into CoO according to [7]

$$Co_3O_4 \longrightarrow 3CoO + \frac{1}{2}O_2 \quad (1)$$

The inserted curves in Figure 1 shows a magnification of both the heating and the cooling curves of the precursor from the 800°C to 940°C temperature range. It illustrates that the cooling of CoO phase formed at 915°C, as a product of the decomposition of spinel $Co_3O_4$ phase, has a weight gain of 1.6%. This weight gain could be attributed to the oxidation of CoO into $Co_3O_4$, and this, in turn, indicates the reversibility of (1).

Figure 2 shows the obtained DTA curve during heating the precursor with F/O molar ratio of 0.5 in air atmosphere degree to 1000°C. Inspection of this figure reveals the presence of an endothermic peak at the 74°C. Such peak, associated with the first observed WL step in Figure 1, could be related to the dehydration of the precursor. In addition, the thermogram shows an exothermic peak at 140°C. The temperature range of such peak is consistent with that of the second and third steps in the TGA pattern. This effect corresponds to simultaneous decomposition of $NO_3^-$ and charring of sucrose. Simultaneously, sucrose is oxidized by nitrate anions (resulting in carbon dioxide, nitrogen, nitrogen dioxide and water [20]) releasing plenty of gases. Thus, the exothermic peak could be attributed to the superposition of precursor decomposition (endothermic) and combustion of the evolved gases (exothermic). It is clear that the combustion process predominates. Also, the thermogram shows another sharp exothermic peak at 275°C. The temperature of such peak is consistent with that of the fourth WL step observed in the TGA pattern. This peak could be ascribed to the combustion of carbon and to the crystallization of spinel $Co_3O_4$. Going to the high temperature range, the obtained DTA thermogram manifests the presence of an endothermic peak at 915°C. Such effect can be ascribed to the decomposition of $Co_3O_4$ into CoO, that is, thermal reduction of $Co^{3+}$ to $Co^{2+}$.

*3.2. X-Ray Diffraction.* X-ray diffraction patterns of the as-prepared $Co_3O_4$ obtained via combustion synthesis at different sucrose/cobalt nitrate molar ratios (i.e., 0.5, 1, 2, 5, 8, and 16) being calcined at 400°C is shown in Figure 3 (curves a–f), respectively. Inspection of this figure revealed that (i) when the F/O molar ratio is 0.5 (pattern a), the powder after combustion contains only well crystallized cubic spinel $Co_3O_4$ (JCPDS card file no. 80-1545), (ii) when the F/O molar ratio is in the range of 1–5 (patterns b–d), the powders are composed of a mixture of $Co_3O_4$ (JCPDS card file No. 80-1545) and CoO (JCPDS 78-0431), and (iii) when the F/O molar ratios are higher than 5, again the powder diffraction patterns are composed only $Co_3O_4$ (JCPDS card file No. 80-1545) as in patterns e and f.

In order to check the role of sucrose/cobalt nitrate molar ratio variation in controlling the structural parameters of the obtained nanocrystalline $Co_3O_4$, the lattice parameter ($a$) and crystallite size ($D$) were calculated. The lattice parameters were computed using the $d$-spacing values and the respective ($hkl$) parameters. The obtained values are plotted in Figure 4(a) as a function of the sucrose/cobalt

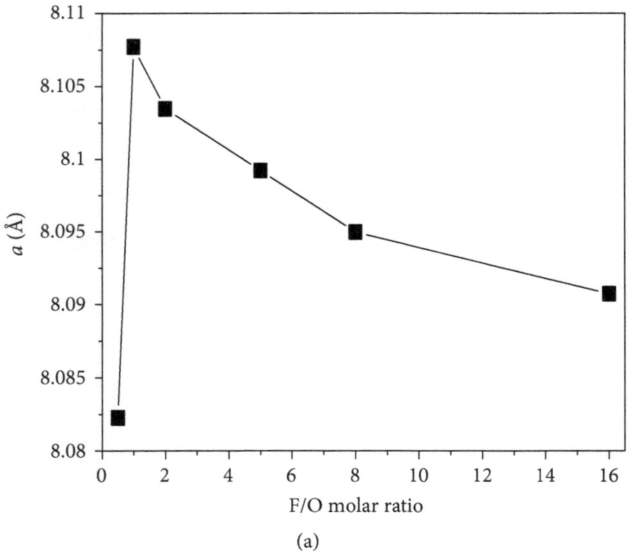


FIGURE 4: Effect of F/O molar ratios (a) and calcination temperature (b) on the lattice parameter values of nanocrystalline $Co_3O_4$.

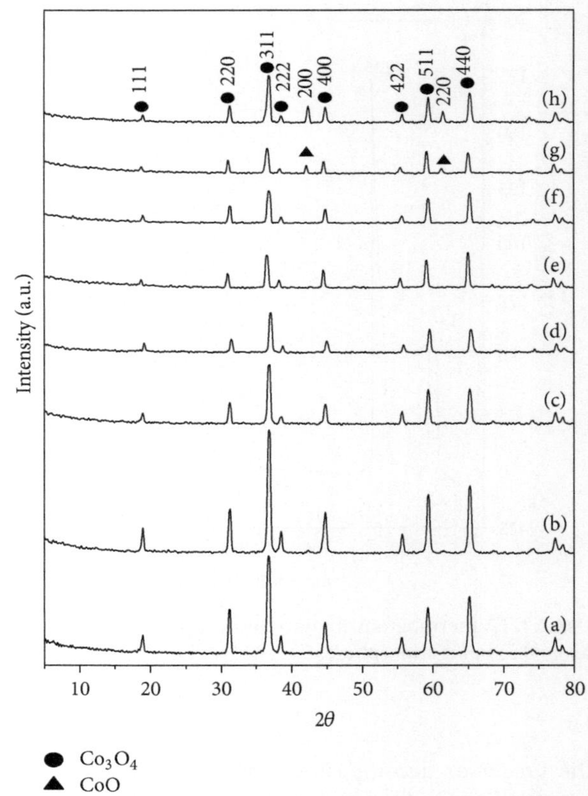


FIGURE 5: XRD patterns of nanocrystalline $Co_3O_4$ samples obtained via F/O molar ratio of 0.5 at different calcination temperature: (a) 350, (b) 400, (c) 500, (d) 600, (e) 700, (f) 800, (g) 900, and (h) 1000°C.

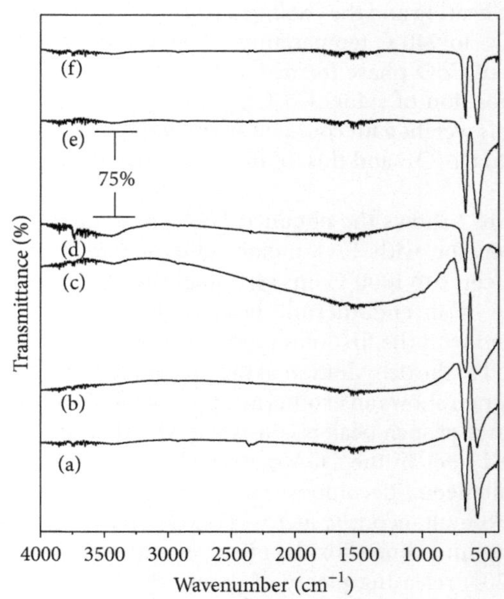


FIGURE 6: FTIR spectra of as-prepared nanocrystalline $Co_3O_4$ at different F/O molar ratios: (a) 0.5, (b) 1, (c) 2 (d) 5, (e) 8, and (f) 16.

nitrate molar ratio. The crystallite size ($D$) was calculated using Scherrer equation [12] from the full-width at half-maximum (FWHM) of the peaks. The relevant values are listed in Table 1. From Table 1, it is evident that the crystallite size of $Co_3O_4$ increased with increasing the sucrose/cobalt nitrate molar ratio from 0.5 till 1, and then it shows a mild decrease upon further F/O increase. This in turn indicates that is the lowest sucrose/cobalt nitrate value gets the smaller the $Co_3O_4$ crystallite size. Regarding the lattice parameter, it is obvious from Figure 4(a) that the variation of the lattice parameter with molar ratio follows the same trend as that observed for the crystallite size.

Since the $Co_3O_4$ obtained by using a molar ratio of 0.5 exhibits the lowest crystallite size, the study was extended to

FIGURE 7: SEM of as-prepared nanocrystalline $Co_3O_4$ at different F/O molar ratios, (a) 0.5, (b) 1, and (c) 5, being calcined at 400°C.

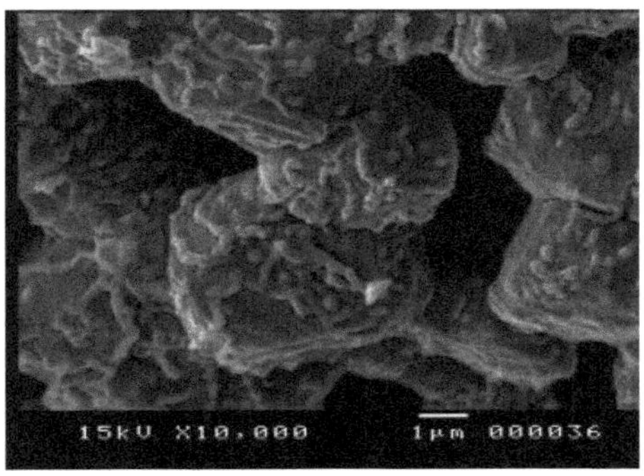


FIGURE 8: SEM of as-prepared nanocrystalline $Co_3O_4$ using sucrose-to-cobalt nitrate molar ratio of 0.5 calcined at 1000°C.

TABLE 1: Crystallite sizes of $Co_3O_4$ at different F/O molar ratios being calcined at 400 °C.

| F/O molar ratio | 0.5 | 1 | 2 | 5 | 8 | 16 |
|---|---|---|---|---|---|---|
| Crystallite size (nm) | 24.1 | 30.8 | 27.5 | 26.1 | 25.5 | 25 |

check the influence of changing the calcination temperature on the morphology and crystallite size of nanocrystalline $Co_3O_4$ at this ratio. XRD patterns of the as-prepared $Co_3O_4$ obtained via calcining sucrose/cobalt nitrate parents having F/O molar ratio of 0.5 at 350–1000°C temperature range are shown in Figure 5. From this figure, two points could be raised: (i) all diffraction peaks belong to one phase only ($Co_3O_4$) at the calcination temperatures range 350–800°C; however, a peak shift to lower angles with increasing the calcination temperature is evident indicating an expansion of the cubic lattice and (ii) the diffraction peaks related to CoO appear at high calcination temperatures (i.e., 900 or 1000°C) which indicates that, CoO formed at high temperature via the thermal reduction of $Co_3O_4$ could not be oxidized completely to $Co_3O_4$ on cooling.

Table 2 gives the average crystallite size of $Co_3O_4$ phase at different calcination temperatures (350–1000°C) calculated from XRD data, whereas the relevant lattice parameters are shown in Figure 4(b). From these data, one can state safely that the sample calcined at 350°C exhibits the lowest average crystallite size ($D = 23.4$ nm). With increasing the calcination temperature, the average crystallite size is

TABLE 2: Crystallite sizes of $Co_3O_4$ at F/O molar ratio of 0.5 : 1 calcined at different calcinations temperature.

| Calcination temperature (°C) | 350 | 400 | 500 | 600 | 700 | 800 | 900 | 1000 |
|---|---|---|---|---|---|---|---|---|
| Crystallite size (nm) | 23.4 | 24.1 | 24.7 | 25.6 | 26.7 | 27 | 27.9 | 30.3 |

increased. This behavior was expected because the heating facilitates the diffusion and agglomeration of the particles [21–23]. Accordingly, we can conclude that the optimum conditions to obtain pure $Co_3O_4$ with the smallest crystallite size using sucrose as a fuel is sucrose/cobalt nitrate molar ratio of 0.5 and a calcination temperature at 350°C.

*3.3. Fourier Transformation Infrared Spectroscopy (FTIR).* Cubic spinel structure of $Co_3O_4$ with $Co^{2+}$ ($3d^7$) and $Co^{3+}$ ($3d^6$) located at tetrahedral and octahedral sites, respectively, belongs to the space group ($Fd3m$). The group theory predicts the following modes in the spinel: $\Gamma = A_{1g}\,(R) + E_g\,(R) + F_{1g}\,(in) + 3F_{2g}\,(R) + 2A_{2u}\,(in) + 2E_u\,(in) + 4F_{1u}\,(IR) + 2F_{2u}\,(in)$, where (R), (IR), and (in) represent Raman active vibrations, infrared-active vibration, and inactive modes, respectively. The FT-IR spectra of as-prepared nanocrystalline $Co_3O_4$ at different F/O molar ratios being calcined at 400°C are shown in Figure 6 (curves a–f). In the investigated region (4000–400 $cm^{-1}$), the entire obtained spectra manifest the presence of two absorption bands at 565 ($\upsilon_1$) and 661 ($\upsilon_2$) $cm^{-1}$ which originate from the stretching vibrations of the metal-oxygen bond and confirm the formation of $Co_3O_4$ spinel oxide [1, 8]. The $\upsilon_1$ band is characteristic of $OB_3$ (where B denotes the $Co^{3+}$ in the octahedral hole) vibration, and the $\upsilon_2$ band is attributable to the $ABO_3$ (where A denotes the $Co^{2+}$ in the tetrahedral hole) vibration in the spinel lattice [12]. In addition, the FTIR spectra show no residual organic compounds and $NO_3^-$ after calcination. It is worth noting that it is very difficult to differentiate between the FT-IR spectra of pure $Co_3O_4$ and that of $Co_3O_4$ with CoO impurities. Thus, the presence of some CoO impurities in the $Co_3O_4$ nanoparticles cannot be excluded based on analysis of FT-IR only. The FTIR spectra of nanocrystalline $Co_3O_4$ samples having sucrose/cobalt nitrate of 0.5 being calcined at 350–1000°C temperature range, not shown, exhibit two

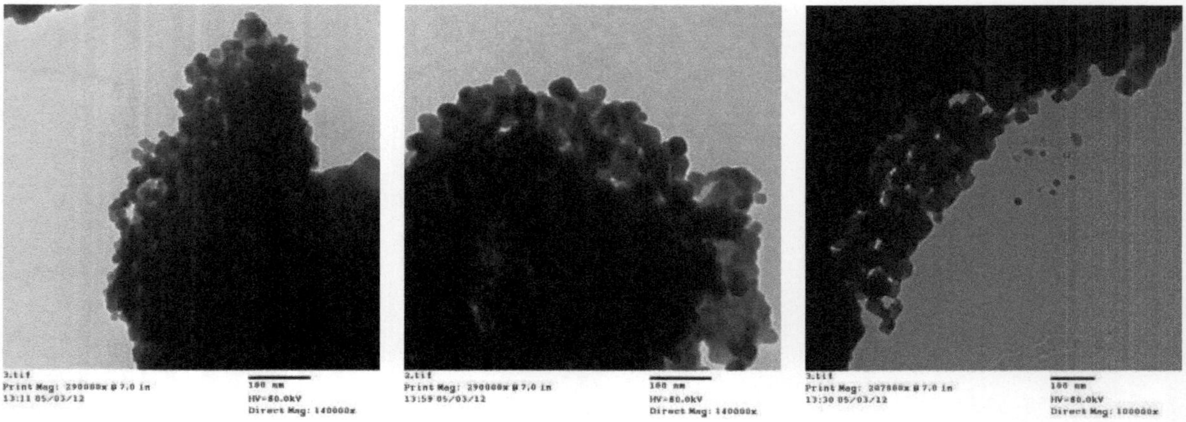

(a) (b) (c)

FIGURE 9: TEM micrographs of as-prepared nanocrystalline $Co_3O_4$ at different sucrose to cobalt molar ratios, (a) 0.5, (b) 1, (c) and 5, being calcined at 400°C.

bands at 565 ($\upsilon_1$) and 661 ($\upsilon_2$) $cm^{-1}$, which originate from the stretching vibrations of the cobalt-oxygen bond and confirm the formation of $Co_3O_4$ spinel oxide.

*3.4. Electron Microscopy (SEM and TEM).* Figure 7 shows SEM of as-prepared nanocrystalline $Co_3O_4$ at different sucrose-to-cobalt nitrate molar ratios from 0.5 to 5 (Figures 7(a)–7(c)), being calcined at 400°C. Inspection of this figure manifests that the prepared spinel $Co_3O_4$ shows a porous network as a consequence of the gases escaping during the combustion process. It can be seen that the samples are extremely uniform with well-defined morphology of each particle. Micrograph of sample at lower F/O value (Figure 7(a)) shows relatively small crystals having few randomly distributed pores with superficial roughening and rounding of the crystal edges. Increasing the molar ratio till 5 is accompanied by the formation of larger agglomerates of crystals having large randomly distributed pores. Figure 8 depicts SEM of as-prepared nanocrystalline $Co_3O_4$ using sucrose-to-cobalt nitrate molar ratio of 0.5 being calcined at 1000°C. In comparison with the micrograph of the sample having F/O ratio of 0.5, Figure 7(a), it is obvious that with increasing the calcination temperature the particles become bigger.

Figure 9 shows TEM nanographs of as-prepared nanocrystalline $Co_3O_4$ at sucrose-to-cobalt nitrate molar ratios of 0.5 (a), 1 (b), and 5 (c) being calcined at 400°C. The TEM images reveal the nanocrystalline nature of the prepared $Co_3O_4$ spinel. It is obvious that the sizes of the particles vary with changing the F/O values. The particle sizes of sample at F/O molar ratio of 0.5, Figure 9(a), are in the range of 13–25 nm. These results are in a good agreement with the sizes determined from XRD analysis by Scherrer's equation (Table 1). In addition, the particles seemed uniform and quasispherical in shape with weak agglomeration. Moreover, increasing the molar ratio from 0.5 to 1 is accompanied by increasing the particle size, Figure 9(b). However, the particles are still quasispherical in shape, but the sizes are in the range of 21–32 nm, and the degree of agglomeration is increased. In addition, further increasing the molar ratio to 5, Figure 9(c), is accompanied by a change from quasispherical to cubic shape and a small decrease in the particle sizes. From Figure 9(c), it is obvious that the particle sizes are in the range 20–28 nm, which again is in a good agreement with the size determined from XRD analysis (Table 1).

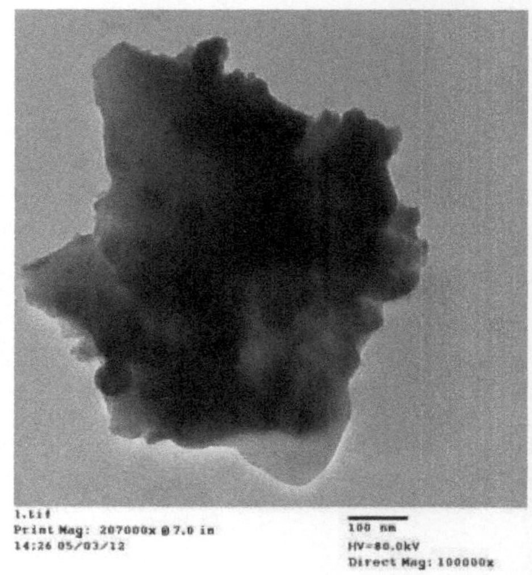

FIGURE 10: TEM of as-prepared nanocrystalline $Co_3O_4$ using sucrose-to-cobalt nitrate molar ratio of 0.5 being calcined at 1000°C.

Figure 10 shows TEM micrographs of as-prepared $Co_3O_4$ at F/O molar ratio of 0.5 calcined at 1000°C. Inspection of Figure 10 revealed that thermal treatment of the precursor at elevated temperatures plays a significant role on the variation of particle size and morphology of $Co_3O_4$ spinel. In this case, it could be seen that the particle size is in the range of 28–31 nm, which is again in close agreement with the sizes determined from XRD analysis (Table 2). Particles of $Co_3O_4$ calcined at elevated temperatures are more agglomerated than that at lower ones (Figure 9). As a conclusion, the F/O ratio has clearly an important role on the structural and morphological properties of the powders. Low F/O ratios are recommended in this case in order to avoid agglomeration of the particles and to obtain powders with small particle size. Also, products with low calcination temperatures are recommended for the same reason.

## 4. Conclusions

It is well known that combustion synthesis is an efficient, quick, simple, low cost, and straightforward method for preparation of nanosized materials at lower temperatures. In this work, we have prepared spinel $Co_3O_4$ employing solution combustion synthesis using sucrose as a fuel. Sucrose content is the key factor that controls the formation of reduced oxidation state compounds such as CoO phase. The ratio of the fuel to nitrates dramatically influenced the phase formation of the final products and the particle size. When the ratio of fuel to nitrates was 2–5, the final products were attributed to $Co_3O_4$ and CoO phases, but at lower (0.5–1) or higher ratios (8–16), the final product was mainly spinel $Co_3O_4$ phase. The morphology of the as-prepared $Co_3O_4$ is quasispherical at lower sucrose/cobalt nitrate ratio, while at higher ratio is cubic. The calcined powders showed the presence of regular particles, with narrow particle size distribution. In addition, this methodology can lead the system to a good chemical homogeneity as the reagents were mixed in an aqueous solution.

## References


[1] B. M. Abu-Zied and S. A. Soliman, "Nitrous oxide decomposition over $MCO_3$-$Co_3O_4$ (M = Ca, Sr, Ba) catalysts," *Catalysis Letters*, vol. 132, no. 3-4, pp. 299–310, 2009.

[2] W. Y. Li, L. N. Xu, and J. Chen, "$Co_3O_4$ nanomaterials in lithium-ion batteries and gas sensors," *Advanced Functional Materials*, vol. 15, no. 5, pp. 851–857, 2005.

[3] T. Maruyama and S. Arai, "Electrochromic properties of cobalt oxide thin films prepared by chemical vapor deposition," *Journal of the Electrochemical Society*, vol. 143, no. 4, pp. 1383–1386, 1996.

[4] P. Poizot, S. Laruelle, S. Grugeon, L. Dupont, and J. M. Tarascon, "Nano-sized transition-metal oxides as negative-electrode materials for lithium-ion batteries," *Nature*, vol. 407, no. 6803, pp. 496–499, 2000.

[5] S. Noguchi and M. Mizuhashi, "Optical properties of CrCo oxide films obtained by chemical spray deposition: substrate temperature effects," *Thin Solid Films*, vol. 77, no. 1–3, pp. 99–106, 1981.

[6] T. Sugimoto and E. Matijevic, "Colloidal cobalt hydrous oxides. Preparation and properties of monodispersed $Co_3O_4$," *Journal of Inorganic and Nuclear Chemistry*, vol. 41, no. 2, pp. 165–172, 1979.

[7] J. Feng and H. C. Zeng, "Size-controlled growth of $Co_3O_4$ nanocubes," *Chemistry of Materials*, vol. 15, no. 14, pp. 2829–2835, 2003.

[8] L. Ren, P. Wang, Y. Han, C. Hu, and B. Wei, "Synthesis of $CoC_2O_4 \cdot 2H_2O$ nanorods and their thermal decomposition to $Co_3O_4$ nanoparticles," *Chemical Physics Letters*, vol. 476, no. 1–3, pp. 78–83, 2009.

[9] Y. Teng, S. Yamamoto, Y. Kusano, M. Azuma, and Y. Shimakawa, "One-pot hydrothermal synthesis of uniformly cubic $Co_3O_4$ nanocrystals," *Materials Letters*, vol. 64, no. 3, pp. 239–242, 2010.

[10] S. Thota, A. Kumar, and J. Kumar, "Optical, electrical and magnetic properties of $Co_3O_4$ nanocrystallites obtained by thermal decomposition of sol-gel derived oxalates," *Materials Science and Engineering B*, vol. 164, no. 1, pp. 30–37, 2009.

[11] J. C. Toniolo, A. S. Takimi, and C. P. Bergmann, "Nanostructured cobalt oxides ($Co_3O_4$ and CoO) and metallic Co powders synthesized by the solution combustion method," *Materials Research Bulletin*, vol. 45, no. 6, pp. 672–676, 2010.

[12] L.-H. Ai and J. Jiang, "Rapid synthesis of nanocrystalline $Co_3O_4$ by a microwave-assisted combustion method," *Powder Technology*, vol. 195, no. 1, pp. 11–14, 2009.

[13] J. Jiu, Y. Ge, X. Li, and L. Nie, "Preparation of $Co_3O_4$ nanoparticles by a polymer combustion route," *Materials Letters*, vol. 54, no. 4, pp. 260–263, 2002.

[14] L. A. Chick, L. R. Pederson, G. D. Maupin, J. L. Bates, L. E. Thomas, and G. J. Exarhos, "Glycine-nitrate combustion synthesis of oxide ceramic powders," *Materials Letters*, vol. 10, no. 1-2, pp. 6–12, 1990.

[15] T. I. Izaak, O. V. Babkina, A. I. Boronin, T. N. Drebushchak, A. I. Stadnichenko, and G. M. Mokrousov, "Constitution and properties of nanocomposites prepared by thermal decomposition of silver salts sorbed by polyacrylate matrix," *Colloid Journal*, vol. 65, no. 6, pp. 720–725, 2003.

[16] L. Xu, B. Wei, Z. Zhang, Z. Lü, H. Gao, and Y. Zhang, "Synthesis and luminescence of europium doped yttria nanophosphors via a sucrose-templated combustion method," *Nanotechnology*, vol. 17, no. 17, pp. 4327–4331, 2006.

[17] R. N. Das, A. Bandyopadhyay, and S. Bose, "Nanocrystalline $\alpha$-$Al_2O_3$ using sucrose," *Journal of the American Ceramic Society*, vol. 84, no. 10, pp. 2421–2423, 2001.

[18] M. G. Lazarraga, L. Pascual, H. Gadjov et al., "Nanosize $LiNi_yMn_{2-y}O_4$ ($0 < y \leq 0.5$) spinels synthesized by a sucrose-aided combustion method. Characterization and electrochemical performance," *Journal of Materials Chemistry*, vol. 14, no. 10, pp. 1640–1647, 2004.

[19] Y. J. Wu, A. Bandyopadhyay, and S. Bose, "Processing of alumina and zirconia nano-powders and compacts," *Materials Science and Engineering A*, vol. 380, no. 1-2, pp. 349–355, 2004.

[20] J. M. Amarilla, R. M. Rojas, and J. M. Rojo, "Understanding the sucrose-assisted combustion method: effects of the atmosphere and fuel amount on the synthesis and electrochemical performances of $LiNi_{0.5}Mn_{1.5}O_4$ spinel," *Journal of Power Sources*, vol. 196, no. 14, pp. 5951–5959, 2011.

[21] N. M. Ghoneim, S. Hanafi, and S. A. Abo El-Enein, "Characteristics and effect of thermal treatment on surface texture of ultrafine zirconia powders," *Journal of Materials Science*, vol. 22, no. 3, pp. 791–797, 1987.

[22] A. Cabot, J. Arbiol, E. Rossinyol, J. R. Morante, F. Chen, and M. Liu, "Synthesis of tin oxide nanostructures with controlled particle size using mesoporous frameworks," *Electrochemical and Solid-State Letters*, vol. 7, no. 5, pp. G93–G97, 2004.

[23] J. Lu and Z. Z. Fang J, "Synthesis and characterization of nanoscaled cerium (IV) oxide via a solid-state mechanochemical method," *Journal of the American Ceramic Society*, vol. 89, no. 3, pp. 842–847, 2006.

# The Influences of CuO/ZnO Ratios on the Crystallization Characteristics Electrical and Magnetic Properties of $Cu_xZn_{1-x}O$ Powders

**Kuan-Jen Chen,[1] Fei-Yi Hung,[2] Truan-Sheng Lui,[2] Cheng-Hung Chen,[2] and Sheng-Po Chang[3]**

[1] *The Instrument Center, National Cheng Kung University, Tainan 70101, Taiwan*
[2] *Department of Materials Science and Engineering, Institute of Nanotechnology and Microsystems Engineering, Center for Micro/Nano Science and Engineering, National Cheng Kung University, Tainan 70101, Taiwan*
[3] *Institute of Microelectronics and Department of Electrical Engineering, Advanced Optoelectronic Technology Center, National Cheng Kung University, Tainan 701, Taiwan*

Correspondence should be addressed to Fei-Yi Hung; fyhung@mail.ncku.edu.tw

Academic Editor: Yang Xu

This study synthesizes $Cu_xZn_{1-x}O$ powders using an aqueous solution method. The $Cu_xZn_{1-x}O$ powders with different content ratios of CuO and ZnO (CuO : ZnO = 1 : 2, 1 : 1, and 2 : 1) were formed. The crystalline characteristics and electrical and magnetic properties depended primarily on the mixing effect and oxygenation. The electrical resistance of $C_{0.5}Z_{0.5}O$ ($1.5 \times 10^5\ \Omega/\square$) powder was lower than that of CuO ($5.82 \times 10^5\ \Omega/\square$) powder after ZnO mixing in CuO. This reduction was attributed to the substitution of $Cu^+$ ions at $Zn^{2+}$ sites or the formation of electron trapping defect centers. The concentration ratio of $Cu_2O$ phase in $Cu_xZn_{1-x}O$ powder mainly dominated the electrical resistance. The $Cu_xZn_{1-x}O$ has a diluted ferromagnetism (DFM) and paramagnetism (PM). The electrical resistance of $Cu_xZn_{1-x}O$ decreased; the magnetic behavior increased instead. This study also analyzes the chemical binding of $Cu_{0.5}Zn_{0.5}O$ powders to confirm the contribution of $Cu^+$ ions to the electrical and magnetic properties.

## 1. Introduction

Metal oxide nanostructures, such as ZnO [1], CuO [2, 3], $SnO_2$ [4], $In_2O_3$ [5], and $TiO_2$ [6], have attracted extensive attention for their optical, electrical and magnetic properties. They are widely applied in various optoelectronic devices, such as photocatalyzers [7], solar cells [8], photodiodes [9], and humidity sensors [10].

Recent reports on CuO-ZnO compound and CuO/ZnO heterojunction structures show the need for efficient characteristics [11, 12]. A few Cu dopants (<10 at.%) doped ZnO samples have been investigated [13, 14]. However, the effects of the higher Cu doping contents (>10 at.%) on electrical properties and magnetic behavior of ZnO were rarely reported. Thus, the electrical properties that dominate devices performance and the influences of higher dopant level on the electrical resistance of $Cu_xZn_{1-x}O$ are topics worthy of further research. In addition, ZnO-diluted magnetic semiconductor (DMS) can enhance its magnetic behavior by Cu doping [15]. The influences of high CuO concentration on magnetic properties of ZnO are not clear.

The study uses an aqueous solution method to synthesize CuO and $Cu_xZn_{1-x}O$ powders. Specifically, this study examines the effects of high ZnO content mixed with CuO on the structural, electrical, and magnetic properties of $Cu_xZn_{1-x}O$ powders. The relation of electrical resistance and

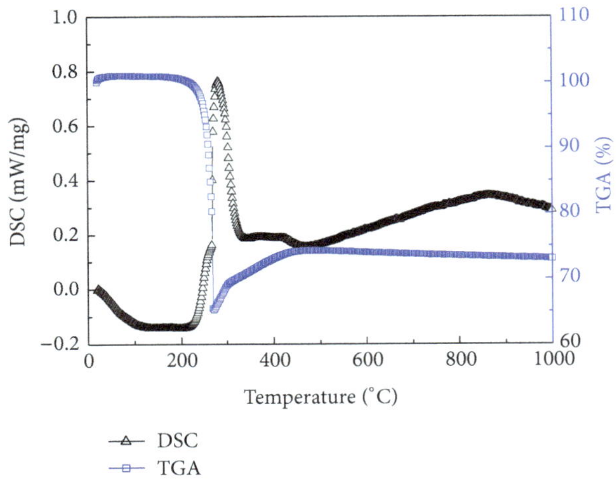


FIGURE 1: DSC-TGA curve of CuO.

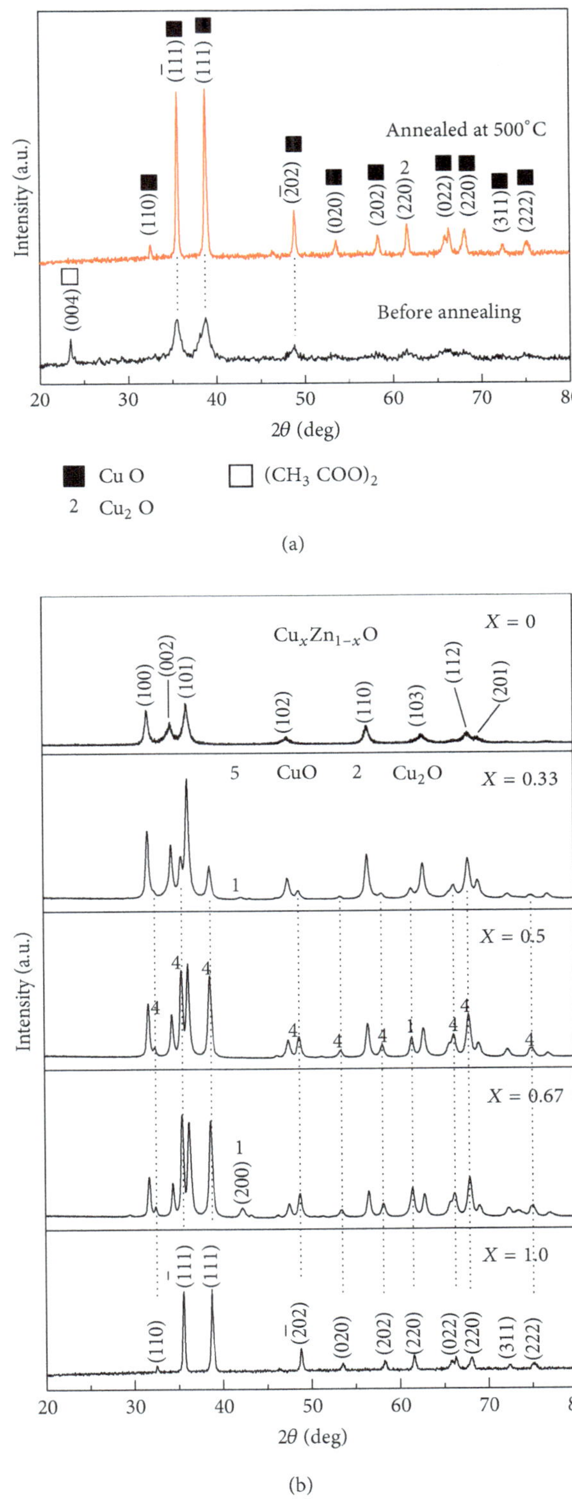


FIGURE 2: XRD patterns of (a) as-grown CuO powder and CuO powder with annealed temperature of 500°C. (b) $Cu_xZn_{1-x}O$ ($x$ = 0, 0.33, 0.5, 0.67, and 1) powders.

magnetic properties for $Cu_xZn_{1-x}O$ powders was carried out to understand the influences of CuO mixing effects and clarify the contribution of CuO and $Cu_2O$ phase.

## 2. Experimental Procedures

The experiments in this study synthesized the CuO powder using an aqueous solution. To acquire the CuO powder, 0.25 M copper acetate [$Cu(CH_3COO)_2$] was synthesized in deionized water. The precursor solution was uniformly stirred at 80°C for 1 h and then dried at 120°C in an oven to evaporate the solvent. The resulting powder was then analyzed by a differential scanning calorimeter (DSC) and a thermogravimetric analyzer (TGA) to determine the thermal properties of fabricated CuO nanoparticles. The DSC and TGA results indicate that CuO powder could be obtained after thermal annealing in a furnace for 1 h. For $Cu_xZn_{1-x}O$ powders, the copper acetate and the zinc acetate were mixed with different ratios (CuO : ZnO = 1 : 2, 1 : 1 and 2 : 1) and the molar ratio of zinc acetate to citric acid ($C_6H_8O_7$) was 2. The $Cu_xZn_{1-x}O$ powders were then designed according to the concentration ratio of CuO and ZnO as $Cu_{0.3}Zn_{0.67}O$, $Cu_{0.5}Zn_{0.5}O$, and $Cu_{0.67}Zn_{0.3}O$.

This study investigates the structural characteristics of powders using X-ray diffraction (XRD, Siemens/D5000) and scanning electron microscopy (SEM, Hitachi/S-4100). To understand the contribution of CuO on the electrical and magnetic characteristics of $Cu_xZn_{1-x}O$ powders, a semiconductor parameter analyzer (Agilent/4155-B) and superconducting quantum interference device vibrating sample magnetometer (SQUID VSM) were used, respectively. The composition and chemical bonding of the crystallization were analyzed using an electron spectroscopy for chemical analysis (ESCA, PHI 5000 VersaProbe).

## 3. Results and Discussions

This study uses DSC and TGA analysis to determine the crystallization conditions of CuO precursors (Figure 1). The

(a) (b) (c)

FIGURE 3: Morphology of $Cu_xZn_{1-x}O$ powders: (a) $Cu_{0.33}Zn_{0.67}O$, (b) $Cu_{0.5}Zn_{0.5}O$, and (c) $Cu_{0.67}Zn_{0.33}O$.

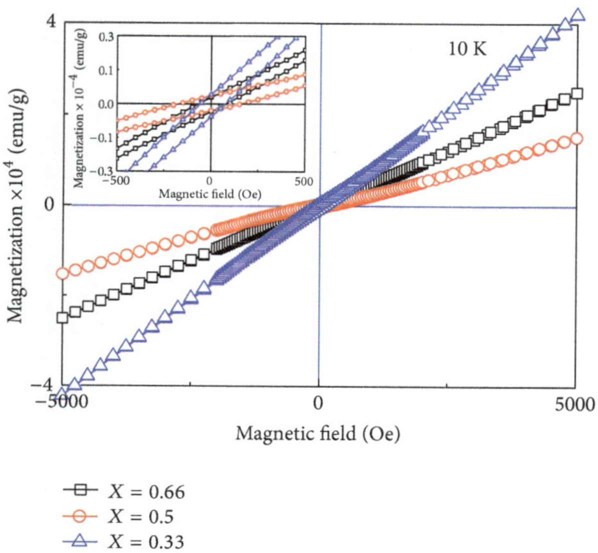


FIGURE 4: Magnetic field dependence of magnetization ($M$-$H$) for $Cu_xZn_{1-x}O$ powders at 10 K. The inset shows the hysteresis loops for lower regions (±500 Oe).

specimens were heated from room temperature to 1000°C at a rate of 10°C/min in air. The TGA data reveals a sharp weight loss in the powder at 270°C because of the evaporation of water and organics [16]. This result is consistent with the DSC curve, which shows an exothermic peak at 280°C. An endothermic peak appears at 480°C and the powder weight (TGA curve) gradually increases and then stabilizes as the temperature exceeds 500°C. The main reason is that the CuO crystallization gradually forms with increasing temperature [17]. Based on these reasons, the CuO powder was annealed at 500°C to estimate the CuO crystallization.

Figure 2(a) shows the XRD patterns of as-grown CuO precursors and CuO powder annealed at 500°C for 1 h. Both samples were polycrystalline and had a monoclinic structure. An additional diffraction peak of $(CH_3COO)_2 \cdot Cu$ appeared in as-grown CuO precursors, indicating that a drying temperature of 120°C was insufficient for its evaporation. After thermal annealing at 500°C, the $Cu(CH_3COO)_2$ phase disappeared and the (111) diffraction peak dominated the CuO crystallization. The intensity of the major diffraction peaks increased, indicating that the sufficient thermal energy improved CuO crystallization. Based on the Scherrer formula [16], the grain size of CuO nanoparticles can be estimated from the full width at half-maximum (FWHM) of the (111) diffraction peak. The average grain size of CuO powder increased from 8 nm to 21 nm after thermal annealing at 500°C. This result is associated with the grains growing more easily under the higher temperature [16]. The ZnO powder was also synthesized and compared to $Cu_xZn_{1-x}O$ powder. XRD analysis of $Cu_xZn_{1-x}O$ ($x$ = 0, 0.33, 0.5, 0.67, and 1) powders were conducted to analyze the effects of ZnO on $Cu_xZn_{1-x}O$ crystallization (Figure 2(b)). A small amount of ZnO mixing with CuO ($Cu_xZn_{1-x}O$, $x$ = 0.67), the monoclinic structure of CuO dominated the $Cu_{0.67}Zn_{0.33}O$ crystallization, and ZnO phases existed in the $Cu_{0.67}Zn_{0.33}O$ matrix. With increasing the ZnO concentration in $Cu_xZn_{1-x}O$ ($x$ value increase from 0.67 to 0.33), the (101) diffraction peak of ZnO was the preferred orientation which indicated that the ZnO phases dominated the $Cu_xZn_{1-x}O$ crystallization. A comparison of the CuO revealed that the diffraction peak of (111) for $Cu_{0.33}Zn_{0.67}O$ shifted to higher degree. In contrast, the diffraction peak of (101) for $Cu_{0.33}Zn_{0.67}O$ shifted to lower degree comparing with ZnO. These results are associated with compressive strain in the crystallization [4]. In addition, the related (220) diffraction peak of the $Cu_2O$ phase was attributed to insufficient oxidation [2]. Notably, the (200) diffraction peak of $Cu_2O$ phase did not appear in $Cu_{0.5}Zn_{0.5}O$ crystallization, indicating that the compositional ratio of $Cu_2O$ phase was less than other $Cu_xZn_{1-x}O$ ($x$ = 0.33 and 0.67) powders. This result should affect the electrical properties of $Cu_xZn_{1-x}O$ powders.

Figure 3 shows SEM images of $Cu_xZn_{1-x}O$ ($x$ = 0.33, 0.5, and 0.67) powders at an annealing temperature of 500°C. All powders displayed a particle-like structure, and the agglomeration of particles was randomly distributed. The morphology of $Cu_{0.5}Zn_{0.5}O$ powder shows an irregular and its size for the agglomeration of particles is larger than that of $Cu_{0.67}Zn_{0.33}O$ and $Cu_{0.33}Zn_{0.67}O$ powders. The electrical properties of powder might be influenced by the particle size [18], therefore, the electrical resistance were examined.

Electrical measurements were conducted to determine the resistance value of CuO powder. The CuO powder was

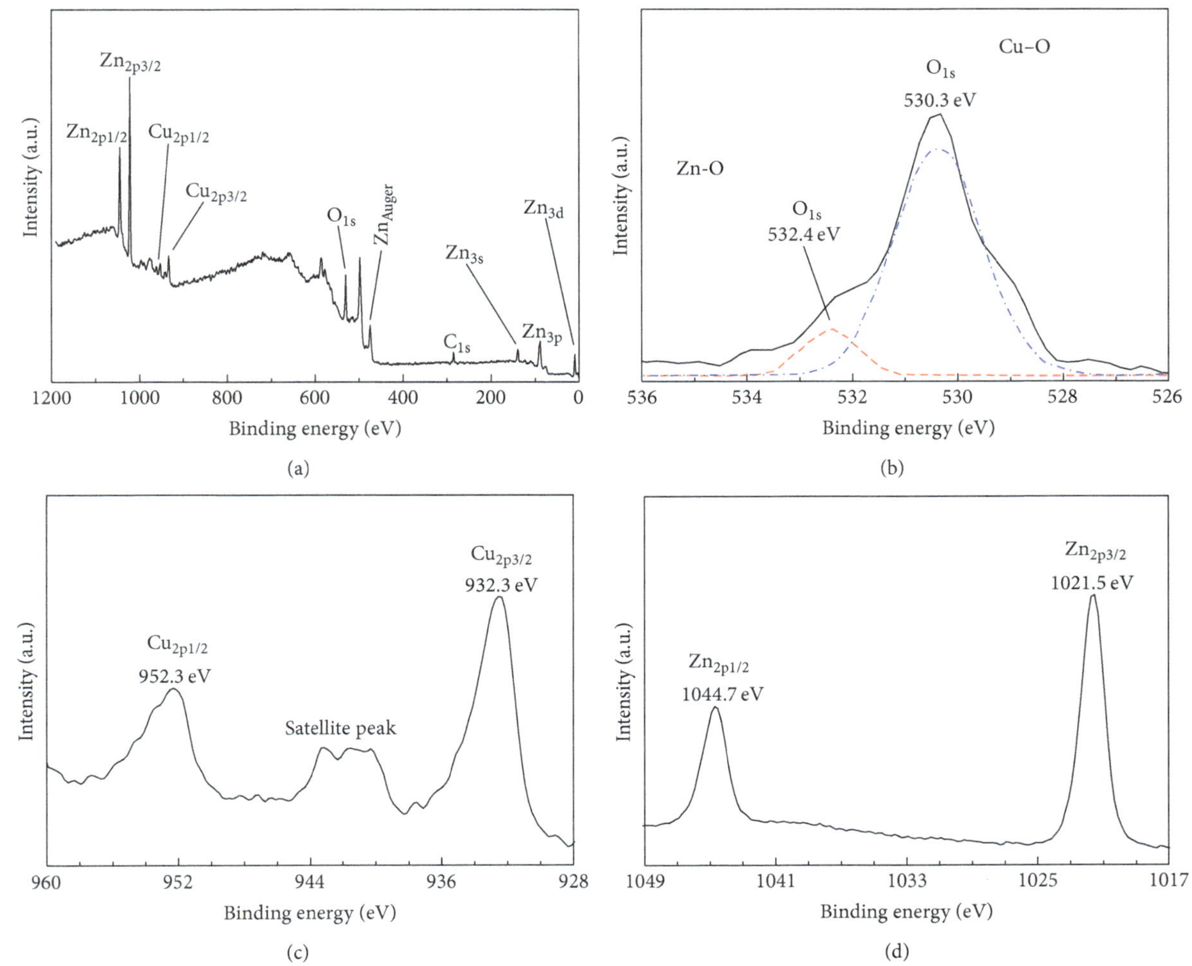


FIGURE 5: XPS spectra of $Cu_{0.5}Zn_{0.5}O$ powder: (a) XPS spectrum of $Cu_{0.5}Zn_{0.5}O$ powder fully scanned from 300 eV to 1200 eV, (b) high-resolution selective spectrum of O, (c) high-resolution selective spectrum of Cu, and (d) high-resolution selective spectrum of Zn.

TABLE 1: The relation of electrical resistances and magnetic properties for $Cu_xZn_{1-x}O$ ($x$ = 0, 0.33, 0.5, 0.67, and 1) powders with different CuO/ZnO ratios.

| Sample | | | Resistivity (Ω/□) | Magnetism |
|---|---|---|---|---|
| $Cu_xZn_{1-x}O$ | Before annealing | $x = 1$ | $8.6 \times 10^{11}$ | — |
| | Annealed at 500°C | | $2.3 \times 10^6$ | AFM |
| | | $x = 0.67$ | $1.0 \times 10^7$ | DFM + PM |
| | | $x = 0.5$ | $1.5 \times 10^5$ | DFM + PM |
| | | $x = 0.33$ | $8.2 \times 10^8$ | DFM + PM |
| | | $x = 0$ | $9.0 \times 10^8$ | DM |

pressed to form ingots (Φ 10 mm/∼2 mm thick) at a pressure of 30 kg/cm$^2$ for 30 s before measuring their resistivity (Table 1). After annealing at 500°C, the resistivity of the CuO sample decreased from $8.6 \times 10^{11}$ Ω/□ to $2.3 \times 10^6$ Ω/□. After CuO mixing with ZnO, the electrical resistance of the $Cu_{0.67}Zn_{0.33}O$ and $Cu_{0.33}Zn_{0.67}O$ samples was higher than that of CuO. This increment of the electrical resistance may be attributed to the substitution of $Cu^+$ ions at $Zn^{2+}$ sites or the formation of electron trapping defect centers [19]. It is noted that the electrical resistance of $Cu_{0.5}Zn_{0.5}O$ sample ($1.5 \times 10^5$ Ω/□) decreased instead comparing with CuO sample. This result indicates that the ratio of the CuO phase to $Cu_2O$ phase in $Cu_xZn_{1-x}O$ system decreased [20] and possessed a stable crystallization comparing with $Cu_{0.67}Zn_{0.33}O$ and $Cu_{0.33}Zn_{0.67}O$ samples.

To understand the contribution of CuO on the magnetic properties, the $Cu_xZn_{1-x}O$ powders with different ratios of CuO and ZnO were measured. Figure 4 shows the magnetization ($M$) as a function of magnetic field ($H$) at the temperature of 10 K for the $Cu_xZn_{1-x}O$ powders. All samples have a linear-like $M$-$H$ variation at a magnetic field of ±5000 Oe without measureable hysteresis, which indicates a paramagnetic behavior [21]. The magnetization of $Cu_{0.33}Zn_{0.67}O$ sample was higher than that of $Cu_{0.5}Zn_{0.5}O$ and $Cu_{0.67}Zn_{0.33}O$ samples, which indicated that the $Cu_{0.33}Zn_{0.67}O$ powder contains a better paramagnetism (PM). The increment of antiferromagnetic interaction possibly reduced the PM that resulted from the formation of CuO crystallization [21]. From XRD data (Figure 2(b)), the CuO phase dominated the $Cu_{0.5}Zn_{0.5}O$, $Cu_{0.67}Zn_{0.33}O$ crystallization and affected their magnetic properties that were consistent with the result of *M-H* curves. In addition, the lower magnetization of $Cu_{0.5}Zn_{0.5}O$, $Cu_{0.67}Zn_{0.33}O$ was also attributed to the secondary phase of $Cu_2O$ [22]. The low-field region of the hysteresis loops of $Cu_xZn_{1-x}O$ powders was clearly observed (inset of Figure 4) which indicated that all $Cu_xZn_{1-x}O$ powders also contained a diluted ferromagnetism (DFM). The FM could be developed by the distortion of ZnO structure by the substitution of remnant $Cu^{2+}$ ions into ZnO lattice [22, 23]. It is found that the coercivity field of the $Cu_{0.33}Zn_{0.67}O$, $Cu_{0.5}Zn_{0.5}O$, and $Cu_{0.67}Zn_{0.33}O$ powders is 75 Oe, 150 Oe, and 78 Oe, respectively. That is to say, the $Cu_{0.5}Zn_{0.5}O$ powder has a good stability for thermal interference [24].

The chemical bonding of the $Cu_{0.5}Zn_{0.5}O$ powder was examined by XPS with full region scanning from 0 eV to 1200 eV (Figure 5). In Figure 5(a), all peaks from oxygen, copper, zinc, and a trace of carbon are apparent, meaning that the $Cu_{0.5}Zn_{0.5}O$ powder is substantially covered by $Cu^{2+}$, $Zn^{2+}$, and $O^{-2}$. The high-resolution scanning information provided in Figures 5(b)–5(d) is for the separate analysis of elements: O, Cu, and Zn, respectively. The high-resolution XPS spectrum of the $O_{1s}$ signal (Figure 5(b)) indicates that the binding energy of 530.3 eV can be attributed to oxidized ions in the CuO particles [25]. Multipeak Gaussian fitting shows another $O_{1s}$ peak located at 532.4 eV, indicating that this binding energy was dominated by $Zn^{2+}$ ion doping [23]. The high-resolution XPS spectrum of the $Cu_{2p3/2}$ mode (Figure 5(c)) appears at 932.3 eV, indicating the presence of $Cu^{2+}$ ion [2]. The binding energy of 952.3 eV ($Cu_{2p1/2}$ mode) can be attributed to $Cu^+$ ions in $Cu_{0.5}Zn_{0.5}O$ that resulted from $Cu_2O$ [26, 27]. This result is consistent with the observation of XRD data (Figure 3(b)). Figure 5(d) shows two strong peaks at 1021.5 eV and 1044.7 eV which correspond to $Zn_{2p3/2}$ and $Zn_{2p1/2}$, respectively. This is consistent with the $Zn^{2+}$ ion binding in previous reports [28]. Based on these results, the $Zn^{2+}$ ions were substituted by $Cu^+$ to form the $Cu_2O$ phase that affects the electrical and magnetic properties.

The correlation of the magnetic properties and resistivity of $Cu_xZn_{1-x}O$ powders were summarized in Table 1. The magnetic property of pure CuO powder varied from antiferromagnetism (AFM) to paramagnetism (PM) when ZnO to CuO. The resistivity of $Cu_xZn_{1-x}O$ powders decreased; the values of the paramagnetic behavior decreased. These results indicate that the paramagnetic behavior in $Cu_{0.33}Zn_{0.67}O$ powder is the highest. The lower carrier concentration may promote paramagnetic ordering in $Cu_{0.33}Zn_{0.67}O$ [29].

## 4. Conclusions

The stability of oxide (CuO or $Cu_2O$) depends primarily on the intensity of oxidation (annealing temperature). When ZnO participated in the CuO system, the crystalline quality of CuO powder deteriorated. The ZnO mixing effect increased the crystallization size and induced a compressive stress in the particle. Although the presence of ZnO phases deteriorated $Cu_{0.5}Zn_{0.5}O$ crystallization, the electrical conductance was improved. A lower $Cu_2O$ phase concentration and stable crystallization reduced the electrical resistance of $Cu_{0.5}Zn_{0.5}O$ powder. The electrical resistance of $Cu_{0.5}Zn_{0.5}O$ powder was the lowest and the magnetic behavior was the smallest because CuO and $Cu_2O$ contents were higher. XPS analysis reveals that the $Zn^{2+}$ ions were substituted by $Cu^{2+}$ and $Cu^+$ ions, forming CuO and $Cu_2O$ phases that confirmed the contribution of $Cu_2O$ on the electrical and magnetic properties.

## Acknowledgments

The authors are grateful to the Instrument Center of National Cheng Kung University, the Center for Micro/Nano Science and Technology (D100-2700), and the National Science Council, Taiwan, for financially supporting this study under Grant nos. 101-2221-E-006-114 and NSC100-2622-E-006-030-CC3.

## References

[1] B. Karthikeyan, T. Pandiyarajan, and K. Mangaiyarkarasi, "Optical properties of sol-gel synthesized calcium doped ZnO nanostructures," *Spectrochimica Acta A*, vol. 82, pp. 97–101, 2011.

[2] H. Qin, Z. Zhang, X. Liu, Y. Zhang, and J. Hu, "Room-temperature ferromagnetism in CuO sol-gel powders and films," *Journal of Magnetism and Magnetic Materials*, vol. 322, no. 14, pp. 1994–1998, 2010.

[3] C. W. Zou, J. Wang, F. Liang, W. Xie, L. X. Shao, and D. J. Fu, "Large-area aligned CuO nanowires arrays: synthesis, anomalous ferromagnetic and CO gas sensing properties," *Current Applied Physics*, vol. 12, pp. 1349–1354, 2012.

[4] J. Wang, N. Du, H. Zhang, J. Yu, and D. Yang, "Large-scale synthesis of $SnO_2$ nanotube arrays as high-performance anode materials of Li-ion batteries," *Journal of Physical Chemistry C*, vol. 115, no. 22, pp. 11302–11305, 2011.

[5] H. G. Na, D. S. Kwak, H. W. Kim, and H. W., "Structural, Raman, and photoluminescence properties of double-shelled coaxial nanocables of $In_2O_3$ core with ZnO and AZO shells," *Crystal Research and Technology*, vol. 47, pp. 79–86, 2012.

[6] L. C. Chen, C. M. Huang, C. S. Gao, G. W. Wang, and M. C. Hsiao, "A comparative study of the effects of In(2)O(3) and SnO(2) modification on the photocatalytic activity and characteristics of TiO(2)," *Chemical Engineering Journal*, vol. 175, pp. 49–55, 2011.

[7] J. P. Li, F. Q. Sun, K. Y. Gu et al., "Preparation of spindly CuO micro-particles for photodegradation of dye pollutants under a halogen tungsten lamp," *Applied Catalysis A*, vol. 406, pp. 51–58, 2011.

[8] R. Motoyoshi, T. Oku, H. Kidowaki et al., "Structure and photovoltaic activity of cupric oxide-based thin film solar cells," *Journal of the Ceramic Society of Japan*, vol. 118, no. 1383, pp. 1021–1023, 2010.

[9] H. T. Hsueh, S. J. Chang, W. Y. Weng et al., "Fabrication and characterization of coaxial *p*-copper oxide/*n*-ZnO nanowire photodiodes," *IEEE Transactions on Nanotechnology.*, vol. 11, pp. 127–133, 2012.

[10] H. T. Hsueh, T. J. Hsueh, S. J. Chang et al., "CuO nanowire-based humidity sensors prepared on glass substrate," *Sensors and Actuators B*, vol. 156, no. 2, pp. 906–911, 2011.

[11] J. X. Wang, X. W. Sun, Y. Yang et al., "Free-standing ZnO–CuO composite nanowire array films and their gas sensing properties," *Nanotechnology*, vol. 22, no. 32, Article ID 325704, 2011.

[12] C. S. Dandeneau, Y. H. Jeon, C. T. Shelton, T. K. Plant, D. P. Cann, and B. J. Gibbons, "Thin film chemical sensors based on *p*-CuO/n-ZnO heterocontacts," *Thin Solid Films*, vol. 517, no. 15, pp. 4448–4454, 2009.

[13] T. Ghosh, M. Dutta, S. Mridha, and D. Basak, "Effect of Cu doping in the structural, electrical, optical, and optoelectronic properties of sol-gel ZnO thin films," *Journal of the Electrochemical Society*, vol. 156, no. 4, pp. H285–H289, 2009.

[14] H. Liu, J. Yang, Z. Hua et al., "The structure and magnetic properties of Cu-doped ZnO prepared by sol-gel method," *Applied Surface Science*, vol. 256, no. 13, pp. 4162–4165, 2010.

[15] S. Y. Zhuo, X. C. Liu, Z. Xiong, J. H. Yang, and E. W. Shi, "Defects enhanced ferromagnetism in Cu-doped ZnO thin films," *Solid State Communications*, vol. 152, pp. 257–260, 2012.

[16] K. J. Chen, T. H. Fang, F. Y. Hung et al., "The crystallization and physical properties of Al-doped ZnO nanoparticles," *Applied Surface Science*, vol. 254, no. 18, pp. 5791–5795, 2008.

[17] J. V. Bellini, M. R. Morelli, and R. H. G. A. Kiminami, "Ceramic system based on ZnO · CuO obtained by freeze-drying," *Materials Letters*, vol. 57, no. 24-25, pp. 3775–3778, 2003.

[18] C. Karunakaran, V. Rajeswari, and P. Gomathisankar, "Optical, electrical, photocatalytic, and bactericidal properties of microwave synthesized nanocrystalline Ag–ZnO and ZnO," *Solid State Sciences*, vol. 13, no. 5, pp. 923–928, 2011.

[19] D. Behera, J. Panigrahi, and B. S. Acharya, "Probing the effect of nitrogen gas on electrical conduction phenomena of ZnO and Cu-doped ZnO thin films prepared by spray pyrolysis," *Ionics*, vol. 17, pp. 741–749, 2011.

[20] X. B. Wang, D. M. Li, F. Zeng, and F. Pan, "Microstructure and properties of Cu-doped ZnO films prepared by dc reactive magnetron sputtering," *Journal of Physics D*, vol. 38, no. 22, pp. 4104–4108, 2005.

[21] M. S. Seehra, P. Dutta, V. Singh, Y. Zhang, and I. Wender, "Evidence for room temperature ferromagnetism in $Cu_xZn_{1-x}O$ from magnetic studies in $Cu_xZn_{1-x}OCuO$ composite," *Journal of Applied Physics*, vol. 101, no. 9, Article ID 09H107, 2007.

[22] J. Qi, D. Gao, L. Zhang, and Y. Yang, "Room-temperature ferromagnetism of the amorphous Cu-doped ZnO thin films," *Applied Surface Science*, vol. 256, no. 8, pp. 2507–2508, 2010.

[23] O. Perales-Perez, A. Parra-Palomino, R. Singhal et al., "Evidence of ferromagnetism in $Zn_{1-x}M_xO$ (M = Ni,Cu) nanocrystals for spintronics," *Nanotechnology*, vol. 18, no. 31, Article ID 315606, 2007.

[24] X. Chen, Z. Zhou, K. Wang et al., "Ferromagnetism in Fe-doped tetra-needle like ZnO whiskers," *Materials Research Bulletin*, vol. 44, no. 4, pp. 799–802, 2009.

[25] R. Al-Gaashani, S. Radiman, N. Tabet, A. R. Daud, and Synthesis a, "nd optical properties of CuO nanostructures obtained via a novel thermal decomposition method," *Journal of Alloys and Compounds*, vol. 509, pp. 8761–8769, 2011.

[26] L. Li, J. G. Lei, and T. H. Ji, "Facile fabrication of *p-n* heterojunctions for Cu(2)O submicroparticles deposited on anatase TiO(2) nanobelts," *Materials Research Bulletin*, vol. 46, pp. 2084–2089, 2011.

[27] H. Kim, B. K. Lee, K. S. An, and S. Ju, "Direct growth of oxide nanowires on CuO(x) thin film," *Nanotechnology*, vol. 23, no. 4, Article ID 830474, 2012.

[28] K. J. Chen, F. Y. Hung, Y. T. Chen, S. J. Chang, and Z. S. Hu, "Surface characteristics, optical and electrical properties on sol-gel synthesized sn-doped ZnO thin film," *Materials Transactions*, vol. 51, no. 7, pp. 1340–1345, 2010.

[29] C. O. Kim, S. Kim, H. T. Oh et al., "Effect of electrical conduction properties on magnetic behaviors of Cu-doped ZnO thin films," *Physica B*, vol. 405, no. 22, pp. 4678–4681, 2010.

# 14

# Antibacterial Activity of Silver Nanoparticles Synthesized by Bark Extract of *Syzygium cumini*

**Ram Prasad and Vyshnava Satyanarayana Swamy**

*Amity Institute of Microbial Technology, Amity University Uttar Pradesh, Sector 125, Noida 201303, India*

Correspondence should be addressed to Ram Prasad; rprasad@amity.edu

Academic Editor: Amir Kajbafvala

The unique property of the silver nanoparticles having the antimicrobial activity drags the major attention towards the present nanotechnology. The environmentally nontoxic, ecofriendly, and cost-effective method that has been developed for the synthesis of silver nanoparticles using plant extracts creates the major research interest in the field of nanobiotechnology. The synthesized silver nanoparticles have been characterized by the UV-visible spectroscopy, atomic force microscopy (AFM), and scanning electron microscopy (SEM). Further, the antibacterial activity of silver nanoparticles was evaluated by well diffusion method, and it was found that the biogenic silver nanoparticles have antibacterial activity against *Escherichia coli* (ATCC 25922), *Staphylococcus aureus* (ATCC 29213), *Pseudomonas aeruginosa* (ATCC 27853), *Azotobacter chroococcum* WR 9, and *Bacillus licheniformis* (MTCC 9555).

## 1. Introduction

The broad spectrum of nanotechnology is important in the major fields of biology, chemistry, physics, and material sciences. Nanotechnology deals with the study of materials at the nanometers [1, 2]. The day to day development of nanotechnology creates a major interest in the development and fabrications of different dimensioned nanoparticles [3]. The nanomaterials can be synthesized by different methods including chemical, physical, irradiation, and biological methods. The development of new chemical or physical methods has resulted in environmental contaminations, since the chemical procedures involved in the synthesis of nanomaterials generate a large amount of hazardous byproducts [4]. Thus, there is a need for "green nanotechnology" that includes a clean, safe, ecofriendly, and environmentally nontoxic method of nanoparticle synthesis, and in this method there is no need to use high pressure, energy, temperature, and toxic chemicals [5, 6]. The biological methods include synthesis of nanomaterial's from the extracts of plant, bacterial, fungal species, and so forth. The synthesis of nanoparticles from the plant extracts is considered to be a process [7]. The preparation and maintenance of fungal and bacterial cultures are time consuming and require aseptic conditions and large manual skills to maintain the cultures [8].

Plant extracts include bark, root, leaves, fruit, flowers, rhizoids, and latex and are used to synthesize the nanoparticles. These nanoparticles show different dimensions including the size, shape, and dispersion which have more efficacy than those synthesized from the chemical and physical procedures. Therefore, the use of green plants for similar nanoparticle biosynthesis methodologies is an exciting possibility which has compatibility for pharmaceutical and other biomedical applications, as they do not use toxic chemicals for the synthesis of nanoparticles [9, 10].

Nanoparticles had a wide variety of application in the major fields of medicine, electronics, therapeutics, and diagnostic agents. Silver nanoparticles have wide application in biomedical science like treatment of burned patients, antimicrobial activity and used the targeted drug delivery, and so forth [11]. Nowadays the nanoparticles are coated on the medical appliances, food covering sheets, and cans for storing the beverages and food [12–14]. However, there are many problems and toxicity of using metal oxide nanoparticles on the human health. Use of plants for the synthesis of nanoparticles does not require high energy, temperatures, and it is easily

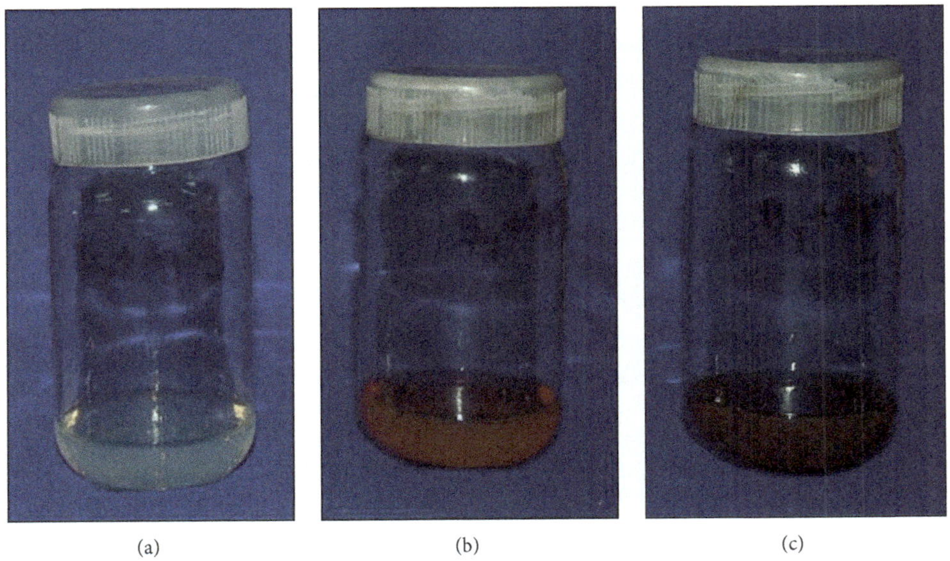


FIGURE 1: *Syzygium cumini* bark extract sample. Change in the color of the solution from brown to dark brown. (a) Silver nitrate solution, (b) reaction mixture, and (c) change in the color of the solution.

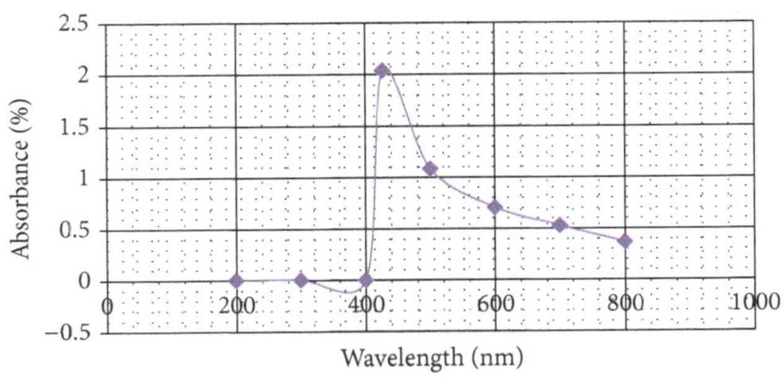


FIGURE 2: UV-visible spectrum of silver nanoparticles.

scaled up for large scale synthesis, and it is cost effective too [15–17].

*Syzygium cumini* is a medicinal plant available in the tropical forests and is used for treatment of diabetes. The leaves and bark are used for controlling blood pressure and gingivitis [14]. The plant contains a variety of phytochemical compounds such as phenols, tannins, alkaloids, glycosides, amino acids, and flavones, and these molecules are expected to self-assemble and cap the metal nanoparticles formed in their presence and thereby induce some shape control during metal ion reduction [18]. In this study we used the silver nanoparticles synthesized from the bark extract of *S. cumini* and its antibacterial effect on the bacteria, namely, *Escherichia coli* (ATCC 25922), *Staphylococcus aureus* (ATCC 29213), *Pseudomonas aeruginosa* (ATCC 27853), *Azotobacter chroococcum* WR 9, and *Bacillus licheniformis* (MTCC 9555).

## 2. Materials and Methods

*2.1. Chemicals.* All analytical reagents and media components were purchased from HiMedia (Mumbai, India) and Sigma Chemicals (St. Louis, MO, USA).

*2.2. Preparation of Plant Extract.* The fresh bark of *Syzygium cumini* was collected and kept in hot air oven for drying at 60°C for six hours. The dried bark was chopped into fine pieces with the help of mixer grinder. It was collected, weighed for 2.5 g, and then mixed in 100 mL of double distilled water. This mixture was boiled at 60°C in the water bath for one hour. The solution was cooled at room temperature and filtered by Whatman filter paper No. 1. The filtrate was collected and stored at 4°C for further experiment.

*2.3. Synthesis of Silver Nanoparticles.* Silver nanoparticles ($AgNO_3$) were synthesized by reducing the freshly prepared 1 mM silver nitrate and stored under dark conditions with the bark extract. The reaction mixture was prepared in ratio of 9 : 1 (V/V) of freshly prepared silver nitrate solution and bark extract, respectively. The initial color of the solution was observed.

*2.4. UV-Visible Spectroscopy.* The silver nanoparticles show the plasmon resonance at 400 to 450 nm in the UV-Visible spectrum. The UV-Visible spectrum of synthesized silver nanoparticles was analysed by spectrophotometer (LAB INDIA UV 300+).

*2.5. Atomic Force Microscopy.* Atomic force microscopy is an advanced characterization technique to identify the size,

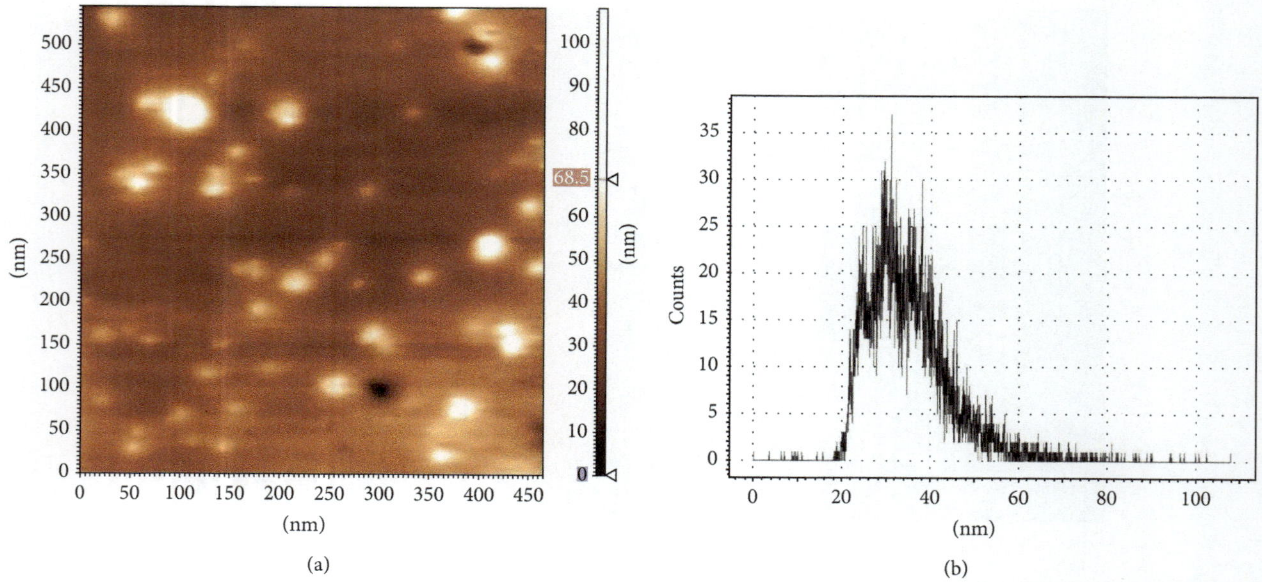


FIGURE 3: Atomic force microscopy. (a) Image of synthesized silver nanoparticles and (b) its histogram.

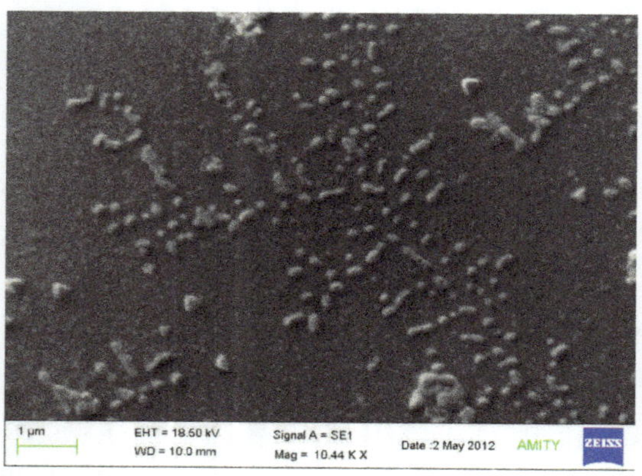


FIGURE 4: SEM images of silver nanoparticles synthesized by *S. cumini* bark extract.

shape, and dispersion of the silver nanoparticles. In order to characterize the silver nanoparticles, the sample was prepared by sonication at room temperature for about 15 minutes in the ultrasonicator. Then the sample solution was dried as a thin layer on mica-based glass slide which was used to view under the AFM Model NT-MDA Solver.

*2.6. Scanning Electron Microscopy.* SEM analysis of the silver nanoparticles provides the information regarding the dimensions including the surface, shape, and size. The sample was prepared by sonicating the sample solution for 15 minutes at room temperature. A small drop of sonicated sample was dried on a glass slide, and it was coated by gold and observed under ZEISS EVO HD SEM.

*2.7. Antibacterial Property.* The antibacterial property of the silver nanoparticles was determined by using the bacterial species including the pathogenic bacteria such as *Escherichia coli* (ATCC 25922), *Staphylococcus aureus* (ATCC 29213), *Pseudomonas aeruginosa* (ATCC 27853), *Azotobacter chroococcum* WR 9, and *Bacillus licheniformis* (MTCC 9555), by the well diffusion method [14]. The different concentrations used were at low concentrations (2, 5, 10, and 15 $\mu$L) and at higher concentrations (25, 50, 75, and 100 $\mu$L) for the identification of antimicrobial activity of the above bacterial species. All the plates were incubated at 37°C for 24 hours, and the zone of inhibition of bacteria was measured.

## 3. Results and Discussion

The green synthesis of silver nanoparticles using *S. cumini* bark extract was successfully carried out, as the change in the color of the solution from yellowish brown to dark brown color exhibits the reduction of the silver nitrate in aqueous solution due to excitation of surface plasmon vibrations in silver nanoparticles [19]. During this reaction process the pH of the solution changes from 5.93 to 5.72, which implies that the reaction occurs under acidic condition. This complete reaction occurs in seven hours. The brown to dark brown color change of the reaction mixture indicated the formation of silver nanoparticles (Figure 1).

The formation of silver nanoparticles was confirmed through measurement of UV-Visible spectrum of the reaction mixture. The UV-Visible spectrophotometric analysis of colloidal reaction mixture of silver nanoparticles synthesized using *S. cumini* bark showed sharp peak at 427 nm in the spectrum, and broadening of peak indicated that the particles are polydispersed [20] (Figure 2). The efficiency of this method was tested for stability also. The reaction mixture was stored for 45 days, and no precipitation in the solution was observed. It was also checked through UV-Vis absorption on regular interval.

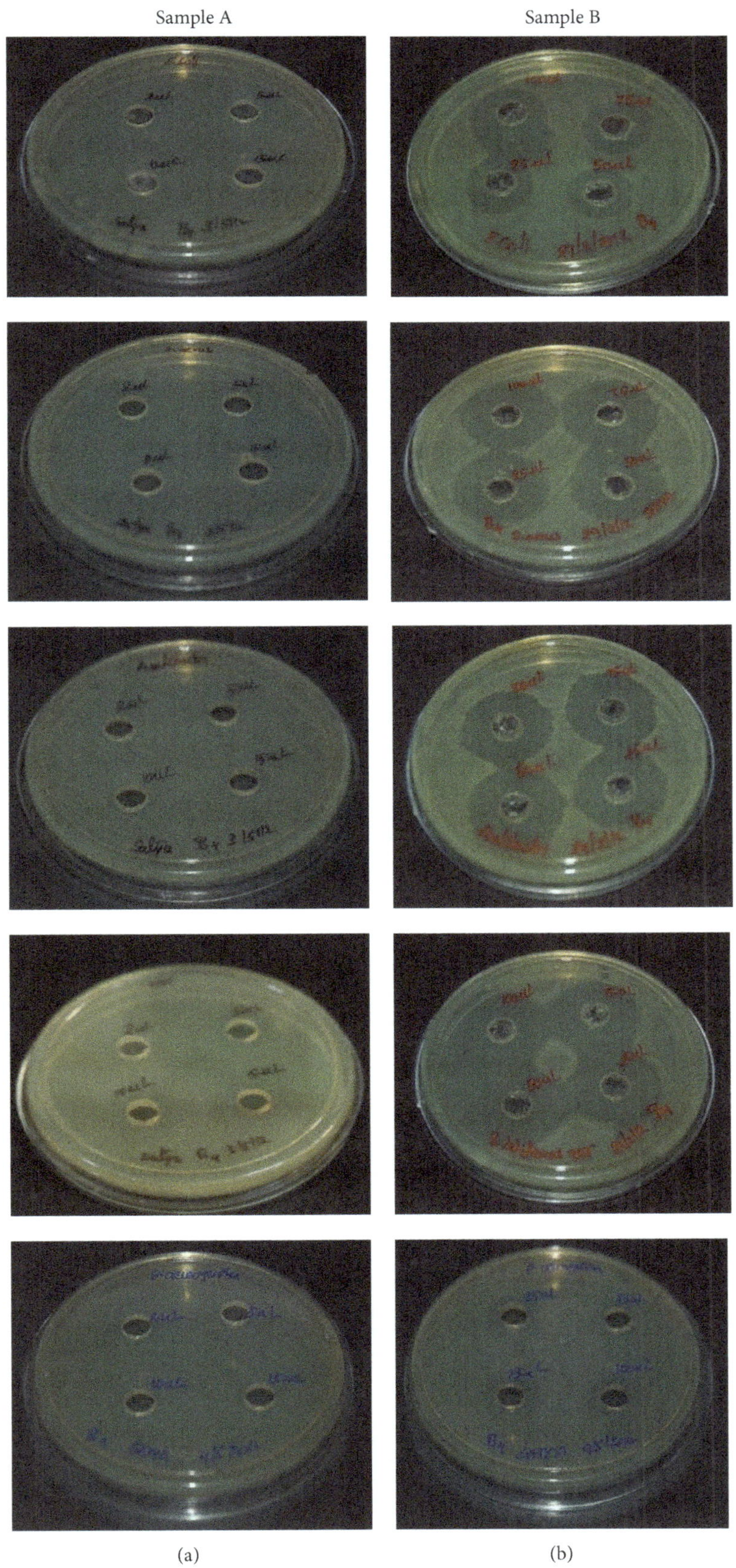


FIGURE 5: Antibacterial effects varying the concentrations of silver nanoparticles samples, (a) lower concentrations (2, 5, 10, and 15 $\mu$L) and (b) Higher concentrations (25, 50, 75, and 100 $\mu$L).

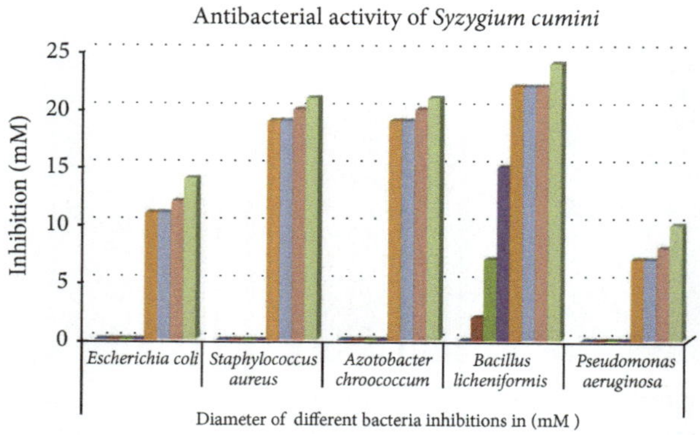


FIGURE 6: Antibacterial activity of *Syzygium cumini*, with different concentrations ranging from 2, 5, 10, 15, 25, 50, 75, and 100 $\mu$L.

The atomic force microscopy (AFM) results display the surface morphology of the monodispersed silver nanoparticles using *S. cumini* bark extract. The particle size of the silver nanoparticles that ranges from 20 to 60 nm was observed. The topographical image of silver nanoparticles indicated that they are agglomerated and formed distinct nanoparticles (Figures 3(a) and 3(b)). The bright spots on the micrograph indicated that the nanoparticles are spherical in shape.

The biosynthesized silver nanoparticles were characterized by scanning electron microscopy for their morphology and size. The SEM micrograph reveals that the synthesized silver nanoparticles have spherical morphology with size range from 20 to 60 nm and also indicated that the particles are well separated showing no agglomeration (Figure 4).

The different species of bacteria show zone of inhibition in the well diffusion method of antimicrobial activity. The different patterns of the zone of inhibitions are observed in Figure 5. Synthesized silver nanoparticles showed antibacterial activity against both Gram positive and negative bacteria (Figure 6). The highest zone of inhibition was observed for *Bacillus licheniformis* even at lower concentration. The exact mechanism of the inhibition of the bacteria is still unknown, but some hypothetical mechanisms show that the inhibition is due to ionic binding of the silver nanoparticles on the surface of the bacteria which creates a great intensity of the proton motive force, and the one hypothesis from the research states that the silver nanoparticles invade the bacterial cell and bind to the vital enzymes containing thiol groups [12, 21, 22]. Also, the findings of Sereemaspun et al. (2008) [23] suggested the inhibition of oxidation-based biological process by penetration of metallic nanosized particles across the microsomal membrane [23, 24]. The molecular basis for the biosynthesis of these silver crystals speculated that the organic matrix contains silver binding properties that provide amino acid moieties that serve as the nucleation sites [25, 26].

## 4. Conclusions

The biological synthesis of the silver nanoparticles is rapid, ecofriendly, cost-effective, and simple method of synthesis. In the present study-silver nanoparticles are synthesized at room temperature within a less span of time. The synthesized silver nanoparticles were characterized by UV-visible spectrometer, AFM, and SEM analysis. The size of the nanoparticles ranges from 20 to 60 nm with spherical shape. AFM and SEM reveal that the synthesized silver nanoparticles are well dispersed showing no agglomeration. These nanoparticles showed a broad spectrum antimicrobial activity against both Gram positive and Gram negative bacteria. Investigation on the antibacterial activity of synthesized silver nanoparticles using *cumini* extract against *Staphylococcus aureus* and *Bacillus licheniformis* reveals high potential as antimicrobial agent in pharmaceutical, food, and cosmetic industries.

## Conflict of Interests

The authors declare that they have no conflict of interests.

## Acknowledgments

The authors are thankful to Drs. Gaurav Raikhy and Ravi Mani Tripathi, Amity University, India, for critically reading the paper and analyzing the data.

## References

[1] E. K. Elumalai, T. N. V. K. V. Prasad, J. Hemachandran, T. S. Viviyan, T. Thirumalai, and E. David, "Extracellular synthesis of silver nanoparticles using leaves of *Euphorbia hirta* and their antibacterial activities," *Journal of Pharmaceutical Sciences and Research*, vol. 2, no. 9, pp. 549–554, 2010.

[2] A. V. Singh, R. Patil, M. B. Kasture, W. N. Gade, and B. L. V. Prasad, "Synthesis of Ag-Pt alloy nanoparticles in aqueous bovine serum albumin foam and their cytocompatibility against human gingival fibroblasts," *Colloids and Surfaces B*, vol. 69, no. 2, pp. 239–245, 2009.

[3] C. Marambio-Jones and E. M. V. Hoek, "A review of the antibacterial effects of silver nanomaterials and potential implications for human health and the environment," *Journal of Nanoparticle Research*, vol. 12, no. 5, pp. 1531–1551, 2010.

[4] M. Zhang, M. Liu, H. Prest, and S. Fischer, "Nanoparticles secreted from ivy rootlets for surface climbing," *Nano Letters*, vol. 8, no. 5, pp. 1277–1280, 2008.

[5] S. Jeong, S. Yeo, and S. Yi, "Antibacterial characterization of silver nanoparticles against *E. coli* ATCC-15224," *Journal of Material Science*, vol. 40, article 5407, 2005.

[6] N. Savithramma, R. M. Linga, K. Rukmini, and D. P. Suvarnalatha, "Antimicrobial activity of silver nanoparticles synthesized by using medicinal plants," *International Journal of ChemTech Research*, vol. 3, no. 3, pp. 1394–1402, 2011.

[7] A. Saxena, R. M. Tripathi, and R. P. Singh, "Biological synthesis of silver nanoparticles by using onion *Allium cepa* extract and their antibacterial activity," *Digest Journal of Nanomaterials and Biostructures*, vol. 5, no. 2, pp. 427–432, 2010.

[8] S. Schultz, D. R. Smith, J. J. Mock, and D. A. Schultz, "Single-target molecule detection with nonbleaching multicolor optical immunolabels," *Proceedings of the National Academy of Sciences of the United States of America*, vol. 97, no. 3, pp. 996–1001, 2000.

[9] K. Vijayaraghavan and S. P. K. Nalini, "Biotemplates in the green synthesis of silver nanoparticles," *Biotechnology Journal*, vol. 5, no. 10, pp. 1098–1110, 2010.

[10] R. M. Crooks, M. Zhao, L. Sun, V. Chechik, and L. K. Yeung, "Dendrimer-encapuslated metal nanoparticles: synthesis, characterization and application to catalysis," *American Chemical Society*, vol. 34, no. 3, pp. 181–190, 2001.

[11] A. Singh, D. Jain, M. K. Upadhyay, N. Khandelwal, and H. N. Verma, "Green synthesis of silver nanoparticles using *Argemone mexicana* leaf extract and evaluation of their antimicrobial activities," *Digest Journal of Nanomaterials and Biostructures*, vol. 5, no. 2, pp. 483–489, 2010.

[12] V. K. Sharma, R. A. Yngard, and Y. Lin, "Silver nanoparticles: green synthesis and their antimicrobial activities," *Advances in Colloid and Interface Science*, vol. 145, no. 1-2, pp. 83–96, 2009.

[13] K. S. Prasad, D. Pathak, A. Patel et al., "Biogenic synthesis of silver nanoparticles using *Nicotiana tobaccum* leaf extract and study of their antibacterial effect," *African Journal of Biotechnology*, vol. 10, no. 41, pp. 8122–8130, 2011.

[14] P. Ram, V. S. Swamy, P. K. Suranjit, and V. Ajit, "Biogenic synthesis of silver nanoparticles from the leaf extract of *Syzygium cumini* (L.)," *International Journal of Pharma and Bio Sciences*, vol. 3, no. 4, pp. 745–752, 2012.

[15] S. Ghosh, S. Patil, M. Ahire et al., "Synthesis of silver nanoparticles using *Dioscorea bulbifera* tuber extract and evaluation of its synergistic potential in combination with antimicrobial agents," *International Journal of Nanomedicine*, vol. 7, pp. 483–496, 2012.

[16] P. S. Vankar and D. Shukla, "Biosynthesis of silver nanoparticles using lemon leaves extract and its applications for antimicrobial finish on fabric," *Applied Nanoscience*, vol. 2, pp. 163–168, 2012.

[17] K. S. Mukunthan, E. K. Elumalai, T. N. Patel, and V. R. Murthy, "*Catharanthus roseus*: a natural source for the synthesis of silver nanoparticles," *Asian Pacific Journal of Tropical Biomedicine*, pp. 270–274, 2011.

[18] N. Ahmad, S. Sharma, M. K. Alam et al., "Rapid synthesis of silver nanoparticles using dried medicinal plant of basil," *Colloids and Surfaces B*, vol. 81, no. 1, pp. 81–86, 2010.

[19] S. S. Shankar, A. Rai, B. Ankamwar, A. Singh, A. Ahmad, and M. Sastry, "Biological synthesis of triangular gold nanoprisms," *Nature Materials*, vol. 3, no. 7, pp. 482–488, 2004.

[20] V. S. Swamy and P. Ram, "Green synthesis of silver nanoparticles from the leaf extract of *Santalum album* and its antimicrobial activity," *Journal of Optoelectronic and Biomedical Materials*, vol. 4, no. 3, pp. 53–59, 2012.

[21] C. Ramteke, T. Chakrabarti, B. K. Sarangi, and R. A. Pandey, "Synthesis of silver nanoparticles from the aqueous extract of leaves of *Ocimum sanctums* for enhanced antibacterial activity," *Journal of Chemistry*, vol. 2013, Article ID 278925, 7 pages, 2013.

[22] S. Kavita, J. Santhanalakshmi, and B. Viswanathan, "Green synthesis of silver nanoparticles using *Polyalthia longifolia* leaf extract along with D-Sorbitol: Study of Antibacterial Activity," *Journal of Nanotechnology*, vol. 2011, Article ID 152970, 5 pages, 2011.

[23] A. Sereemaspun, P. Hongpiticharoen, R. Rojanathanes, P. Maneewattanapinyo, S. Ekgasit, and W. Warisnoicharoen, "Inhibition of human cytochrome P450 enzymes by metallic nanoparticles: a preliminary to nanogenomics," *International Journal of Pharmacology*, vol. 4, no. 6, pp. 492–495, 2008.

[24] R. M. Linga and N. Savithramma, "Antimicrobial activity of silver nanoparticles synthesized by using stem extract of *Svensonia hyderobadensis* (Walp.) mold-a rare medicinal plant," *Research in Biotechnology*, vol. 3, no. 3, pp. 41–47, 2012.

[25] N. Prabhu, T. R. Divya, and G. Yamuna, "Synthesis of silver phyto nanoparticles and their antibacterial efficacy," *Digest Journal of Nanomaterials and Biostructures*, vol. 5, no. 1, pp. 185–189, 2010.

[26] N. Savithramma, M. Linga Rao, K. Rukmini, and P. Suvarnalatha Devi, "Antimicrobial activity of silver nanoparticles synthesized by using medicinal plants," *International Journal of ChemTech Research*, vol. 3, no. 3, pp. 1394–1402, 2011.

# 15

# Sol-Gel Synthesis of $TiO_2/SiO_2$ and $ZnO/SiO_2$ Composite Films and Evaluation of Their Photocatalytic Activity towards Methyl Green

**V. L. Chandraboss, B. Karthikeyan, J. Kamalakkannan, S. Prabha, and S. Senthilvelan**

*Department of Chemistry, Annamalai University, Annamalai Nagar, Tamil Nadu 608 002, India*

Correspondence should be addressed to S. Senthilvelan; dr_senthilvel@yahoo.co.in

Academic Editor: Amir Kajbafvala

The $TiO_2/SiO_2$ and $ZnO/SiO_2$ composite films were prepared by sol-gel dip coating method. The surface morphology and crystal structure of thin films were characterized by means of scanning electron microscopy (SEM) with elementary dispersive X-ray analysis (EDX) and X-ray diffractometer (XRD). Optical properties of films have been investigated using ultraviolet and visible spectroscopy (UV-visible spectroscopy). The photocatalytic activity was established by testing the degradation and decolorization of methyl green (MG) from aqueous solution with artificial UV-light.

## 1. Introduction

Nanomaterials may provide solutions to scientific and ecological challenges in the areas of catalysis, medicine, solar energy conversion, and water treatment [1, 2]. This increasing demand must be accompanied by "green" synthesis methods. In the global efforts to decrease generated hazardous waste, "green" chemistry and chemical processes are progressively integrating with contemporary developments in science and industry. Implementation of these sustainable processes should adopt the 12 fundamental principles of green chemistry [3]. These principles are gear to guide in minimizing the use of dangerous products and maximizing the efficiency of chemical processes. Hence, any synthetic route or chemical process should address these principles by using environmentally benign solvents and nontoxic chemicals [4].

From a biological and physiological point of view, the removal of poisonous chemicals from waste water is currently one of the most important subjects in pollution control. MG is a basic triphenylmethane-type dicationic dye, usually used for staining solutions in medicine and biology [5] and as a photochromophore to sensitize gelatinous films [6]. Triphenylmethane dyes are used widely in the textile industry for dyeing of nylon, wool, cotton, and silk as well as for coloring of waxes, varnish, oil, plastics, and fats.

The application of illuminated semiconductors has been effectively working for the decomposition of variety of organic contaminants in water [7]. The major organic compounds that constitute the industrial wastewater include dyes, phenols, chlorophenols, aliphatic alcohols, aromatics, polymers, and carboxylic acids. Among these, toluene, salicylic acid, and 4-chlorophenol have been identified as a water pollutant arising from numerous sources including paper milling, textile, and cosmetic industries [8], causing bad odor to the water. Hence, the destruction of organic compounds is of considerable interest. Over the years, a large number of semiconductors have been utilized as photocatalysts. The most commonly studied photocatalysts are $TiO_2$ and ZnO [9, 10].

$TiO_2$ films have been broadly studied in photocatalytic degradation of the dyes, because of their low solubility in wastewater, high oxidation efficiency, ecofriendly properties, and avoiding the need for posttreatment separation in a system. Moreover, the coatings propose a benefit of a strong particle-substrate coupling that allows long-term performance. A variety of methods are obtainable for the preparation of nanostructured $TiO_2$-based photocatalytic films,

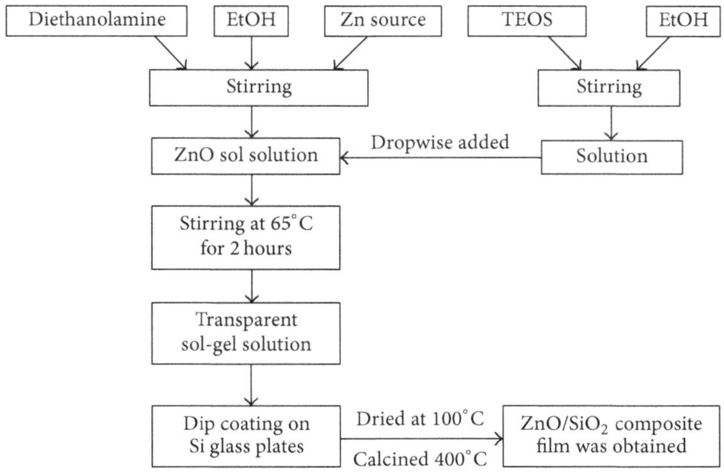


SCHEME 1: Preparation route of $ZnO/SiO_2$ composite film.

such as vacuum arc plasma evaporation [11], electrochemical method [12–14], classical and modified sol-gel methods [15–18], and chemical vapor deposition (CVD) [19, 20].

ZnO has a competitive photocatalytic activity (PA) greater in some cases than $TiO_2$, for example, in the discoloration of Reactive Blue19, a textile anthraquinone dye, in aqueous suspension [21], and in the oxidation of protocatechuic acid [22]. Furthermore, ZnO thin films have been found to decompose aqueous solutions of reactive dyes [23, 24], as well as phenol and chlorophenol [25], and other environmental pollutants [26, 27].

ZnO is available at low cost, which gives it an important advantage. However, the solar UV-light reaching the surface of the earth and available to excited $TiO_2$ is relatively small (around 4%), and artificial UV-light sources are somewhat expensive. The biggest advantage of ZnO is that it absorbs over a larger fraction of the solar spectrum than $TiO_2$ [28, 29]. The biggest advantage of ZnO in comparison with $TiO_2$ is that it absorbs over a larger fraction of UV spectrum and the corresponding threshold of ZnO is 425 nm [30]. For this reason, ZnO is the most suitable photocatalyt for photocatalytic degradation of dyes under UV-light irradiation.

Recently, some researchers synthesized $TiO_2$, ZnO thin films, and composite materials for photocatalytic activity of some industrial pollutants and dyes [31–34]. In this context our research group has successfully prepared $TiO_2/SiO_2$ and $ZnO/SiO_2$ composite films by sol-gel method. Then, the photocatalytic activity of these composites towards is studied.

## 2. Materials and Methods

*2.1. Materials.* Zinc acetate dihydrate [$Zn(CH_3COO)_2\cdot 2H_2O$], titanium tetraisopropoxide [$Ti(iso\text{-}OC_3H_7)_4$], tetraethyl orthosilicate [TEOS], methyl green, diethanolamine, and ethanol were purchased and used as such. The aqueous solutions were prepared by using double distilled water. All glassware was cleaned with chromic acid followed by thorough washing with distilled water. Chemical structure of MG is shown in Figure 1.

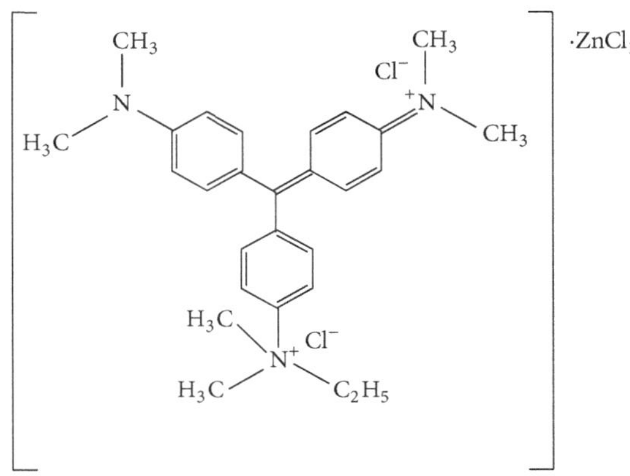


FIGURE 1: Chemical structure of methyl green.

*2.2. Preparation of $TiO_2/SiO_2$ and $ZnO/SiO_2$ Composite Films.* Initially diethanolamine in ethanol was stirred well and then zinc acetate dihydrate was added to the previous solution (solution I). Tetraethyl orthosilicate dissolved in ethanol is taken as another (solution II). The solution II is added to the solution I and stirred (600 rpm) at 65°C for 2 hour. The dip coating was achieved when cleaned glass plates were gradually immersed into the solution. The films were dried at 100°C for 4 hours to evaporate the solvent and to remove the organic residuals. Finally the films were calcinated at 400°C to get uniform $ZnO/SiO_2$ composite film (Scheme 1). The above procedure was also carried out for $TiO_2$-doped $SiO_2$ film.

*2.3. Analytical Methods.* Scanning electron microscopy (SEM) with elementary dispersive X-ray analysis (EDX) experiments was carried out on an FEI Quanta FEG 200 instrument with EDX analyzer facility at 25°C. X-ray diffraction (XRD) spectra were recorded on the X'PERT PRO model X-ray diffractometer from Pan Analytical instruments operated at a voltage of 40 kV and a current of 30 mA with Cu $K\alpha$ radiation. UV-visible absorbance spectra were measured over a range of 200–800 nm with a Shimadzu

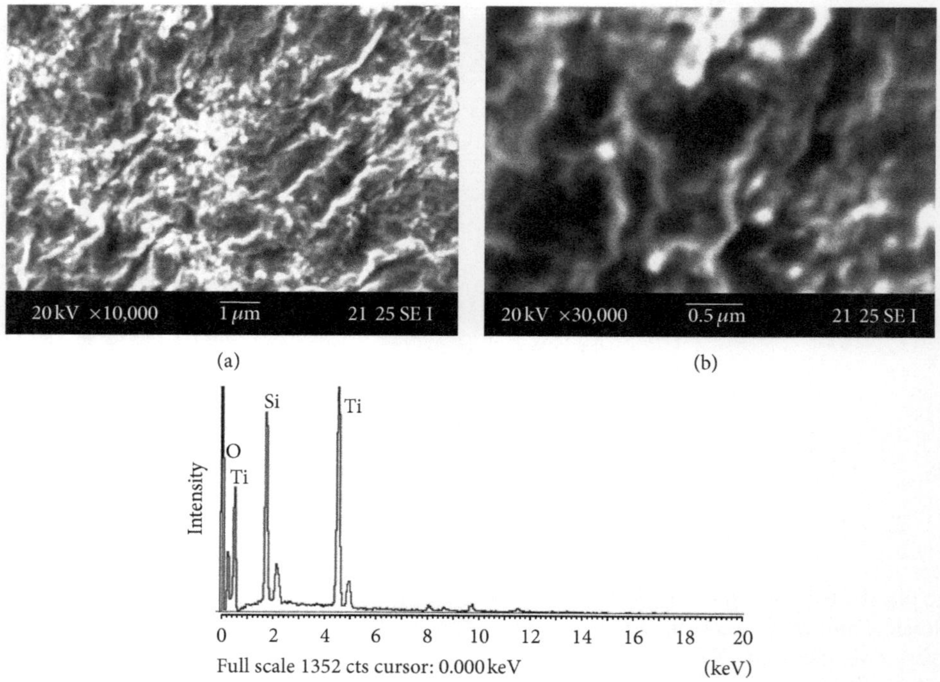


Binding energy

| Element | App. conc. | Intensity corrn. | Weight% |
|---|---|---|---|
| O K | 05.50 | 0.4412 | 12.48 |
| Si K | 02.22 | 0.8276 | 02.69 |
| Ti K | 05.68 | 0.8590 | 06.61 |
| Total | | | 21.78 |

(c)

FIGURE 2: SEM images of $TiO_2/SiO_2$ composite film with scale bars of (a) 1 μm, (b) 0.5 μm, and (c) EDX spectrum of $TiO_2/SiO_2$ composite film.

UV-1650PC recording spectrometer using a quartz cell with 10 mm of optical path length.

*2.4. Photocatalytic Activity.* Photocatalytic studies were carried out in a multilamp photoreactor (HML MP88, supplied by Heber Scientific, India) fitted with eight 8 W mercury UV lamps of wavelength 365 nm. The volume of the reaction solution was maintained as 30 mL. Air was bubbled through the reaction solution for effective stirring. The degradation of MG was followed spectrophotometrically. The solution was diluted to keep the absorbance within the Beer-Lambert law limit.

## 3. Results and Discussion

### *3.1. Characterization of $TiO_2/SiO_2$ and $ZnO/SiO_2$ Composite Films*

*3.1.1. SEM with EDX Analysis.* The morphology of the $TiO_2/SiO_2$ and $ZnO/SiO_2$ composite films was determined by scanning electron microscopy. Figure 2 shows SEM micrograph of $TiO_2/SiO_2$ film annealed at 400°C (scale bars of 1 and 0.5 μm) (Figures 2(a) and 2(b)). EDX analysis confirmed the presence of Ti, Si, and O (Figure 2(c)). Figure 3 shows SEM micrograph of $ZnO/SiO_2$ film annealed at 400°C (scale bars of 1 and 0.5 μm) (Figures 3(a) and 3(b)). EDX analysis confirmed the presence of Zn, Si, and O (Figure 3(c)). Uniform morphology of the prepared composite materials has been confirmed from the previous observation.

*3.1.2. XRD Analysis.* Figure 4(a) shows XRD of $TiO_2$-doped $SiO_2$ composite film. The peaks at 23.5°, 27.0°, 38.0°, 42.4°, and 48.8° are the diffractions of the $TiO_2$ (1 0 1), $SiO_2$ (1 0 1), $TiO_2$ (1 1 2), $SiO_2$ (2 0 0), and $TiO_2$ (2 0 0). The diffractogram consists of broad diffraction peaks of $TiO_2$ anatase phase (JCPDS file no. 84-1286). Figure 4(b) shows XRD of ZnO-doped $SiO_2$ composite film. The peaks at 26.5°, 31.7°, 33.3°, 36.0°, 49.4°, and 68.9° are the diffractions of the $SiO_2$ (1 0 1), ZnO (1 0 0), ZnO (0 0 2), ZnO (1 0 1), $SiO_2$ (1 1 2), and ZnO (2 0 1). The diffractogram consists of broad diffraction peaks of ZnO (JCPDS file no. 36-1451).

*3.1.3. Optical Transmittance Analysis.* Optical transmittance spectra of $TiO_2/SiO_2$ and $ZnO/SiO_2$ composite films

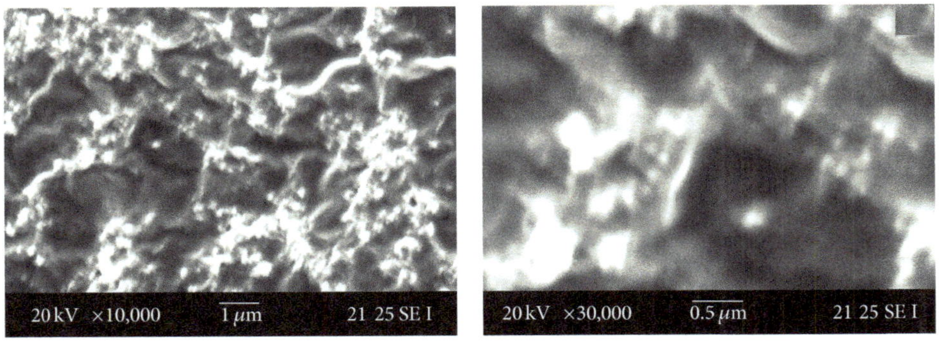


(a) (b)

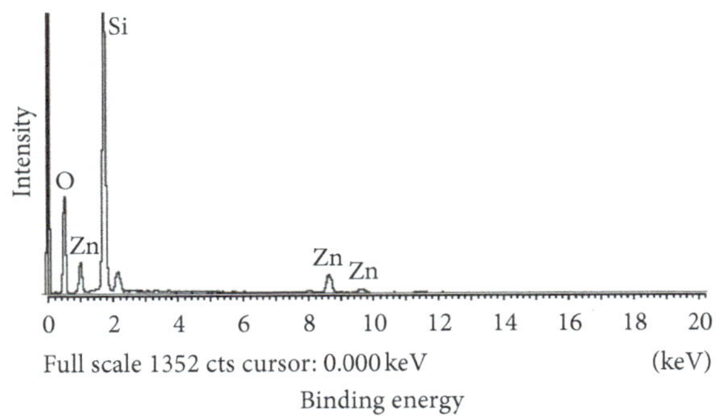


| Element | App. conc. | Intensity corrn. | Weight% |
|---|---|---|---|
| O K | 10.33 | 0.8796 | 11.74 |
| Si K | 07.67 | 0.7576 | 10.13 |
| Zn K | 04.70 | 0.8259 | 05.70 |
| Total | | | 27.56 |

(c)

FIGURE 3: SEM images of $ZnO/SiO_2$ composite film with scale bars of (a) 1 μm, (b) 0.5 μm, and (c) EDX spectrum of $ZnO/SiO_2$ composite film.

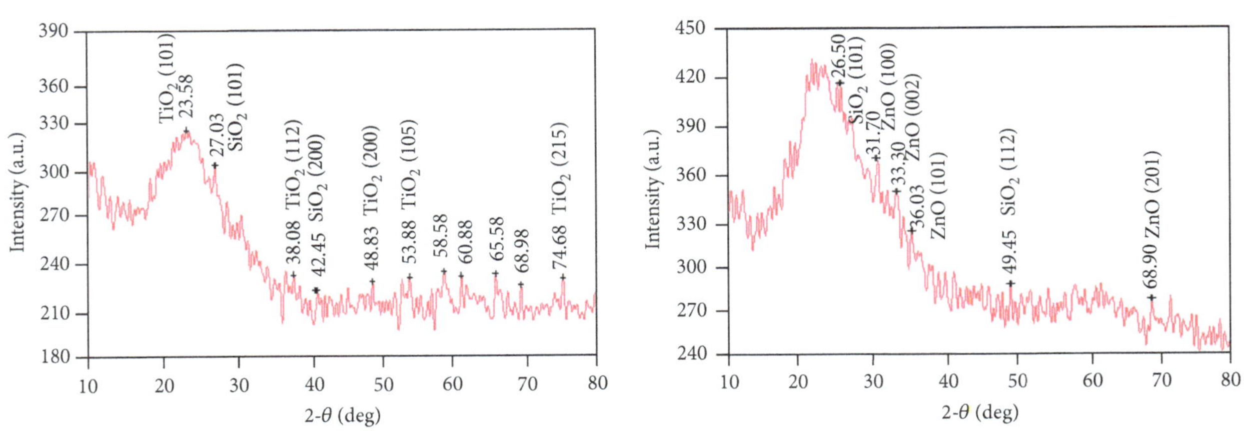


(a) (b)

FIGURE 4: XRD amorphous data of (a) $TiO_2/SiO_2$ composite film and (b) $ZnO/SiO_2$ composite film.

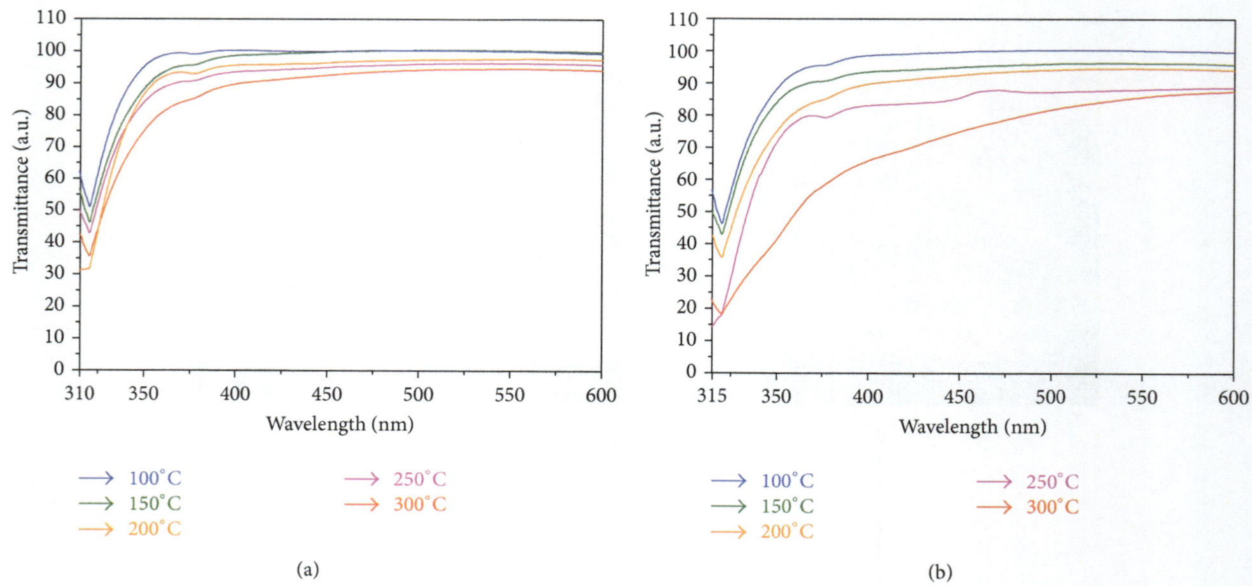


FIGURE 5: Optical transmittance spectra of (a) $TiO_2/SiO_2$ composite film and (b) $ZnO/SiO_2$ composite film.

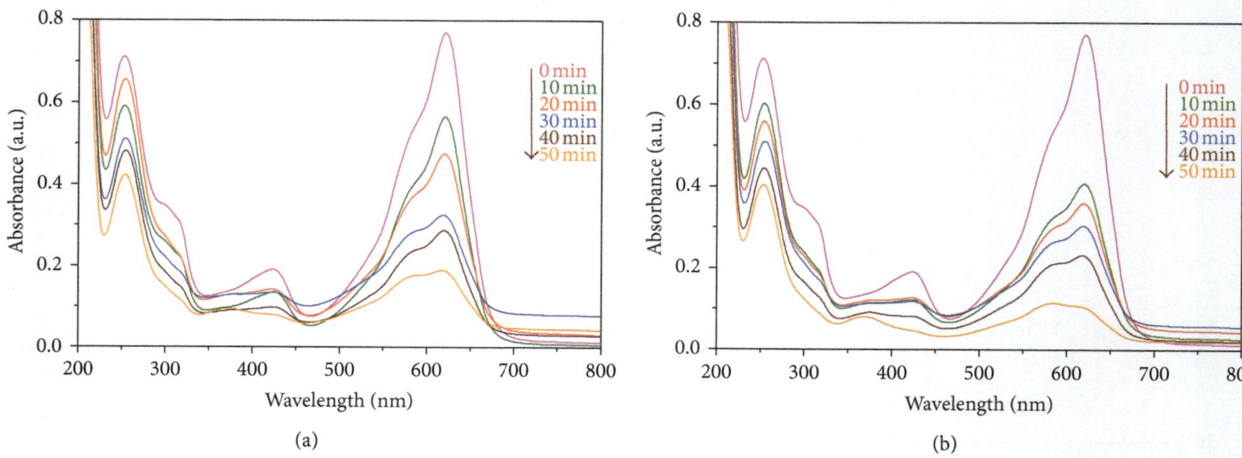


FIGURE 6: UV-visible absorption spectra of methyl green at 10-minute interval: (a) $TiO_2/SiO_2$ composite film and (b) $ZnO/SiO_2$ composite film.

annealed at different temperatures (100°C, 150°C, 200°C, 250°C, and 300°C) are shown in Figure 5. When the films were heated at different temperatures, the transmittance of the films becomes lower as the temperature increases. Transmittance of the $ZnO/SiO_2$ film was slightly lower than the $TiO_2/SiO_2$ with increasing temperature (Figures 5(a) and 5(b)).

### 3.2. Photodegradation and Decolorization of MG

*3.2.1. Photodegradation of Methyl Green with Artificial UV Light.* The photodegradation of MG in aqueous medium in the presence of atmospheric air on $TiO_2/SiO_2$ and $ZnO/SiO_2$ composite films was studied using multilamp photoreactor with mercury UV lamps of wavelength 365 nm. The reference wavelength of MG reaction solution is 630 nm. Initially the dye solution is dark green in color; after the photodegradation and decolorization the color of the solution becomes pale green. The reaction time affords the photodegradtation and decolorization of MG.

*3.2.2. Photocatalytic Activity of $TiO_2/SiO_2$ and $ZnO/SiO_2$ Composite Films.* The absorption spectrum of MG in the presence of $TiO_2/SiO_2$ and $ZnO/SiO_2$ is shown in Figures 6(a) and 6(b), respectively. The absorption spectrum shows the decrease in intensity with respect to time. The rapid decrease in the intensity of $ZnO/SiO_2$ is noted in the photocatalysis compared with $TiO_2/SiO_2$. The enhanced

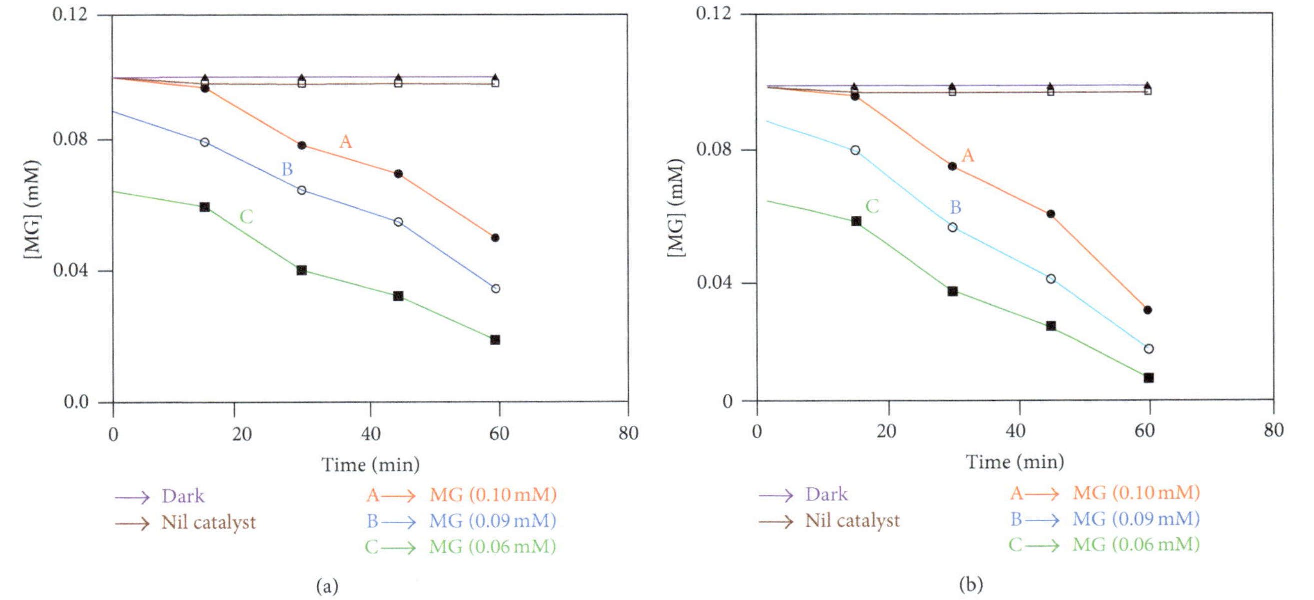


Figure 7: Photodegradation at different concentration of methyl green: (a) $TiO_2/SiO_2$ composite film and (b) $ZnO/SiO_2$ composite film.

photocatalytic activity shown by $ZnO/SiO_2$ is compared with $TiO_2/SiO_2$ composites.

*3.2.3. Effect of Dye Concentration.* The progress of the photodegradation of MG represents increases with light intensity in the presence of $TiO_2/SiO_2$ and $ZnO/SiO_2$ composite films. Different concentrations of MG were prepared and used for the photodegradation process. At lower concentration, the degradation and decolorization of MG are high but at higher concentration, the photocatalytic activity is low (Figure 7). The photodegradation of $TiO_2/SiO_2$ and $ZnO/SiO_2$ is shown in Figures 7(a) and 7(b), respectively. From these figures, it is concluded that the degradation efficiency of $ZnO/SiO_2$ composite film is higher than $TiO_2/SiO_2$ composites only at lower concentration of MG.

## 4. Conclusions

$TiO_2/SiO_2$ and $ZnO/SiO_2$ composite films were prepared by the sol-gel method. The prepared films were characterized by SEM with EDX, XRD, and optical transmittance studies. $TiO_2/SiO_2$ and $ZnO/SiO_2$ films were used for the photodegradation of MG under UV-light irradiation. The photodegradation of MG on $TiO_2/SiO_2$ and $ZnO/SiO_2$ composite films in aqueous medium has been studied as a function of dye concentration. $ZnO/SiO_2$ composite film exhibited reasonable activity under UV-light and had much better activity when compared to that of $TiO_2/SiO_2$ composite films.

## Acknowledgment

S. Senthilvelan is highly thankful to UGC, New Delhi, for granting a major research project (F. no. 39-771/2010(SR)).

## References

[1] J. A. Dahl, B. L. S. Maddux, and J. E. Hutchison, "Toward greener nanosynthesis," *Chemical Reviews*, vol. 107, no. 6, pp. 2228–2269, 2007.

[2] J. E. Hutchison, "Greener nanoscience: a proactive approach to advancing applications and reducing implications of nanotechnology," *ACS Nano*, vol. 2, no. 3, pp. 395–402, 2008.

[3] J. M. DeSimone, "Practical approaches to green solvents," *Science*, vol. 297, no. 5582, pp. 799–803, 2002.

[4] P. Raveendran, J. Fu, and S. L. Wallen, "Completely "Green" synthesis and stabilization of metal nanoparticles," *Journal of the American Chemical Society*, vol. 125, no. 46, pp. 13940–13941, 2003.

[5] F. J. Green, *The Sigma-Aldrih Handbook of Stains, Dyes, and Indicators*, Aldrich Chemical, Milwaukee, Wis, USA, 1990.

[6] T. Geethakrishnan and P. K. Palanisamy, "Degenerate four-wave mixing experiments in Methyl green dye-doped gelatin film," *Optik*, vol. 117, no. 6, pp. 282–286, 2006.

[7] M. A. Behnajady, N. Modirshahla, and R. Hamzavi, "Kinetic study on photocatalytic degradation of C.I. Acid Yellow 23 by ZnO photocatalyst," *Journal of Hazardous Materials*, vol. 133, no. 1–3, pp. 226–232, 2006.

[8] A. Mills, C. E. Holland, R. H. Davies, and D. Worsely, "Photo mineralization of salicylic acid: a kinetic study," *Journal of Photochemistry and Photobiology A*, vol. 83, pp. 257–263, 1994.

[9] S. K. Kansal, M. Singh, and D. Sud, "Studies on photodegradation of two commercial dyes in aqueous phase using different photocatalysts," *Journal of Hazardous Materials*, vol. 141, no. 3, pp. 581–590, 2007.

[10] Y. Wei, Y. Huang, J. Wu et al., "Synthesis of hierarchically structured ZnO spheres by facile methods and their photocatalytic deNOx properties," *Journal of Hazardous Materials*, vol. 248, pp. 202–210, V.

[11] T. Miyata, S. Tsukada, and T. Minami, "Preparation of anatase $TiO_2$ thin films by vacuum arc plasma evaporation," *Thin Solid Films*, vol. 496, no. 1, pp. 136–140, 2006.

[12] S. Wang, X. Wu, W. Qin, and Z. Jiang, "$TiO_2$ films prepared by micro-plasma oxidation method for dye-sensitized solar cell," *Electrochimica Acta*, vol. 53, no. 4, pp. 1883–1889, 2007.

[13] X. H. Wu, Z. H. Jiang, L. H. Liu, X. D. Li, and X. G. Hu, "$TiO_2$ ceramic films prepared by micro-plasma oxidation method for photodegradation of rhodamine B," *Materials Chemistry and Physics*, vol. 80, no. 1, pp. 39–43, 2003.

[14] S. Karuppuchamy, N. Suzuki, S. Ito, and T. Endo, "Enhanced efficiency of dye sensitized solar cells by UV-O3 treatment of $TiO_2$ layer," *Current Applied Physics*, vol. 9, pp. 243–248, 2009.

[15] D. L. Liao, C. A. Badour, and B. Q. Liao, "Preparation of nanosized $TiO_2$/ZnO composite catalyst and its photocatalytic activity for degradation of methyl orange," *Journal of Photochemistry and Photobiology A*, vol. 194, no. 1, pp. 11–19, 2008.

[16] X. Zhang and Q. Liu, "Preparation and characterization of titania photocatalyst co-doped with boron, nickel, and cerium," *Materials Letters*, vol. 62, no. 17-18, pp. 2589–2592, 2008.

[17] L. Andronic and A. Duta, "The influence of $TiO_2$ powder and film on the photodegradation of methyl orange," *Materials Chemistry and Physics*, vol. 112, no. 3, pp. 1078–1082, 2008.

[18] I. Stambolova, V. Blaskov, I. N. Kuznetsova, N. Kostova, and S. Vassilev, "Effect of the thermal treatment on the morphology and optical properties of nanosized $TiO_2$ films," *Journal of Optoelectronics and Advanced Materials*, vol. 13, pp. 381–386, 2011.

[19] B. H. Kim, J. Y. Lee, Y. H. Choa, M. Higuchi, and N. Mizutani, "The effect of target density and its morphology on $TiO_2$ thin films," *Materials Science and Engineering B*, vol. 107, pp. 289–294, 2004.

[20] T. K. Ghorai, D. Dhak, S. K. Biswas, S. Dalai, and P. Pramanik, "Photocatalytic oxidation of organic dyes by nano-sized metal molybdate incorporated titanium dioxide ($M_xMo_xTi_{1-x}O_6$) (M = Ni, Cu, Zn) photocatalysts," *Journal of Molecular Catalysis A*, vol. 273, no. 1-2, pp. 224–229, 2007.

[21] I. Poulios, D. Makri, and X. Prohaska, "Photocatalytic treatment of olive milling wastewater: oxidation of protocatechuic acid," *GLOBAL NEST*, vol. 1, pp. 55–62, 1999.

[22] S. Sakthivel, B. Neppolian, M. V. Shankar, B. Arabindoo, M. Palanichamy, and V. Murugesan, "Solar photocatalytic degradation of azo dye: comparison of photocatalytic efficiency of ZnO and $TiO_2$," *Solar Energy Materials and Solar Cells*, vol. 77, no. 1, pp. 65–82, 2003.

[23] C. A. K. Gouvêa, F. Wypych, S. G. Moraes, N. Durán, N. Nagata, and P. Peralta-Zamora, "Semiconductor-assisted photocatalytic degradation of reactive dyes in aqueous solution," *Chemosphere*, vol. 40, no. 4, pp. 433–440, 2000.

[24] C. Lizama, J. Freer, J. Baeza, and H. D. Mansilla, "Optimized photodegradation of reactive blue 19 on $TiO_2$ and ZnO suspensions," *Catalysis Today*, vol. 76, no. 2–4, pp. 235–246, 2002.

[25] S. Lathasree, A. N. Rao, B. Sivasankar, V. Sadasivam, and K. Rengaraj, "Heterogeneous photocatalytic mineralisation of phenols in aqueous solutions," *Journal of Molecular Catalysis A*, vol. 223, no. 1-2, pp. 101–105, 2004.

[26] L. B. Khalil, W. E. Mourad, and M. W. Rophael, "Photocatalytic reduction of environmental pollutant Cr(VI) over some semiconductors under UV/visible light illumination," *Applied Catalysis B*, vol. 17, no. 3, pp. 267–273, 1998.

[27] Y. Huang, Y. Wei, J. Wu et al., "Low temperature synthesis and photocatalytic properties of highly oriented ZnO/$TiO_2$–xNy coupled photocatalysts," *Applied Catalysis B*, vol. 123, pp. 9–17, 2012.

[28] S. Sakthivel, B. Neppolian, M. Palanichamy, B. Arabindoo, and V. Murugesan, "Photocatalytic degradation of leather dye, Acid green 16 using ZnO in the slurry and thin film forms," *Indian Journal of Chemical Technology*, vol. 6, no. 3, pp. 161–165, 1999.

[29] I. Poulios and I. Tsachpinis, "Photodegradation of the textile dye Reactive Black 5 in the presence of semiconducting oxides," *Journal of Chemical Technology and Biotechnology*, vol. 74, pp. 349–357, 1999.

[30] S. S. Shinde, P. S. Shinde, C. H. Bhosale, and K. Y. Rajpure, "Zinc oxide mediated heterogeneous photocatalytic degradation of organic species under solar radiation," *Journal of Photochemistry and Photobiology B*, vol. 104, no. 3, pp. 425–433, 2011.

[31] I. Stambolova, M. Shipochka, V. Blaskov, A. Loukanovb, and S. Vassilevc, "Sprayed nanostructured $TiO_2$ films for efficient photocatalytic degradation of textile azo dye," *Journal of Photochemistry and Photobiology B*, vol. 117, pp. 19–26, 2012.

[32] G. Torres Delgado, C. I. Zúñiga Romero, S. A. Mayén Hernández, R. Castanedo Péreza, and O. Zelaya Angel, "Optical and structural properties of the sol- gel-prepared ZnO thin films and their effect on the photocatalytic activity," *Solar Energy Materials and Solar Cells*, vol. 93, no. 1, pp. 55–59, 2009.

[33] X. Zhang and H. Zheng, "Synthesis of $TiO_2$-doped $SiO_2$ composite films and its applications," *Bulletin of Materials Science*, vol. 32, pp. 787–790, 2008.

[34] A. Mahyar, M. A. Behnajady, and N. Modirshahla, "Enhanced photocatalytic degradation of C.I. Basic violet 2 using $TiO_2$-$SiO_2$ composite nanoparticles," *Photochemistry and Photobiology*, vol. 87, no. 4, pp. 795–801, 2011.

# 16

# Electrochemical Method for Ag-PEG Nanoparticles Synthesis

**M. Virginia Roldán, Nora Pellegri, and Oscar de Sanctis**

*Laboratorio de Materiales Cerámicos, FCEIyA, IFIR, Universidad Nacional de Rosario (UNR), Avenida Pellegrini 250, S2000BTP Rosario, Argentina*

Correspondence should be addressed to M. Virginia Roldán; vroldan@fceia.unr.edu.ar

Academic Editor: Mohammad H. Maneshian

In this work we present an electrochemical method to successfully prepare silver nanoparticles using only polyethylene glycol as stabilizer and without any other reactive. Here we study the use of the polymeric stabilizer to allow the introduction of a potential tool to reinforce the control of the size and shape of the nanoparticles throughout the synthesis process. The evolution of the reactions was followed by UV-Vis spectroscopy. The electrode processes were characterized by cyclic voltammetric measurements and the final product was studied by Atomic Force Microscopy, Transmission Electron Microscopy, and X-Ray Diffraction. The influences of the current density, polymer length, and concentration media were analyzed.

## 1. Introduction

Noble metal nanoparticles have been intensely investigated due to their amazing properties such as optical, catalytic, and electric ones that can be controlled depending on the particles size, the size distribution, and shape [1]. Metallic silver particles in particular are technologically important because they show unique properties normally related to noble metals (excellent conductivity [2], chemical stability, nonlinear optical behavior [3, 4], etc.) besides other specific ones (catalytic activity [5], antibacterial action [6]). These characteristics make them suitable for a variety of potential applications on several devices [7–9].

Several methods have been reported for Ag nanoparticles synthesis, including Ag ions chemical reduction in aqueous solutions with or without stabilizing agents [10–13], thermal decomposition in organic solvents [14, 15], biochemical reduction [16, 17], chemical and photo reduction in reverse micelles [18–20], "nanosphere lithography" (NSL) [21], electrochemical reduction [22–27], irradiation reduction [28], microwave assisted [29–31] and chemical reduction in nonaqueous solvents with surface modifiers [32, 33]. Each method has typical advantages and disadvantages.

Particularly, the electrochemical techniques are quite interesting because they allow obtaining particles with a high purity using fast and simple procedures and controlling the particle size easily by adjusting the current density [22, 23]. Through these techniques particles have been obtained with determined size and shape [24, 25] of several compositions [26, 27, 34–36]. Also, this method is eco-friendly because it avoids the use of reducer agents that usually are toxic.

Different stabilizers have been used in electrochemical techniques, which include organic monomers as electrostatic stabilizers [26] and polymeric compounds as steric stabilizers [37]. Ionic organic compounds can act as stabilizers and support electrolytes simultaneously avoiding the use of additional chemicals [26]. However, some authors showed that steric stabilizers allow obtaining more stable nanoparticles than electrostatic stabilizers [38]. Also, chemical reduction techniques assisted by polymers allow controlling the shape and size of nanoparticles by changing the polymer chain length [39, 40]. On the other hand, some authors showed that polyethylene glycol-stabilized silver nanoparticles are highly stable and sufficiently permeable to allow the diffusion of small molecules [41]. Thus, they are useful to several applications as Surface-enhanced Raman Scattering techniques and others. Then, we propose the use of the polymeric stabilizer polyethylene glycol to obtain stable nanoparticles and we study their ability to control the nanoparticles morphology.

In this work we present an electrochemical method to prepare silver nanoparticles using polyethylene glycol as stabilizer. We study the effect of the stabilizer concentration, polymer chain length, and current density over the particle size and shape. These are common strategies for the control of the particle morphology, but they have dissimilar behavior at different systems, mainly the stabilizer concentration. Also, the use of a polymeric stabilizer allows the introduction of a potential tool to reinforce the control of the size and shape of the nanoparticles by using different polymer chains lengths and so we study this strategy too.

## 2. Experimental Section

*2.1. Electrochemical Synthesis of Silver Nanoparticles.* Silver nanoparticles were prepared by electrochemical reduction within a simple two-electrode type cell. The volume of the electrolysis cell was 50 ml. A platinum sheet (0.5 × 7 cm) was employed as cathode and a platinum rod (0.1 × 7 cm) as anode, the two being 3 cm apart. The cell was maintained into an ultrasonic bath. Silver nitrate ($AgNO_3$, Merck) was used as silver ion precursor and polyethylene glycol (PEG, Aldrich) with different chains lengths $M_w$ = 400 (PEG-400), 600 (PEG-600), 1450 (PEG-1450), 2000 (PEG-2000), and 6000 (PEG-6000) which was employed as stabilizing agent.

In a typical synthesis, the reaction medium was prepared to obtain an aqueous solution of silver nitrate 2.5 mM with 0.5 to 2% w/v of PEG. The solution was mixed and purged with $N_2$ during 20 minutes. Then, the electrolysis was carried out under ultrasonication at constant current and $N_2$ atmosphere during 30 minutes. The current chosen (7 mA, 10 mA, or 13 mA) was given by adjusting the applied potential.

*2.2. Characterization of Silver Nanoparticles.* In order to study the optical behavior of silver colloid, ultraviolet-visible (UV-Vis) absorption spectra of different samples were recorded by a Jasco V-530 UV-Vis spectrophotometer using as reference a corresponding blank sample.

The shapes and sizes of the nanoparticles were determined by Atomic Force Microscopy (AFM) and Transmission Electron Microscopy (TEM). AFM images were taken using NanoTec ELECTRONICA equipment in tapping mode configuration with a $Si_3N_4$ tip in air at room temperature. In order to prepare the sample for the AFM study, the particles were redissolved in ethanol and drop-cast onto a freshly peeled HOPG substrate. TEM images were taken with Philips EM 301 equipment. Samples were prepared by placing a drop over a TEM grid and leaving it to evaporate the solvent. To study the crystalline structure of the nanoparticles, X-ray Diffraction patterns were recorded by grazing incidence with a 2° incidence angle with an X'Pert Phillips PW 1700 diffractometer using CuK radiation (1.5405 Å) and a graphite monochromator (the step size was of $2\theta = 0.2°$; 10 s/step) for samples prepared over $SiO_2$ by drop-cast.

The cyclic voltammetric measurements (CV) were performed with an EG&G potentiostat/galvanostat in a conventional three-electrode cell at room temperature. A platinum rod embedded in epoxy resin was used as a working electrode and only its cross section was allowed to contact the electrolytes; a platinum microelectrode and a saturated Ag/AgCl electrode were used as the auxiliary electrode and the reference electrode, respectively.

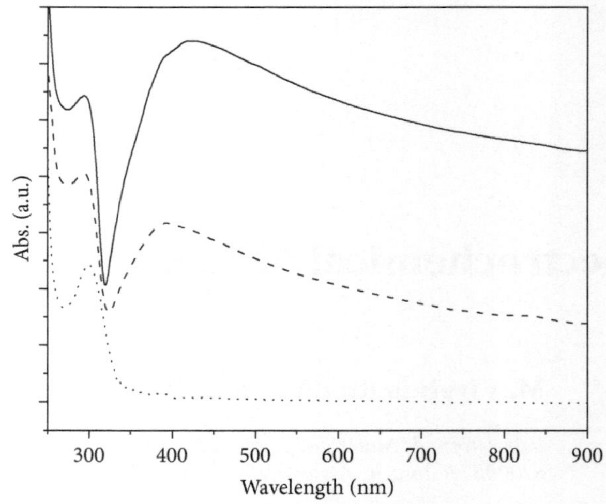


FIGURE 1: Absorption spectra of silver nanoparticles prepared using PEG-2000 at a concentration of 1% w/v with a current equal to 10 mA. Initial spectrum (dotted line) and spectra after 15 (dashed line) and 30 minutes of reaction (solid line).

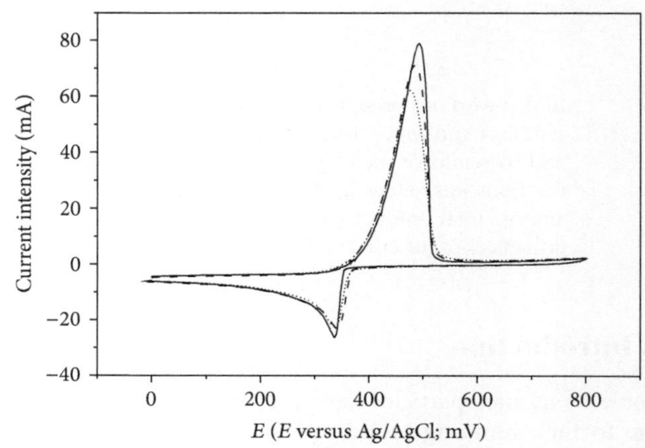


FIGURE 2: Cyclic voltammograms of 2.5 mM $AgNO_3$ aqueous solutions with 1% w/v (dashed line), 2% w/v (dotted line), and without PEG-2000 (solid line).

## 3. Results and Discussion

*3.1. Characterization of the Reaction Progress.* The reaction mixture was characterized in different stages of the reaction progress by UV-Vis spectroscopy. Figure 1 shows the evolution of the optical properties for silver nanoparticles prepared by electrochemical reduction using PEG-2000 as stabilizer at a concentration of 1% w/v and applying a current equal to 10 mA. At the initial time, the absorption spectrum exhibits a single shoulder at a wavelength near to 300 nm, which was attributed to ions $NO_3^-$ in aqueous solution [42] with a lower contribution of $Ag^+$ [43]. After 15 minutes of reaction, a broad band appeared at 392 nm and its intensity increased until an

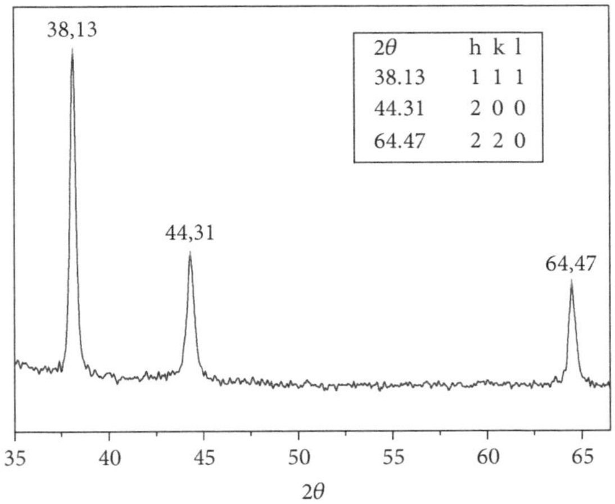


| 2θ | h k l |
|---|---|
| 38.13 | 1 1 1 |
| 44.31 | 2 0 0 |
| 64.47 | 2 2 0 |

FIGURE 3: X-Ray Diffraction pattern of silver nanoparticles. The principal position peaks are shown in the figure.

absorption maximum centered at 425 nm which was obtained after 30 minutes of reaction. This latter absorption band corresponds to the surface plasmon resonance of spherical silver nanoparticles [44]. An absorption minimum appears at 320 nm, which corresponds to a minimum in the imaginary part of the refractive index for bulk silver [45]. In a control experiment a similar solution was placed in the ultrasonic bath with nitrogen atmosphere, but no one current was applied. After 50 minutes no spectral change was observed (not shown here); then reducing action by PEG-2000 was ruled out in these conditions.

In the electrochemical synthesis of silver nanoparticles, there is a competition between two different cathode surface processes: the silver particle formation and the silver deposition on cathode [46]. The key to electrochemical synthesis of silver nanoparticles in aqueous medium is to avoid formation of silver plating and to force reduced particles to leave the cathode surface. In order to test the PEG influence over these processes, CV experiments were carried out [37]. Figure 2 shows the CV in silver nitrate solutions with different amounts of PEG. All CV experiments present one reduction peak near 337.4 mV and one oxidation peak near 492.0 mV approximately. The reduction peak is attributed to silver ions reduction and their electrodeposition on the platinum surface; the oxidation peak is related to oxidation of the silver electrodeposition on the electrode [42]. CV show that the reduction peak intensity is the same for all the cases. The oxidation peak intensity decreases when the PEG amount increases and this is attributed to the PEG influence over the electrode processes. In the PEG-free solution the reduced silver was almost completely deposited on the surface of working electrode and then, when the scan was swept in the oxidation sense, almost all the reduced silver was oxidized. On the other hand, in the PEG containing solution, the silver ions reduced were divided: a part was deposited on the electrode surface whereas another part was reduced to nanoparticles and stabilized by PEG. Thus, when the scan was swept in the oxidation sense, only the deposited silver was oxidized again. So, CV results show that PEG effectively enhances the nanoparticles formation and reduces the electrodeposition on the electrode surface.

*3.2. Characterization of Silver Nanoparticles.* The nanoparticles were characterized by XRD, AFM microscopy, and UV-Vis spectroscopy. In Figure 3 the XRD pattern of nanoparticles, is shown where three distinct diffraction peaks were observed at $2\theta$ values of 38.13°, 44.31°, and 64.47°, corresponding to the (111), (200), and (220) crystalline planes of cubic Ag, respectively. The broad nature of the XRD peaks is attributed to the nanosize of the particles [47].

In order to determine the size and shape of the particles, AFM images were registered. Figure 4 shows a typical AFM image. The particles are uniformly spherical in shape with an average diameter of 30.36 nm. The average diameter was determined by statistical analysis from histograms of several AFM images of the nanoparticles. This average diameter is near to the value obtained from the XRD spectrum. An average nanoparticle size of 45.85 nm was calculated from the breadth of (111) diffraction line at its half-intensity (Figure 3), with an instrumental broadening of 0.25°, using the Scherrer equation [48]. If we compare the mean sizes of the Ag nanoparticles estimated by microscopic technique and XRD, it seems that the Scherrer calculus overestimated the radius. This fact is due to the fact that the sample is polydisperse and the XRD is particularly sensitive to the largest particles or crystallites [49]; then, this result seems to be reasonable.

TEM images (Figure 5) show mainly spherical particles between 10 and 30 nm of size and a few nanospheres bigger, with diameter of 200 nm approximately. These last ones would be responsible for the dispersion effect that is observed as a mild decline of the signal towards longer wavelengths in the UV-Vis extinction spectra.

Furthermore, the morphological characteristics of the particles during different experiments stages were followed by the study of the UV-Vis spectra taking into account the knowing relations between the number and position of the absorption bands and the size and shape of the particles [50]. In silver nanospheres with diameter shorter than 100 nm only one peak in the UV-Vis spectra is observed between 400 and 500 nm [51].

In metallic nanoparticles, the size dependence of the surface plasmon absorption is not easily explained as in the case of semiconductor nanoparticles, where a blue shift or a red shift of the absorption onset undoubtedly results from decreasing size or increasing size, respectively.

The peak position of surface resonance is not well suited for discussion of size effect within the intrinsic size region in metallic nanoparticles [52]. However, the size dependence of the optical spectra of large nanoparticles ($2R > 25$ nm) is an extrinsic particle-size effect governed only by dimension of the particle with respect to the wavelength of the light, and therefore a red shift of the peak position of surface resonance undoubtedly results from increasing size [43]. This relationship is used to obtain qualitative information about relative media size between different samples.

*3.3. Influence of Synthesis Conditions.* In order to determine the influence of the current density, PEG concentration and

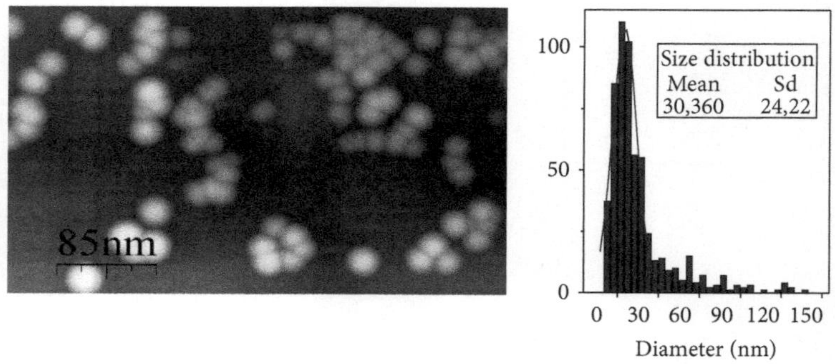


(a) (b)

FIGURE 4: AFM image of PEG protected silver nanoparticles (a) and the corresponding size distribution analysis (b).

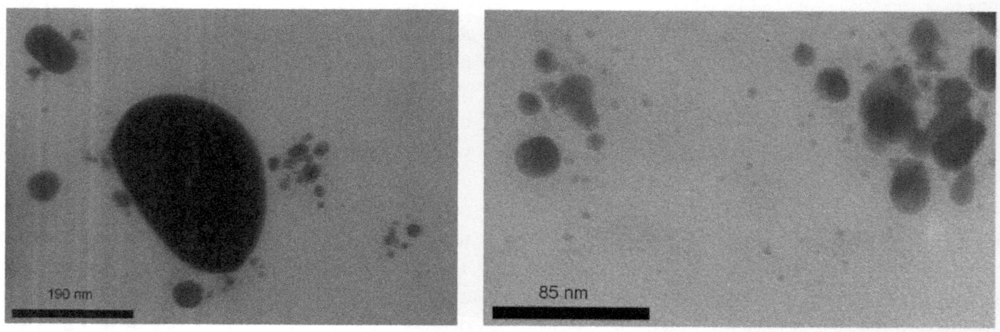


(a) (b)

FIGURE 5: TEM images of PEG protected silver nanoparticles.

polymer chain length on the size and shape of the Ag nanoparticles several experiments were performed to study one by one each parameter of the experimental conditions. After each experiment, a UV-Vis spectrum was taken to analyze the number and position of the absorption bands and then to infer the nanoparticles shape and relative size.

*3.3.1. Current Density.* Reetz and Helbig determined that the particle size of Pd clusters obtained by electrochemical reduction can be controlled by variation of the current density [53]. In order to determine if the current density has some effect on silver particle size, the synthesis was repeated using currents of 7, 10, and 13 mA, respectively, whereas the other conditions were the same for these samples: 2.5 mM $AgNO_3$ and 1% w/v PEG-2000. Figure 6 shows the UV-Vis spectra of the obtained colloids. A blue shift in the plasmon peak can be observed as the current is increased, from 454 nm for the sample prepared at 7 mA to 405 nm for the sample prepared at 13 mA. This infers that for our samples the use of higher current density during the synthesis produces small particles. This behavior is similar to other reports by several authors for different systems with other metallic ions or stabilizers [45, 53].

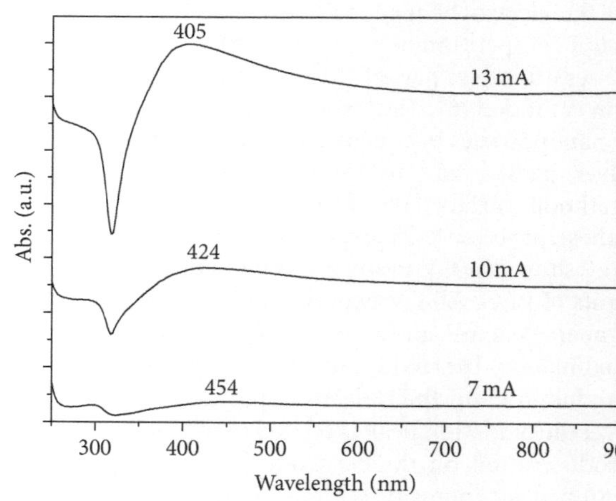


FIGURE 6: UV-Vis spectra of Ag nanoparticles obtained by using different current density during the synthesis process. From the bottom to the top: 7 mA, 10 mA and 13 mA. In the figure, the plasmon position is indicated for each spectrum.

*3.3.2. PEG Concentration.* The influence of PEG-2000 concentration on the silver particle size was studied through UV-Vis spectroscopy. The synthesis was repeated using PEG-2000 concentration of 0.5, 1, 1.5, and 2% w/v. Figure 7 shows the UV-Vis spectra for the colloids obtained with different PEG-2000 amounts.

A displacement of the maximum of absorption was observed towards the blue with the increase of the concentration of PEG-2000 in the range between 0.5 and 2% w/v. Then, increasing the PEG concentration in the reaction medium,

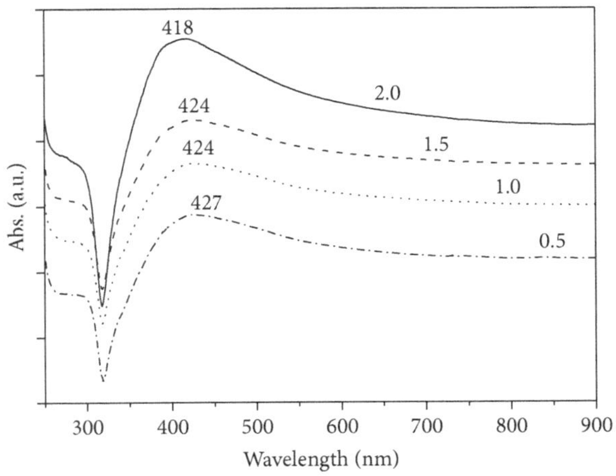


FIGURE 7: UV-Vis spectra of Ag nanoparticles obtained by using different polymer concentrations of PEG-2000 during the synthesis process. From the bottom to the top: 0.5, 1.0, 1.5, and 2.0% w/v. In the figure, the plasmon position is indicated for each spectrum.

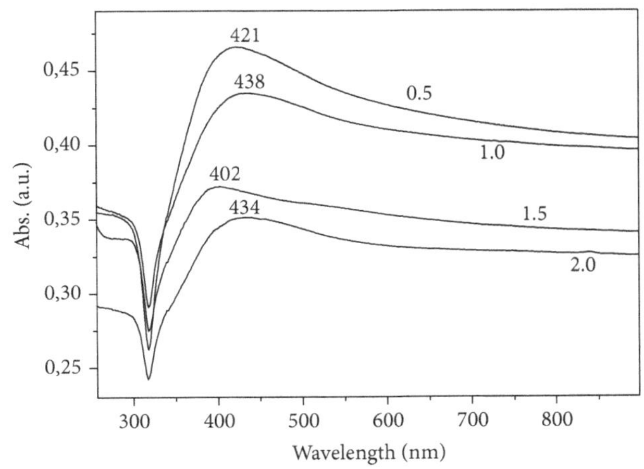


FIGURE 9: UV-Vis spectra of Ag nanoparticles obtained by using different polymer concentration of PEG-600 during the synthesis. From the bottom to the top: 0.5, 1.0, 1.5, and 2.0% w/v. In the figure, the plasmon position is indicated for each spectrum.

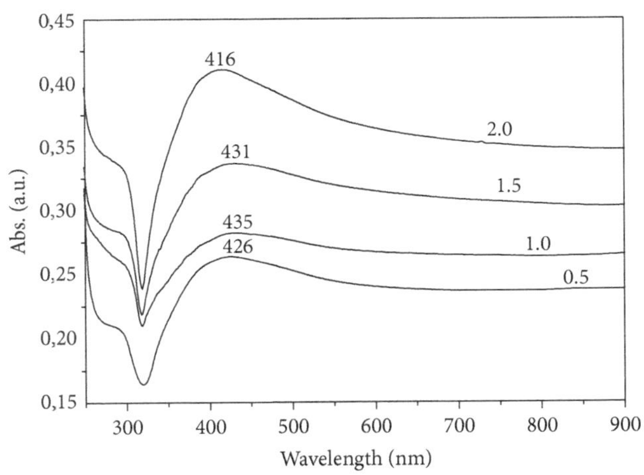


FIGURE 8: UV-Vis spectra of Ag nanoparticles obtained by using different polymer concentration of PEG-6000 during the synthesis process. From the bottom to the top: 0.5, 1.0, 1.5, and 2.0% w/v. In the figure, the plasmon position is indicated for each spectrum.

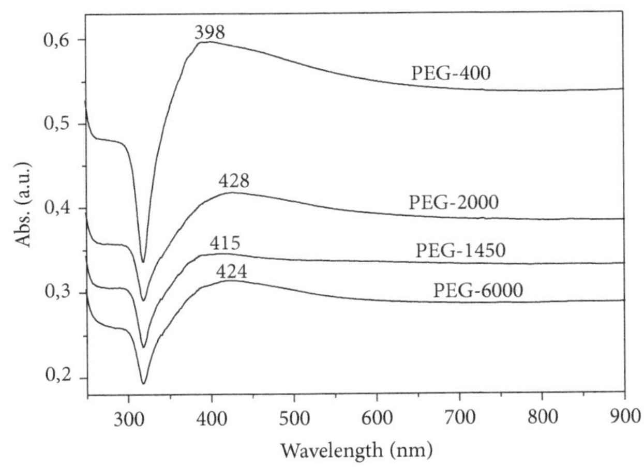


FIGURE 10: UV-Vis spectra of Ag nanoparticles obtained by using different polymer length during the synthesis process. From the bottom to the top: PEG-6000, PEG-1450, PEG-2000, and PEG-400. In the figure, the plasmon position is indicated for each spectrum.

the particle size obtained decreases. Then, it is concluded that it is possible to reinforce the control over the particle size by adjusting the PEG-2000 concentration. This behavior was similar when the same experiment was repeated using PEG-6000 in the range between 1 and 2% w/v (Figure 8). Meanwhile with PEG-600 an unsystematic behavior of the position of the maximum of absorption in function of the wavelength was found (Figure 9), and the same to PEG-400 (not shown here). Then, it is concluded that it is possible to control the particle size by adjusting the PEG-2000 and PEG-6000 concentration between 1 and 2% w/v.

*3.3.3. PEG Chain Length.* In some cases, when polymeric stabilizers are used, just by changing the chain length of the stabilizer, variations in the form of particles can be produced [39]. The effect of PEG chain length on the particle size was also investigated by UV-Vis spectroscopy. Polymers with several molecular weights were used: $M_w$ = 400, 1450, 2000, and 6000. Nevertheless, in all our experiments only one maximum of absorption between 400 and 430 nm appears to be assigned to the spherical appearance of nanoparticles (Figure 10). Then, it is concluded that the $M_w$ did not influence the form of particles and nanospheres are obtained for all the cases.

With all the studies shown during these investigations we obtained stable nanoparticles using PEG as stabilizer and we found the correct experimental parameters that control the morphology. These nanoparticles have the possibility to be used for several purposes, such as bactericide, because they could be manipulated as a powder.

## 4. Conclusions

Spherical silver nanoparticles were synthesized by an electrochemical method with PEG as stabilizer. This method is a rapid and eco-friendly technique to obtain silver nanoparticles. PEG is an efficient stabilizer that favours the particle formation over silver deposition on the cathode. The influence of synthesis conditions was studied. The current density is an adequate parameter to control particle size. The PEG chain length does not show effects on the shape of the nanoparticles obtained, which are spherical for all the cases. Nevertheless, with the longer polymer chain length it is possible to control the particle size through the PEG concentration. Thus, the use of polymeric stabilizer of long chain adds an additional tool to control the particle size with respect to other electrochemistry techniques.

## Acknowledgments

The authors thank UNR and CONICET for their institutional and financial support. This work was performed principally with the aid of Grant PICT 20466 from Agencia Nacional de Promoción Científica y Tecnológica, Argentina. The authors thank Dr. Claudia Palopoli (Ph.D.) for the CV measurements and discussion.

## References

[1] D. Feldheim and C. Foss, *Metal Nanoparticles: Synthesis, Characterization and Applications*, Marcel Dekker, New York, NY, USA, 2002.

[2] J. Lue, "A review of characterization and physical property studies of metallic nanoparticles," *Journal of Physics and Chemistry of Solids*, vol. 62, pp. 1599–1612, 2001.

[3] N. Faraji, W. Mahomood Mat Yunus, A. Kharazmi, E. Saion, M. Shahmiri, and N. Tamchek, "Synthesis, characterization and nonlinear optical properties of silver/PVA nanocomposites," *Journal of the European Optical Society*, vol. 7, p. 12040, 2012.

[4] T. He, C. Wang, X. Pan, and Y. Wang, "Nonlinear optical response of Au and Ag nanoparticles doped polyvinylpyrrolidone thin films," *Physics Letters A*, vol. 373, pp. 592–595, 2009.

[5] R. Vadakkekara, M. Chakraborty, and P. Parikh, "Catalytic performance of silica-supported silver nanoparticles for liquid-phase oxidation of ethylbenzene," *Industrial and Engineering Chemistry Research*, vol. 51, no. 16, pp. 5691–5698, 2012.

[6] M. Radzig, V. Nadtochenko, O. Koksharova, J. Kiwi, V. Lipasva, and I. Khmel, "Antibacterial effects of silver nanoparticles on gram-negative bacteria: influence on the growth and biofilms formation, mechanisms of action," *Colloids and Surfaces B*, vol. 102, pp. 300–306, 2013.

[7] M. V. Roldán, A. Frattini, O. de Sanctis, H. Troiani, and N. Pellegri, "Characterization and applications of Ag nanoparticles in waveguides," *Applied Surface Science*, vol. 254, no. 1, pp. 281–285, 2007.

[8] Q. Wu, H. Cao, Q. Luan et al., "Biomolecule-assisted synthesis of water-soluble silver nanoparticles and their biomedical applications," *Inorganic Chemistry*, vol. 47, no. 13, pp. 5882–5888, 2008.

[9] S. Prabhu and E. Poulose, "Silver nanoparticles: mechanism of antimicrobial action, synthesis, medical applications, and toxicity effects," *International Nano Letters*, vol. 2, p. 32, 2012.

[10] L. M. Liz Marzan and I. Lado-Tourino, "Reduction and stabilization of silver nanoparticles in ethanol by nonionic surfactants," *Langmuir*, vol. 12, pp. 3585–3589, 1996.

[11] M. Guzmán, J. Dille, and S. Godet, "Synthesis of silver nanoparticles by chemical reduction method and their antibacterial activity," *International Journal of Chemical and Biological Engineering*, vol. 2, pp. 104–111, 2009.

[12] G. N. Glavee, K. J. Klabunde, C. M. Sorensen, and G. C. Hadjipanayis, "Borohydride reduction of cobalt ions in water. Chemistry leading to nanoscale metal, boride, or borate particles," *Langmuir*, vol. 9, no. 1, pp. 162–169, 1993.

[13] N. Leopold and B. Lendl, "A new method for fast preparation of highly surface-enhanced raman scattering (SERS) active silver colloids at room temperature by reduction of silver nitrate with hydroxylamine hydrochloride," *The Journal of Physical Chemistry B*, vol. 107, no. 24, pp. 5723–5727, 2003.

[14] P. Jeevanandam, C. Srikanth, and S. Dixit, "Synthesis of monodisperse silver nanoparticles and their self-assembly through simple thermal decomposition approach," *Materials Chemistry and Physics*, vol. 122, no. 2-3, pp. 402–407, 2010.

[15] K. Akhbari, A. Morsali, and P. Retailleau, "Silver nanoparticles from the thermal decomposition of a two-dimensional nano-coordination polymer," *Polyhedron*, vol. 29, no. 18, pp. 3304–3309, 2010.

[16] T. Edison and M. Sethuraman, "Instant green synthesis of silver nanoparticles using *Terminalia chebula* fruit extract and evaluation of their catalytic activity on reduction of methylene blue," *Process Biochemistry*, vol. 47, pp. 1351–1357, 2012.

[17] V. Kotakadi, Y. Rao, S. Gaddam, T. Prasad, A. Reddy, and D. Sai Gopal, "Simple and rapid biosynthesis of stable silver nanoparticles using dried leaves of *Catharanthus roseus*. Linn. G. Donn and its anti microbial activity," *Colloids and Surfaces B*, vol. 105, pp. 194–198, 2013.

[18] K. Shameli, M. Ahmad, E. Al-Mulla et al., "Green biosynthesis of silver nanoparticles using *Callicarpa maingayi* stem bark extraction," *Molecules*, vol. 17, no. 7, pp. 8506–8517, 2012.

[19] R. Trbojevich, N. Pellegri, A. Frattini, O. de Sanctis, P. J. Morais, and R. M. Almeida, "Preparation and isolation of gold nanoparticles coated with a stabilizer and sol-gel compatible agent," *Journal of Materials Research*, vol. 17, no. 8, pp. 1973–1980, 2002.

[20] D. Zhang, X. Liu, X. Wang, X. Yang, and L. Lu, "Optical properties of monodispersed silver nanoparticles produced via reverse micelle microemulsion," *Physica B*, vol. 406, no. 8, pp. 1389–1394, 2011.

[21] K. H. Baek, J. H. Kim, K. B. Lee, H. S. Ahnn, and C. S. Yoon, "Surface plasmon resonance tuning of silver nanoparticle array produced by nanosphere lithography through ion etching and thermal annealing," *Journal of Nanoscience and Nanotechnology*, vol. 10, no. 5, pp. 3118–3122, 2010.

[22] M. L. Rodríguez-Sánchez, M. J. Rodríguez, M. C. Blanco, J. Rivas, and M. A. López-Quintela, "Kinetics and mechanism of the formation of Ag nanoparticles by electrochemical techniques: a plasmon and cluster time-resolved spectroscopic study," *The Journal of Physical Chemistry B*, vol. 109, pp. 1183–1191, 2005.

[23] M. M. Wadkar, V. R. Chaudhari, and S. K. Haram, "Synthesis and characterization of stable organosols of silver nanoparticles by electrochemical dissolution of silver in DMSO," *Journal of Physical Chemistry B*, vol. 110, no. 42, pp. 20889–20894, 2006.

[24] J. Zhu, X. Liao, X. Zhao, and H. Chen, "Preparation of silver nanorods by electrochemical methods," *Materials Letters*, vol. 49, no. 2, pp. 91–95, 2001.

[25] M. Mazur, "Electrochemically prepared silver nanoflakes and nanowires," *Electrochemistry Communications*, vol. 6, no. 4, pp. 400–403, 2004.

[26] Y. Yu, S. Chang, C. Lee, and C. Wang, "Gold nanorods: electrochemical synthesis and optical properties," *The Journal of Physical Chemistry B*, vol. 101, no. 34, pp. 6661–6664, 1997.

[27] M. Zhou, S. Chen, S. Zhao, and H. Ma, "RETRACTED: one-step synthesis of Au-Ag alloy nanoparticles by a convenient electrochemical method," *Physica E*, vol. 33, pp. 28–34, 2006.

[28] M. Bordenave, A. Scarpettini, M. Roldán, N. Pellegri, and A. Bragas, "Plasmon-induced photochemical synthesis of silver triangular prisms and pentagonal bipyramids by illumination with light emitting diodes," *Materials Chemistry and Physics*, vol. 139, no. 1, pp. 100–106, 2013.

[29] H. Peng, A. Yang, and J. Xiong, "Green, microwave-assisted synthesis of silver nanoparticles using bamboo hemicelluloses and glucose in an aqueous medium," *Carbohydrate Polymers*, vol. 91, pp. 348–355, 2013.

[30] A. Pal, S. Shah, and S. Devi, "Microwave-assisted synthesis of silver nanoparticles using ethanol as a reducing agent," *Materials Chemistry and Physics*, vol. 114, pp. 530–532, 2009.

[31] H. Yin, T. Yamamoto, Y. Wada, and S. Yanagida, "Large-scale and size-controlled synthesis of silver nanoparticles under microwave irradiation," *Materials Chemistry and Physics*, vol. 83, no. 1, pp. 66–70, 2004.

[32] A. Frattini, N. Pellegri, D. Nicastro, and O. de Sanctis, "Effect of amine groups in the synthesis of Ag nanoparticles using aminosilanes," *Materials Chemistry and Physics*, vol. 94, no. 1, pp. 148–152, 2005.

[33] L. Huang, Y. Zhai, S. Dong, and J. Wang, "Efficient preparation of silver nanoplates assisted by non-polar solvents," *Journal of Colloid and Interface Science*, vol. 331, no. 2, pp. 384–388, 2009.

[34] A. Rodríguez-López, A. Paredes-Arroyo, J. Mojica-Gomez et al., "Electrochemical synthesis of magnetite and maghemite nanoparticles using dissymmetric potential pulses," *Journal of Nanoparticle Research*, vol. 14, p. 993, 2012.

[35] C. Y. Cheng, S. Thiagarajan, and S. M. Chen, "Electrochemical fabrication of AuRh nanoparticles and their electroanalytical applications," *International Journal of Electrochemical Science*, vol. 6, no. 5, pp. 1331–1341, 2011.

[36] D. F. Yancey, E. V. Carino, and R. M. Crooks, "Electrochemical synthesis and electrocatalytic properties of Au@Pt dendrimer-encapsulated nanoparticles," *Journal of the American Chemical Society*, vol. 132, no. 32, pp. 10988–10989, 2010.

[37] B. Yin, H. Ma, S. Wang, and S. Chen, "Electrochemical synthesis of silver nanoparticles under protection of poly(*N*-vinylpyrrolidone)," *The Journal of Physical Chemistry B*, vol. 107, no. 34, pp. 8898–8904, 2003.

[38] M. Tejmaya, I. Römer, R. Merrifield, and J. Lead, "Stability of citrate, PVP, and PEG coated silver nanoparticles in ecotoxicology media," *Environmental Science and Technology*, vol. 46, no. 13, pp. 7011–7017, 2012.

[39] C. Luo, Y. Zhang, X. Zeng, Y. Zeng, and Y. Wang, "The role of poly(ethylene glycol) in the formation of silver nanoparticles," *Journal of Colloid and Interface Science*, vol. 288, no. 2, pp. 444–448, 2005.

[40] R. G. Shimmin, A. B. Schoch, and P. V. Braun, "Polymer size and concentration effects on the size of gold nanoparticles capped by polymeric thiols," *Langmuir*, vol. 20, no. 13, pp. 5613–5620, 2004.

[41] A. Shkilnyy, M. Soucé, P. Dubois, F. Warmont, M. L. Saboungi, and I. Chourpa, "Poly(ethylene glycol)-stabilized silver nanoparticles for bioanalytical applications of SERS spectroscopy," *Analyst*, vol. 134, no. 9, pp. 1868–1872, 2009.

[42] D. Sutton, *Electronic Spectra of Transition Metal Complexes*, McGraw-Hill, London, UK, 1968.

[43] H. Wang, X. Qiao, J. Chen, X. Wang, and S. Ding, "Mechanisms of PVP in the preparation of silver nanoparticles," *Materials Chemistry and Physics*, vol. 94, pp. 449–453, 2005.

[44] U. Kreibig and M. Vollmer, *Optical Properties of Metal Clusters*, Springer, Berlin, Germany, 1995.

[45] X. K. Meng, S. C. Tang, and S. Vongehr, "A review on diverse silver nanostructures," *Journal of Materials Science & Technology*, vol. 26, pp. 487–522, 2010.

[46] L. Rodríguez-Sánchez, M. C. Blanco, and M. A. López-Quintela, "Electrochemical synthesis of silver nanoparticles," *The Journal of Physical Chemistry B*, vol. 104, pp. 9683–9688, 2000.

[47] R. He, X. Qian, J. Yin, and Z. Zhu, "Preparation of polychrome silver nanoparticles in different solvents," *Journal of Materials Chemistry*, vol. 12, pp. 3783–3786, 2002.

[48] R. Jenkins and R. L. Snyder, *Introduction to X-Ray Powder Diffractometry*, John Wiley & Sons, New York, NY, USA, 1996.

[49] S. Calvin, S. X. Luo, C. Caragianis-Broadbridge et al., "Comparison of extended x-ray absorption fine structure and Scherrer analysis of x-ray diffraction as methods for determining mean sizes of polydisperse nanoparticles," *Applied Physics Letters*, vol. 87, Article ID 233102, 3 pages, 2005.

[50] K. Kelly, E. Coronado, L. Zhao, and G. Schatz, "The optical properties of metal nanoparticles: the influence of size, shape, and dielectric environment," *The Journal of Physical Chemistry B*, vol. 107, pp. 668–677, 2003.

[51] D. Evanoff and G. Chumanov, "Size-controlled synthesis of nanoparticles. 2. Measurement of extinction, scattering, and absorption cross sections," *The Journal of Physical Chemistry B*, vol. 108, no. 37, pp. 13957–13962, 2004.

[52] S. Link and M. A. El-Sayed, "Spectral properties and relaxation dynamics of surface plasmon electronic oscillations in gold and silver nanodots and nanorods," *The Journal of Physical Chemistry B*, vol. 103, pp. 8410–8426, 1999.

[53] M. Reetz and W. Helbig, "Size-selective synthesis of nanostructured transition metal clusters," *Journal of the American Chemical Society*, vol. 116, no. 16, pp. 7401–7402, 1994.

# 17

# Effect of Nanoalumina on the Electrochemical and Mechanical Properties of Waterborne Polyurethane Composite Coatings

**Shailesh K. Dhoke, Narayani Rajgopalan, and A. S. Khanna**

*Department of Metallurgical Engineering and Materials Science, Indian Institute of Technology Bombay, Powai, Mumbai 400076, India*

Correspondence should be addressed to Shailesh K. Dhoke; shailesh_zeo@rediffmail.com

Academic Editor: Vijaya Rangari

A nanocomposite coating was formed by incorporating nanoalumina pigment in a waterborne polyurethane dispersion (WPUD) to different loading levels (0.1% and 1.0% by weight). Electrochemical performance of the nanocomposite coating was evaluated by applying these nanomodified coatings on mild steel substrate and exposing them to salt-spray, humidity, and accelerated UV weathering. The surface morphology of the composite coating was evaluated using various analytical techniques. SEM and AFM were used to investigate the dispersion of nanoalumina pigment and surface morphological changes of the nanomodified coating, before and after exposure to the test environment. Mechanical properties like scratch resistance were studied by using nanoscratch technique (Nanoindenter TI-900, Hysitron Inc, USA) and hardness using pencil hardness test method. The results showed an improvement in the corrosion, UV weathering, and mechanical properties of the coatings at lower concentration (0.1% by wt), indicating the positive effect of addition of nanoalumina pigment to the coating.

## 1. Introduction

An aqueous polyurethane dispersion is a binary colloid system in which polyurethane particles are dispersed in a continuous aqueous medium [1, 2]. The basic building blocks of solvent borne polyurethanes like diisocyanates, polyols, amines catalysts, and additives are common to aqueous-based system as well [3]. Though, the development of aqueous polyurethane dispersions was motivated primarily by environmental considerations, one technical advantage of aqueous polyurethane dispersion is that the viscosity of dispersion is independent of the molecular weight of polymer. Thus, polyurethane dispersion can be prepared at a high solid content with a molecular weight high enough to form films with excellent performance solely by physical drying [2]. Economical aspect is another reason as they do not contain the expensive solvents [3]. Compared to their solvent counterparts, polyurethane dispersions can offer the following advantages: viscosity and flow properties independent of molecular weight, the absence of external emulsifiers, environmental safety, good adhesion, and rheology characteristics [4]. Other advantages which make them attractive as waterborne coatings are the outstanding properties of solvent, stain, chemical resistance, and toughness with flexibility [5].

Also, it has been found that additive and pigments not only provides esthetics to the waterborne coatings, but also helps in improving many properties of the coatings such as UV resistance, corrosion resistance. With the quest for new developed coating systems with better performance, aqueous-based polyurethane coatings are modified with various nanoparticles [6–13]. Nanotechnology presents a wide range of opportunities to improve performance of coatings and promises to deliver breakthrough performance specifically with respect to scratch and mar resistance, barrier properties including corrosion resistance and mechanical properties [14, 15]. Optical clarity is one of the many features of nanoparticles, extremely important in expanding nanoparticle applications in coatings. They can be added to a clear coating formulation with little or no adverse impact on visual characteristics. Nanoparticles most commonly used in coatings are $SiO_2$ [7, 11–13], $TiO_2$ [16], ZnO [6, 8, 9, 17], $Al_2O_3$ [18–21], $Fe_2O_3$ [22], and $CaCO_3$ [23], $ZrO_2$ [24]. Use of nanoparticles is most commonly based on the inherent properties they possess. For example, nanotitania and nanozinc oxide are most commonly used as UV blocking agents, whereas nanoalumina and nanosilica are used to improve scratch and abrasion resistance of the coating. The enhanced properties are result of the much greater surface to volume ratio of the nanopigment that is often characterized by very high aspect ratios [25].

Present study refers to the effect of addition of nanoalumina on water-borne polyurethane coating system at two different loading levels (0.1% and 1.0% by weight). Alumina is a harder material (hardness on Mohs's scale is 9) and is mostly used to modify coatings in order to improve its scratch and abrasion resistance. The nanomodified coatings were applied on pretreated mild steel panels by dip coating techniques. Dip coating techniques can be described as a process where the substrate to be coated is immersed in a liquid and then withdrawn with a well-defined withdrawal speed under controlled temperature and atmospheric conditions. The coating thickness is mainly defined by the withdrawal speed, by the solid content, and by the viscosity of the liquid. Dip coating was used keeping into view the industrial viability of the coating. Using this technique, uniform film can be applied onto flat and cylindrical substrates with ease and high production rates and high transfer efficiency. The process is cost effective and requires low labor as compared to other application process. The performance behavior of the nanoparticle modified coating systems with respect to electrochemical, UV resistance, and mechanical properties was investigated.

## 2. Experimental

*2.1. Material Used.* Nanoalumina (Nano DurTM), of nanoparticle size ranges from 30–47 nm with a mean size of 35 nm and specific area of 35 $m^2 g^{-1}$, was procured from Johnson Mathey Company. A one-pack translucent waterborne polyurethane dispersion (WPUD) based on aliphatic urethane with approximately 34% of total solid content was procured from Dooall Corporation Pvt. Ltd, India.

*2.2. Preparation of Nanoparticle Modified Polyurethane Waterborne Coating.* Waterborne polyurethane dispersion (WPUD) was stabilized with nanoalumina pigments in different concentrations by ultrasonic probe dispersion equipment (100 W, 40 khz) for about 45 minutes until a clear coating was obtained. To minimize temperature driven side effects during sonication, the suspension container was immersed in a cooling bath. The nanomodified coating system thus formed was applied on pretreated cold roll steel panels by dip coating. Dip coating techniques can be described as a process where the substrate to be coated is immersed in a liquid and then withdrawn with a well-defined withdrawal speed under controlled temperature and atmospheric conditions. The coating thickness is mainly defined by the withdrawal speed, by the solid content, and by the viscosity of the liquid. Dip coating was used keeping into view the industrial viability of the coating. Using this technique, uniform film can be applied onto flat and cylindrical substrates with ease and high production rates and high transfer efficiency. The process is cost effective and requires low labor as compared to other application process. Pretreated mild steel substrates (3″× 6″) were immersed in the solution of the coating material at a constant speed and held in coating bath for a while. The substrate was pulled up from the bath, while the thin layer of the coating deposits itself on the substrate. The excess coating dripping from the substrate was wiped off and the coating was allowed for oxidative curing at room temperature for 24 hours. The average coating thickness was found to be 10 $\mu$m. The coated panels were used for electrochemical and mechanical characterizations.

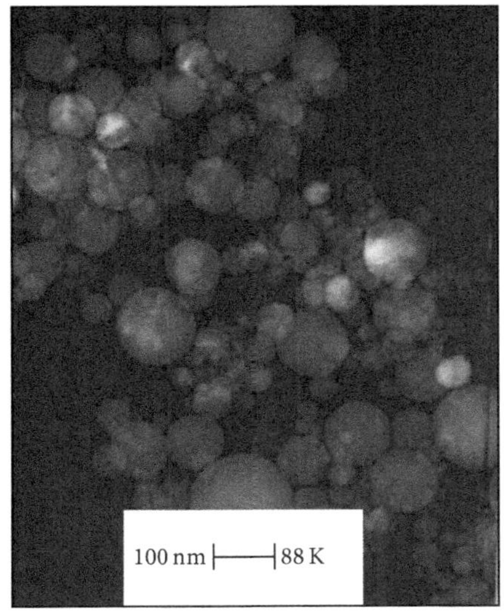


FIGURE 1: TEM photograph of nano-$Al_2O_3$ particles.

*2.3. Characterization.* The powder nanoparticles were investigated by transmission electron microscopy (TEM, Philips CM200) for its shape and size. Electrochemical behavior of the coating was studied by exposing the coated test panels to Salt Spray (ASTM B117), Humidity (ASTM D 2247), and UV weathering (ASTM D 4587) tests. Surface morphology of the coatings after electrochemical studies was determined using scanning electron microscopy (SEM, Model no. S3400, Hitachi) and atomic force microscopy (AFM, Digital Instrument Nanoscope IV). For SEM and AFM study a 1 cm × 1 cm size peace was cut from the sample. Mechanical properties like scratch resistance were studied by using nanoscratch technique (Nanoindenter TI-900, Hysitron Inc, USA) and hardness using pencil hardness test method (ASTM D-3363-05(2011)e2).

## 3. Results and Discussion

*3.1. Characterization of Nanoalumina Oxide Powders.* Figure 1 shows the TEM photograph of nano-$Al_2O_3$ particles. It shows that the particles are of spherical shape.

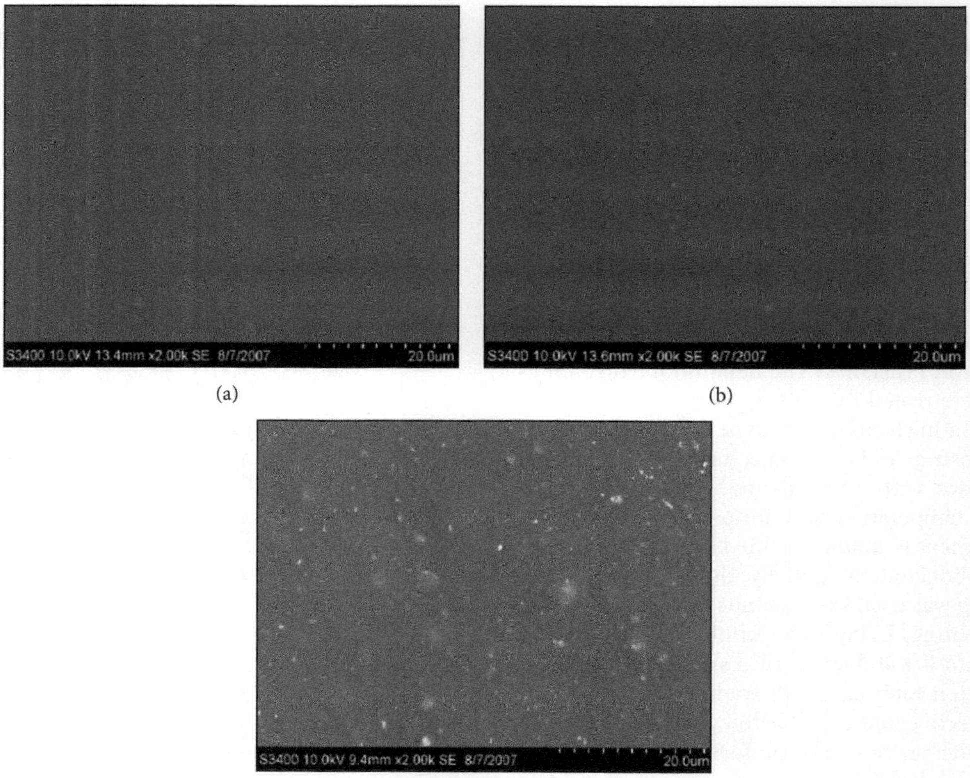

(c)

FIGURE 2: SEM micrographs of unexposed samples: (a) neat WPUD, (b) WPUD + 0.1% nano-$Al_2O_3$, and (c) WPUD + 1.0% nano-$Al_2O_3$.

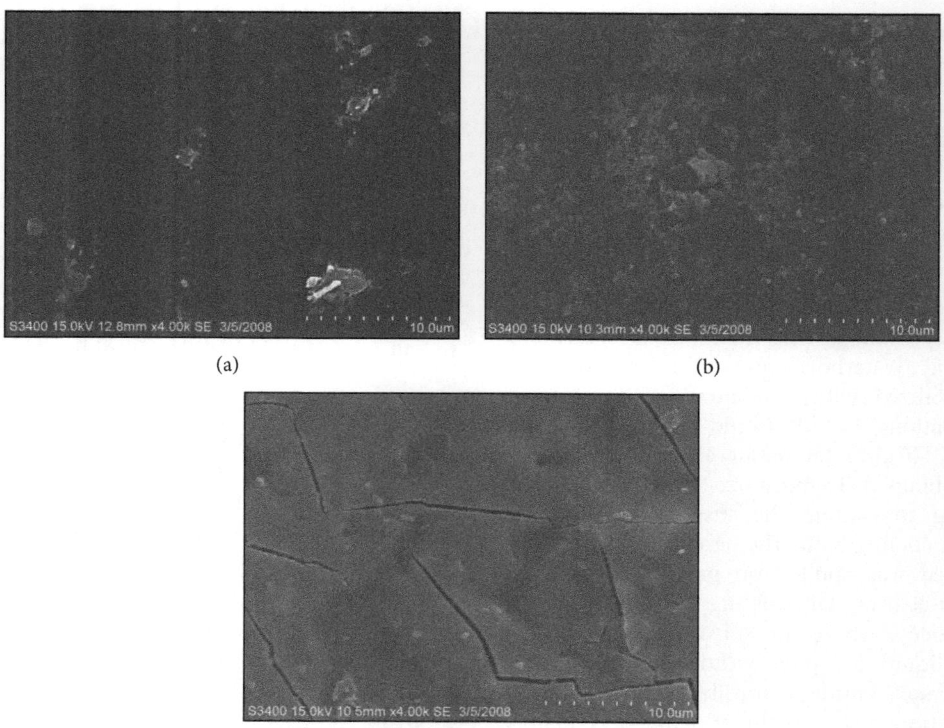

(c)

FIGURE 3: SEM micrographs of salt-spray-exposed samples: (a) neat WPUD, (b) WPUD + 0.1% nanoalumina, and (c) WPUD + 1.0% nanoalumina.

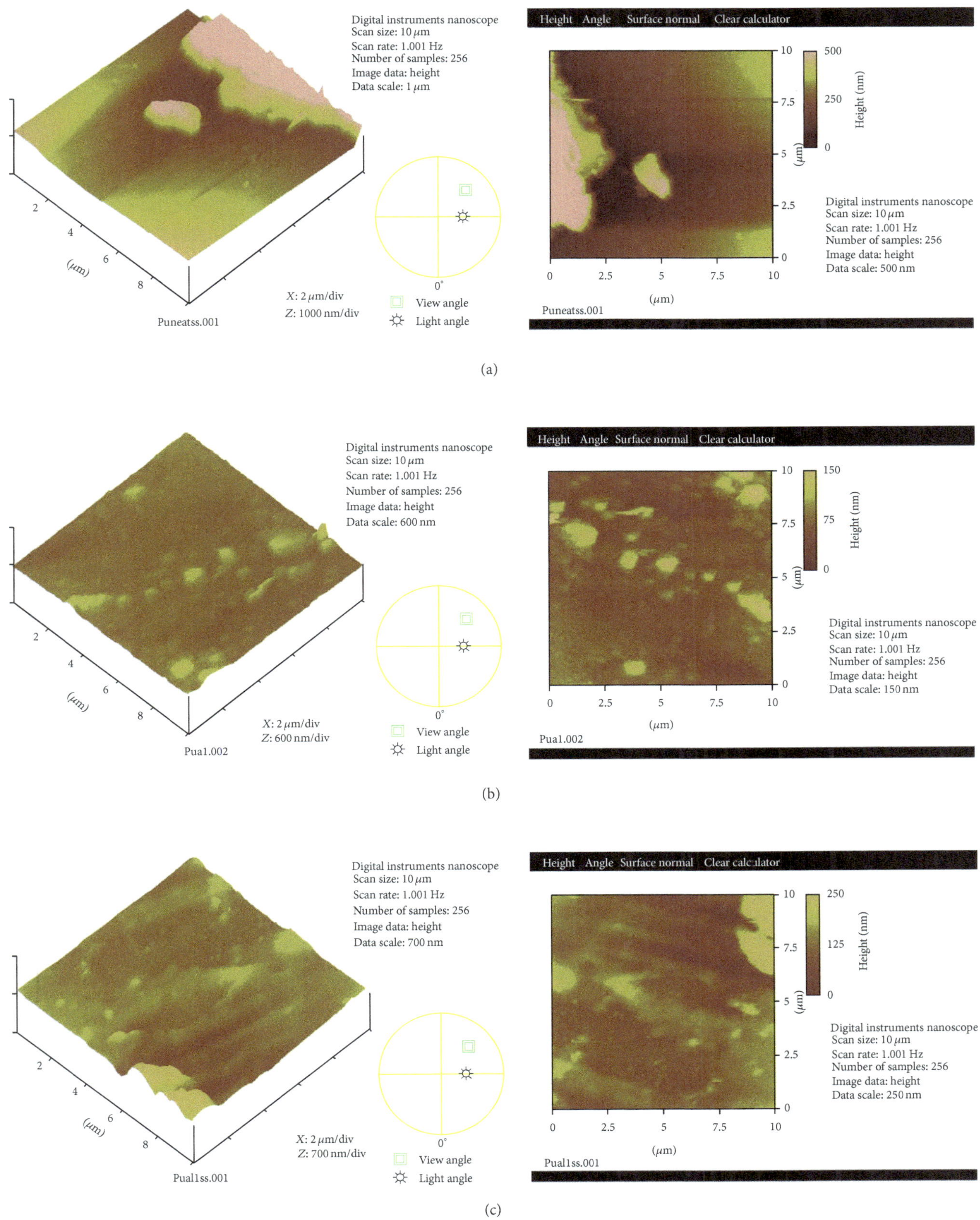


FIGURE 4: AFM topographic height images of salt-spray-exposed coating surface at a scan size of 10 μm × 10 μm: (a) neat WPUD, (b) WPUD + 0.1% nanoalumina, and (c) WPUD + 1.0% nanoalumina.

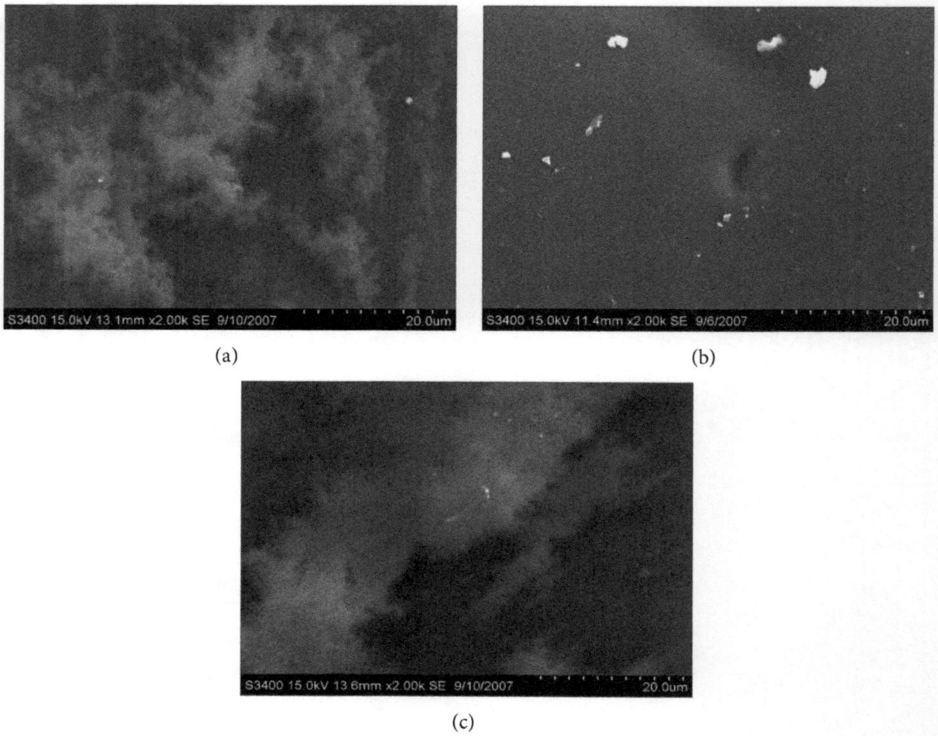


FIGURE 5: SEM micrographs of humidity-exposed samples: (a) neat WPUD, (b) WPUD + 0.1% nanoalumina, and (c) WPUD + 1.0% nanoalumina.

The nanoparticle size ranges from 30 to 47 nm with a mean size of 35 nm.

### 3.2. Characterization of Nanocomposite Coating

*3.2.1. Electrochemical Analysis by Using SEM and AFM.* Surface properties of the coating before and after modification are discussed on the basis of the change of surface topography using SEM and AFM. Figure 2 shows the SEM micrographs of neat WPUD coating and coating modified with nano-$Al_2O_3$ particles in different loading levels, before exposure to the test environments. From Figure 2(a), it can be clearly seen that the coating formed is uniform and free from heterogeneity on the surface. After modifying WPUD with nano-$Al_2O_3$ particles, the coating modified with 0.1% nano-$Al_2O_3$ particles (Figure 2(b)) shows uniform surface morphology indicating proper dispersion of nanoparticles, while the surface micrograph of coating modified with 1.0% nano-$Al_2O_3$ particles (Figure 2(c)) shows white spots all over the surface indicating agglomeration of the nanoparticles.

*Salt-Spray Exposure.* Figure 3 shows the surface micrographs of the neat WPUD and nanocomposite coatings after 800 hours of exposure in salt spray chamber. From Figure 3, it can be seen that the coating degradation after exposure to salt spray is more distinct on neat sample (Figure 3(a)) and on coating modified with 1.0% nanoalumina (Figure 3(c)) where appearance of microcracks is observed. However, the performance was as comparatively better for the coating modified with 0.1% nanoalumina (Figure 3(b)). The degradation of neat WPUD coating resulting into chemical changes during exposure is attributed to the fact that coating without barrier pigments is permeable to water and other corrosion initiators such as chloride and oxygen. However, after modification of WPUD with nanoalumina, it is suggested that the concentration of nanoparticles has dramatic effect on corrosion protection properties of these nanocomposites. At lower loading level (0.1% nanoalumina), the concentration of nanoalumina particle is so small that the well-disperse particles in the coating restrict the diffusion of the corrosive electrolyte through the coating film acting as an effective barrier. It may be attributed that for lower loading level of nanoalumina, the well-dispersed particle having large surface area and small size absorbs more resin on its surface which enhances the density of the coatings thereby reducing the transport paths for the corrosive electrolyte to pass through the coating and consequently reducing the corrosion process [9, 26, 27]. At higher loading level (1.0% nanoalumina), the increase in the number of particles in the coating surpasses the critical loading level which may lead to agglomeration of the particles resulting in microsize defects at the agglomerated site. These agglomerates because of their high surface activity introducing defect like crack and increasing brittleness in nanocomposite coatings. These defects act as site for electrochemical reaction affecting the coating performance. AFM topographic images for salt-spray-exposed samples for 800 hours also provide the same information (Figure 4). As can be seen from Figure 4(a),

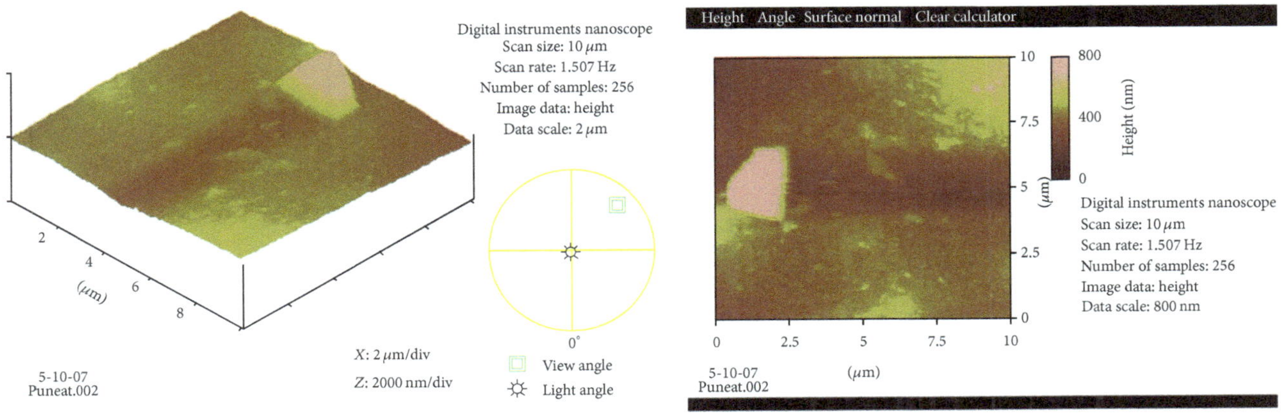


(a)

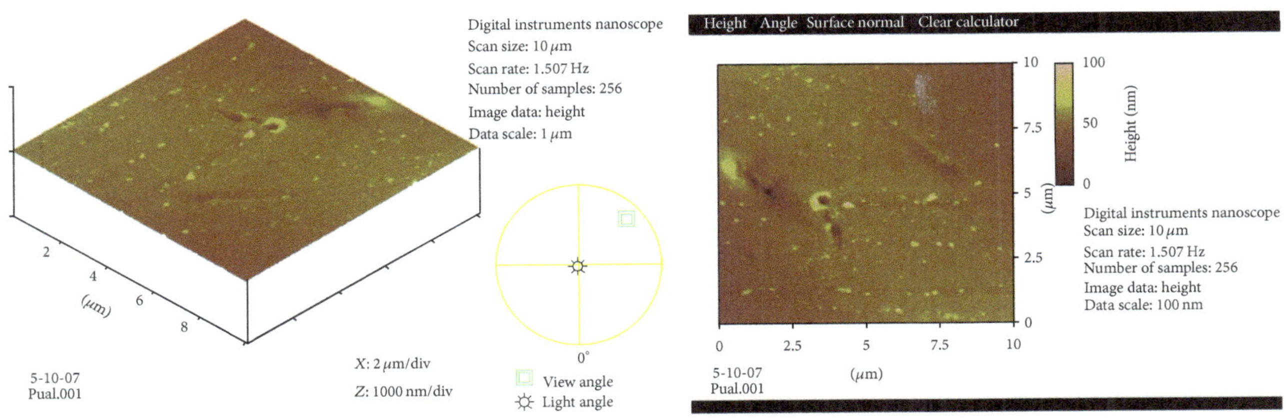


(b)

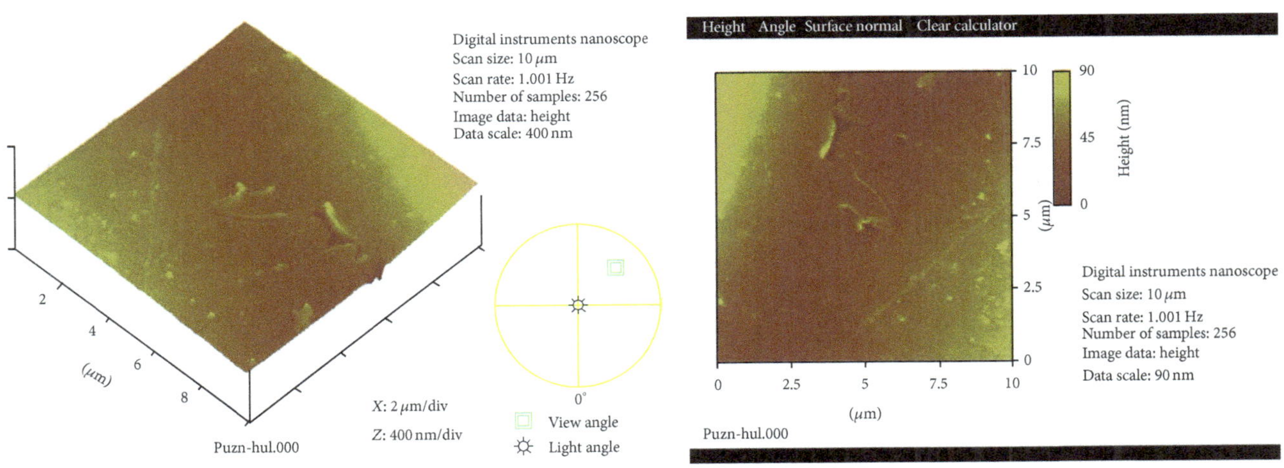


(c)

FIGURE 6: AFM topographic height images of humidity-exposed coating surface at a scan size of 10 $\mu$m × 10 $\mu$m: (a) neat WPUD, (b) WPUD + 0.1% nanoalumina, and (c) WPUD + 1.0% nanoalumina.

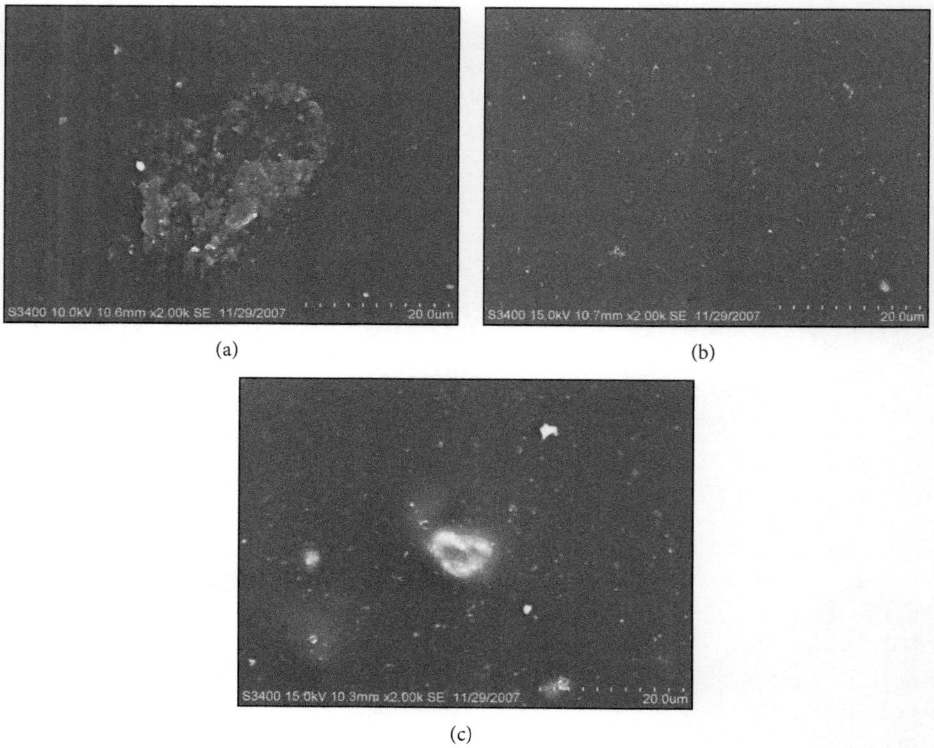


FIGURE 7: SEM micrographs of UV-exposed samples: (a) neat WPUD, (b) WPUD + 0.1% nanoalumina, and (c) WPUD + 1.0% nanoalumina.

the surface of neat sample and coating modified with 1.0% nanoalumina (Figure 4(c)) is completely heterogeneous due to breakdown of polymer matrix as compared to coating modified with 0.1% nanoalumina (Figure 4(b)), which shows a little change on the surface appearance, indicating better protection. It is appeared that the neat coating had undergone the breakdown of polymer matrix due to long exposure to the electrolyte, whereas the addition of critical amount of nano-$Al_2O_3$ had retarded the diffusion of corrosive species.

*Humidity Exposure.* Figure 5 shows the SEM micrographs of humidity-exposed samples. From Figure 5, the effect of humidity was more pronounced on neat sample (Figure 5(a)) and on coating modified with 1.0% nanoalumina (Figure 5(c)) as compared to coating modified with 0.1% nanoalumina (Figure 5(b)). The surface micrograph of the neat sample (Figure 5(a)) shows presence of blisters all over the surface. This indicates that the coating has undergone drastic changes during exposure leading to the failure. With the incorporation of 0.1% of nanoalumina pigment in the coating system, no such defects were observed. As can be seen from the SEM micrographs (Figure 5(b)) coating at lower loading level (0.1%) appears more uniform and devoid of areas of delamination, degradation, and blisters. This again can be attributed to proper pigment concentration and dispersion in the polymer matrix. Also, the interaction of the nanoalumina with the polymer is strong, whereby the strongly bounded interface does not allow the permeation of water [20, 21]. However, at higher loading level (Figure 5(c)), heterogeneities are clearly observed similar to that of neat coating. This can be attributed to improper pigment to polymer ratio resulting in the agglomerated site consequently leading to lose of their desired functional properties [12, 27]. AFM topographic height images (Figure 6) for humidity-exposed samples also provide the parallel observations. Figure 6(a) represents AFM image of neat samples showing completely distorted surface, indicating changes in the surface roughness and microstructure of the coating. Coating system with lower concentration of nanoalumina particles (0.1% by wt) shows little change in the surface roughness and microstructure of the coating (Figure 6(b)) as compared to the coating system (Figure 6(c)) with higher loading level (1.0% by wt), suggesting the improvement in the humidity resistance of the coating.

*UV Weathering Test.* Figure 7 shows the surface micrographs of UV-exposed samples after 1000 hours, where the source of irradiation used was UV-B lamp which emits wavelength of 313 nm. The degradation of coating due to photo-chemical interaction is distinctly observed in neat-WPUD-coated sample. The presence of surface heterogeneity clearly indicates the degradation of coating due to long-term exposure to UV radiation (Figure 7(a)). The coating modified with nanoalumina showed an improvement in the UV resistance. No changes were observed for the coating modified with 0.1% of nanoalumina (Figure 7(b)). However, for coating modified with 1.0% of nanoalumina (Figure 7(c)), appearance of pin holes is clearly observed. This suggests that though the induction time for surface degradation is delayed, the performance of the coating was inferior as

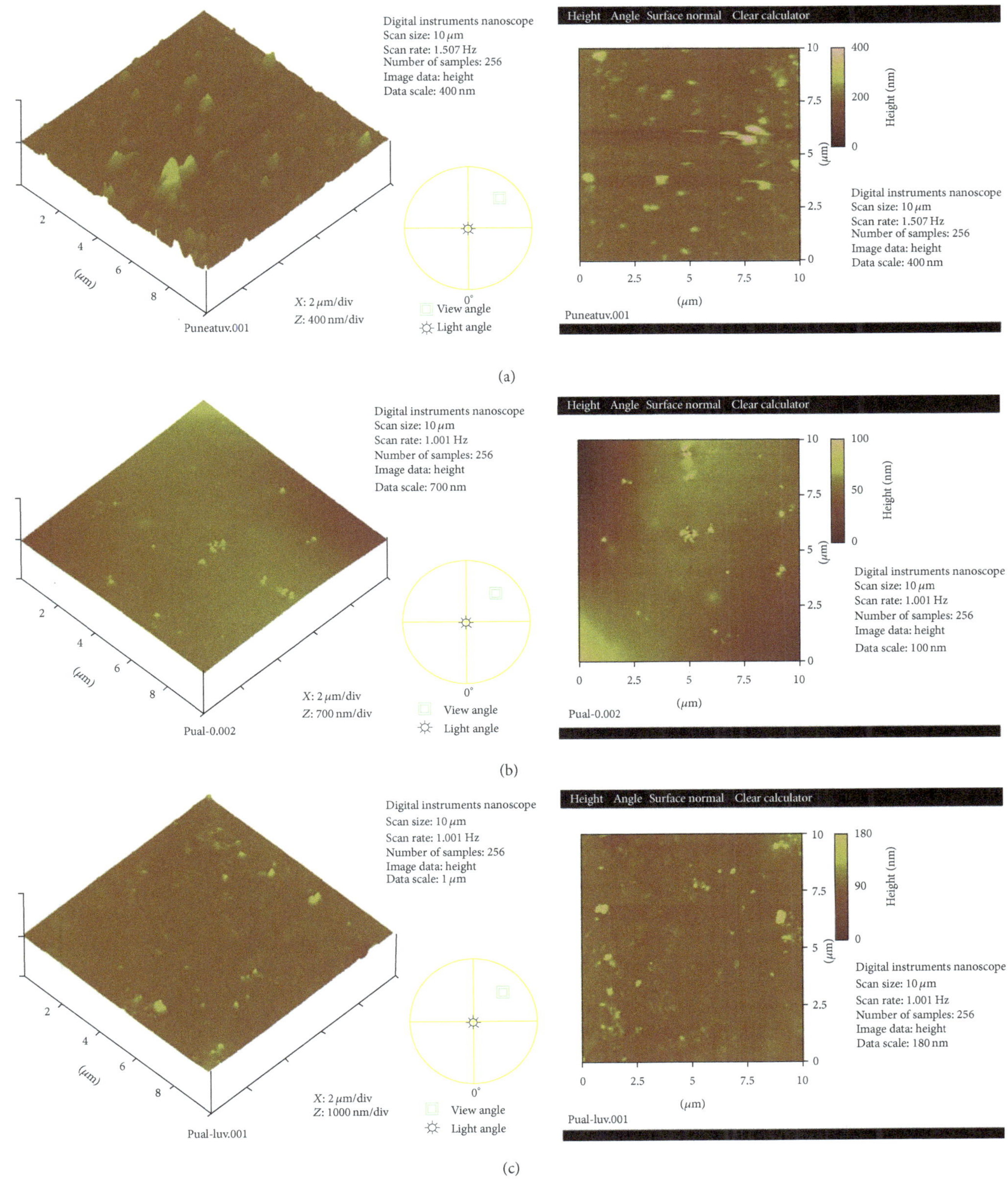


FIGURE 8: AFM topographic height images of UV-exposed coating surface at a scan size of 10 $\mu m \times 10$ $\mu m$: (a) neat WPUD, (b) WPUD + 0.1% nanoalumina, and (c) WPUD + 1.0% nanoalumina.

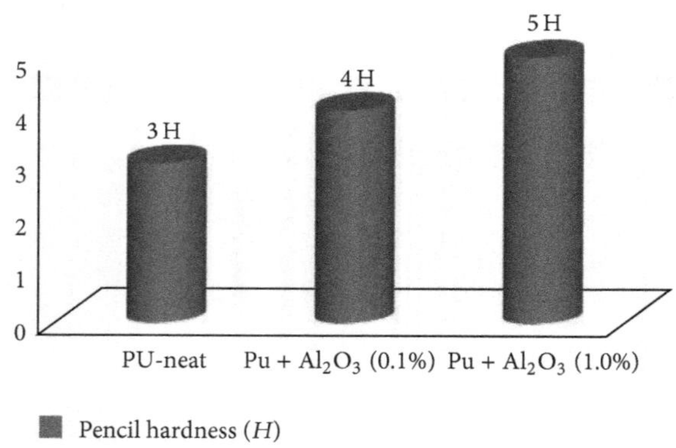


FIGURE 9: Pencil hardness test.

compared to the coating modified with lower concentration of nanoalumina. The improvement of UV resistance of the modified coating can be attributed to the fact that nanoalumina has a wide band gap (3.37 eV) and large excitation binding energy of 60 meV; therefore, it can absorb light that matches or exceeds their band gap energy and which lies in the UV range of the solar spectrum [28, 29]. Thus, a well-dispersed nanoalumina particles attenuate UV radiations and protect the polymer matrix from degradation. Figure 8 shows the AFM topographic height image of UV-exposed samples. For Neat WPUD coating the surface appears to be more heterogeneous with increased roughness, indicating surface degradation (Figure 8(a)). Coating with 0.1% nanoalumina shows effective improvement in UV blocking properties (Figure 8(b)). It can be attributed that the small surface area per unit mass and volume increases the effectiveness of the blocking UV radiation and prevents photodegradation of base polymer matrix. Also, nanoalumina due to its ceramic nature is inherently stable [30] and at this dimension possesses lower photochemical activity [31] thereby providing good UV blocking property to the coating system without interacting with the UV radiation. However, the performance of nanoalumina at higher loading level is not satisfactory, as presence of pinholes and cracks is clearly seen (Figure 8(c)) due to formation of agglomeration in the coating matrix.

*3.2.2. AFM Surface Roughness Analysis.* Table 1 represents surface roughness values for neat WPUD and nano-$Al_2O_3$-modified WPUD coating systems, before and after exposure to the test environment. After exposure to various test environments, it was found that the surface roughness of neat sample is higher due to breakdown of polymer matrix as compared to the surface roughness of nano-$Al_2O_3$-modified coating. This suggests the improvement in the properties of the coating after modification. However, the surface roughness of coating modified with 0.1% nano-$Al_2O_3$ was relatively less than the coating modified with 1.0% nano-$Al_2O_3$, suggesting that for a higher loading level of nanoparticles the resin available is not sufficient to fill all interpigmentary interstices leading to porous film with inferior corrosion resistance.

*3.3. Mechanical Characterization*

*3.3.1. Pencil Hardness.* The surface hardness is an important property of coatings, because a harder surface resists scratch and abrasive forces better. Pencil hardness test was carried out for neat WPUD and nanoalumina (0.1% by wt and 1.0% by wt) modified with WPUD-coated samples (Figure 9). Pencil hardness for the neat-WPUD-coated sample was found to be 3 H. With addition of nanoalumina particles, the hardness of the coating was enhanced, and the grade is 5 H at the higher loading level (1.0%) and 4 H at lower loading level (0.1%).

*3.3.2. Nanoscratch Testing.* The AFM-scanned pictures for the scratched coatings are shown in Figure 10. A distinct difference in the scratch morphology is observed on neat-WPUD (Figure 10(a)) and nanoalumina modified with WPUD coating (Figures 10(b) and 10(c)). A straight, wider, and completely clear scratch was obtained for the neat-WPUD coating (Figure 10(a)). On the other hand, the profile of the scratch was not straight and uniform for the nanoalumina modified coating (Figures 10(b) and 10(c)), suggesting the resistance towards scratch damage after modification. The homogeneous distribution of nanoparticles toughens the coating system by hindering the mobility of macromolecular chains at the interface around the nanoparticles thereby improving the scratch resistance property [32, 33]. Also, the width of the scratch profile for neat-WPUD coating is narrow, whereas the width of the scratch profile for nanoalumina modified coatings is broad, suggesting that the lateral force required to create scratch to the nanoalumina-modified coating is higher as compared to the neat-WPUD coating. Thus, addition of nanoalumina (hardness of 9 Mohs) particles inhibits the crack formation and propagation during the scratch process. It is also apparent that the addition of nanoalumina in higher concentrations decreases the hardness because of nanoparticle aggregation [34]. This may be the probable reason for relatively less resistance offered by the coating modified with 1.0% nanoalumina as compared to coating modified with 0.1% nanoalumina.

## 4. Conclusions

In the present study, two different concentrations of nanoalumina particles (0.1 wt% and 1.0 wt%) were employed to prepare nanocomposite polyurethane coatings. Although, the neat-WPUD coating acts as a barrier against corrosive environment, the protection seems to be reduced due to the diffusion of corrosive electrolyte through the coating film after long-term exposure to the corrosive environment. However, after modification with nanoalumina the coatings showed delay in corrosion process. The increase in the induction time of coating degradation may be attributed as a result of uniform distribution of nanoparticles which restricts the diffusion of the corrosive electrolyte through the coating film acting as an effective barrier. It was observed that UV degradation of the neat WPUD was also reduced after modification with nanoalumina, indicating improvement in UV resistance property. This may be attributed to the fact that at nanoscale the particles themselves effectively absorb

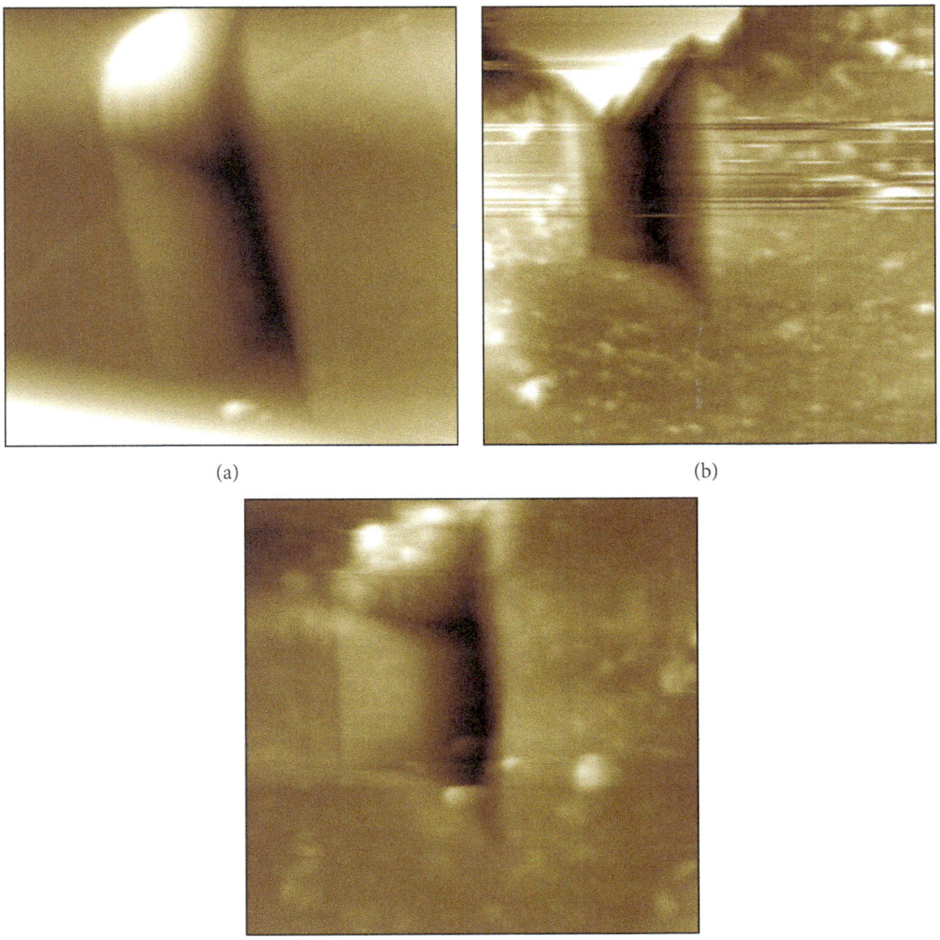

(a) (b) (c)

FIGURE 10: AFM scan pictures of scratched films of (a) neat WPUD, (b) WPUD + 0.1% nanoalumina, and (c) WPUD + 1.0% nanoalumina.

TABLE 1: AFM Surface roughness values (nm) measured for coatings exposed to salt spray, humidity, and UV weathering.

| Test | Neat WPUD | WPUD + 0.1% Nanoalumina | WPUD + 1.0% Nanoalumina |
|---|---|---|---|
| Salt spray | 128.66 nm | 18.64 nm | 47.72 nm |
| Humidity | 132.71 nm | 6.79 nm | 17.93 nm |
| UV weathering | 20.54 nm | 10.70 nm | 13.94 nm |

rather than scatter the UV radiation providing UV protection to the polymer. A considerable improvement in the electrochemical, UV weathering, and mechanical (hardness and scratch) properties of the WPUD coating modified with 0.1 wt% nanoalumina particles was observed. However, optimizing the concentration and improving the dispersion of nanoalumina in the polymer matrix can further improve the performance properties of the WPUD.

## References


[1] B. K. Kim, "Aqueous polyurethane dispersions," *Colloid Polymer Science*, vol. 274, pp. 599–611, 1996.

[2] M. G. Lu, J. Y. Lee, M. J. Shim, and S. W. Kim, "Synthesis and properties of anionic aqueous polyurethane dispersions," *Journal of Applied Polymer Science*, vol. 86, no. 14, pp. 3461–3465, 2002.

[3] K. L. Noble, "Waterborne polyurethanes," *Progress in Organic Coatings*, vol. 32, no. 1-4, pp. 131–136, 1997.

[4] R. G. Coogan, "Post-crosslinking of water-borne urethanes," *Progress in Organic Coatings*, vol. 32, no. 1–4, pp. 51–63, 1997.

[5] H. S. Bender, "Overview of polyurethane dispersion chemistry," in *FSCT Coating Fundamental: Waterborne Coating Technology*, pp. 41–47, 2004.

[6] S. Heberer, R. Maier, R. Mertsch, G. Michael, and M. Pridohl, *Novel UV Protection Coatings Based on Nano-ZnO Particles*, Smart Coatings III, Berlin, Germany, 2004.

[7] S. Zhou, L. Wu, J. Sun, and W. Shen, "The change of the properties of acrylic-based polyurethane via addition of nano-silica," *Progress in Organic Coatings*, vol. 45, no. 1, pp. 33–42, 2002.

[8] J. Zheng, R. Ozisik, and R. W. Siegel, "Disruption of self-assembly and altered mechanical behavior in polyurethane/zinc oxide nanocomposites," *Polymer*, vol. 46, no. 24, pp. 10873–10882, 2005.

[9] L. H. Yang, F. C. Liu, and E. H. Han, "Effects of P/B on the properties of anticorrosive coatings with different particle size," *Progress in Organic Coatings*, vol. 53, no. 2, pp. 91–98, 2005.

[10] J. González-Irun Rodrigúez, P. Carreira, A. García et al., "Nanofiller effect on the glass transition of a polyurethane," *Journal of Thermal Analysis and Calorimetry*, vol. 87, no. 1, pp. 45–47, 2004.

[11] S. X. Zhou, L. M. Wu, J. Sun, and W. D. Shen, "Effect of nanosilica on the properties of polyester-based polyurethane," *Journal of Applied Polymer Science*, vol. 88, no. 1, pp. 189–193, 2003.

[12] M. M. Jalili, S. Moradian, H. Dastmalchian, and A. Karbasi, "Investigating the variations in properties of 2-pack polyurethane clear coat through separate incorporation of hydrophilic and hydrophobic nano-silica," *Progress in Organic Coatings*, vol. 59, no. 1, pp. 81–87, 2007.

[13] Z. S. Petrović, Y. J. Cho, I. Javni et al., "Effect of silica nanoparticles on morphology of segmented polyurethanes," *Polymer*, vol. 45, no. 12, pp. 4285–4295, 2004.

[14] R. Fernando, "Nanomateriál technology applications in coatings," *JCT CoatingsTech*, vol. 1, no. 5, pp. 32–38, 2004.

[15] R. H. Cayton and T. Swaitowski, *The Impact of Nano-Materials on Coating Technologies*, vol. 2, NSTI-Nanotech, 2005.

[16] N. S. Allen, M. Edge, A. Ortega et al., "Degradation and stabilisation of polymers and coatings: nano versus pigmentary titania particles," *Polymer Degradation and Stability*, vol. 85, no. 3, pp. 927–946, 2004.

[17] A. Ammala, A. J. Hill, P. Meakin, S. J. Pas, and T. W. Turney, "Degradation studies of polyolefins incorporating transparent nanoparticulate zinc oxide UV stabilizers," *Journal of Nanoparticle Research*, vol. 4, no. 1-2, pp. 167–174, 2002.

[18] R. H. Cayton and R. W. Brotzman Jr., "Nanocomposite coatings-applications and properties," in *Materials Research Society Symposium Proceedings*, vol. 703, 2002.

[19] L. J. Brickweg, B. R. Floryancic, E. D. Sapper, and R. H. Fernando, "Shear-induced 1-D alignment of alumina nanoparticles in coatings," *Journal of Coating Technology and Research*, vol. 4, no. 1, pp. 107–110, 2007.

[20] S. K. Dhoke, T. J. Mangal Sinha, and A. S. Khanna, "Effect of nano-$Al_2O_3$ particles on the corrosion behavior of alkyd based waterborne coatings," *Journal of Coatings Technology Research*, vol. 6, no. 3, pp. 353–368, 2009.

[21] S. K. Dhoke and A. S. Khanna, "Electrochemical impedance spectroscopy (EIS) study of nano-alumina modified alkyd based waterborne coatings," *Progress in Organic Coatings*, vol. 74, pp. 92–99, 2012.

[22] Q. Wang, M. Yang, and Y. Chen, "Effects of nanosized iron oxide with different morphology on nanomechanical properties of nanocomposite coating," *Key Engineering Materials*, vol. 336–338, pp. 2218–2220, 2007.

[23] A. S. Hamdy, "Advanced nano-particles anti-corrosion ceria based sol gel coatings for aluminum alloys," *Materials Letters*, vol. 60, no. 21-22, pp. 2633–2637, 2006.

[24] S. Kozhukharov, G. Tsaneva, V. Kozhukhuarov et al., "Corrosion protection properties of composite hybrid coatings with involved nano-particles of zirconia and ceria," *Journal of the University of Chemical Technology and Metallurgy*, vol. 43, no. 1, pp. 73–80, 2008.

[25] D. R. Baer, P. E. Burrows, and A. A. El-Azab, "Enhancing coating functionality using nanoscience and nanotechnology," *Progress in Organic Coatings*, vol. 47, no. 3-4, pp. 342–356, 2002.

[26] A. Kalendová, "Effects of particle sizes and shapes of zinc metal on the properties of anticorrosive coatings," *Progress in Organic Coatings*, vol. 46, no. 4, pp. 324–332, 2003.

[27] H. Shi, F. Liu, E. Han, and Y. Wei, "Effects of nano pigments on the corrosion resistance of alkyd coating," *Journal of Materials Science and Technology*, vol. 23, no. 4, pp. 551–558, 2007.

[28] H. F. Lu, B. Fei, J. H. Xin, R. Wang, and L. Li, "Fabrication of UV-blocking nanohybrid coating via miniemulsion polymerization," *Journal of Colloid and Interface Science*, vol. 300, no. 1, pp. 111–116, 2006.

[29] S. M. Al-Hilli and M. Willander, "Optical properties of zinc oxide nano-particles embedded in dielectric medium for UV region: numerical simulation," *Journal of Nanoparticle Research*, vol. 8, no. 1, pp. 79–97, 2006.

[30] B. Pacaud, J. N. Bousseau, and J. Lemaire, "Nano-titania as UV blockers in stains," *European Coatings Journal*, no. 11, pp. 842–848, 1998.

[31] S. Heberer, R. Maier, R. Mertsch, G. Michael, and M. Pridohl, "Novel UV protection coatings based on Nano-ZnO particles," in *Proceedings of the European Coatings Conference: Smart Coatings III*, Berlin, Germany, June 2004.

[32] S. K. Dhoke, R. Bhandari, and A. S. Khanna, "Effect of nano-ZnO addition on the silicone-modified alkyd-based waterborne coatings on its mechanical and heat-resistance properties," *Progress in Organic Coatings*, vol. 64, no. 1, pp. 39–46, 2009.

[33] P. Kardar, M. Ebrahimi, and S. Bastani, "Study the effect of nano-alumina particles on physical-mechanical properties of UV cured epoxy acrylate via nano-indentation," *Progress in Organic Coatings*, vol. 62, no. 3, pp. 321–325, 2008.

[34] C. Sow, B. Riedl, and P. Blanchet, "UV-waterborne polyurethane-acrylate nanocomposite coatings containing alumina and silica nanoparticles for wood: Mechanical, optical, and thermal properties assessment," *Journal of Coatings Technology Research*, vol. 8, no. 2, pp. 211–221, 2011.

# 18

# Naturally Self-Assembled Nanosystems and Their Templated Structures for Photonic Applications

**K. Pradeesh, Nageswara Rao Kotla, Shahab Ahmad, Vindesh K. Dwivedi, and G. Vijaya Prakash**

*Nanophotonics Lab, Department of Physics, Indian Institute of Technology Delhi, Hauz Khas, New Delhi 110016, India*

Correspondence should be addressed to G. Vijaya Prakash; prakash@physics.iitd.ac.in

Academic Editor: Amir Kajbafvala

Self-assembly has the advantage of fabricating structures of complex functionalities, from molecular levels to as big as macroscopic levels. Natural self-assembly involves self-aggregation of one or more materials (organic and/or inorganic) into desired structures while templated self-assembly involves interstitial space filling of diverse nature entities into self-assembled ordered/disordered templates (both from molecular to macro levels). These artificial and engineered new-generation materials offer many advantages over their individual counterparts. This paper reviews and explores the advantages of such naturally self-assembled hybrid molecular level systems and template-assisted macro-/microstructures targeting simple and low-cost device-oriented fabrication techniques, structural flexibility, and a wide range of photonic applications.

## 1. Introduction

Fabrication of nano/mesa photonic architectures from *top-down* technology involve precise growth techniques like molecular beam epitaxy (MBE), chemical vapor deposition (CVD) and also involve patterning techniques such as photolithography, particle beam lithography, scanning probe lithography, and nanoimprint lithography. While the above mentioned processes are laborious, time-consuming, and costly, the *"bottom-up"* technology based on *self-assembly* approach is the simplest, cost effective technique. Self-assembly is one of the most important *"molecular engineering"* strategies used in fabricating complex functional structures, from micro to the molecular levels, utilising the advantage of self-interaction of molecules. *Molecular self-assembly* is a strategy for nanofabrication that involves designer molecules and supramolecular entities so that molecules naturally aggregate into specific desired structures [1, 2]. This method reduces many difficult steps in nanofabrication, those involving atomic-level modifications of the structures. Moreover, molecular self-assembly tends to produce structures that are relatively defect-free and self-healing, because the target structures are selective with thermodynamically stable assembly between the possible configurations. Several self-assembly methods have been developed in the recent past, such as phase-separation of copolymers, formation of pores in alumina, liquid crystals, zerogels, and polymer spheres templating [3–5]. At molecular levels, one of the examples of self-assembly is the *intercalation strategy* wherein the organic entities are space-filled within naturally self-assembled crystalline inorganic semiconductor hosts, with an opportunity to produce a very special and tailor-made semiconductor, known as *inorganic-organic hybrids* [6–8]. In a macro level, mono dispersed mesa sized spherical colloids are self-assembled to form three-dimensionaly periodic lattices and are famously known as synthetic opals (3D photonic crystals) [9–17]. Another relatively easy and cost-effective methodology to produce nano to wavelength-scaled photonic structures with long-range order is through self-organized systems, which can be used to create periodic patterns, followed by material filling into the interstitial spaces through techniques like electrochemical deposition. This approach is called templated self-assembly. Overall, for large-scale production, the self-assembly and template assisted self-assembly techniques are better alternatives to *top-down* technology.

## 2. Naturally Self-Assembled Nanosystems

Though inorganic semiconductors are still a material of choice for high-performance semiconductor devices, in recent years organic semiconductors have attracted considerable interest for applications as active materials in low-cost semiconductor devices. For example, displays based on organic light-emitting diodes have already made their way into commercial products, and semiconductor devices like organic FETs have been demonstrated and are being rapidly improved [18–22]. Analogous to low-dimensional inorganic semiconductors are low-dimensional organic nanolayers, which are basically a self-assembly of organic molecules (e.g., *J*-aggregates) [23–25]. Significant advantages of these self-organized molecular nanostructures (*J*-aggregates) are the ultra-sharp absorption line widths with giant oscillator strengths, especially room-temperature performance. *J*-aggregates exhibit a very narrow and red-shifted electronic absorption band (*J* band) and strong fluorescence with a small Stokes shift. These characteristic optical properties, which are the origins of their expected functions, are explained by the interaction between the transition dipole moments [26, 27]. However, while organic semiconductors have obvious advantages, due to simple fabrication and high performance, the main draw-backs are a slow-optical response, lack of thermal/mechanical durability, and most importantly limited life span.

## 3. Naturally Self-Assembled Inorganic-Organic (IO) Hybrid Systems and Applications

Inorganic-organic (IO) hybrid nanostructures have recently emerged as highly-promising systems for applications as optoelectronic devices, opening up a new dimension to nanotechnology, as unique replacement to their inorganic and organic counterparts. These hybrid systems have great advantage to combine distinct properties of inorganic and organic components within a single-molecular material. The art of combining dissimilar components to yield improved materials is not actually new: ancient building construction material, adobe [28], was made from a mixture of clay (inorganic) and straw (organic). Another classical example is *"blue Maya color"* developed in around 900 A.D., is an IO-hybrid composite. In fact the *Maya blue* is an inorganic-organic hybrid composed of *palygorskite* clay and organic indigo dye $C_{16}H_{10}N_2O_2$ [29, 30]. The most characteristic beauty is its unusual stability; even after centuries of exposure to heat, humidity, and extreme atmospheric conditions, the color hardly faded.

Among various IO-hybrids, one of the most interesting and well-studied materials is a *perovskite* type hybrid [31]. These self-organized materials are derived from the general structure form of $AMX_3$ where A is an organic moiety, M is a divalent metal (such as $Pb^{2+}$, $Sn^{2+}$, $Ge^{2+}$, $Cu^{2+}$, $Ni^{2+}$, $Mn^{2+}$, $Fe^{2+}$, $Co^{2+}$, and $Eu^{2+}$), and X is a halide (such as I, Br, and Cl). This simple 3D structure $AMX_3$ consists of corner-sharing $MX_6$ octahedra extended in three dimensions, where the "A" cations are located in the larger

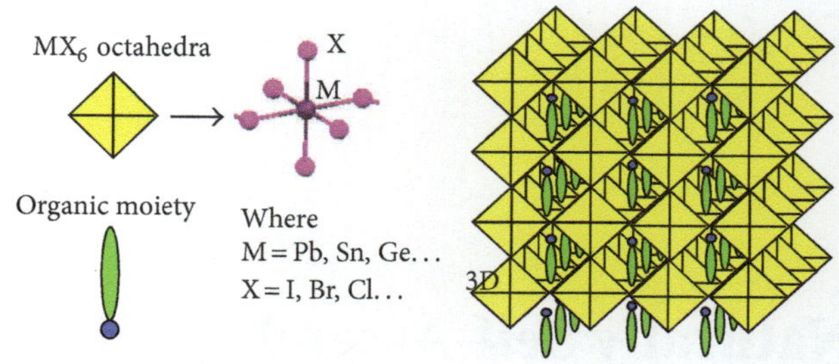


FIGURE 1: Schematic of $MX_6$ octahedra and the organic moiety of the basic $AMX_3$ perovskite unit cell and three-dimensional network formed by $AMX_3$ perovskite unit cells.

12-fold coordinated voids between the octahedra (Figure 1). These self-assembling inorganic-organic perovskites adopt an alternating framework of semiconducting inorganic sheets and organic layers. The increasing interest is because of the ability to derive low-dimensional crystals, which show unique crystal structure and physical and optical properties, from parent 3D networks of $AMX_3$ from simple and effective natural self-assembly [32–43]. These materials involve different types of interactions allowing the assembly of complex and highly-ordered structures with various bonding schemes. The chemical bonding involved in these IO-hybrid assembled systems are generally described as

(i) covalent/ionic bonding within the inorganic network which favors the formation of sheets of corner-sharing metal halide octahedra,

(ii) hydrogen/ionic bonding between the organic cations and the halogens in the inorganic sheets,

(iii) various weak interactions like *Van der Waals* interactions between the organic R-groups.

Many structural, electrical, thermochromic, and magnetic studies were carried out from almost a decade, to explore the advantages of IO-hybrids over organic and inorganic counterparts [44–46], These hybrids have an advantage of structural flexibility to choose suitable organic spacers (usually monofunctional or difunctional amines). The crystallographic orientation and the thickness of the perovskite sheets can be tailored as per the choice of appropriate organic cations. In other words, inorganic units can be self-organized into low-dimensional crystals of zero-(0D), one-(1D) and two-(2D) dimensional networks (Figure 2). In 0D networks, $MX_6$ octahedra are isolated and are surrounded with the organic spacers. In 1D networks, metal halides ($MX_6$ octahedra) are extended as a chain along one direction with corner/edge/face shared to form 1D hybrid. Similarly, inorganic network can be extended as layered sheets with corner shared $MX_6$ octahedra to form 2D networks. In 2D, inorganic and organic layers are stacked alternatively with orientation along a specific crystal direction.

During synthesis, the organic moieties, apart from controlling the dimensionality, can also alter the crystallographic orientation of the parent network. For example, based on the organic moiety interaction with metal halide network,

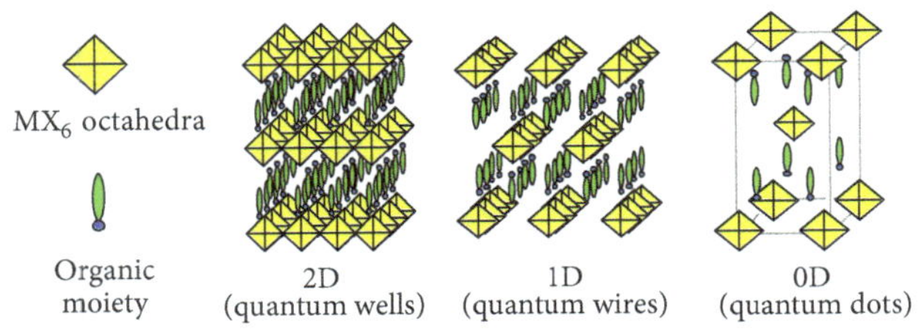


FIGURE 2: Schematic of 2D, 1D, and 0D IO-hybrid derived from parent $AMX_3$ type 3D IO-hybrid.

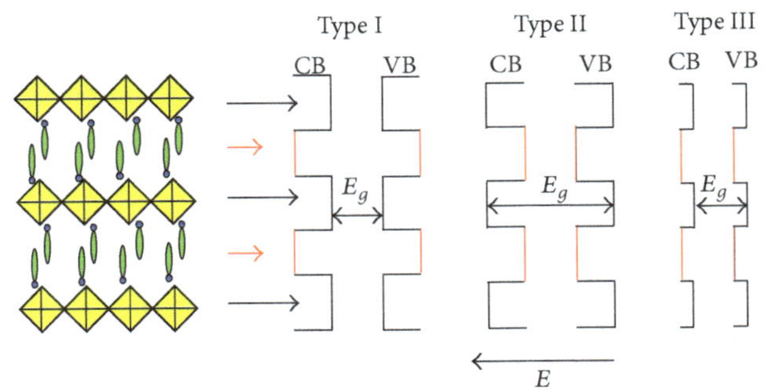


FIGURE 3: 2D IO-hybrid structure and several possible energy level schemes.

the resultant network can deviate from ⟨100⟩ oriented 2D network [47] to 1D dimensional ⟨110⟩ oriented hybrid structures also [48, 49]. Under special circumstances, based on organic moiety conformation, the metal halides may self-assemble into either edge sharing or face sharing of metal halide octahedra, forming various low dimensional inorganic halide networks of different orientations. Based on such different networks, these hybrids show marked variation in their structural and optical features [50–59].

Several studies were carried out in the recent past to prove the potential ability of IO-hybrids in photonic applications: electro-absorption and electroluminescence [60–64], photoconductive devices, [65, 66], optical nonlinear devices [67, 68], stark effect [69] magneto- absorption, and spontaneous magnetization [70, 71]. Apart from linear optical studies, high-optical excitation effects such as ultrafast dynamics of excitons [50, 72], observation of higher-order excitons (biexciton and triexciton) [51, 52, 73–75], and even an attempt of biexciton lasing [76] were also reported. Photonic devices such as thin film transistors (TFTs), inorganic-organic field-effect transistors (IOFETs), inorganic-organic light emitting diodes (IOLEDs), and scintillators were also been successfully demonstrated [77–79].

## 4. Naturally Self-Assembled Two-Dimensional IO-Hybrid Systems

Among the several low-dimensional hybrids mentioned before, 2D hybrids are of special interest. The 2D (⟨100⟩ oriented) hybrids are analogous to natural *multiple quantum wells* (MQWs), where inorganic and organic sheets (of molecular level sizes) are alternatively stacked. This is of special attraction because such natural MQWs are easily achieved from solution-processing techniques, without any involvement of laborious instrumentation like molecular beam epitaxy deposition. Various possible layered schemes of these MQWs are shown in Figure 3. Depending on respective inorganic and organic bandgaps, these MQWs can be classified [80] into Type I, II, or III (Figure 3).

In Type I, the conduction band of the inorganic layer is generally below that of the organic layer, and the valence band is above that of the organic layer (Figure 3). Therefore, inorganic sheets act as "quantum wells" for both electrons and holes leading to Type I heterostructure. Similarly, if larger bandgap inorganic sheets are integrated with more complex, conjugated organic cations (with bandgap less than inorganic layer), the well and barrier layer roles can be reversed [77] forming Type II heterostructure (Figure 3). In special cases, by appropriate modifications of the chemistry of the organic and inorganic layers, the bandgaps for the organic and inorganic layers can also be offset, leading to Type III heterostructure, in which the wells for the electrons and holes are in different layers. The present study deals with the IO-hybrid system of R-$PbI_4$ (where R is organic), having Type I structure, where the inorganic (PbI network ~3 eV) bandgap is much less than the organic bandgap (4 to 6 eV), therefore, the electron-hole confinement is solely within inorganic network.

The most influencing factors in 2D IO-hybrids are (1) the choice of organic moiety, (2) how the organic moiety intercalates into the inorganic network (*vice versa*), thereby altering the structural rearrangement and the consequent energy band structure, (3) quantum confinement due to quantum-limit widths of the individual organic and inorganics, and (4) the dielectric contrast between the organic and inorganic sheets. Technically the last two are dependent on first two factors.

## 5. Structure and Fabrication Strategies of Naturally Self-Assembled IO-Hybrid Systems

*5.1. Structure of Naturally Self-Assembled IO-Hybrids.* Generic way of visualization of these IO-hybrids is intercalation of organic guest moieties into a parent crystalline host. Recent efforts in the crystal engineering resulted into the reduction of structures into 0, 1, 2, or 3- low-dimensional hybrid networks [32, 33, 77, 83–87]. The dimensionality of these IO-hybrids, based on the bridging of organic moiety between the MX one-dimensional planes, is critically dependent on (1) the choice of hydrogen bonding scheme between protonated amine terminal group(s) of organic moiety and the MX network and (2) the driving force, size, and shape of the organic molecule [80]. The simplest $MX_4^{2-}$ network consists of corner-sharing metal halide octahedra oriented along ⟨100⟩ plane and, based on how the organic is

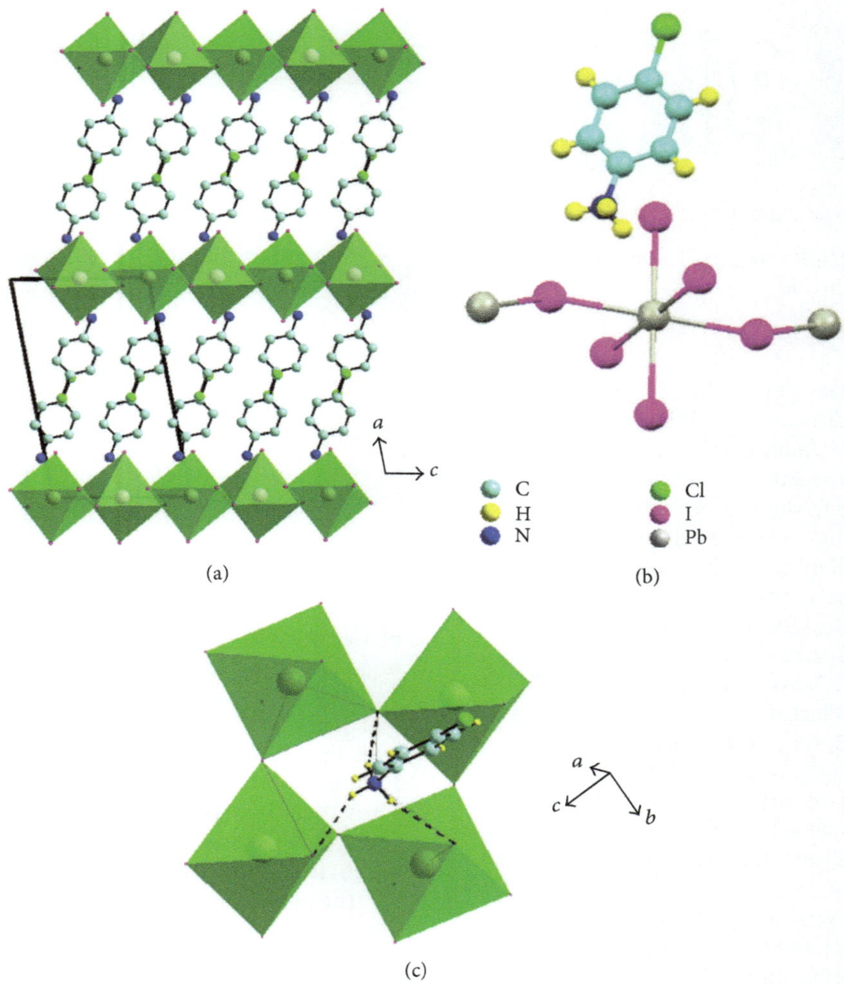


FIGURE 4: (a) Packing structure, (b) asymmetric unit, and (c) NH-I terminal halide configuration (equilateral triangle configuration) of CAPI [81].

intercalated into the parental network, intercalated structures of 0D, 1D, 2D, and 3D can be expected [32, 33, 48, 86, 88, 89].

In the two-dimensional R-$MX_4$ type hybrids, $(MX_4)^{2-}$ octahedral network sheets are stacked up along $\langle 100 \rangle$ direction with alternate layers of organic moieties. The crystallographic information for one of the IO-hybrid (4-$ClC_6H_4NH_3)_2PbI_4$(CAPI) has been presented in Figure 4. CAPI IO-hybrid crystallizes in the monoclinic space group $P2_1/c$ in which the asymmetric unit consists of half a $(PbI_4)^{2-}$ anion and one $(Cl\text{-}C_6H_4NH_3)^+$ cation. These structures comprises of well-ordered organic and inorganic layers, arranged alternately stacked along the *a*-direction with layers infinitely extended in the *bc* plane.

*5.2. General View of Synthesis, Fabrication, and Implications.* Many device applications demand simple and effective fabrication protocols specially the techniques to make highly uniform device-quality thin films. IO-hybrids are generally fabricated from conventional solution processing methods, and single crystals are harvested by slow evaporation technique [90–94]. Though several synthesis recipes are available, a simpler, generalized, high product yield, and commercially viable process is as follows. Stoichiometric quantities of organic moiety and inorganic ($PbI_2$) were mixed with concentrated aqueous HI at 60°C. The resultant solution was allowed to rest at 60°C for an hour and then cooled slowly to room temperature without stirring. The precipitate, thus, obtained was filtered off and dried.

The general synthesis is as follows:

$$2\left(R-NH_2\right) + PbI_2 + HI \longrightarrow \left(R-NH_3\right)PbI_4. \quad (1)$$

Single crystals of the respective compounds were harvested from slow evaporation process by dissolving the compound in a sparingly soluble solvent. However, the synthesis procedure slightly varies from its generic route depending on the nature of organic moiety.

From the application perspective, thin film processing demands to achieve easy and controlled thickness/morphology over large areas and most importantly highly oriented IO-hybrids. The applications of IO-hybrids will be immense only if the fabrication parameters are precisely controlled. For the same reason, one has to develop methods of fabrication that can be carefully predicted and controlled for a predetermined technological application. Usual way of fabricating these thin films is from spin-coating of IO-hybrid solutions onto a desired substrate. Although other techniques like single- and double-source thermal vapor deposition, Langmuir Blodgett (LB) method, layer by layer deposition, spray pyrolysis, and low-temperature melting process had been employed to obtain films, it is always difficult to find empirical conditions and processes to obtain well-ordered thin films of these IO-hybrids [95–100]. Especially the applicability of thermal vapor deposition technique is limited due to stability and contamination issues and to balance organic and inorganic evaporation rates simultaneously. Recently, Rikukawa group [95] had developed layer-by-layer self-assembly method to fabricate ultra-thin films of bifunctional amino end-group based IO-hybrid. This method is based on alternate dipping of hydrophilic substrates in organic iodide and lead halides solutions followed by repeated washing to remove unreacted residuals, and this procedure was repeated several times to obtain required self-assembly films, up to 12 layers.

We have recently explored one of the much simpler, but efficient technique, so-called intercalation process to fabricate highly-ordered IO-hybrids over centimeter size lateral dimensions. In fact, the word *intercalation* in general refers to insertion of guest into self- assembled 2D/3D solids. Gieseking [101] and MacEwan [102] showed, for the first time, the ability of the formation of IO-hybrid by intercalation of organic cations into layered and charged inorganic hosts and further extended to neutral guest/hosts by Bradley [103]. Owing to the intense interest in new nanocomposite functional hybrid materials for fundamental and device-oriented research, new *intercalation chemistry* has been established [104]. For 2D layered hybrids it is essential to focus on both new hybrids as well as highly ordered films, however, the later one has not been widely considered.

As mentioned before, the kinetics and layer formation during intercalation are critically dependent on the nature and shape of the guest moiety, concentration of guest molecules, thickness of parent films, intercalation time, and the solvent used [7]. The schematic of intercalation strategy of IO-hybrid for a high-quality thin film fabrication is demonstrated in Figure 5. When predeposited layered $PbI_2$ film is intercalated with presynthesized organic iodide, the structural network of $PbI_2$ and the conformation of organic chain are changed to form IO-hybrids. Uniform 2D IO-hybrids films can, thus, be fabricated by an appropriate choice of organic moieties. A brief description of the intercalation process is explained here by taking an example of the IO-hybrid 2(1-cyclohexenyl) ethylammonium tetraiodoplumbate ($(C_6H_9C_2H_4NH_3)_2PbI_4$, CHPI) [7]. The-organic iodide, 2-(1-cyclohexenyl) ethylammonium iodide ($C_6H_9C_2H_4NH_3I$) (CHI) is formed when 1 mL of 2-(1-cyclohexenyl) ethylamine was added to 2.1 mL of HI (47%). The obtained light yellow precipitate CHI eventually has been filtered and dried for further use. Similar is the method for other organic iodide synthesis. Suitable solvent (such as

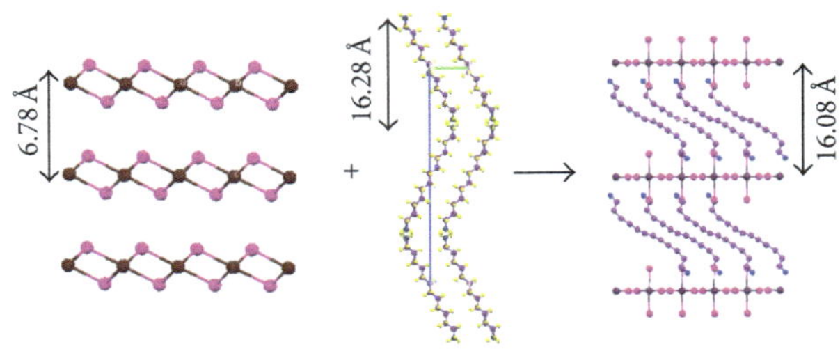


FIGURE 5: Schematic representation of intercalation process for 2D layered IO-hybrid thin films [7]. Essentially the intercalation method involves the intercalation of organic moiety into intestinal spaces of layered inorganic host to obtain desired inorganic-organic exfoliated layered hybrid.

toluene or a combination of toluene and isopropanol) is taken to dissolve organic iodide. Special care has to be taken on the solvent ratio so as to dissolve only the organic iodide but not $PbI_2$ or the resultant hybrid. Finally, the deposited $PbI_2$ thin films are dipped into organic iodide films, with a controlled speed and for specific time, to obtain desired IO-hybrid films. The resultant films, fabricated by the intercalation process, were smooth and uniform over a large area and had shown relatively well-stacked (*00l* oriented) inorganic and organic monolayers [7] (Figure 6).

## 6. Room-Temperature Optical Exciton Features

Exciton absorption/emission features in these low-dimensional IO-hybrids, especially in 2D hybrids, are significantly enhanced as compared to the 3D counterpart, due to the low-dimensionality. In the PbI based 2D IO-hybrids, electrons are excited from the valence band (VB), consisting of a mixture of Pb (6s) and I (5p) states, to the conduction band (CB), derived mainly from the Pb (6p) states, leaving holes in the VB. An electron and hole pair up to form an exciton via coulomb interaction, and the resulted excitons produce photoluminescence by radiative recombination [105, 106]. The enhancement of exciton features in 2D systems is a well-known phenomenon because of the spatial electron and hole confinement in a very thin and deep quantum wells and, hence, multiples the exciton binding energy, enabling *quantum confinement effect*. Apart from the usual quantum confinement in these natural *MQWs*, the excitons binding energies are further enhanced due to large contrast in dielectric constants of organic and inorganic layers. Such large binding energy enhancement, leading to strong room-temperature exciton features, often is referred as *dielectric confinement* effects [107, 108]. Ishihara et al. [105] in 1989, reported for the first time the exciton binding energy of $(C_{10}H_{21}NH_3)_2PbI_4$ (C10PI), which is 370 meV and is much higher than that of bulk $PbI_2$, (~30 meV) [106]. This was well-accounted from *dielectric confinement* assumption, where the dielectric difference between "well" and "barrier" induces strong columbic interaction between an electron and a hole

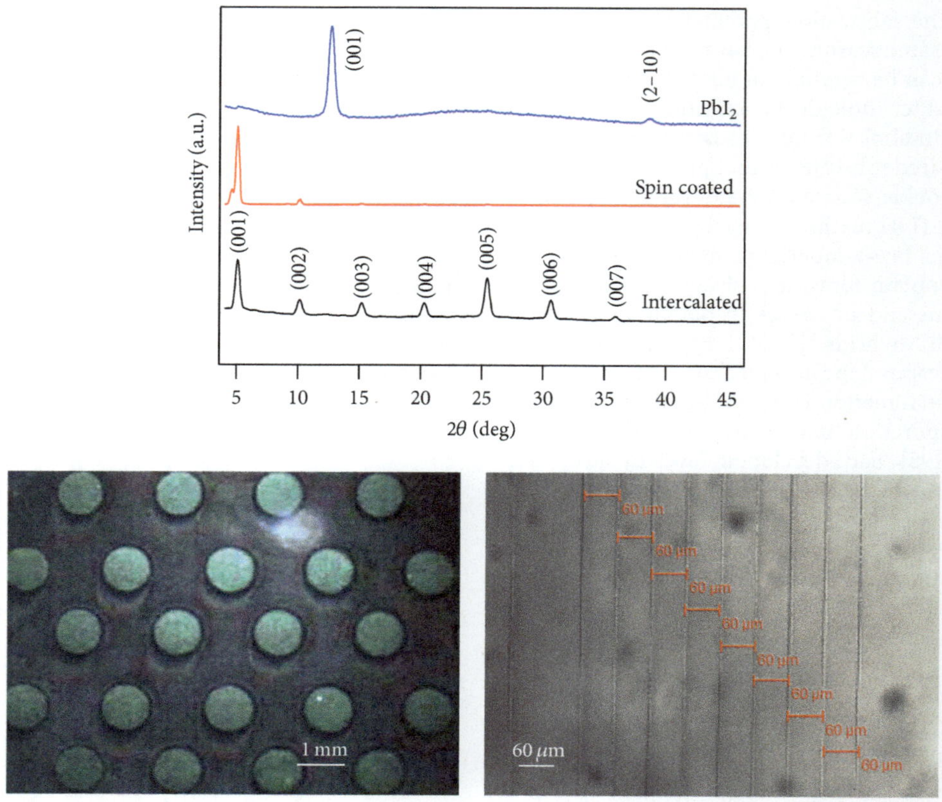


(a) (b)

FIGURE 6: X-ray diffraction patterns of pure $PbI_2$ film, spin-coated CHPI (from synthesized CHPI), and intercalated CHPI films (for 10 seconds) [7]. Bottom microscopic reflection images represent the obtained periodic photonic structures using top-down technology: (a) patterned structures from direct deposition of materials from templates using intercalation technique and (b) femtosecond laser written structures on CHPI thin films.

and as a consequence the binding energy of the exciton is 12 times larger than that of $PbI_2$ [84, 105, 107–109].

For example, in CAPI (Figure 4), the inorganic and organic layer thicknesses are estimated to be 6 Å and 10 Å. This layered structure resembles *MQWs*, where inorganic layer has a bandgap of ~3 eV forming "well," and organic layer has a bandgap ~6 eV forming "barrier." Figure 7 shows the typical photoluminescence (PL) and absorption spectra of CAPI thin film.

The room-temperature absorption of CAPI shows two principal absorptions: a broad absorption at ~400 nm and a strong narrow peak at ~480 nm (Figure 7 (black)). While the former is attributed to the charge-transfer transition between the organic and inorganic layers, the narrow absorption peak at about 480 nm is attributed to the lowest exciton within the inorganic layers [84, 105, 110]. CAPI thin films show strong room-temperature photoluminescence (PL) at ~485 nm upon UV excitation (Figure 7 (red)). The PL spectrum of CAPI has narrow line shape with spectral width ~15 nm. The oscillator strength of exciton absorption peak obtained for the hybrid CAPI thin film (of thickness 100 nm)

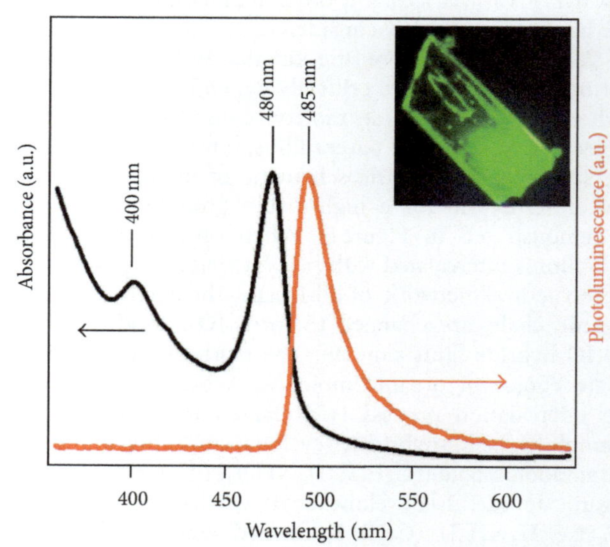


FIGURE 7: Absorption and PL spectra of the 2D layered IO-hybrid, CAPI, thin film [81]. Inset shows typical single crystal PL microscope image.

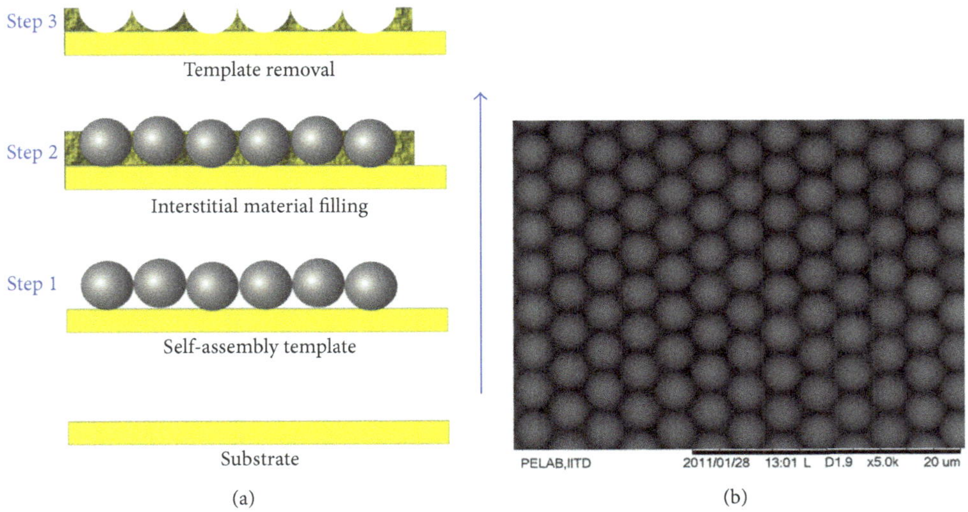


FIGURE 8: (a) Schematic representation of self-assembly template assisted fabrication of photonic structures and (b) polymer microsphere templates.

is $f = 6.5 \times 10^{15}\,cm^{-2}$. Since, for the film thickness of 100nm, there are ~70 quantum wells, the oscillator strength per quantum-well $f_{qw} = 9.2 \times 10^{13}\,cm^{-2}$. This value is an order of magnitude higher than the conventional inorganic quantum wells such as InGaAs structure [111].

Despite some understanding on the dependence of exciton energies of 2D layered IO-hybrids on various parameters like the inorganic *well* width and organic *barrier* separation, the dielectric contrast and the inorganic layer geometrical arrangement, quantitative calculations of exciton binding energies remain out of reach. A systematic correlation between the exciton energies and a specific structural feature has recently been established and discussed [110, 112]. However, the structural features are strongly dependent on several factors such as (1) disorder or conformation of the organic moiety, (2) crystal packing, (3) arrangement of inorganic layers, and (4) position of tagging of ammonium group in organic moiety to the PbI network. As a consequence the optical excitons features are also strongly dependent on the studied IO-hybrid thin films based on organic moiety conformation, solvent used, temperature, and the film thickness [110, 113–115]. Special control over thickness of the IO-hybrid film is critical for device oriented 3D structure fabrication, to avoid the exciton deformation and defect related emission/absorption features.

## 7. Templated Self-Assembled Microstructures

Carving useful materials itself into the nano to micron-sized structures using simple bottom-up technology is economically viable. This new approach is based on the natural self-assembly of templates and subsequently space filling the voids either by precipitation via chemical routes or by the electrochemical reduction of materials (Figure 8). Therefore, nano-/microstructuring using templates such as artificial polymer opals and liquid crystal is a whole new class of research. These micro-/nanostructures have potential applications such as tunable plasmonic bandgaps [116], novel types of liquid crystal displays [117], and nanolaser cavities [118]. These macroporous materials are already available in the market as surface enhanced Raman scattering (SERS) devices [119]. Despite the fact that the methodology offers highly ordered structures with very large single-crystalline domains, it has so far been restricted only to metals.

However, fabrication of self-assembled semiconductor photonic structures is still scarce, particularly understanding the chemistry for interstitial filling and deposition for semiconductors and adequate interconnectivity between the pores [120–123]. Electrochemical deposition is one of the low-cost deposition techniques which is an optimal bottom up technique for complete interstitial space filling of desired material through various types of templates. Over the years, conducting thin films (Ag, Au, etc.) were deposited on various semiconducting and conducting substrates utilizing this technique, but the deposition of semiconductors, organic materials, and polymers have always been a formidable task, due to low-conductivity issues. Constant research work on electrodeposition finally emerged out to be more useful than expected when deposition of semiconductors and several organic materials had been possible. However, structural and optoelectronic properties of semiconductors are critically affected by the preparation conditions such as fabrication method, types of substrates, thickness, and annealing conditions. Nevertheless, it is necessary to optimize the fabrication conditions and postprocessing methods for desired applications.

Here the template assisted method to fabricate periodic structures from 50 nm to as big as 20 micron is exemplified. After having optimized electrodeposition conditions, such as electrolyte recipe, surface quality, structural, and optical properties, we have extensively investigated to fabricate 2D and 3D periodic and quasiperiodic nano/micro-structure

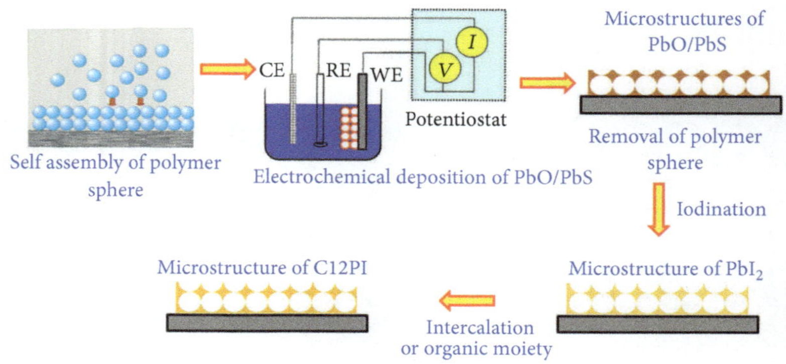


FIGURE 9: Schematic representation of carving naturally self-assembled hybrid systems into 2D/3D microstructures from template self-assembly method.

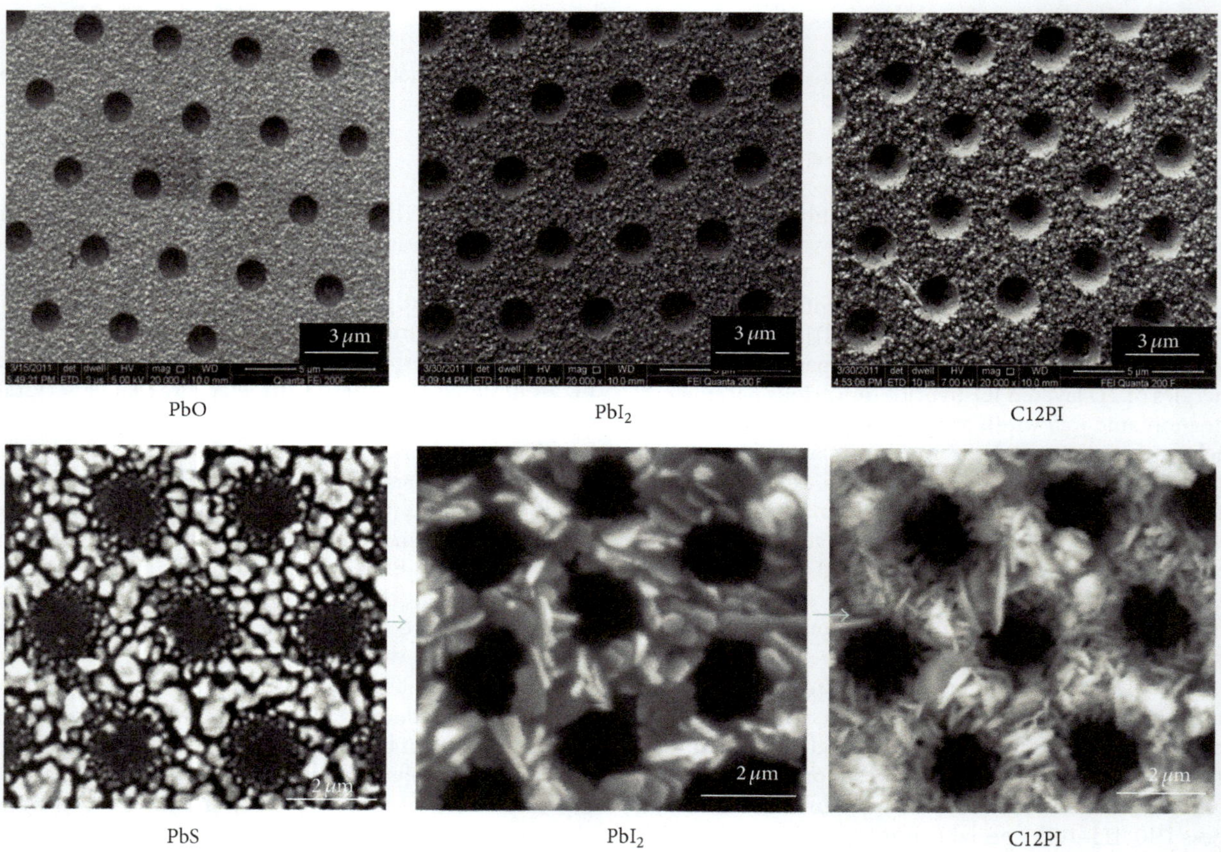


FIGURE 10: Scanning electron microscope images of template-assisted PbO, $PbI_2$, PbS, and C12PI microstructures [82].

from template-assisted growth techniques [11, 12, 17, 82]. During this investigation, several composite semiconductors such as CdSe, CdTe, ZnO, PbO, PbS, and $PbI_2$ are successfully fabricated and their optoelectronic properties are thoroughly investigated. We further demonstrated micron-scale 2D periodic highly emitting IO-hybrid structures, using template-assisted electrochemical growth followed by three-step processing, which can be easily extended to wave-length scale and nanoscale structures.

The systematic procedure has been explained in the schematic diagram (Figure 9). Essentially the electrodeposited PbO/PbS microstructures (Figure 10) are iodinised to obtain $PbI_2$ microstructures, then the presynthesized organic iodide is intercalated into $PbI_2$ to obtain desired IO-hybrid microstructures. During the process, the thickness of IO-hybrid film has to be fixed and such thickness optimization is required because these type of IO-hybrids are sensitive to thickness induced stacking imperfections, which directly

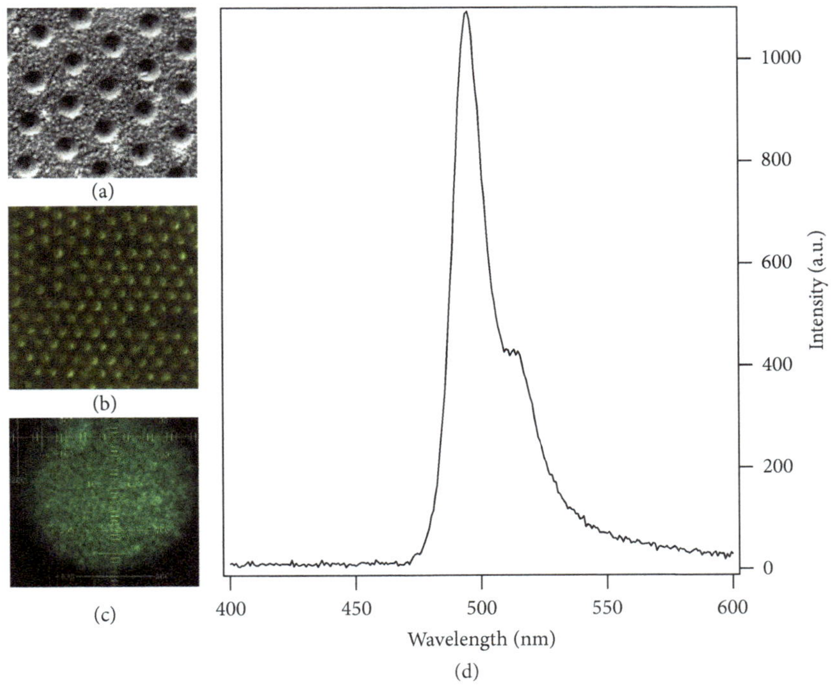


Figure 11: (a) SEM, (b) dark field microscope, and (c) PL microscope images ($\lambda_{ex} \sim 410$ nm), (d) corresponding to the PL spectra from the fabricated C12PI microstructures [82].

results into rapid change in their exciton-related emission/absorption behavior. The thickness-dependent disorder produce uneven crystalline planes, and as a consequence, the shift in the exciton PL peak and/or broad defect emission were observed [81, 113].

The fabricated IO-hybrid microstructures (Figures 10 and 11) are uniform over large areas and are highly luminescent. In general, the fabrication of 3D structures from infiltration methods, using conventionally synthesized hybrids is difficult due to surface morphology issues [81]. In contrast to that, the hybrid structures fabricated from this novel method pave the way for new directions in the fabrication of different photonic structures of IO-hybrids. As a hybrid nanosystem, low-dimensional IO-hybrid systems have shown potential applications and those applications were reviewed in previous sections. Photonic structures of these nano systems could further improve the optical properties and hence would find device applications in the area of optoelectronics.

## 8. Conclusions

Fabrication, structural and optical exciton features of naturally self-assembled low-dimensional IO-hybrid nano systems were discussed. While the fabrication of these self-assembled systems are usually from solution chemistry techniques, a novel device-compatible thin film fabrication from very inexpensive method, that is, intercalation, was reviewed. Finally, the designing and fabrication of optoelectronic-compatible photonic architectures from these IO-hybrids, especially from template-assisted method, have been clearly discussed.

## Acknowledgments

This work is funded from High-impact Research initiative of IIT Delhi, UK-India Education Research Initiative (UKIERI), Nano Research Facility (MCIT, Govt. India funded) of IIT Delhi, and the Department of Science and Technology (DST, Govt. of India). Authors are thankful to Professor Jeremy Baumberg, University of Cambridge, UK, for his valuable support and fruitful discussions.

## References

[1] G. M. Whitesides and M. Boncheva, "Beyond molecules: self-assembly of mesoscopic and macroscopic components," *Proceedings of the National Academy of Sciences of the United States of America*, vol. 99, no. 8, pp. 4769–4774, 2002.

[2] J. M. Lehn, "Toward self-organization and complex matter," *Science*, vol. 295, no. 5564, pp. 2400–2403, 2002.

[3] P. Yang, D. Zhao, D. I. Margolese, B. F. Chmelka, and G. D. Stucky, "Block copolymer templating syntheses of mesoporous metal oxides with large ordering lengths and semicrystalline framework," *Chemistry of Materials*, vol. 11, no. 10, pp. 2813–2826, 1999.

[4] S. Mahima, R. Kannan, I. Komath, M. Aslam, and V. K. Pillai, "Synthesis of platinum Y-junction nanostructures using hierarchically designed alumina templates and their enhanced electrocatalytic activity for fuel-cell applications," *Chemistry of Materials*, vol. 20, no. 3, pp. 601–603, 2008.

[5] G. S. Attard, C. G. Göltner, J. M. Corker, S. Henke, and R. H. Templer, "Liquid-crystal templates for nanostructured metals," *Angewandte Chemie (International Edition in English)*, vol. 36, no. 12, pp. 1315–1317, 1997.

[6] K. J. Bachmann and J. L. Shay, "An InGaAs detector for the 1.0–1.7-$\mu$m wavelength range," *Applied Physics Letters*, vol. 32, p. 446, 1978.

[7] K. Pradeesh, J. J. Baumberg, and G. Vijaya Prakash, "*In situ* intercalation strategies for device-quality hybrid inorganic-organic self-assembled quantum wells," *Applied Physics Letters*, vol. 95, no. 3, Article ID 033309, 2009.

[8] I. Saikumar, Shahab Ahmad, J. J. Baumberg, and G. Vijaya Prakash, "Fabrication of excitonic luminescent inorganic-organic hybrid nano- and microcrystals," *Scripta Materialia*, vol. 67, no. 10, pp. 834–837, 2012.

[9] Y. Xia, J. A. Rogers, K. E. Paul, and G. M. Whitesides, "Unconventional methods for fabricating and patterning nanostructures," *Chemical Reviews*, vol. 99, pp. 1823–1848, 1999.

[10] M. Trupke, F. Ramirez-Martinez, E. A. Curtis et al., "Pyramidal micromirrors for microsystems and atom chips," *Applied Physics Letters*, vol. 88, no. 7, Article ID 071116, 2006.

[11] G. Vijaya Prakash, R. Singh, A. Kumar, and R. K. Mishra, "Fabrication and characterisation of CdSe photonic structures from self-assembled templates," *Materials Letters*, vol. 60, no. 13-14, pp. 1744–1747, 2006.

[12] G. Vijaya Prakash, K. Pradeesh, A. Kumar et al., "Fabrication and optoelectronic characterisation of ZnO photonic structures," *Materials Letters*, vol. 62, no. 8-9, pp. 1183–1186, 2008.

[13] R. V. Nair and R. Vijaya, "Three-dimensionally ordered photonic crystal heterostructures with a double photonic stop band," *Journal of Applied Physics*, vol. 102, Article ID 056102, 3 pages, 2007.

[14] R. V. Nair and R. Vijaya, "Structural and optical characterization of photonic crystals synthesized using the inward growing self-assembling method," *Applied Physics A*, vol. 90, pp. 559–563, 2008.

[15] M. C. Gonçalves, L. M. Fortes, R. M. Almeida, A. Chiasera, A. Chiappini, and M. Ferrari, "3-D rare earth-doped colloidal photonic crystals," *Optical Materials*, vol. 31, no. 9, pp. 1315–1318, 2009.

[16] L. Irimpan, V. P. N. Nampoori, P. Radhakrishnan, A. Deepthy, and B. Krishnan, "Size dependent fluorescence spectroscopy of nanocolloids of ZnO," *Journal of Applied Physics*, vol. 102, no. 6, Article ID 063524, 2007.

[17] S. Coyle, G. Vijaya Prakash, J. J. Baumberg, M. Abdelsalem, and P. N. Bartlet, "Spherical micromirrors from templated self-assembly: polarization rotation on the micron scale," *Applied Physics Letters*, vol. 83, no. 4, p. 767, 2003.

[18] V. Bulovic, G. Gu, P. E. Burrows, M. E. Thompson, and S. R. Forrest, "Transparent light-emitting devices," *Nature*, vol. 380, p. 29, 1996.

[19] G. Gu, V. Bulovic, P. E. Burrows, S. R. Forrest, and M. E. Thompson, "Transparent organic light emitting devices," *Applied Physics Letters*, vol. 68, p. 2606, 1996.

[20] V. Bulović, V. G. Kozlov, V. B. Khalfin, and S. R. Forrest, "Transform-limited, narrow-linewidth lasing action in organic semiconductor microcavities," *Science*, vol. 279, no. 5350, pp. 553–555, 1998.

[21] V. G. Kozlov, V. Bulović, P. E. Burrows, and S. R. Forrest, "Laser action, in organic semiconductor waveguide and double-heterostructure devices," *Nature*, vol. 389, no. 6649, pp. 362–364, 1997.

[22] W. Brütting, Ed., *Physics of Organic Semiconductors*, Wiley-VCH, Weinheim, Germany, 2005.

[23] M. Furuki, M. Tian, Y. Sato et al., "Observation of sub-100-fs optical response from spin-coated films of squarylium dye *J* aggregates," *Applied Physics Letters*, vol. 78, no. 18, pp. 2634–2636, 2001.

[24] K. Misawa, H. Ono, K. Minoshima, and T. Kobayashi, "New fabrication method for highly oriented *J* aggregates dispersed in polymer films," *Applied Physics Letters*, vol. 63, no. 5, pp. 577–579, 1993.

[25] F. Sasaki, S. Kobayashi, and S. Haraichi, "Enhancement of the optical nonlinearity in pseudoisocyanine *J* aggregates embedded in distributed feedback microcavities," *Applied Physics Letters*, vol. 81, p. 391, 2002.

[26] A. Eilmes, "Excited-state polarizability of J-aggregates," *Chemical Physics Letters*, vol. 347, no. 1–3, pp. 205–210, 2001.

[27] N. Fukutake, S. Takasaka, and T. Kobayashi, "Energy transfer between two kinds of J-aggregates studied by near-field absorption-fluorescence spectroscopy," *Chemical Physics Letters*, vol. 361, no. 1-2, pp. 42–48, 2002.

[28] P. Gomez-Romero, "Hybrid organic-inorganic materials-in search of synergic activity," *Advanced Materials*, vol. 13, no. 3, pp. 163–174, 2001.

[29] E. Fois, A. Gamba, and A. Tilocca, "On the unusual stability of Maya blue paint: molecular dynamics simulations," *Microporous and Mesoporous Materials*, vol. 57, no. 3, pp. 263–272, 2003.

[30] D. E. Arnold, J. R. Branden, P. R. Williams, G. M. Feinman, and J. P. Brown, "The first direct evidence for the production of Maya Blue: rediscovery of a technology," *Antiquity*, vol. 82, no. 315, pp. 151–164, 2008.

[31] K. S. Aleksandrov and V. V. Beznosikov, "Hierarchies of perovskite-like crystals," *Physics of the Solid State*, vol. 39, no. 5, pp. 695–715, 1997.

[32] D. B. Mitzi, "Synthesis, structure, and properties of organic-inorganic perovskites and related materials," *Progress in Inorganic Chemistry*, vol. 48, p. 1, 1999.

[33] D. B. Mitzi, "Organic-inorganic perovskites containing trivalent metal halide layers: the templating influence of the organic cation layer," *Inorganic Chemistry*, vol. 39, no. 26, pp. 6107–6113, 2000.

[34] A. Poglitsch and D. Weber, "Dynamic disorder in methylammoniumtrihalogenoplumbates (II) observed by millimeter-wave spectroscopy," *The Journal of Chemical Physics*, vol. 87, no. 11, pp. 6373–6378, 1987.

[35] D. Weber, "$CH_3NH_3PBX_3$, a Pb(II)-system with cubic perovskite structure," *Zeitschrift für Naturforschung B*, vol. 33, pp. 1443–1445, 1978.

[36] S. Wang, D. B. Mitzi, G. A. Landrum, H. Genin, and R. Hoffmann, "Synthesis and solid state chemistry of $CH_3BiI_2$: a structure with an extended one-dimensional organometallic framework," *Journal of the American Chemical Society*, vol. 119, no. 4, pp. 724–732, 1997.

[37] D. B. Mitzi, "Synthesis and crystal structure of the alkylbismuth diiodides: a family of extended one-dimensional organometallic compounds," *Inorganic Chemistry*, vol. 35, no. 26, pp. 7614–7619, 1996.

[38] E. O. Schlemper and W. C. Hamilton, "The crystal structure of dimethyltin dilfluoride. An example of octahedral coordination of tin," *Inorganic Chemistry*, vol. 5, no. 6, pp. 995–998, 1966.

[39] X. Huang, J. Li, and H. Fu, "The first covalent organic-inorganic networks of hybrid chalcogenides: structures that may lead to a new type of quantum wells," *Journal of the American Chemical Society*, vol. 122, no. 36, pp. 8789–8790, 2000.

[40] F. Chiarella, R. Mosca, M. Pavesi, A. Zappettini, P. Ferro, and F. Licci, "Enhanced luminescence of CuCl microcrystals in a organic-inorganic hybrid matrix," *Applied Physics A*, vol. 88, no. 2, pp. 235–237, 2007.

[41] M. Eraa, K. Miyakea, Y. Yoshidac, and K. Yase, "Orientation of azobenzene chromophore incorporated into metal halide-based layered perovskite having organic-inorganic superlattice structure," *Thin Solid Films*, vol. 393, pp. 24–27, 2001.

[42] N. F. Stephens, A. M. Z. Slawin, and P. Lightfoot, "A novel scandium fluoride, $[C_2N_2H_{10}]_{(0.5)}[ScF_4]$, with an unprecedented tungsten bronze-related layer structure," *Chemical Communications*, no. 5, pp. 614–615, 2004.

[43] M. Szafrafiski, "Investigation of phase instabilities in guanidinium halogenoplumbates(II)," *Thermochimica Acta*, vol. 307, no. 2, pp. 177–183, 1997.

[44] R. Kind, "Phase transitions and incommensurability in crystalline model bilayers," *Berichte der Bunsengesellschaft für physikalische Chemie*, vol. 87, pp. 248–254, 1983.

[45] J. Etxebarria, J. Fernandez, M. A. Arriandiaga, and M. J. Tello, "Influence of the thermal expansion on the piezoelectric photoacoustic detection of ferro-paraelastic phase transition in $(CH_3CH_2NH_3)_2CuCl_4$," *Journal of Physics C*, vol. 18, no. 1, pp. L13–L17, 1985.

[46] T. Goto, B. Lthi, R. Geick, and K. Strobel, "Elastic soft mode in perovskite-type layer-structure materials," *Physical Review B*, vol. 22, no. 7, pp. 3452–3458, 1980.

[47] N. Mercier, N. Louvain, and W. Bi, "Structural diversity and retro-crystal engineering analysis of iodometalate hybrids," *CrystEngComm*, vol. 11, no. 5, pp. 720–734, 2009.

[48] D. B. Mitzi, "Templating and structural engineering in organic-inorganic perovskites," *Journal of the Chemical Society, Dalton Transactions*, no. 1, pp. 1–12, 2001.

[49] D. B. Mitzi, C. D. Dimitrakopoulos, and L. L. Kosbar, "Structurally tailored organic-inorganic perovskites: optical properties and solution-processed channel materials for thin-film transistors," *Chemistry of Materials*, vol. 13, no. 10, pp. 3728–3740, 2001.

[50] T. Fujita, H. Nakashima, M. Hirasawa, and T. Ishihara, "Ultrafast photoluminescence from $(C_6H_5C_2H_4NH_3)_2PbI_4$," *Journal of Luminescence*, vol. 87, pp. 847–849, 2000.

[51] M. Shimizu, J. Fujisawa, and T. Ishihara, "Photoluminescene of the inorganic-organic layered semiconductor $(C_6H_5C_2H_4NH_3)_2PbI_4$: observation of triexciton formation," *Physical Review B*, vol. 74, Article ID 155206, 6 pages, 2006.

[52] M. Shimizu, J. I. Fujisawa, and T. Ishihara, "Nonlinear luminescence from an inorganic-organic layered semiconductor," *Journal of Luminescence*, vol. 122-123, no. 1-2, pp. 485–487, 2007.

[53] T. Ishihara, "Optical properties of PbI-based perovskite structures," *Journal of Luminescence*, vol. 60-61, pp. 269–274, 1994.

[54] G. A. Mousdis, G. C. Papavassiliou, C. P. Raptopouloub, and A. Terzis, "Preparation and characterization of $[H_3N(CH_2)_6NH_3]PbI_4$ and similar compounds with a layered perovskite structure," *Journal of Materials Chemistry*, vol. 10, pp. 515–518, 2000.

[55] G. C. Papavassiliou, I. B. Koutselas, A. Terzis, and M. H. Whangbo, "Structural and electronic properties of the natural quantum-well system $(C_6H_5CH_2CH_2NH_3)_2SnI_4$," *Solid State Communicationsications*, vol. 91, no. 9, pp. 695–698, 1994.

[56] A. Lemmerer and D. G. Billing, "Two packing motifs based upon chains of edge-sharing $PbI_6$ octahedra," *Acta Crystallographica Section C*, vol. 62, no. 12, p. m597, 2006.

[57] C. P. Raptopoulou, A. Terzis, G. A. Mousdis, and G. C. Papavassiliou, "Preparation, structure and optical properties of $[CH_3SC(NH_2)_2]_3SnI_5$, $[CH_3SC(NH_2)_2][HSC(NH_2)_2]SnBr_4$, $(CH_3C_5H_4NCH_3)PbBr_3$, and $[C_6H_5CH_2SC(NH_2)_2]_4Pb_3I_{10}$," *Zeitschrift für Naturforschung B*, vol. 57, pp. 645–650, 2002.

[58] G. C. Papavassiliou, G. A. Mousdis, A. Terzis, and C. P. Raptopoulou, "Crystal structure and optical properties of 4-[4-(dimethylamino)-styryl]-1-methyl-pyridinium lead tribromide," *Zeitschrift für Naturforschung B*, vol. 58, pp. 815–816, 2003.

[59] G. C. Papavassiliou, G. A. Mousdis, I. Koutselas et al., "Some new synthetic low-dimensional semiconductors based on inorganic units," *Advanced Materials for Optics and Electronics*, vol. 8, no. 5, pp. 263–267, 1998.

[60] C. Xu, S. Fukuta, H. Sakakura et al., "Anomalous electro-absorption in the low-temperature phase of $(C_{10}H_{21}NH_3)_2PbI_4$," *Solid State Communications*, vol. 77, pp. 923–926, 1991.

[61] M. Era, S. Morimoto, T. Tsutsui, and S. Saito, "Organic-inorganic heterostructure electroluminescent device using a layered perovskite semiconductor $(C_6H_5C_2H_4NH_3)_2PbI_4$," *Applied Physics Letters*, vol. 65, pp. 676–3, 1994.

[62] K. Morii, M. Ishida, T. Takashima et al., "Encapsulation-free hybrid organic-inorganic light-emitting diodes," *Applied Physics Letters*, vol. 89, no. 18, Article ID 183510, 3 pages, 2006.

[63] K. Chondroudis and D. B. Mitzi, "Electroluminescence from an organic-inorganic perovskite incorporating a quaterthiophene dye within lead halide perovskite layers," *Chemistry of Materials*, vol. 11, no. 11, pp. 3028–3030, 1999.

[64] T. Hattori, T. Taira, M. Era, T. Tsutsui, and S. Saito, "Highly efficient electroluminescence from a heterostructure device combined with emissive layered-perovskite and an electron-transporting organic compound," *Chemical Physics Letters*, vol. 254, pp. 103–108, 1996.

[65] K. Saruwatari, H. Sato, T. Idei et al., "Photoconductive properties of organic-inorganic hybrid films of layered perovskite-type niobate," *Journal of Physical Chemistry B*, vol. 109, no. 25, pp. 12410–12416, 2005.

[66] W. E. Mahmoud, "A novel photodiode made of hybrid organic/inorganic nanocomposite," *Journal of Physics D*, vol. 42, no. 15, Article ID 155502, 2009.

[67] A. M. Guloy, Z. Tang, P. B. Miranda, and V. I. Srdanov, "A new luminescent organic-inorganic hybrid compound with large optical nonlinearity," *Advanced Materials*, vol. 13, pp. 833–837, 2001.

[68] M. Shimizu, J. Fujisawa, and J. Ishi-Hayase, "Influence of the dielectric confinement on excitonic nonlinearity in inorganic-organic layered semiconductors," *Physical Review B*, vol. 71, no. 20, Article ID 205306, 9 pages, 2005.

[69] M. Shimizu, N. A. Gippius, S. G. Tikhodeev, and T. Ishihara, "Coulomb correction to the dressed exciton in an inorganic-organic layered semiconductor: detuning dependence of the Stark shift," *Physical Review B*, vol. 69, no. 15, Article ID 155201, 5 pages, 2004.

[70] C. Q. Xu, H. Sakakura, T. Kondo et al., "Magneto-optical effects of excitons in $(C_{10}H_{21}NH_3)_2PbI_4$ under high magnetic fields up to 40 T," *Solid State Communicationsications*, vol. 79, no. 3, pp. 249–253, 1991.

[71] T. Sekine, T. Okuno, and K. Awaga, "Observation of spontaneous magnetization in the layered perovskite ferromagnet, (*p*-Chloroanilinium)$_2CuBr_4$," *Inorganic Chemistry*, vol. 37, no. 9, pp. 2129–2133, 1998.

[72] K. Tanaka, T. Takahashi, T. Kondo et al., "Electronic and excitonic structures of inorganic-organic perovskite-type quantum-well crystal $(C_4H_9NH_3)_2PbBr_4$," *Japanese Journal of Applied Physics*, vol. 44, no. 8, pp. 5923–5932, 2005.

[73] J. Fujisawa and T. Ishihara, "Excitons and biexcitons bound to a positive ion in a bismuth-doped inorganic-organic layered lead iodide semiconductor," *Physical Review B*, vol. 70, no. 20, Article ID 205330, 6 pages, 2004.

[74] Y. Kato, D. Ichii, K. Ohashi et al., "Extremely large binding energy of biexcitons in an organic-inorganic quantum-well material $(C_4H_9NH_3)_2PbBr_4$," *Solid State Communicationsications*, vol. 128, no. 1, pp. 15–18, 2003.

[75] T. Goto, H. Makino, T. Yao et al., "Localization of triplet excitons and biexcitons in the two-dimensional semiconductor $(CH_3C_6H_4CH_2NH_3)_2PbBr_4$," *Physical Review B*, vol. 73, no. 11, Article ID 115206, pp. 1–5, 2006.

[76] T. Kondo, T. Azuma, T. Yuasa, and R. Ito, "Biexciton lasing in the layered perovskite-type material $(C_6H_{13}NH_3)_2PbI_4$," *Solid State Communicationsications*, vol. 105, no. 4, pp. 253–255, 1998.

[77] D. B. Mitzi, K. Chondroudis, and C. R. Kagan, "Organic-inorganic electronics," *IBM Journal of Research and Development*, vol. 45, no. 1, pp. 29–45, 2001.

[78] C. R. Kagan, D. B. Mitzi, and C. D. Dimitrakopoulos, "Organic-inorganic hybrid materials as semiconducting channels in thin-film field-effect transistors," *Science*, vol. 286, no. 5441, pp. 945–947, 1999.

[79] K. Shibuya, M. Koshimizu, Y. Takeoka, and K. Asai, "Scintillation properties of $(C_6H_{13}NH_3)_2PbI_4$: exciton luminescence of an organic/inorganic multiple quantum well structure compound induced by 2.0 MeV protons," *Nuclear Instruments and Methods in Physics Research B*, vol. 194, pp. 207–212, 2002.

[80] D. B. Mitzi, *Functional Hybrid Materials*, Wiley, Weinheim, Germany, 2004, edited by P. G. Romero and C. Sanchez.

[81] G. Vijaya Prakash, K. Pradeesh, R. Ratnani, K. Saraswat, M. E. Light, and J. J. Baumberg, "Structural and optical studies of local disorder sensitivity in natural organic-inorganic self-assembled semiconductors," *Journal of Physics D*, vol. 42, no. 18, Article ID 185405, 2009.

[82] V. K. Dwivedi, J. J. Baumberg, and G. Vijaya Prakash, "Direct deposition of inorganic-organic hybrid semiconductors and their template-assisted microstructures," *Materials Chemistry and Physics*, vol. 137, no. 3, pp. 941–946, 2013.

[83] Y. Kawabata, M. Yoshizawa-Fujita, Y. Takeoka, and M. Rikukawa, "Relationship between structure and optoelectrical properties of organic-inorganic hybrid materials containing fullerene derivatives," *Synthetic Metals*, vol. 159, no. 9-10, pp. 776–779, 2009.

[84] T. Ishihara, J. Takahanshi, and T. Goto, "Optical properties due to electronic transitions in two-dimensional semiconductors $(C_nH_{2n+1}NH_3)_2PbI_4$," *Physical Review B*, vol. 42, pp. 11099–11107, 1990.

[85] S. Noda, M. Fujita, and T. Asano, "Spontaneous-emission control by photonic crystals and nanocavities," *Nature Photonics*, vol. 1, no. 8, pp. 449–458, 2007.

[86] D. B. Mitzi, C. A. Feild, W. T. A. Harrison, and A. M. Guloy, "Conducting tin halides with a layered organic-based perovskite structure," *Nature*, vol. 369, no. 6480, pp. 467–469, 1994.

[87] D. B. Mitzi, S. Wang, C. A. Feild, C. A. Chess, and A. M. Guloy, "Conducting layered organic-inorganic halides containing ⟨110⟩-oriented perovskite sheets," *Science*, vol. 267, no. 5203, pp. 1473–1476, 1995.

[88] Z. Xu, D. B. Mitzi, C. D. Dimitrakopoulos, and K. R. Maxcy, "Semiconducting perovskites $(2\text{-}XC_6H_4C_2H_4NH_3)_2SnI_4$ (X = F, Cl, Br): steric interaction between the organic and inorganic layers," *Inorganic Chemistry*, vol. 42, pp. 2031–2039, 2003.

[89] D. B. Mitzi, D. R. Medeiros, and P. R. L. Malenfant, "Intercalated organic-inorganic perovskites stabilized by fluoroaryl-aryl interactions," *Inorganic Chemistry*, vol. 41, no. 8, pp. 2134–2145, 2002.

[90] K. Tanakaa, T. Takahashia, T. Bana, T. Kondoa, K. Uchidab, and N. Miura, "Comparative study on the excitons in lead-halide-based perovskite-type crystals $CH_3NH_3PbBr_3$ $CH_3NH_3PbI_3$," *Solid State Communicationsications*, vol. 127, p. 619, 2003.

[91] Z. Xu, D. B. Mitzi, and D. R. Medeiros, "$[(CH_3)_3NCH_2CH_2NH_3]$ $SNI_4$: a layered perovskite with quaternary/primary ammonium dications and short interlayer iodine-iodine contacts," *Inorganic Chemistry*, vol. 42, no. 5, pp. 1400–1402, 2003.

[92] A. Lemmerer and D. G. Billing, "P-phenylenediammonium tetraiodozincate(II) dihydrate," *Acta Crystallographica Section E*, vol. 62, no. 4, pp. m779–m781, 2006.

[93] A. Lemmerer and D. G. Billing, "Two packing motifs based upon chains of edge-sharing $PbI_6$ octahedra," *Acta Crystallographica Section C*, vol. 62, no. 12, pp. m597–m601, 2006.

[94] H. Krautscheid, C. Lode, F. Vielsack, and H. Vollmer, "Synthesis and crystal structures of iodoplumbate chains, ribbons and rods with new structural types," *Journal of the Chemical Society, Dalton Transactions*, no. 7, pp. 1099–1104, 2001.

[95] T. Matsui, A. Yamaguchi, Y. Takeoka, M. Rikukawa, and K. Sanui, "Fabrication of two-dimensional layered perovskite $[NH_3(CH_2)_{12}NH_3]PbX_4$ thin films using a self-assembly method," *Chemical Communications*, no. 10, pp. 1094–1095, 2002.

[96] T. Matsushima, K. Fujita, and T. Tsutsui, "High field-effect hole mobility in organic-inorganic hybrid thin films prepared by vacuum vapor deposition technique," *Japanese Journal of Applied Physics*, vol. 43, pp. L1199–L1201, 2004.

[97] K. Ikegami, "Spectroscopic study of *J* aggregates of amphiphilic merocyanine dyes formed in their pure Langmuir films," *Journal of Chemical Physics*, vol. 121, p. 2337, 2004.

[98] D. B. Mitzi, M. T. Prikas, and K. Chondroudis, "Thin film deposition of organic-inorganic hybrid materials using a single source thermal ablation technique," *Chemistry of Materials*, vol. 11, no. 3, pp. 542–544, 1999.

[99] Z. Y. Cheng, H. F. Wang, Z. W. Quan, C. K. Lin, J. Lin, and Y. C. Han, "Layered organic-inorganic perovskite-type hybrid materials fabricated by spray pyrolysis route," *Journal of Crystal Growth*, vol. 285, no. 3, pp. 352–357, 2005.

[100] D. B. Mitzi, D. R. Medeiros, and P. W. DeHaven, "Low-temperature melt processing of organic-inorganic hybrid films," *Chemistry of Materials*, vol. 14, no. 7, pp. 2839–2841, 2002.

[101] J. E. Gieseking, "The mechanism of cation exchange in the montmorillonite-beidellite-nontronite type of clay minerals," *Soil Science*, vol. 47, no. 1, pp. 1–14, 1939.

[102] D. M. C. MacEwan, "Identification of the montmorillonite group of minerals by X-rays," *Nature*, vol. 154, no. 3914, pp. 577–578, 1944.

[103] W. F. Bradley, "Molecular associations between montmorillonite and some polyfunctional organic liquids," *Journal of the American Chemical Society*, vol. 67, no. 6, pp. 975–981, 1945.

[104] E. Ruiz-Hitzky, *Functional Hybrid Materials*, Wiley, Weinheim, Germany, 2004, edited by P. G. Romero and C. Sanchez.

[105] T. Ishihara, J. Takahashi, and T. Goto, "Exciton state in two-dimensional perovskite semiconductor $(C_{10}H_{21}NH_3)_2PbI_4$," *Solid State Communicationsications*, vol. 69, no. 9, pp. 933–936, 1989.

[106] L. C. Thanh, C. Depeursinge, F. Levy, and E. Mooser, "The band gap excitons in $PbI_2$," *Journal of Physics and Chemistry of Solids*, vol. 36, no. 7-8, pp. 699–702, 1975.

[107] L. V. Keldysh, "Coulomb interaction in thin semiconductor and semimetal films," *Journal of Experimental and Theoretical Physics*, vol. 29, p. 658, 1979.

[108] E. Hanamura, N. Nagaosa, M. Kumagai, and T. Takagahara, "Quantum wells with enhanced exciton effects and optical non-linearity," *Materials Science and Engineering B*, vol. 1, no. 3-4, pp. 255–258, 1988.

[109] A. Piryatinski, S. A. Ivanov, S. Tretiak, and V. I. Klimov, "Effect of quantum and dielectric confinement on the exciton-exciton interaction energy in type II core/shell semiconductor Nanocrystals," *Nano Letters*, vol. 7, pp. 108–115, 2007.

[110] K. Pradeesh, K. Nageswara Rao, and G. Vijaya Prakash, "Synthesis, structural, thermal and optical studies of inorganic-organic hybrid semiconductors, R-$PbI_4$," *Journal of Applied Physics*, vol. 113, no. 8, Article ID 083523, 9 pages, 2013.

[111] M. S. Skolnick, T. A. Fisher, and D. M. Whittaker, "Strong coupling phenomena in quantum microcavity structures," *Semiconductor Science and Technology*, vol. 13, p. 645, 1998.

[112] K. Pradeesh, J. J. Baumberg, and G. Vijaya Prakash, "Exciton switching and Peierls transitions in hybrid inorganic-organic self-assembled quantum wells," *Applied Physics Letters*, vol. 2, no. 10, Article ID 173305, 3 pages, 2009.

[113] K. Pradeesh, J. J. Baumberg, and G. Vijaya Prakash, "Strong exciton-photon coupling in inorganic-organic multiple quantum wells embedded low-Q microcavity," *Optics Express*, vol. 17, no. 24, pp. 22171–22178, 2009.

[114] K. Pradeesh, M. Agarwal, K. K. Rao, and G. Vijaya Prakash, "Synthesis, crystal structure and optical properties of quasi-one-dimensional lead (II) iodide: $C_{14}H_{18}N_2Pb_2I_6$," *Solid State Sciences*, vol. 12, no. 1, pp. 95–98, 2010.

[115] K. Pradeesh, G. S. Yadav, M. Singh, and G. Vijaya Prakash, "Synthesis, structure and optical studies of inorganic-organic hybrid semiconductor, $NH_3(CH_2)_{12}NH_3PbI_4$," *Materials Chemistry and Physics*, vol. 124, no. 1, pp. 44–47, 2010.

[116] S. Barman, N. V. Venkataraman, S. Vasudevan, and R. Seshadri, "Phase transitions in the anchored organic bilayers of long-chain alkylammonium lead iodides $(C_nH_{2n+1}NH_3)_2PbI_4$; $n$ = 12, 16, 18," *The Journal of Physical Chemistry B*, vol. 107, no. 8, pp. 1875–1883, 2003.

[117] K. Gauthron, J. S. Lauret, L. Doyennette et al., "Optical spectroscopy of two-dimensional layered $(C_6H_5C_2H_4\text{-}NH_3)_2\text{-}PbI_4$ perovskite," *Optics Express*, vol. 18, no. 6, pp. 5912–5919, 2010.

[118] D. G. Billing and A. Lemmerer, "Synthesis, characterization and phase transitions of the inorganic-organic layered perovskite-type hybrids $[(C_nH_{2n+1}NH_3)_2PbI_4]$ ($n$ = 12, 14, 16 and 18)," *New Journal of Chemistry*, vol. 32, pp. 1736–1746, 2008.

[119] S. Zhang, G. Lanty, J. S. Lauret, E. Deleporte, P. Audebert, and L. Galmiche, "Synthesis and optical properties of novel organic-inorganic hybrid nanolayer structure semiconductors," *Acta Materialia*, vol. 57, no. 11, pp. 3301–3309, 2009.

[120] I. Zhitomirsky, L. Gal-Or, A. Kohn, and H. W. Hennicke, "Electrochemical preparation of PbO films," *Journal of Materials Science Letters*, vol. 14, no. 11, pp. 807–810, 1995.

[121] T. K. Chaudhuri and H. N. Acharya, "Preparation of lead iodide films by iodination of chemically deposited lead sulphide films," *Materials Research Bulletin*, vol. 17, no. 3, pp. 279–286, 1982.

[122] G. D. Currie, J. Mudar, and O. Risgin, "Photoconductive and photovoltaic spectral response in $PbI_2$ crystals," *Applied Optics*, vol. 6, no. 6, pp. 1137–1138, 1967.

[123] A. E. Dugan and H. K. Hknisch, "Defect energy-level structure of $PbI_2$ single crystals," *Physical Review*, vol. 171, no. 3, pp. 1047–1051, 1968.

# 19

# Gold Nanoparticles and Nanocomposites in Clinical Diagnostics Using Electrochemical Methods

**Pranjal Chandra,[1,2,3] Jai Singh,[4] Amardeep Singh,[2] Ananya Srivastava,[5] Rajendra N. Goyal,[3,6] and Yoon Bo Shim[2,3]**

[1] *Amity Institute of Biotechnology, Amity University Uttar Pradesh, Sector 125, Noida 201303, India*
[2] *Department of Chemistry, Pusan National University, Busan 609-735, Republic of Korea*
[3] *Institute of BioPhysio Sensor Technology, Pusan National University, Busan 609-735, Republic of Korea*
[4] *Faculty of Nanotechnology and Advanced Materials, HMC, and Graphene Research Institute, Sejong University, Seoul 143-747, Republic of Korea*
[5] *Department of Chemistry, DDU Gorakhpur University, Gorakhpur 273009, India*
[6] *Department of Chemistry, Indian Institute of Technology Roorkee, Roorkee 247667, India*

Correspondence should be addressed to Pranjal Chandra; pchandral@amity.edu and Yoon Bo Shim; ybshim@pusan.ac.kr

Academic Editor: Young-Seok Shon

Progress and development in clinical diagnostics certainly focus upon the advances in the nanomaterials, particularly gold nanoparticles (AuNPs) that offer promise to solve the biocompatible and sensitive detection systems. This paper focuses on the recent application of AuNPs in clinical diagnosis. Various important methods of AuNPs synthesis and their application in clinical detection of various biomolecules using electrochemical detection methods have been described. AuNPs alone and in various composites are also described based on the various biosensors design recently published for the detection of cancer biomarkers, proteins, bacteria, and cancer cells. The effect of AuNPs type and size in clinical detection has also been briefly illustrated.

## 1. Introduction

The last decade has witnessed an exponential progress of activities in the field of nanoscience and nanotechnology worldwide, motivated both by the anticipation of considerate new science and by the impending trust for applications and financially feasible impacts. The prime action in this field has been in the production and characterization of new materials consisting of particles with dimensions in the order of a few nanometers, purported nanocrystalline materials. These nanosized materials have properties that are often significantly different from their counterparts with the bulk size [1, 2].

Inorganic, organic, and biological nanomaterials may have existed in nature since the evolution of life started on earth [3]. Some evident examples are microorganism and fine grained minerals in rocks [4]. In addition, nanostructures include quantum dots, quantum wires, grains, particles, nanotubes, nanorods, nanofibers, nanofoams, nanocrystals, nanoprecision self-assemblies, and thin films of metals, intermetallics, semiconductors, ferroelectrics, dielectrics, composites, alloys, blends, organics, organominerals, biomaterials, biomolecules, oligomers, polymers, functional structures and devices [5–8]. These novel materials made up of nanosized grains or building blocks offer unique and entirely different electrical, optical, mechanical, and magnetic properties compared to conventional micro- or millimeter-size materials owing to their distinctive size, shape, surface chemistry, and topology [7, 8]. Nanostructured materials and their base technologies have opened new exciting possibilities for future applications in medical, aerospace, catalysts, batteries, nonvolatile memories, sensors, insulators, color imaging, printing, flat panel displays, waveguides, modulators, computer chips, magneto-optical discs, transducers, photodetectors, optoelectronics, solar cells, lithography, holography, photoemitters, molecular sized transistors and switches, drug delivery, medicine, medical implants, pharmacy, cosmetics, and several others [9–12].

In general, noble metal used in the field of nanobiotechnology range in particle size between 10 and 100 nm, rarely exceeding above than 600 nm [13]. Biocompatible metal nanoparticles have gained considerable attention in recent years for potential applications in nanomedicine due to their interesting size dependent chemical, electronic, and optical properties [14]. In addition, silver (Ag) and gold (Au) nanoparticles (NPs) have fascinated scientist for over a century and are now heavily utilized in biomedical sciences and engineering [15–17]. The nanosize of these particles allows various messages with biomolecules on the cell surfaces and within the cells in way that can be decoded and designated to various biochemical and physiochemical properties of these cells [18]. Similarly, its potential application in drug delivery and noninvasive imaging offered various advantages over conventional pharmaceutical agents. In an effort to utilize nanoparticles at their full choke, it is important that the nanoparticulate systems should be stable, biocompatible, and selectively directed to specific sites in the body after systemic administration [19]. More specific targeting systems are designed to recognize the targeted cells such as cancer cells. AuNPs have attracted huge scientific and technological interest due to their ease of synthesis, chemical stability, and unique properties. The applications of AuNPs in different field are summarized in Table 1.

## 2. Synthetic Routes for the Preparation of AuNPs

Inorganic nanomaterials have received remarkable attention in the past few years due to their optical and well-defined size related to electronic properties [29–31]. Development of new materials in the nanometer scale called nanoparticles (NPs) is expected to form the basis of many of the technological and biological innovations of this century [32–35]. Particularly, AuNPs are being considered in chemistry, biological, and medical due to their oxide-free surface, bioconjugation properties, good biocompatibility, and unique optical properties. Specifically, because of optical activity, AuNPs are extensively used in colorimetric biosensors [36, 37], drug delivery [38–40], cancer imaging [41–44], and cancer therapies [45, 46]. They also have the potential to help establish specific beneficial processes and achieve selectivity within biological settings. These applications have sparked great interest in the development of synthetic methods for preparing different gold-based nanostructures. The general synthesis methods of AuNPs are briefly described in this section.

*2.1. General Synthetic Routes.* AuNPs can be manufactured into a variety of shapes including Au nanospheres, nanorods, nanoshells, nanocubes, nanoclusters, and nanostars [46–48]. The development of simple and versatile methods for the preparation of NPs in a size- or shape-selected and controlled manner has been a challenging but intellectually rewarding task (Figure 2). Most commonly, a Au(III) salt is reduced to Au(0) to form an activated species, either in a single step or via a Au(I) intermediate followed by reduction to Au(0). These activated Au(0) species are thermodynamically unstable and rapidly aggregate to form nuclei and eventually the desired gold nanoparticle, as presented in Figure 1.

This is typically achieved not only by altering the relative growth rates of different facets by the selective localization of surface-modifying or capping agents, but also by the modulation of nucleation and reaction parameters such as time, temperature, reagent concentration, and pH [49–52]. On the basis of reported literature, [46–68] the following methods have been widely used in the synthesis of nanoparticles.

*2.2. Physical Methods.* Many of the physical methods have been involved to the synthesis of Au nanoparticles [53–55]. In these methods, the size of the NP is controlled by temporarily inactivating the source of evaporation or by slowing the rate by introducing gas molecules to collide with the particles. The growth of nanoparticles generally occurs rapidly, from milliseconds to seconds, requiring precise control over experimental parameters.

*2.2.1. Photochemistry (UV, Near-IR), Sonochemistry, and Radiolysis.* Photochemical synthesis of nanoparticles is carried out by the light-induced decomposition or the reduction of a metal salt by photogenerated reducing agents such as solvated electrons (radiolysis). UV-irradiation is another method that can improve the quality of the AuNPs especially when it is used in synergy with micelles or seeds [55]. Near-IR laser irradiation facilitates an enormous size growth of thiol-stabilized AuNPs [56]. Sonochemistry can also be used for the synthesis of AuNPs within the pores of silica [57]. Radiolysis has been used to control the AuNPs size or to synthesize them in the presence of specific radicals [58], and the mechanism of AuNPs formation upon $\gamma$-irradiation has been carefully examined [59]. Laser photolysis has been used to produce AuNPs in block copolymer micelles [60]. Laser ablation is another technique for AuNPs synthesis that has been used under various conditions, whereby size control can be induced by a laser.

*2.3. Chemical Methods.* Chemical methods have emerged to be indispensable for synthesizing nanocrystals of various types of materials. These methods are generally carried out under mild conditions and are relatively straightforward. One of the important factors that determine the quality of a synthetic procedure is the monodispersity of the nanocrystals obtained. It is desirable to have nanoparticles of nearly the same size, in order to be able to relate the size and the property under study. A variety of reducing agents, such as sodium borohydride, presence of sodium hydroxide, ascorbic acid, sodium citrate, trisodium citrate, and hydrogen peroxide have been used in the reduction of Au(III) ions to Au(0) [61–63]. Such reduction takes place in the presence of one or more water-soluble polymers, surfactants, or capping agents

TABLE 1: Expected applications of AuNPs in various areas.

| Serial number | Fields | Applications |
|---|---|---|
| 1 | Electronics | AuNPs are considered for use as conductors from printable inks to electronic chips. As the world of electronics becomes smaller in size, nanoparticles are important components in the electronic chip design. AuNPs are being used to connect resistors, conductors, and other elements of electronic devices [20]. |
| 2 | Photodynamic therapy | Photodynamic therapy that is a mode of treatment by laser light at a particular wavelength is combined with exogenous chromophores to accomplish a therapeutic effect. Near-IR absorbing AuNPs (including Au nanoshells and nanorods) produce heat when excited by light at wavelengths from 700 to 800 nm. This enables these nanoparticles to destroy targeted tumors. When light is applied to a tumor-containing AuNPs, the particles rapidly heat up, killing tumor cells in a treatment also known as hyperthermia therapy [21]. |
| 3 | Therapeutic agent delivery | Therapeutic agents can be covered with AuNPs. Whereas the NIR-resonant Au/silica nanoshells used in therapeutic and imaging applications have an average diameter of 120–140 nm [22]. |
| 4 | Sensors | AuNPs can be used in different sensors. For example, a colorimetric sensor based on AuNPs can identify if foods are suitable for consumption [23]. Other methods, such as surface-enhanced Raman spectroscopy, exploit AuNPs as substrates to enable the measurement of vibrational energies of chemical bonds. This strategy could also be used for the detection of proteins, pollutants, and other molecules label-free [24]. |
| 5 | Probes | AuNPs also scatter light and can create an array of interesting colors under dark-field microscopy. The scattered colors of AuNPs are currently used for biological imaging. Also, AuNPs are relatively dense, allowing them helpful as probes for transmission electron microscopy [25]. |
| 6 | Diagnostics | AuNPs are also used to detect biomarkers in clinical and medical diagnosis. They are also useful in the lateral flow immunoassays, a common household example being the home pregnancy test [26]. |
| 7 | Catalysis | AuNPs are used as catalysts in a number of chemical reactions [27]. The surface of AuNPs can be used for oxidation or reduction of certain chemicals. AuNPs are being developed for fuel cell applications. These technologies would be useful in the automotive and display industry. |
| 8 | Fluorescence properties | Fluorescence-based detection techniques are among the most highly sensitive and popular biological tests in the clinical analysis. AuNPs show outstanding behavior of antiphoto bleaching under the presence of powerful light [28]. |

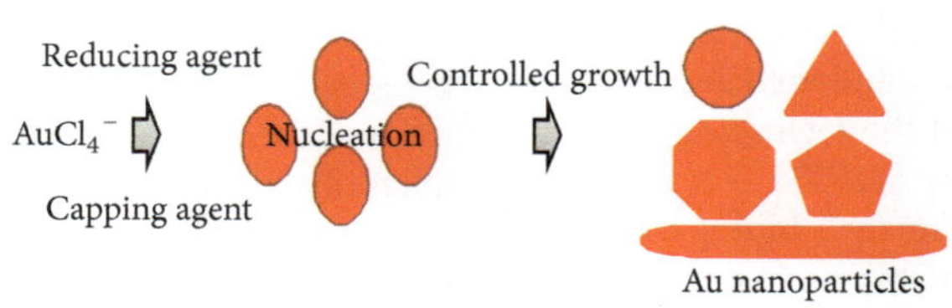


FIGURE 1: Synthetic route of growth of gold nanoparticles.

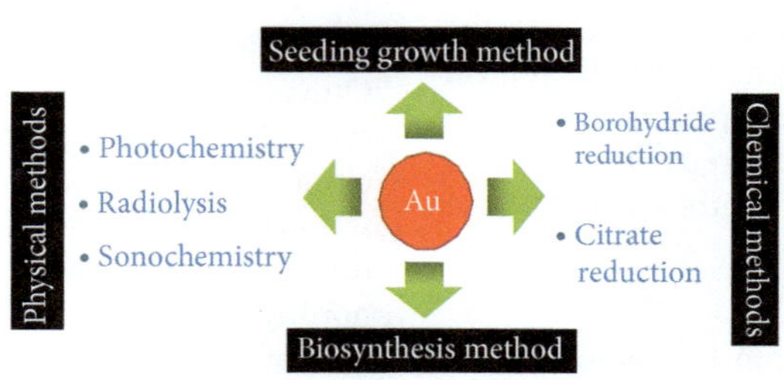


FIGURE 2: Various methods for AuNPs preparation.

and with introducing external supplied energy such as photo irradiation, ultrasound irradiation, or heating. These methods allow for adequate control of the size and concentration of the dispersed particles. Moreover, the surface-modifying or capping agents help to make colloidal stabilization and prevent NPs aggregation [62]. The most common approach to achieve colloidal stability proceeds *via* chemical binding of ligands at the surface of the NPs and a covalent linkage between the ligand. It may change the properties of the NPs through a modification of their electronic density and the dielectric constant of the surrounding medium. A strategy based on the physical adsorption of ligands on the surface of the nanoparticles may be preferable, in order to maintain the intended properties of the nanomaterials. The two reduction methods are outlined as follows.

*2.3.1. Borohydride Reduction.* The preparation of metal NPs in solution is most commonly based on the chemical reduction of metallic ions to be metalized in organic/aqueous solvent. The reduction by borohydride has been in existence for a number of years [64, 65]. Although, various metal nanocrystals have been synthesized *via* the borohydride reduction method, but this method is widely used in the synthesis of AuNPs. Superhydride and hexadecylaniline have been used as alternative reagents to $NaBH_4$ for the reduction of Au(III) in the synthesis of thiol-stabilized AuNPs [66]. Problems associated with borohydride reduction include its irreproducibility, especially in the aqueous medium and the incorporation of boron in the product [67].

*2.3.2. Citrate Reduction.* The most common method of synthesizing gold nanoparticles is through the use of citrate. Jana et al. reported that adding sodium citrate to a boiling solution of chlorauric acid led to the formation of gold nanoparticles [68]. Later work showed that variations in temperature and the ratio of reactants allowed for control over the core size from 10 to 100 nm [69]. More recent work has shown that Au/citrate solutions reduced with $NaBH_4$ allow the synthesis of nanoparticles <10 nm in average core size. Despite being the most common method of producing gold nanoparticles, citrate-stabilized nanoparticles are disadvantageous for several reasons. First, they cannot be isolated from solution, making it difficult to store or study them in the solid state. Second, their stability with changes in pH or ionic strength is minimal. Finally, and most importantly, their functionalization, either through ligand exchange derivatization of carboxylic acids in the ligand shell, is extremely limited.

*2.4. Seeding-Growth Method.* The seeding-growth procedure is another popular technique that has been used for a number of years. Recent studies have successfully led to control of the size distribution in the range 5–40 nm, whereas the sizes can be manipulated by varying the ratio of seed to metal salt [70]. The step-by-step particle enlargement is more effective than a one-step-seeding method to avoid secondary nucleation. Gold nanorods have been conveniently fabricated using the seeding-growth method [71]. In a typical seeded growth reaction, a previously synthesized "seed" nanoparticle is introduced to a growth solution containing an Au(III) salt, a weak reducing agent, and a directing agent (most commonly a surfactant). Initially, the seed particles grow through the slow diffusion of gold atoms onto their surface.

*2.5. Biosynthesis Methods.* As far as the synthesis of nanoparticles is concerned there is a great and growing need to develop clean, nontoxic, and environmentally friendly synthetic procedures. Biological methods have recently been considered as possible environmentally friendly nanofactories. Both living and dead microorganisms are gaining importance by virtue of their facile assembly of nanoparticles. Prokaryotic bacteria have primarily attracted the most attention in the area of biosynthesis of metal nanoparticles [72]. Different research groups continued to accomplish some excellent work based on uni- and multicellular organisms, Algae, and Fungi to synthesize inorganic materials, both intra- and extracellularly. However, the use of plants and their extracts for NP synthesis is a comparatively new and underresearched technique. The majority of research on the use of plants to synthesize AuNPs has been investigated without the need for their removal from the plant biomatrix [73]. More recently synthesized mesoporous material SBA-16 are used synthesis of AuNPs for preventing the aggregation or growth of metal NPs [74].

*2.6. Overview of Synthesized AuNPs.* Despite the progress achieved, concerns, and problems with the preparation of metal NPs remain, such as the byproducts from the reducing agent, the multiple steps often required, and the high concentration of protective agents. In addition, a simple modification of the surface functionalization of the citrate-stabilized AuNPs with a variety of chemical and biochemical vectors is also another important factor to be considered. The ability to synthesize smaller AuNPs while preserving their biocompatibility and derivability would present new alternatives for these areas of research. To date, isotopic replacement of the solvent has solely been explored in the case of Au nanorods seeding-growth. Finally, the modification of the reducing and reduced species due to isotope replacement offers unique insights into the AuNP formation mechanism [75]. Recently, AuNPs are synthesized using monosodium glutamate as a reducing agent. However, the particles possess the tendency to aggregate in high molar ratios of gold capped with glutamates and are therefore unstable. Consequently, PMMA is introduced as a polymer-capping agent to stabilize the AuNPs [76]. Thus, nature of the particle stabilizer, solvent, reaction condition, namely, pH, temperature, and so forth, plays crucial role in determining the final size of the particles. It is also important to note that the citrate/aqueous synthesis of AuNPs is the most employed recipe in today nanotechnology explosion. All these novel procedures offer an attractive alternative to the existing ones and open a full range of possibilities for biological studies.

## 3. Nanosensors Based on Nano-AuNPs in Clinical Chemistry

AuNPs are one of the highly used particles in sensors used for clinical chemistry. In this section, the detection of some important clinical molecules has been summarized using AuNPs or AuNPs-composite sensor. A review on the use of AuNPs and AgNPs for clinical diagnosis has also appeared in the literature [77]. It has been demonstrated that the use of nanoparticles not only improves the performance of sensors but also permits arrays of sensors to be constructed for parallel sensing [78]. The chemical methods have also been found to produce AuNPs of different shapes such as spherical, nanoplates, and flower shapes [79, 80]. A comparison of spherical AuNPs and nanogold plates-based indium tin oxide (ITO) sensor has been reported for the determination of dopamine and ascorbic acid [81]. The nanoplates based sensor has been found to exhibit higher electrocatalytic activity for both compounds due to their (1 1 1) lattice as the basal plane and due to more available surface edges. In addition the increased surface area of nanoplates, due to their large dimension (400–500 nm), also appears to play a significant role. A review on the biopharmaceutical applications of AuNPs has also appeared in the literature describing the detection of diseases biomarker for the diagnosis of patients with germ cell tumors and hepatocellular carcinoma [82]. The biosensing and therapeutic applications of AuNPs have been found due to their catalytic properties. An electrochemical immunoassay method has also been reported for the detection of tumor markers based on electrochemical stripping of AuNPs, which were deposited by immunosensor array with a reduction reaction catalyzed by AuNPs [83]. The method was sensitive, and a detection limit of 3.5 pg/mL was observed with acceptable reproducibility. Significant methods for the determination of brain neurotransmitters have been developed in the past few years using AuNPs. One commonly known neurotransmitter is dopamine (DA), chemically known as 4-(2-aminoethyl) benzene-1,2-diol, widely present in the mammalian brain tissues and plays a pivotal role in the function of the cardiovascular, hormonal, renal, and central nervous systems. Abnormal levels of DA have been linked with Parkinson's disease, Tourette's syndrome, Schizophrenia, attention deficit hyperactive disorder, and generation of pituitary tumours. Similarly, serotonin (5-hydroxytryptamine or 5-HT) is a monoamine neurotransmitter widely distributed in the brain tissues and plays a significant role in temperature regulation, muscle contraction, liver regeneration, endocrine regulation, depression, and so forth. It has also been found that DA and 5-HT influence each other in their respective releasing, and hence simultaneous determination of DA and 5-HT has been carried out using AuNPs-based nanosensors. These nanosensors have been found rapid and convenient for the simultaneous determination of DA and 5-HT in human serum and urine with detection limit of 0.5 and 3.0 nM, respectively [84]. The large excess of ascorbic acid (AA) in biological fluids did not interfere in the determination. Norepinephrine (NE) is also a significant neurotransmitter present in many parts of the central nervous system, where it is engrossed in emotional arousal, blood pressure regulation, and mood disorders. The determination of NE in biological fluids has also been carried out at AuNPs nanosensor using square wave voltammetry. A stable layer of AuNPs was deposited on the surface of ITO, and the nanosensor was used for the detection of NE. The reduction of NE was observed in two pH dependent peaks, and the linear dynamic range was found between 100 nM and 25 $\mu$M with a detection limit of 87 nM [85]. Attempts have also been made to determine DA in human urine using AuNPs-coated poly(3,4-ethylene-dioxythiophene) nanosensor using sodium dodecyl sulfate [86]. The DA concentration versus peak current plot showed two linear regions: 0.5–20 $\mu ML^{-1}$ and 25–140 $\mu ML^{-1}$. The method was found highly sensitive and reproducible with good selectivity and long term stability. AuNPs have also been used in the composites for the determination of neurotransmitters. One of such examples is the fabrication of nano-Au/cysteamine/GC sensor for the determination of DA in presence of high concentration of ascorbic acid (AA). The cysteamine was bound at the surface of glassy carbon (GC) using cyclic voltammetry [87]. The oxidation of DA occurred at 0.175 V, and linear dynamic range for the determination was observed between $1.0 \times 10^{-8}$ and $2.5 \times 10^{-5}$ $ML^{-1}$ with a detection limit of 4 $nML^{-1}$. The method was simple, and satisfactory results were observed. The electrodeposition of AuNPs clusters over insulating overoxidized polypyrrole film modified GC has been used by Guo et al. [88] for the determination of DA and 5-HT in presence of 100-fold concentration of AA in human blood serum. The oxidation peaks of the three compounds were observed at 0.37, 0.20 and 0.01 V versus SCE for 5-HT, DA, and AA, respectively. The method was sensitive, and a detection limit of 1 nM for 5-HT and 15 nM for DA was observed.

Another important clinical target is tryptophan (2-amino-3-(1H-indol-3-yl)-propionic acid, Trp), an important and essential amino acid for humans and herbivores. It is also a potent precursor of several metabolites such as 5-HT, melatonin, and niacin. It is an indispensable ingredient of various types of proteins, therefore, must be added in human nutrition for establishing and maintaining positive balance of nitrogen. A nanosensor based on the electrodeposition of AuNPs onto carbon nanotube film has been used for the amperometric determination of Trp at the physiological pH [89]. The linear dynamic range was found between 30 nM and 2.5 $\mu$M, and a detection limit of 10 nM was observed. The method was successfully used to determine tryptophan in pharmaceutical samples. In another attempt, AuNPs attached multiwall carbon nanotubes (MWCNT) on ITO have been used for the determination of Trp [90]. The use of AuNPs-doped MWCNT shifted the peak potential of Trp to 0.669 V as compared to 0.760 V in absence of AuNPs, which clearly indicated the high electrocatalytic activity of the fabricated nanosensor. The determination of Trp was carried out in human urine and blood serum using standard addition method, and good reproducibility was observed. AuNPs have also seen their applications in clinical analysis of glucose.

A glucose nanosensor based on AuNPs decorated MWCNT at GC was used by Lim et al. [91]. The sensor was further modified with a layer of glucose oxidase and chitosan and used as electrochemiluminescence nanosensor. The sensor showed excellent performance for glucose detection and a wide linear range of 1–1000 $\mu$M was observed with a detection limit of 0.5 $\mu$M. A colorimetric bioassay for the glucose determination in human serum has been developed using self-assembled AuNPs with a thiol derivatized mannose [92]. The change in simple surface plasmon absorption band resulting from this method provides a selective method to quantify glucose. A linear concentration range of 500–4000 $\mu$g/mL with a detection limit of 363 $\mu$g/mL was observed. The practical utility of the method was examined by applying the colorimetric assay for the determination of glucose in human serum samples. The reported method is simple, and results can be achieved in 10 min and can be applied for the determination of human blood glucose levels without the use of enzymes.

Morphine (MO), a well-known alkaloid, is frequently used to relieve severe pain for patients, especially for those who undergo a surgical procedure. However, when overdosed or abused, MO is toxic and can cause disruption in the central nervous system. Therefore, to prevent overdose-induced toxication, it is necessary to sensitively determine the concentrations of MO in patient's blood or urine. An easy-to-use approach for directly electrodepositing AuNPs onto carbon paste electrode (CPE) to construct AuNPs modified carbon paste electrode (AuNPCPE) is performed [93]. The electrochemistry of MO is investigated by cyclic voltammetry, differential pulse voltammetry, and electrochemical impedance measurements. The results indicate that this strategy provides a promising approach for highly sensitive morphine sensing that offers an excellent response for morphine in the concentration range of $4.0 \times 10^{-7} - 2.0 \times 10^{-4}$ mol L$^{-1}$, with a detection limit of $4.21 \times 10^{-9}$ mol L$^{-1}$. AuNPCPE has also been successfully applied to the determination of morphine in urine samples with a low detection limit and satisfactory recovery. The good results indicate that AuNPCPE holds great promise in practical applications. Apart from the direct application of AuNPs in clinical chemistry of electrochemically active compounds, AuNPs have also attracted attention in the immunoassay of proteins and hormones. For example, an enzyme-catalyzed Ag deposition on irregular shaped AuNPs has been used for the electrochemical immunoassay of alpha-fetoprotein [94]. The assay was carried out using a sandwich-type protocol, and a wide dynamic linear range from 0.1 to 200 ng mL$^{-1}$ was observed. The interesting feature of the nanobiosensor was its disposable nature and acceptable reproducibility, as no significant difference was observed in the analysis of 10 clinical serum samples. An electrochemical assay was also developed by using AuNPs-polyaniline-nanogold microspheres (GPGs) for detecting thyroid stimulating hormone [95]. The GPGs were first synthesized and functionalized with horseradish peroxidase-conjugated thyroid-stimulating hormone antibody (HRP-Ab2), and then thyroid-stimulating hormone (TSH) was detected. The strong attachment of HRP-Ab2 was observed on GPGs and resulted in good reproducibility, and a detection limit of 0.005 $\mu$IU mL$^{-1}$ TSH was observed. The analysis of 15 spiked serum samples exhibited no significant difference between the developed method and the commercially available enzyme-linked immunosorbent assay (ELISA).

## 4. Biosensors Based on AuNPs and Conjugates

Biosensors are analytical devices that are based on the coupling of an immobilized biologically active compound called a receptor with a signal transducer and an electronic amplifier [96]. After the discovery of nanomaterials, various interdisciplinary strategies have been developed for the sensitive detection of various analytes present in biological and environmental matrix. Among all nanomaterials, Au nanoparticles (AuNPs) have widely been used because of their size-dependent electrical properties, high surface area-to-volume ratios, high electrocatalytic activities, and ease of chemical modification. Various biorecognision elements (antibodies, aptamers, peptides, DNA, etc.) have been attached in various ways to achieve a low detection limits. In this section some of the important fabrications strategies have been discussed based on various types of biorecognision elements.

The general working principle of the immunosensors relying on the specific immunochemical identification of antibodies antigens immobilized on a transducer to antigens in the sample media can produce analytical signals dynamically varying with the concentrations of analytes of interest. In the recent years construction of immunosensors using AuNPs have received significant attention. Numerous immunosensors are developed so far where antibodies are directly attached with the AuNPs. The ability of AuNPs to provide a stable immobilization of biomolecules, biocompatibility, and ease of surface functionalization provides a major advantage of AuNPs in the development of immunosensors. Various approaches like layer by layer self-assembly using nafion [97], dithiothreitol 3, cysteamine [98], 4-aminothiophenol [99], polyvinyl butyral [100] have been used in the fabrication of immunosensors. In one study, a spacer arm antibody was covalently attached onto the AuNPs and applied for the electrochemical immunoassay [101]. AuNPs have also been widely applied for the detection of cancer biomarkers. This study described that AuNPs were efficient in preserving the action and orientation of the antibody, and it can form a key platform in several medical immunodiagnostics. Very recently, inkjet-printed AuNPs arrays coupled with electrochemical detection cancer biomarker interleukin-6 (IL-6) in serum have been developed. Biorecognision antibodies for IL-6 were linked onto the eight-electrode array and used in sandwich immunoassays. A biotinylated secondary antibody with 16–18 horseradish peroxidase labels was used, and detection was achieved by hydroquinone-mediated amperometry with a clinically applicable detection limit of 20 pg mL$^{-1}$ [102]. Apart of cancer biomarker detection, strategies using AuNPs have also been developed to understand the cancer immunology. In this regard, a quick and sensitive protocol for the detection of alpha-enolase (ENO1) is developed using an electrochemical sandwich immunosensor. Since ENO1 has been correlated with small cell lung cancer, nonsmall cell lung cancer, and head and neck cancer, it has been used as a potential diagnostic marker for lung cancer. The anti-ENO1 monoclonal antibody was adsorb on polyethylene glycol-modified disposable screen-printed electrode as the detection platform. The polyclonal secondary anti-ENO1-tagged AuNPs congregates as electrochemical signal probes. This AuNPs congregate-based assay provided an amplification approach for detecting ENO1 at trace levels, leading to a detection limit as low as 11.9 fg (equivalent to 5 microL of a 2.38 pg/mL solution [103]. Another example of AuNPs for signal amplification was shown by Zhu et al. [104]. In this case, an amperometric immunosensor with an enhanced sensitivity for the detection of neomycin (Neo) was prepared by covalently immobilizing a monoclonal Neo antibody onto a new nanoconducting surface, as a sensor probe. The probe was used to detect Neo in a sandwich-type approach, where the secondary antibody was attached to AuNPs-decorated multiwall carbon nanotubes labeled with hydrazine. Hydrazine on the conjugate served as a catalyst for the reduction of hydrogen peroxide, and the catalytic current was monitored at −0.45 V versus Ag/AgCl. The performance of the immunosensor with and without AuNPs on the probe and the conjugate was compared. A linear dynamic range for Neo analysis was obtained between 10 ng/mL and 250 ng/mL with a detection limit of 6.76 ± 0.17 ng/mL. In another signal amplification strategy, multifunctionalized AuNPs (MFAuNPs) was developed and applied for the electrochemical impedance immunosensor. The sensor was utilized for the detection of protein A, through the stepwise immobilization of 1, 6-hexanedithiol, AuNPs, and IgG on the Au electrode via self-assembling technique. The interfacial properties of the modified electrodes were evaluated in the presence of $Fe(CN)_6^{4-/3-}$ couple redox as a probe by cyclic voltammetry and electrochemical impedance spectroscopy. The electrochemical behavior of the redox probe was affected by the accumulation of treated substances on the electrode surface. It was interesting to note that there is no significant increase in the electron transfer resistance after the binding of protein A to IgG on the electrode surface without amplification. The changes in the electron transfer resistance on the IgG-modified electrodes became more sensitive after the MFAuNPs were introduced for signal amplification. The increments of amplified impedance showed good correlation of the detection of protein A in the range of 5–1000 pg/mL with a detection limit of 1 pg/mL. In this strategy, the MFAuNPs were successfully applied for the highly sensitive and selective amplification of biomolecules in the picogram range [105]. Very recently, a new electrochemical detection methodology has been developed describing the practical application and integration of nanoparticle-enhanced detection into electrochemical biosensing technologies [106]. Immunoglobulin E (IgE) was used as a model protein biomarker, which possesses two distinct epitopes for antibody (anti-IgE) and DNA aptamer binding. A surface sandwich assay format was utilized involving the specific adsorption of IgE onto a gold electrode surface that was premodified with a monolayer of aptamer-nanoparticle conjugates followed by the specific interaction of alkaline phosphatase- (ALP-) conjugated anti-IgE. A femtomolar detection of IgE has been achieved. The signal enhancement associated with the nanoparticle was clearly demonstrated using analytical methods and control experiments. Another focus in recent time is to detect important cardiac biomarker such as troponin-I using AuNPs. In one study, troponin-I has been detected using an immunoassay where a specific monoclonal antibody against troponin-I was modified on the GNP-coated ITO electrode surface by self-assembly and characterized by electrochemical techniques. The immunoreaction was monitored by measuring open circuit potential (OCP). This method was fairly simple and stable. A linear dependence of OCP changes according to troponin-I concentrations is observed in the range of concentration from 1 to 100 ng/mL [107]. AuNPs have also been used for the evaluation of protein phosphorylation using a simple electrochemical method. This approach of protein phosphorylation is expected to be much simpler and sensitive compared to the conventional methods using mass spectrometry. The method relies on the labeling of a specific phosphorylation event with Au nanoparticles, followed by electrochemical detection [108]. This study demonstrates the use of AuNPs as electroactive indicators for kinase activity and inhibition has been demonstrated. Furthermore, the electrochemical protein phosphorylation assay will be helpful for the development of cost-effective biosensors for signal transduction studies. Apart from these approaches, the effect of the AuNPs size, ranging from 5 nm to 80 nm, on the electrochemical response of screen-printed carbon electrodes (SPCEs) used as electrochemical transducers is investigated for the first time recently [109]. To confirm this size effect, a simple hydrodynamic modelling and calculation at the nanoscale level is applied so as to find the effect of the size of AuNPs upon the electrochemical response. The results show that the greatest signal for AuNPs suspension for the same concentration of total Au is obtained for the 20 nm sized nanoparticles. The size effect is also studied for AuNPs acting as electroactive labels in an immunosensor that employs magnetic beads as platforms of the bioreactions. The best response for the 5 nm AuNPs in this case is due to the fact that in the immunosensing conditions the Brownian motions are minimized because the AuNPs contact with the electrotransducer surface is induced by the immunoreaction, and the fast magnetic collection of the nanoparticles used as antibody labels upon application of a magnetic field. These findings can be extended towards the tuning sensor fabrication for other clinical molecule detection [110]. The successful interaction between Au and SH has been utilized by Shim's laboratory for the development of electropolymerized self-assembled layer on AuNPs for the detection of inducible nitric oxide synthase (i-NOS) in neuronal cell culture [111]. The results show that AuNP-encapsulated conductive polymer are good nanostructured materials as biosensor probes and have a potential application in cell biosensors. The self-assembled layer modified electrode was found to have excellent properties in terms of better conductivity and lower charge transfer resistance. The main advantage of this sensor probe is its simplicity compared to the classical immunoassay methods or optical detection-based immunosensing systems. This method was also extended for the toxicity analysis of endocrine disruptors through the *in vitro* monitoring of i-NOS concentrations using an AuNP-encapsulated conductive polymer-modified immunosensor [112].

AuNPS have also been well utilized in the detection of various kinds of cells such as bacteria and cancer cells. For instance, a disposable immunosensor for *Salmonella-enterica* subsp. enterica serovar Typhimurium LT2 (S) detection using a magneto-immunoassay and AuNPs as label for electrochemical detection is developed [113]. The immunosensor relies on the application of a screen-printed carbon electrode that incorporates a stable magnet underneath. *Salmonella* containing samples (i.e., skimmed milk) were examined by using anti-*Salmonella* magnetic beads (MBs-pSAb) as detector and sandwiching afterwards with AuNPs-modified antibodies (sSAb-AuNPs) detected using differential pulse voltammetry with the detection limit of 143 cells $mL^{-1}$. This AuNPs detection tools together with magnetic field application report a limit of detection lower than the usual commercial method carried out for comparison purposes in skimmed milk samples.

A fast cancer cell detection and quantification method, based on the electrocatalytic properties of AuNPs towards the hydrogen evolution reaction has been reported recently [114]. This method effectively detects the rare circulating tumor cells (CTC) through the epithelial cell adhesion molecule (EpCAM), a 30–40 kDa type I glycosylated membrane protein expressed at low levels in a variety of human epithelial tissues and overexpressed in most solid cancers [115]. The method establishes a selective CTC assay capable of detecting $4 \times 10^3$ cancer cells in suspension that can be extended to several other cells detection scenarios. In another study, a novel strategy for the rapid identification and high sensitive detection of different kinds of cancer cells by means of electrochemical and contact angle measurements has been performed. A simple, label-free method based on the functionalized AuNPs-modified interface has been utilized to distinguish the different cancer cells and drug resistant blood cancer cells. The detection limit was $\sim 10^3$ cells $mL^{-1}$ which is highly applicable for clinical applications of cells [116]. Conducting polymer nanofibers-AuNPs nanocomposite has also been applied for the detection of cancer cells. In this regard, polyaniline nanofibers- (PANI-NF-) AuNPs composites were prepared and based on the simple electrostatic reaction between the positive charges on the surface of PANI-NF and the negatively charged AuNPs [117]. This nanocomposite was utilized to develop an electrochemical cytosensor based on the self-assembly of foliate on PANI-NF/AuNP nanocomposite. This cytosensor was applied for the *in vitro* detection of cancer cells based on the selective binding between the folate receptor on cancer cell surface and a folate immobilized on PANI-NF/AuNPs nanocomposite. This cytosensor has several advantages, such as easy construction, short response time, high sensitivity, easy regeneration, good stability, and freedom of additional labeling, which can be thus used for a highly sensitive detection of cancer cells.

## 5. Conclusions

AuNPs such fascinating features as ease of synthesis and surface functionalization that are highly biocompatibility, sensitive, and decent optical properties make AuNPs still the most attractive nanomaterial and one of the most studied in the bioanalytical field. AuNPs have emerged as promising nanoplatforms for efficient diagnostics and therapeutics by merging the characteristic properties they possess at the nanometric scale with the practical immobilization of specific ligands on the surface. Therefore, they have become ideal candidates for molecularly sensitive detection, highly efficient contrast agents for molecular imaging, and carriers for targeted drug and gene delivery. Nonetheless, a better fundamental understanding of AuNPs behavior in biological systems needs to be addressed in future, as well as the engineering of novel AuNPs, which can overcome the drawbacks related to currently developed AuNPs. In future AuNPs possessing enhance biodistribution, and enhanced electrical and optical characteristics will provide better way towards clinical diagnostics.

## Acknowledgments

P. Chandra is thankful to Amity Institute of Biotechnology (AIB), Amity University for the AIB research facilities. R. N. Goyal is thankful to Council of Scientific and Industrial Research, New Delhi, and also Korean Federation of Science and Technology, South Korea, for awarding Brain Pool Fellowship. Y. B. Shim is thankful to the support of a 2-Year research Grant of Pusan National University (2011-2012).

## References

[1] C. N. R. Rao, F. L. Deepak, G. Gundiah, and A. Govindaraj, "Inorganic nanowires," *Progress in Solid State Chemistry*, vol. 31, no. 1-2, pp. 5–147, 2003.

[2] C. Suryanarayana and C. C. Koch, "Nanocrystalline materials—current research and future directions," *Hyperfine Interactions*, vol. 130, no. 1–4, pp. 5–44, 2000.

[3] H. S. Nalwa, *Handbook of Nanostructured Materials and Nanotechnology*, Elsevier, 2000.

[4] G. M. Gadd, "Metals, minerals and microbes: geomicrobiology and bioremediation," *Microbiology*, vol. 156, pp. 609–643, 2010.

[5] J. D. Bryan and D. R. Gamelin, "Doped semiconductor nanocrystals: synthesis, characterization, physical properties, and applications," *Progress in Inorganic Chemistry*, vol. 54, pp. 47–126, 2005.

[6] E. R. Hitzky, K. Ariga, and Y. Lvov, *Bio-Inorganic Hybrid Nanomaterials—Strategies, Syntheses, Characterization and Application*, Wiley-VCH, 2008.

[7] Y. Yin and A. Alivisatos, "Colloidal nanocrystal synthesis and the organic-inorganic interface," *Nature*, vol. 437, pp. 664–670, 2005.

[8] C. M. Lieber, "Semiconductor nanowires: a platform for nanoscience and nanotechnology," *MRS Bulletin*, vol. 36, pp. 1052–1063, 2011.

[9] T. Hillie and M. Hlophe, "Nanotechnology and the challenge of clean water," *Nature Nanotechnology*, vol. 2, no. 11, pp. 663–664, 2007.

[10] M. M. Khin, A. S. Nair, V. J. Babu, R. Murugan, and S. Ramakrishna, "A review on nanomaterials for environmental remediation," *Energy & Environmental Science*, vol. 5, pp. 8075–8109, 2012.

[11] P. J. Vikesland and K. R. Wigginton, "Nanomaterial enabled biosensors for pathogen monitoring—a review," *Environmental Science & Technology*, vol. 44, pp. 3656–3669, 2010.

[12] A. Rodriguez, "Activation of gold nanoparticles on titania: a novel $DeSO_x$ catalyst," *ACS Symposium Series*, vol. 890, pp. 205–209, 2004.

[13] X. Wang, C. Wang, L. Cheng, S.-T Lee, and Z. Liu, "Noble metal coated single-walled carbon nanotubes for applications in surface enhanced Raman scattering imaging and photothermal therapy," *Journal of the American Chemical Society*, vol. 134, pp. 7414–7422, 2012.

[14] Y. Su, X. Wei, F. Peng et al., "Gold nanoparticles-decorated silicon nanowires as highly efficient near-infrared hyperthermia agents for cancer cells destruction," *Nano Letters*, vol. 12, pp. 1845–1850, 2012.

[15] M. Faraday, "Experimental relations of gold (and other metals) to light," *Philosophical Transactions of the Royal Society of London*, vol. 147, pp. 145–181, 1857.

[16] G. T. Beilby, "The effects of heat and of solvents on thin films of metal," *Proceedings of the Royal Society A*, vol. 72, pp. 226–235, 1903.

[17] V. V. Mody, R. Siwale, A. Singh, and H. R. Mody, "Introduction to metallic nanoparticles," *Journal of Pharmacy and Bioallied Sciences*, vol. 2, pp. 282–289, 2010.

[18] D. A. Giljohann, D. S. Seferos, W. L. Daniel, M. D. Massich, P. C. Patel, and C. A. Mirkin, "Gold nanoparticles for biology and medicine," *Angewandte Chemie*, vol. 49, pp. 3280–3294, 2010.

[19] M. A. Hayat, *Colloidal Gold: Principles, Methods, and Applications*, Academic Press, San Diego, Calif, USA, 1989.

[20] D. Huang, F. Liao, S. Molesa, D. Redinger, and V. Subramanian, "Plastic-compatible low resistance printable gold nanoparticle conductors for flexible electronics," *Journal of the Electrochemical Society*, vol. 150, no. 7, pp. G412–G417, 2003.

[21] T. Stuchinskaya, M. Moreno, M. J. Cook, D. R. Edwards, and D. A. Russell, "Targeted photodynamic therapy of breast cancer cells using antibody-phthalocyanine-gold nanoparticle conjugates," *Photochemical and Photobiological Sciences*, vol. 10, no. 5, pp. 822–831, 2011.

[22] S. D. Brown, P. Nativo, J. A. Smith et al., "Gold nanoparticles for the improved anticancer drug delivery of the active component of oxaliplatin," *Journal of the American Chemical Society*, vol. 132, no. 13, pp. 4678–4684, 2010.

[23] C. S. S. R. Kumar, *Nanomaterials for Biosensors*, John Wiley & Sons, 2007.

[24] K. Saha, S. S. Agasti, C. Kim, X. Li, and V. M. Rotello, "Gold nanoparticles in chemical and biological sensing," *Chemical Reviews*, vol. 112, pp. 2739–2779, 2012.

[25] S. D. Perrault and W. C. W. Chan, "In vivo assembly of nanoparticle components to improve targeted cancer imaging," *Proceedings of the National Academy of Sciences of the United States of America*, vol. 107, no. 25, pp. 11194–11199, 2010.

[26] G. Peng, U. Tisch, O. Adams et al., "Diagnosing lung cancer in exhaled breath using gold nanoparticles," *Nature Nanotechnology*, vol. 4, pp. 669–673, 2009.

[27] D. T. Thompson, "Using gold nanoparticles for catalysis," *Nano Today*, vol. 2, no. 4, pp. 40–43, 2007.

[28] Z.-J. Zhang, C.-X. Wan, and Y. Wang, "Fluorescent property of gold nanoparticles with different surface structures," *Chinese Journal of Chemical Physics*, vol. 20, p. 796, 2007.

[29] D. L. Feldheim and C. A. Foss, *Metal Nanoparticles: Synthesis, Characterization, and Applications*, Marcel Dekker, New York, NY, USA, 2002.

[30] M. C. Daniel and D. Astruc, "Gold nanoparticles: assembly, supramolecular chemistry, quantum-size-related properties, and applications toward biology, catalysis, and nanotechnology," *Chemical Reviews*, vol. 104, no. 1, pp. 293–346, 2004.

[31] J. A. Dahl, B. L. S. Maddux, and J. E. Hutchison, "Toward greener nanosynthesis," *Chemical Reviews*, vol. 107, no. 6, pp. 2228–2269, 2007.

[32] S. K. Ghosh and T. Pal, "Interparticle coupling effect on the surface plasmon resonance of gold nanoparticles: from theory to applications," *Chemical Reviews*, vol. 107, no. 11, pp. 4797–4862, 2007.

[33] N. L. Rosi and C. A. Mirkin, "Nanostructures in biodiagnostics," *Chemical Reviews*, vol. 105, no. 4, pp. 1547–1562, 2005.

[34] X. Xu, M. S. Han, and C. A. Mirkin, "A gold-nanoparticle-based real-time colorimetric screening method for endonuclease activity and inhibition," *Angewandte Chemie*, vol. 46, no. 19, pp. 3468–3470, 2007.

[35] J. S. Lee, P. A. Ulmann, M. S. Han, and C. A. Mirkin, "A DNA-gold nanoparticle-based colorimetric competition assay for the detection of cysteine," *Nano Letters*, vol. 8, no. 2, pp. 529–533, 2008.

[36] P. Podsiadlo, V. A. Sinani, J. H. Bahng, N. W. S. Kam, J. Lee, and N. A. Kotov, "Gold nanoparticles enhance the anti-leukemia action of a 6-mercaptopurine chemotherapeutic agent," *Langmuir*, vol. 24, pp. 568–574, 2008.

[37] Y. H. Chen, C. Y. Tsai, P. Y. Huang et al., "Methotrexate conjugated to gold nanoparticles inhibits tumor growth in a syngeneic lung tumor model," *Molecular Pharmaceutics*, vol. 4, no. 5, pp. 713–722, 2007.

[38] G. F. Paciotti, L. Myer, D. Weinreich et al., "Colloidal gold: a novel nanoparticle vector for tumor directed drug delivery," *Drug Delivery*, vol. 11, no. 3, pp. 169–183, 2004.

[39] D. J. Javier, N. Nitin, M. Levy, A. Ellington, and R. Richards-Kortum, "Aptamer-targeted gold nanoparticles as molecular-specific contrast agents for reflectance imaging," *Bioconjugate Chemistry*, vol. 19, no. 6, pp. 1309–1312, 2008.

[40] J. Aaron, N. Nitin, K. Travis et al., "Plasmon resonance coupling of metal nanoparticles for molecular imaging of carcinogenesis in vivo," *Journal of Biomedical Optics*, vol. 12, no. 3, Article ID 034007, 2007.

[41] S. Lee, E. J. Cha, K. Park et al., "A near-infrared-fluorescence-quenched gold-nanoparticle imaging probe for in vivo drug screening and protease activity determination," *Angewandte Chemie*, vol. 47, no. 15, pp. 2804–2807, 2008.

[42] I. H. El-Sayed, X. Huang, and M. A. El-Sayed, "Surface plasmon resonance scattering and absorption of anti-EGFR antibody conjugated gold nanoparticles in cancer diagnostics: applications in oral cancer," *Nano Letters*, vol. 5, no. 5, pp. 829–834, 2005.

[43] G. von Maltzahn, J.-H. Park, A. Agrawal et al., "Computationally guided photothermal tumor therapy using long-circulating gold nanorod antennas," *Cancer Research*, vol. 69, no. 9, pp. 3892–3900, 2009.

[44] Y. Cheng, A. C. Samia, J. D. Meyers, I. Panagopoulos, B. Fei, and C. Burda, "Highly efficient drug delivery with gold nanoparticle

vectors for in vivo photodynamic therapy of cancer," *Journal of the American Chemical Society*, vol. 130, no. 32, pp. 10643–10647, 2008.

[45] F. Kim, J. H. Song, and P. Yang, "Photochemical synthesis of gold nanorods," *Journal of American Chemical Society*, vol. 124, pp. 14316-14317, 2002.

[46] Y. Sun and Y. Xia, "Mechanistic study on the replacement reaction between silver nanostructures and chloroauric acid in aqueous medium," *Journal of the American Chemical Society*, vol. 126, no. 12, pp. 3892–3901, 2004.

[47] S. H. Yu, H. Cölfen, and Y. Mastai, "Formation and optical properties of gold nanoparticles synthesized in the presence of double-hydrophilic block copolymers," *Journal of Nanoscience and Nanotechnology*, vol. 4, no. 3, pp. 291–298, 2004.

[48] T. Sakai and P. Alexandridis, "Single-step synthesis and stabilization of metal nanoparticles in aqueous pluronic block copolymer solutions at ambient temperature," *Langmuir*, vol. 20, no. 20, pp. 8426–8430, 2004.

[49] X. Sun, S. Dong, and E. Wang, "One-step polyelectrolyte-based route to well-dispersed gold nanoparticles: synthesis and insight," *Materials Chemistry and Physics*, vol. 96, no. 1, pp. 29–33, 2006.

[50] W. Chen, W. P. Cai, C. H. Liang, and L. D. Zhang, "Synthesis of gold nanoparticles dispersed within pores of mesoporous silica induced by ultrasonic irradiation and its characterization," *Materials Research Bulletin*, vol. 36, no. 1-2, pp. 335–342, 2001.

[51] A. Housni, M. Ahmed, S. Liu, and R. Narain, "Monodisperse protein stabilized gold nanoparticles via a simple photochemical process," *Journal of Physical Chemistry C*, vol. 112, no. 32, pp. 12282–12290, 2008.

[52] K. L. McGilvray, M. R. Decan, D. Wang, and J. C. Scaiano, "Facile photochemical synthesis of unprotected aqueous gold nanoparticles," *Journal of the American Chemical Society*, vol. 128, no. 50, pp. 15980–15981, 2006.

[53] S. Meltzer, R. Resch, B. E. Koel et al., "Fabrication of nanostructures by hydroxylamine seeding of gold nanoparticle templates," *Langmuir*, vol. 17, no. 5, pp. 1713–1718, 2001.

[54] K. Mallick, Z. L. Wang, and T. Pal, "Seed-mediated successive growth of gold particles accomplished by UV irradiation: a photochemical approach for size-controlled synthesis," *Journal of Photochemistry and Photobiology A*, vol. 140, no. 1, pp. 75–80, 2001.

[55] V. G. Pol, A. Gedanken, and J. Calderro-Moreno, "Deposition of gold nanoparticles on silica spheres: a sonochemical approach," *Chemistry of Materials*, vol. 15, pp. 1111–1123, 2003.

[56] A. Dawson and P. V. Kamat, "Complexation of gold nanoparticles with radiolytically generated thiocyanate radicals ($(SCN)_2^{-}$)," *Journal of Physical Chemistry B*, vol. 104, no. 50, pp. 11842–11846, 2000.

[57] E. Gachard, H. Remita, J. Khatouri, B. Keita, L. Nadjo, and J. Belloni, "Radiation-induced and chemical formation of gold clusters," *New Journal of Chemistry*, vol. 22, no. 11, pp. 1257–1265, 1998.

[58] F. Mafune and T. Kondow, "Formation of small gold clusters in solution by laser excitation of interband transition," *Chemical Physics Letters*, vol. 372, pp. 199–204, 2003.

[59] T. Ishii, H. Otsuka, K. Kataoka, and Y. Nagasaki, "Preparation of functionally PEGylated gold nanoparticles with narrow distribution through autoreduction of auric cation by α-Biotinyl-PEG-block-[poly(2-(N,N-dimethylamino)ethyl methacrylate)]," *Langmuir*, vol. 20, no. 3, pp. 561–564, 2004.

[60] N. R. Jana, L. Gearheart, and C. J. Murphy, "Seeding growth for size control of 5–40 nm diameter gold nanoparticles," *Langmuir*, vol. 17, no. 22, pp. 6782–6786, 2001.

[61] Y. Xiao, B. Shlyahovsky, I. Popov, V. Pavlov, and I. Willner, "Shape and color of Au nanoparticles follow biocatalytic processes," *Langmuir*, vol. 21, no. 13, pp. 5659–5662, 2005.

[62] H. I. Schlesinger, H. C. Brown, A. E. Finholt, J. R. Gilbreath, H. R. Hoekstra, and E. K. Hyde, "Sodium borohydride, its hydrolysis and its use as a reducing agent and in the generation of hydrogen," *Journal of the American Chemical Society*, vol. 75, no. 1, pp. 215–219, 1953.

[63] H. C. Brown and C. A. Brown, "New, Highly active metal catalysts for the hydrolysis of borohydride," *Journal of American Chemical Society*, vol. 84, pp. 1493–1494, 1962.

[64] P. R. Selvakannan, S. Mandal, R. Pasricha, S. D. Adyanthaya, and M. Sastry, "One-step synthesis of hydrophobized gold nanoparticles of controllable size by the reduction of aqueous chloroaurate ions by hexadecylaniline at the liquid-liquid interface," *Chemical Communications*, no. 13, pp. 1334–1335, 2002.

[65] S. Chen, K. Huang, and J. A. Stearns, "Alkanethiolate-protected palladium nanoparticles," *Chemistry of Materials*, vol. 12, pp. 540–547, 2000.

[66] J. Turkevich, P. C. Stevenson, and J. Hillier, "The formation of colloidal gold," *Journal of Physical Chemistry*, vol. 57, pp. 670–673, 1953.

[67] G. Frens, "Controlled nucleation for the regulation of the particle size in monodisperse gold suspensions," *Nature*, vol. 241, pp. 20–22, 1973.

[68] N. R. Jana, L. Gearheart, and C. J. Murphy, "Evidence for seed-mediated nucleation in the chemical reduction of gold salts to gold nanoparticles," *Chemistry of Materials*, vol. 13, no. 7, pp. 2313–2322, 2001.

[69] G. Carrot, J. C. Valmalette, C. J. G. Plummer et al., "Gold nanoparticle synthesis in graft copolymer micelles," *Colloid and Polymer Science*, vol. 276, no. 10, pp. 853–859, 1998.

[70] D. Mandal, M. E. Bolander, D. Mukhopadhyay, G. Sarkar, and P. Mukherjee, "The use of microorganisms for the formation of metal nanoparticles and their application," *Applied Microbiology and Biotechnology*, vol. 69, no. 5, pp. 485–492, 2006.

[71] R. Bali and A. T. Harris, "Biogenic synthesis of Au nanoparticles using vascular plants," *Industrial and Engineering Chemical Research*, vol. 49, pp. 12762–12772, 2010.

[72] Y. Hao, Y. Chong, S. Li, and H. Yang, "Controlled synthesis of Au nanoparticles in the nanocages of SBA-16: improved activity and enhanced recyclability for the oxidative esterification of alcohols," *Journal of Physical Chemistry C*, vol. 116, pp. 6512–6519, 2012.

[73] H. A. Keul, M. Moeller, and M. R. Bockstaller, "Effect of solvent isotopic replacement on the structure evolution of gold nanorods," *Journal of Physical Chemistry C*, vol. 112, no. 35, pp. 13483–13487, 2008.

[74] J. Liu, M. Anand, and C. B. Roberts, "Synthesis and extraction of β-D-glucose-stabilized au nanoparticles processed into low-defect, wide-area thin films and ordered arrays using $CO_2$-expanded liquids," *Langmuir*, vol. 22, no. 9, pp. 3964–3971, 2006.

[75] M. Larguinho and P. V. Baptisa, "Gold and silver nano particles for clinical diagnostics—from genomics and proteomics," *Journal of Proteomics*, vol. 75, pp. 2811–2823, 2012.

[76] P. D'Orazio, "Biosensors in clinical chemistry—2011 update," *Clinica Chimica Acta*, vol. 412, pp. 1749–1761, 2011.

[77] P. Chandra, H. B. Noh, M. S. Won, and Y. B. Shim, "Detection of daunomycin using phosphatidylserine and aptamer co-immobilized on Au nanoparticles deposited conducting polymer," *Biosensors and Bioelectronics*, vol. 26, no. 11, pp. 4442–4449, 2011.

[78] M. Kambayashi, J. Zhang, and M. Oyama, "Crystal growth of gold nanoparticles on indium tin oxides in the absence and presence of 3-mercaptopropyl-trimethoxysilane," *Crystal Growth and Design*, vol. 5, no. 1, pp. 81–84, 2005.

[79] A. A. Umar and M. Oyama, "Formation of gold nanoplates on indium tin oxide surface: two-dimensional crystal growth from gold nanoseed particles in the presence of poly(vinylpyrrolidone)," *Crystal Growth and Design*, vol. 6, no. 4, pp. 818–821, 2006.

[80] R. N. Goyal, A. Aliumar, and M. Oyama, "Comparison of spherical nanogold particles and nanogold plates for the oxidation of dopamine and ascorbic acid," *Journal of Electroanalytical Chemistry*, vol. 631, no. 1-2, pp. 58–61, 2009.

[81] F. K. Alanazi, A. A. Radwan, and A. A. Alsarra, "Biopharmaceutical applications of nanogold," *Saudi Pharmaceutical Journal*, vol. 18, pp. 179–193, 2010.

[82] G. Lai, L. Wang, J. Wu, H. Ju, and F. Yan, "Electrochemical stripping analysis of nanogold label-induced silver deposition for ultrasensitive multiplexed detection of tumor markers," *Analytica Chimica Acta*, vol. 721, pp. 1–6, 2012.

[83] R. N. Goyal, V. K. Gupta, M. Oyama, and N. Bachheti, "Gold nanoparticles modified indium tin oxide electrode for the simultaneous determination of dopamine and serotonin: application in pharmaceutical formulations and biological fluids," *Talanta*, vol. 72, no. 3, pp. 976–983, 2007.

[84] R. N. Goyal, M. A. Aziz, M. Oyama, S. Chatterjee, and A. R. S. Rana, "Nanogold based electrochemical sensor for determination of norepinephrine in biological fluids," *Sensors and Actuators B*, vol. 153, no. 1, pp. 232–238, 2011.

[85] N. F. Atta, A. Galal, and E. H. El-Ads, "Gold nanoparticles-coated poly(3,4-ethylene-dioxythiophene) for the selective determination of sub-nano concentrations of dopamine in presence of sodiumdodecyl sulfate," *Electrochimica Acta*, vol. 69, pp. 102–111, 2012.

[86] G. Z. Hu, D. P. Zhang, W. L. Wu, and Z. S. Yang, "Selective determination of dopamine in the presence of high concentration of ascorbic acid using nano-Au self-assembly glassy carbon electrode," *Colloids and Surfaces B*, vol. 62, no. 2, pp. 199–205, 2008.

[87] J. Li and X. Lin, "Simultaneous determination of dopamine and serotonin on gold nanocluster/overoxidized-polypyrrole composite modified glassy carbon electrode," *Sensors and Actuators B*, vol. 124, no. 2, pp. 486–493, 2007.

[88] Y. Guo, S. Guo, Y. Fang, and S. Dong, "Gold nanoparticle/carbon nanotube hybrids as an enhanced material for sensitive amperometric determination of tryptophan," *Electrochimica Acta*, vol. 55, no. 12, pp. 3927–3931, 2010.

[89] R. N. Goyal, S. Bishnoi, H. Chasta, A. Md. Aziz, and M. Oyama, "Effect of surface modification of indium tin oxide by nanoparticles on the electrochemical determination of tryptophan," *Talanta*, vol. 85, pp. 2626–2631, 2011.

[90] B. Haghighi, S. Bozorgzadeh, and L. Gorton, "Fabrication of a novel electrochemiluminescence glucose biosensor using Au nanoparticles decorated multiwalled carbon nanotubes," *Sensors and Actuators B*, vol. 155, no. 2, pp. 577–583, 2011.

[91] K. R. Lim, J.-M. Park, H. N. Choi, and W.-Y. Lee, "Gold glyconanoparticle-based colorimetric bioassay for the determination of glucose in human serum," *Microchemical Journal*, vol. 106, pp. 154–159, 2013.

[92] W. O. Ho, S. Krause, C. J. McNeil et al., "Electrochemical sensor for measurement of urea and creatinine in serum based on ac impedance measurement of enzyme-catalyzed polymer transformation," *Analytical Chemistry*, vol. 71, no. 10, pp. 1940–1946, 1999.

[93] W. Lai, D. Tang, X. Que, J. Zhuang, L. Fu, and G. Chen, "Enzyme-catalyzed silver deposition on irregular-shaped gold nanoparticles for electrochemical immunoassay of alpha-fetoprotein," *Analytica Chimica Acta*, vol. 755, pp. 62–68, 2012.

[94] Y. Cui, H. Chen, L. Hou et al., "Nanogold-polyaniline-nanogold microspheres-functionalized molecular tags for sensitive electrochemical immunoassay of thyroid-stimulating hormone," *Analytica Chimica Acta*, vol. 738, pp. 76–84, 2012.

[95] A. P. F. Turner, I. Karube, and G. S. Wilson, *Biosensors: Fundamentals and Applications*, Oxford University Press, New York, NY, USA, 1987.

[96] D. P. Tang, R. Yuan, Y. Q. Chai et al., "Novel potentiometric immunosensor for hepatitis B surface antigen using a gold nanoparticle-based biomolecular immobilization method," *Analytical Biochemistry*, vol. 333, no. 2, pp. 345–350, 2004.

[97] M. P. Chatrathi, J. Wang, and G. E. Collins, "Sandwich electrochemical immunoassay for the detection of Staphylococcal enterotoxin B based on immobilized thiolated antibodies," *Biosensors and Bioelectronics*, vol. 22, no. 12, pp. 2932–2938, 2007.

[98] M. Wang, L. Wang, H. Yuan et al., "Immunosensors based on layer-by-layer self-assembled Au colloidal electrode for the electrochemical detection of antigen," *Electroanalysis*, vol. 16, no. 9, pp. 757–764, 2004.

[99] M. Wang, L. Wang, H. Yuan et al., "Application of impedance spectroscopy for monitoring colloid Au-enhanced antibody immobilization and antibody-antigen reactions," *Biosensors and Bioelectronics*, vol. 19, pp. 575–582, 2004.

[100] D. Tang, R. Yuan, Y. Chai, J. Dai, X. Zhong, and Y. Liu, "A novel immunosensor based on immobilization of hepatitis B surface antibody on platinum electrode modified colloidal gold and polyvinyl butyral as matrices via electrochemical impedance spectroscopy," *Bioelectrochemistry*, vol. 65, no. 1, pp. 15–22, 2004.

[101] G. K. Ahirwal and C. K. Mitra, "Gold nanoparticles based sandwich electrochemical immunosensor," *Biosensors and Bioelectronics*, vol. 25, no. 9, pp. 2016–2020, 2010.

[102] G. C. Jensen, C. E. Krause, G. A. Sotzing, and J. F. Rusling, "Inkjet-printed gold nanoparticle electrochemical arrays on plastic. Application to immunodetection of a cancer biomarker protein," *Physical Chemistry Chemical Physics*, vol. 13, pp. 4888–4894, 2011.

[103] J. A. Ho, H. C. Chang, N. Y. Shih et al., "Diagnostic detection of human lung cancer-associated antigen using a gold nanoparticle-based electrochemical immunosensor," *Analytical Chemistry*, vol. 82, pp. 5944–5950, 2010.

[104] Y. Zhu, J. I. Son, and Y. B. Shim, "Amplification strategy based on gold nanoparticle-decorated carbon nanotubes for neomycin immunosensors," *Biosensors and Bioelectronics*, vol. 26, no. 3, pp. 1002–1008, 2010.

[105] C.-C. Lin, L.-C. Chen, C.-H. Huang, S.-J. Ding, C.-C. Chang, and H.-C. Chang, "Development of the multi-functionalized

gold nanoparticles with electrochemical-based immunoassay for protein A detection," *Journal of Electroanalytical Chemistry*, vol. 619, pp. 39–45, 2008.

[106] E. J. Nam, E. J. Kim, A. W. Wark, S. Rho, H. Kim, and H. J. Lee, "Highly sensitive electrochemical detection of proteins using aptamer-coated gold nanoparticles and surface enzyme reactions," *Analyst*, vol. 137, pp. 2011–2016, 2012.

[107] A. J. Saleh Ahammad, Y. H. Choi, K. Koh, J. H. Kim, J. J. Lee, and M. Lee, "Electrochemical detection of cardiac biomarker troponin I at gold nanoparticle-modified ITO electrode by using open circuit potential," *International Journal of Electrochemical Science*, vol. 6, no. 6, pp. 1906–1916, 2011.

[108] K. Kerman, M. Chikae, S. Yamamura, and E. Tamiya, "Gold nanoparticle-based electrochemical detection of protein phosphorylation," *Analytica Chimica Acta*, vol. 588, no. 1, pp. 26–33, 2007.

[109] A. de la Escosura-Muñiz, C. Parolo, F. Maran, and A. Mekoçi, "Size-dependent direct electrochemical detection of gold nanoparticles: application in magnetoimmunoassays," *Nanoscale*, vol. 3, pp. 3350–3356, 2011.

[110] E. Vasilyeva, B. Lam, Z. Fang, M. D. Minden, E. H. Sargent, and S. O. Kelley, "Direct genetic analysis of ten cancer cells: tuning sensor structure and molecular probe design for efficient mRNA capture," *Angewandte Chemie*, vol. 50, pp. 4137–4141, 2011.

[111] W. Choon, A. Koh, P. Chandra, D.-M. Kim, and Y.-B. Shim, "Electropolymerized self-assembled layer on gold nanoparticles: detection of inducible nitric oxide synthase in neuronal cell culture," *Analytical Chemistry*, vol. 83, pp. 6177–6183, 2011.

[112] P. Chandra, W. C. Koh, H.-B. Noh, and Y.-B. Shim, "In vitro monitoring of i-NOS concentrations with an immunosensor: the inhibitory effect of endocrine disruptors on i-NOS release," *Biosensors and Bioelectronics*, vol. 32, pp. 278–282, 2012.

[113] A. S. Afonso, B. Pérez-Lo'pez, R. C. Faria et al., "Electrochemical detection of Salmonella using gold nanoparticles," *Biosensors and Bioelectronics*, vol. 40, pp. 121–126, 2013.

[114] M. Maltez-da Costa, A. de la Escosura-Muñiz, C. Nogués, L. Barrios, E. Ibáñez, and A. Merkoçi, "Detection of circulating cancer cells using electrocatalytic gold nanoparticles," *Small*, vol. 8, no. 23, pp. 3605–3612.

[115] P. T. Went, A. Lugli, S. Meier et al., "Frequent EpCam protein expression in human carcinomas," *Human Pathology*, vol. 35, no. 1, pp. 122–128, 2004.

[116] F. He, Q. Shen, H. Jiang et al., "Rapid identification and high sensitive detection of cancer cells on the gold nanoparticle interface by combined contact angle and electrochemical measurements," *Talanta*, vol. 77, pp. 1009–1014, 2009.

[117] W. Hui, W. Tian, Y. Yan-Xia et al., "Construction of an electrochemical cytosensor based on polyaniline nanofiber/gold nanoparticle interface and application to detection of cancer cells," *Chinese Journal of Analytical Chemistry*, vol. 40, pp. 184–190, 2012.

# $CuO$-$CeO_2$ Nanocomposite: An Efficient Recyclable Catalyst for the Synthesis of Aryl-14*H*-dibenzo[a-j]xanthenes

**Jalal Albadi,[1] Abdolhossein Razeghi,[2] Hossein Abbaszadeh,[3] and Azam Mansournezhad[3]**

[1] *College of Science, Behbahan Khatam Alanbia University of Technology, Behbahan 736254, Iran*
[2] *Catalysis and Reaction Engineering, Tehran University, Tehran, Iran*
[3] *Departmaent of Chemistry, Gachsaran Branch, Islamic Azad University, Gachsaran, Iran*

Correspondence should be addressed to Jalal Albadi; jalal.albadi@gmail.com

Academic Editor: Hamed Bahmanpour

$CuO$-$CeO_2$ nanocomposite is reported as a highly efficient recyclable catalyst that is applied for the synthesis of Aryl-14*H*-dibenzo[a-j]xanthenes under solvent-free conditions. The catalyst was synthesized by coprecipitation method and characterized by X-ray powder diffraction (XRD), BET specific surface area, field emission scanning electron microscopy (FESEM), and energy dispersive spectroscopy (EDS) analysis. The copper nanoparticles are dispersed as fine and amorphous phases on the surface of ceria and made nanoclusters with average size of about 33 nm. This catalyst can be recovered by simple filtration and recycled up to 8 consecutive runs without any losing of its efficiency. This procedure provides several advantages such as simple workup, mild reaction conditions, short reaction times, and high yields of the products.

## 1. Introduction

In recent years, nanocatalysis has emerged as a sustainable and competitive alternative to conventional catalyst since the nanoparticles possess a high-surface-to-volume ratio, which enhances their activity and selectivity, while, at the same time, maintaining the intrinsic features of a heterogeneous catalyst [1]. In particular, nanocrystalline oxides have proved to be useful to chemists in the laboratory and industry due to the good activation of adsorbed compounds and reaction rate enhancement, selectivity, easier workup and recyclability of the supports, and the eco-friendly green reaction conditions [2–6]. Also, the practical applications of nanocomposite metal oxides as the catalysts in organic synthesis have been increased due to their high catalytic activity because of the high surface area [7, 8]. The recyclability of the catalyst is the added advantage in the case of these catalysts. The catalytic activity of $CuO$-$CeO_2$ nanocomposite is well known for the oxidation of CO in $H_2$-rich streams [9]. Xanthene derivatives are important class of compounds that received significant attention from many pharmaceutical and organic chemists because of the broad spectrum of their biological and pharmaceutical properties such as antibacterial [10], anti-inflammatory [11], and antivirial properties [12]. Furthermore, these compounds are used as dyes and fluorescent material for visualization of biomolecules and in laser technologies due to their useful spectroscopic properties [13, 14]. Aryl-14*H*-dibenzo[a-j]xanthenes are among the most important classes in the family of xanthenes due to their distinctive structures and great potential for the further transformations [15]. Various methods have been reported for the synthesis of Aryl-14*H*-dibenzo[a-j]xanthenes; among these methods, the one-pot condensation of $\beta$-naphthol with aldehydes is the most common procedure. Therefore, various catalysts have been developed for the improvement of this reaction [16–25]. In the continuation of our research program to develop the efficient, and green catalysts in organic synthesis [26–29], herein, we wish to report a green, efficient and recyclable catalyst for the synthesis of Aryl-14*H*-dibenzo[a-j]xanthenes under solvent-free conditions (Scheme 1).

## 2. Experimental

All products were identified by the comparison of their spectral and physical data with authentic samples. Chemicals (2-naphthol, aromatic aldehydes, cerium, and copper nitrates)

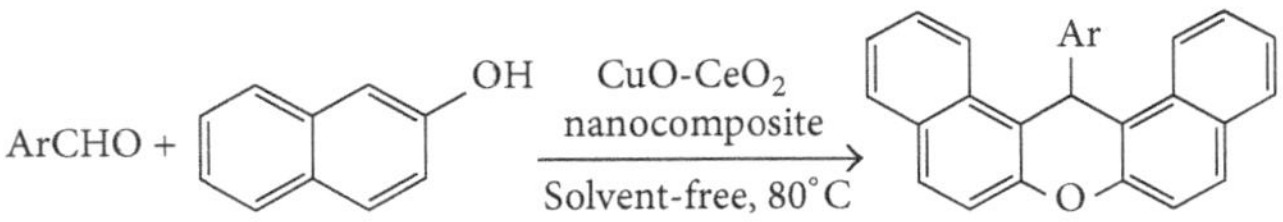


SCHEME 1: Synthesis of Aryl-14*H*-dibenzo[a-j]xanthenes catalyzed by CuO-$CeO_2$ nanocomposite oxide.

were purchased from Merck Chemical company in Germany. Yields refer to isolated pure products.

*2.1. Catalyst Preparation.* CuO-$CeO_2$ nanocomposite was prepared by coprecipitation method using aqueous solution of cerium and copper nitrates and drop-wise adding KOH as precipitant agent under vigorous mixing, while temperature and pH were fixed at unique values. Then, acquired sample was filtered, washed, and calcined to obtain final catalyst for using atsynthesis of Aryl-14*H*-dibenzo[a-j]xanthenes.

*2.2. General Procedure for Synthesis of Aryl-14H-dibenzo[a-j]xanthenes.* A mixture of 2-naphthol (2 mmol), aldehyde (1 mmol), and CuO-$CeO_2$ nanocomposite (0.05 g) was added to a round-bottomed flask. The reaction mixture was placed in an oil bath at 80°C and magnetically stirred for the appropriate times according to Table 1. After completion of the reaction (monitored by TLC), the mixture reaction was cooled to room temperature, next acetone (10 mL) was added, and the mixture stirred for 5 min. Then, the catalyst was recovered by filtration to be reused subsequently. The filtrate was evaporated to dryness, and the solid residue recrystallized from hot ethanol to give pure products in high yields. The experimental procedure with this catalyst is very simple and the catalyst can be removed easily by filtration. The solid products were easily recrystallized from hot ethanol and were obtained in good to high yields during the short reaction times. Very low amount of the catalyst is needed. Moreover, our procedure is environmentally friendly as it does not use any toxic auxiliary or solvent.

*2.3. Catalyst Characterization.* The catalyst structure characterization was performed by X-ray powder diffraction (XRD), using a Bruker AXS D8 advanced diffractometer equipment with CuK$\alpha$ radiation ($\lambda$ = 1.5406 A°). The Debye-Scherrer equation is used to determine average crystallite size of nanoparticles. Field emission scanning electron microscopy (FESEM) and energy dispersive spectroscopy (EDS) carried out by a Hitachi S4160 instrument to see the morphology, the evaluation of cluster size, and the metal composition of the catalyst. BET specific surface area was estimated from the $N_2$ adsorption/desorption isotherms, measured at 77°K using a Quantachrome CHEMBET-3000 instrument.

## 3. Results and Discussions

XRD pattern of fresh CuO-$CeO_2$ nanocomposite calcined at 450°C is illustrated (Figure 1). The main reflection placed at $2\theta$ range of 15–80° indicates a cubic fluorite structure with highest intensity at (1, 1, 1) plane that belongs to the XRD standard JCPDS card number of 43-1002. According to the pattern of pure $CeO_2$ (not shown in the figure), only the diffraction peaks of ceria crystallites are detected, and there is no peaks related to copper oxide nanoparticles. It is supposed that fine dispersion and/or amorphous structure of CuO with low content on the surface of ceria clusters is responsible for absence of copper oxide peaks at the XRD pattern [9, 30, 31]. However, there is some difference between values of unit cell parameters for pure $CeO_2$ and CuO-$CeO_2$ due to lattice shrinkage in surface ceria structure. The calculation showed 5.45 and 5.42° A for $CeO_2$ and CuO-$CeO_2$, respectively. The observed shrinkage may be consequence of partial incorporation of CuO into the $CeO_2$ lattice structure that has been reported for coprecipitation synthesis [30, 31]. The calculated average crystal size of the ceria in the

TABLE 1: Synthesis of Aryl-14*H*-dibenzo[a-j]xanthenes catalyzed by CuO-$CeO_2$ nanocomposite.

| Entry | Aldehyde | Time (min) | Yield (%)[a] | M. P. (°C) |
|---|---|---|---|---|
| 1 | PhCHO | 15 | 93 | 185–186 |
| 2 | 2-$NO_2C_6H_4$CHO | 12 | 91 | 213–215 |
| 3 | 3-$NO_2C_6H_4$CHO | 15 | 90 | 210–211 |
| 4 | 4-$NO_2C_6H_4$CHO | 8 | 90 | 310–312 |
| 5 | 2-Cl$C_6H_4$CHO | 12 | 90 | 213–215 |
| 6 | 4-Cl$C_6H_4$CHO | 10 | 91 | 289–290 |
| 7 | 2,4-$Cl_2C_6H_3$CHO | 12 | 90 | 251–252 |
| 8 | 2,6-$Cl_2C_6H_3$CHO | 15 | 89 | 267–269 |
| 9 | 4-F$C_6H_4$CHO | 10 | 90 | 237–238 |
| 10 | 2-Me$C_6H_4$CHO | 18 | 89 | 258–260 |
| 11 | 3-Me$C_6H_4$CHO | 12 | 90 | 199–201 |
| 12 | 4-Me$C_6H_4$CHO | 20 | 90 | 228–229 |
| 13 | 4-MeO$C_6H_4$CHO | 30 | 89 | 202–204 |
| 14 | 4-Br$C_6H_4$CHO | 15 | 91 | 296–298 |
| 15 | 4-OH$C_6H_4$CHO | 45 | 88 | 139–141 |

[a]Isolated yield. [b]Products were characterized by the comparison of their spectroscopic data (NMR and IR) and melting points with those reported in the literature [17, 18].

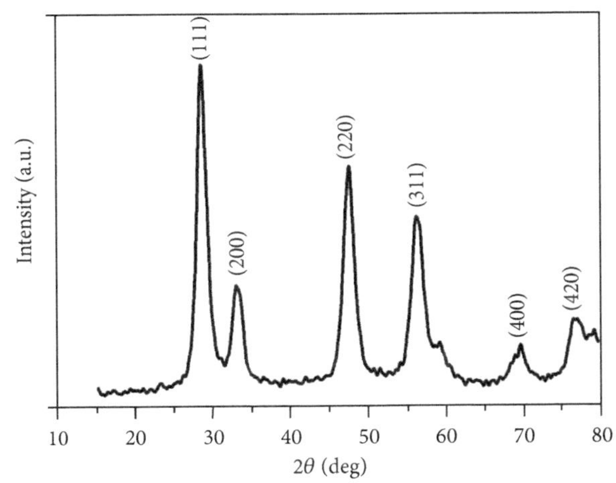


FIGURE 1: XRD pattern of the CuO-$CeO_2$ nanocomposite catalyst.

(a) (b)

FIGURE 2: FESEM image of the $CuO\text{-}CeO_2$ nanocomposite catalyst.

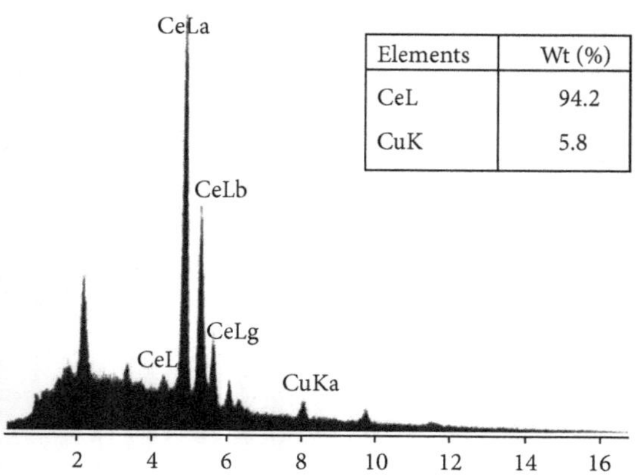


FIGURE 3: EDS analysis of the $CuO\text{-}CeO_2$ nanocomposite catalyst.

catalyst based on dominant peak at $2\theta$ of 28.5° using Debye-Scherrer equation was 7.5 nm. BET specific surface area result was 131 $m^2/g$ that confirms estimated value for average nanoparticle size of the catalyst by XRD test.

Figures 2(a) and 2(b) present FESEM micrograph of the catalyst at two different magnifications to study the morphology and the evaluation of nanoclusters size of catalyst. It seems that 7 nm particles calculated based on the BET surface area are agglomerated to approximately 33 nm sized clusters and made bigger nanostructure $CuO\text{-}CeO_2$ catalyst.

EDS analysis (Figure 3) showed the presence of 5.8 and 94.2 wt.% for Cu and Ce elements in the catalyst, respectively.

Optimization of the reaction conditions showed that the best results in the presence of 0.05 g of $CuO\text{-}CeO_2$ nanostructured catalyst at 80°C were obtained under solvent-free conditions, when the relative ratio of the aldehyde and 2-naphthol was 2 : 1, respectively. Using these optimized conditions, the reaction of various aromatic aldehydes, containing electron-donating and electron-withdrawing groups, was explored (Table 1). Aliphatic aldehydes remain intact under the same reaction conditions. Therefore, the method can be useful for the chemoselective synthesis of Aryl-14*H*-dibenzo[a-j]xanthenes from aromatic aldehydes in the presence of aliphatic ones.

TABLE 2: Recyclability of $CuO\text{-}CeO_2$ nanocomposite.

| Run | 1 | 2 | 3 | 4 | 5 | 6 | 7 | 8 |
|---|---|---|---|---|---|---|---|---|
| Time (min) | 15 | 15 | 15 | 18 | 18 | 20 | 20 | 25 |
| Yield (%)[a] | 93 | 93 | 92 | 92 | 92 | 90 | 88 | 88 |

[a] Isolated yield.

The activity of the recycle catalyst was also examined under the optimized conditions, and the desired product was obtained in high yields after 1–8 runs (Table 2). To investigate these property for our introduced catalyst, the reaction of benzaldehyde with 2-naphthol was selected as the model (Table 2). After each reaction, we washed and dried the catalyst with acetone and reused for the next run. This process was repeated for 8 runs, and no appreciable yield decrease was observed.

## 4. Conclusions

We have developed a new, efficient, and green procedure for the synthesis of biologically important xanthene derivatives catalyzed by recyclable $CuO\text{-}CeO_2$ nanostructured catalyst, under solvent-free conditions. This catalyst can promote the yields and reaction times over 8 runs without noticeable loss in its efficiency. Moreover, high yields of products, short reaction times, ease of workup, and clean operation are the most important advantages of this method which makes the procedure beneficial compared to conventional methods.

## Acknowledgment

The authors are thankful to the Behbahan Khatam Al-Anbia, University of Technology, for the partial support of this work.

## References

[1] D. Astruc, *Nanoparticles and Catalysis*, Wiley-VCH, Weinheim, Germany, 2008.

[2] H. Sharghi and M. Hosseini, "Solvent-free and one-step Beckmann rearrangement of ketones and aldehydes by zinc oxide," *Synthesis*, no. 8, pp. 1057–1059, 2002.

[3] M. H. Sarvari and H. Sharghi, "Reactions on a solid surface. A simple, economical and efficient Friedel-Crafts acylation reaction over zinc oxide (ZnO) as a new catalyst," *Journal of Organic Chemistry*, vol. 69, no. 20, pp. 6953–6956, 2004.

[4] S. F. Yin, B. Q. Xu, S. J. Wang, C. F. Ng, and C. T. Au, "Magnesia-Carbon nanotubes (MgO-CNTs) nanocomposite: novel support of Ru Catalyst for the generation of COx-free Hydrogen from Ammonia," *Catalysis Letters*, vol. 96, pp. 113–116, 2004.

[5] M. T. Drexler and M. D. Amiridis, "Kinetic investigation of the heterogeneous synthesis of flavanone over MgO," *Catalysis Letters*, vol. 79, no. 1–4, pp. 175–181, 2002.

[6] R. S. Varma, "Solvent-free organic syntheses: using supported reagents and microwave irradiation," *Green Chemistry*, vol. 1, no. 1, pp. 43–55, 1999.

[7] S. Samantaray, D. K. Pradhan, G. Hota, and B. G. Mishra, "Catalytic application of $CeO_2$-CaO nanocomposite oxide synthesized using amorphous citrate process toward the aqueous phase one pot synthesis of 2-amino-2-chromenes," *Chemical Engineering Journal*, vol. 193, pp. 1–9, 2012.

[8] S. Li, W. Zhang, M. H. So, C. M. Che, R. Wang, and R. Chen, "One-pot solvothermal synthesis of Pd/$F_{e3O4}$ nanocomposite and its magnetically recyclable and efficient catalysis for Suzuki reactions," *Journal of Molecular Catalysis A*, vol. 359, pp. 81–87, 2012.

[9] A. Razeghi, A. Khodadadi, H. Ziaei-Azad, and Y. Mortazavi, "Activity enhancement of Cu-doped ceria by reductive regeneration of CuO-$CeO_2$ catalyst for preferential oxidation of CO in $H_2$-rich streams," *Chemical Engineering Journal*, vol. 164, no. 1, pp. 214–220, 2010.

[10] S. Chatterjee, M. Iqbal, J. C. Kauer et al., "Xanthene derived potent nonpeptidic inhibitors of recombinant human calpain I," *Bioorganic and Medicinal Chemistry Letters*, vol. 6, no. 13, pp. 1619–1622, 1996.

[11] J. P. Poupelin, G. Saint Ruf, O. Foussard Blanpin, G. Narcisse, G. Uchida-Ernouf, and R. Lacroix, "Synthesis and antiinflammatory properties of bis (2 hydroxy 1 naphthyl) methane derivatives," *European Journal of Medicinal Chemistry*, vol. 13, no. 1, pp. 67–71, 1978.

[12] O. Sirkecioglu, N. Tulinli, and A. Akar, "Synthesis of 14-Aryl-14*H*-dibenzo[a-j]xanthenes," *Journal of Chemical Research*, pp. 502–506, 1995.

[13] R. M. Ion, A. Planner, K. Wiktorowicz, and D. Frackowiak, "The incorporation of various porphyrins into blood cells measured via flow cytometry, absorption and emission spectroscopy," *Acta Biochimica Polonica*, vol. 45, no. 3, pp. 833–845, 1998.

[14] R. J. Sarma and J. B. Baruah, "One step synthesis of dibenzoxanthenes," *Dyes and Pigments*, vol. 64, no. 1, pp. 91–92, 2005.

[15] K. Chibale, M. Visser, D. Van Schalkwyk, P. J. Smith, A. Saravanamuthu, and A. H. Fairlamb, "Exploring the potential of xanthene derivatives as trypanothione reductase inhibitors and chloroquine potentiating agents," *Tetrahedron*, vol. 59, no. 13, pp. 2289–2296, 2003.

[16] L. Wu, J. Zhang, L. Fang, C. Yang, and F. Yan, "Silica chloride catalyzed synthesis of 14-aryl-14*H*-dibenzo[a,i]xanthene-8,13-diones," *Dyes and Pigments*, vol. 86, no. 1, pp. 93–96, 2010.

[17] S. Allameh, A. Davoodnia, and A. Khojastehnezhad, "An efficient and eco-friendly synthesis of 14-Aryl-14*H*-dibenzo[a-j] xanthenes using $H_4[SiW_{12}O_{40}]$ as a heterogeneous and reusable catalyst under solvent-free conditions," *Chinese Chemical Letters*, vol. 23, pp. 17–20, 2012.

[18] I. Mohammadpoor-Baltork, M. Moghadam, V. Mirkhani, S. Tangestaninejad, and H. R. Tavakoli, "Highly efficient and green synthesis of 14-aryl(alkyl)-14*H*-dibenzo[a,j] xanthene and 1,8-dioxooctahydroxanthene derivatives catalyzed by reusable zirconyl triflate $[ZrO(OTf)_2]$ under solvent-free conditions," *Chinese Chemical Letters*, vol. 22, no. 1, pp. 9–12, 2011.

[19] A. R. Hajipour, Y. Ghayeb, N. Sheikhan, and A. E. Ruoho, "Brønsted acidic ionic liquid as an efficient and reusable catalyst for synthesis of 14-aryl-or 14-Aryl-14*H*-dibenzo[a-j]xanthenes under solvent-free conditions," *Synlett*, no. 5, pp. 741–744, 2010.

[20] S. Ko and C. F. Yao, "Heterogeneous catalyst: amberlyst-15 catalyzes the synthesis of 14-substituted-14*H*-dibenzo[a,j] xanthenes under solvent-free conditions," *Tetrahedron Letters*, vol. 47, no. 50, pp. 8827–8829, 2006.

[21] T. S. Rivera, A. Sosa, G. P. Romanelli, M. N. Blanco, and L. R. Pizzio, "Tungstophosphoric acid/zirconia composites prepared by the sol-gel method: an efficient and recyclable green catalyst for the one-pot synthesis of 14-Aryl-14*H*-dibenzo[a-j]xanthenes," *Applied Catalysis A*, vol. 443, pp. 207–213, 2012.

[22] M. Dabiri, M. Baghbanzadeh, M. Shakouri Nikcheh, and E. Arzroomchilar, "Eco-friendly and efficient one-pot synthesis of alkyl- or Aryl-14*H*-dibenzo[a-j]xanthenes in water," *Bioorganic and Medicinal Chemistry Letters*, vol. 18, no. 1, pp. 436–438, 2008.

[23] G. B. Dharma Rao, M. P. Kaushik, and A. K. Halve, "An efficient synthesis of naphtha[1, 2-e]oxazinone and 14-substituted-14*H*-dibenzo [a, j] xanthene derivatives promoted by zinc oxide nanoparticle under thermal and solvent-free conditions," *Tetrahedron Letters*, vol. 53, pp. 2741–2744, 2012.

[24] F. Shirini and N. Ghaffari Khaligh, "Succinimide-N-sulfonic acid: an efficient catalyst for the synthesis of xanthene derivatives under solvent-free conditions," *Dyes and Pigments*, vol. 95, pp. 789–794, 2012.

[25] R. Kumar, G. C. Nandi, R. K. Verma, and M. S. Singh, "A facile approach for the synthesis of 14-aryl- or alkyl-14*H*-dibenzo[a,j]xanthenes under solvent-free condition," *Tetrahedron Letters*, vol. 51, no. 2, pp. 442–445, 2010.

[26] J. Albadi, M. Keshavarz, F. Shirini, and M. Vafaie-nezhad, "Copper iodide nanoparticles on poly(4-vinyl pyridine): a new and efficient catalyst for multicomponent click synthesis of 1,4-disubstituted-1,2,3-triazoles in water," *Catalysis Communications*, vol. 27, pp. 17–20, 2012.

[27] J. Albadi, N. Iravani, F. Shirini, and F. Dehghan, "A green recyclable poly(4-vinylpyridine)-supported copper iodide nanoparticle catalyst for the chemoselective synthesis of pentaerythritol diacetals from aromatic aldehydes," *Journal of Chemical Research*, vol. 36, no. 10, pp. 610–611, 2012.

[28] J. Albadi, N. Iravani, and M. Khoshakhlagh, "A new, green and recyclable poly(4-vinylpyridine)-supported copper iodide nanoparticles catalyst for the synthesis of aryl -14*H*-dibenzo[a-o]xanthenes," *Iranian Journal of Catalysis*, vol. 2, pp. 85–89, 2012.

[29] J. Albadi, M. Keshavarz, M. Abedini, and M. Vafaie-nezhad, "Copper iodide nanoparticles on poly(4-vinyl pyridine) as new and green catalyst for multicomponent click synthesis of 1, 4-disubstituted-1, 2, 3-triazoles in water," *Chinese Chemical Letters*, vol. 23, pp. 797–800, 2012.

[30] P. Ratnasamy, D. Srinivas, C. V. V. Satyanarayana et al., "Influence of the support on the preferential oxidation of CO in hydrogen-rich steam reformates over the CuO-$CeO_2$-$ZrO_2$ system," *Journal of Catalysis*, vol. 221, pp. 455–465, 2004.

[31] Z. Liu, R. Zhou, and X. Zheng, "Preferential oxidation of CO in excess hydrogen over a nanostructured $CuO-CeO_2$ catalyst with high surface areas," *Catalysis Communications*, vol. 9, pp. 2183–2186, 2008.

# 21

# Green Chemistry Approach for Efficient Synthesis of Schiff Bases of Isatin Derivatives and Evaluation of Their Antibacterial Activities

**Jnyanaranjan Panda,[1] V. Jagannath Patro,[1] Biswa Mohan Sahoo,[1] and Jitendriya Mishra[2]**

[1] *Department of Pharmaceutical Chemistry, Roland Institute of Pharmaceutical Sciences and Berhampur, Odisha 760010, India*
[2] *University Institute of Pharmaceutical Sciences, UGC Centre of Advanced Study (CAS), Panjab University, Chandigarh 160014, India*

Correspondence should be addressed to Jnyanaranjan Panda; jnyana@rediffmail.com

Academic Editor: Amir Kajbafvala

Microwave-assisted organic synthesis, a green chemistry approach, is nowadays widely used in the drug synthesis. Microwave-assisted synthesis improves both throughput and turnaround time for medicinal chemists by offering the benefits of drastically reduced reaction times, increased yields, and pure products. Schiff bases are the important class of organic compounds due to their flexibility, and structural diversities due to the presence of azomethine group which is helpful for elucidating the mechanism of transformation and rasemination reaction in biological system. This novel compound could also act as valuable ligands for the development of new chemical entities. In the present work, some Schiff bases of Isatin derivatives was synthesized using microwave heating method. Schiff base of Isatin were synthesized by condensation of the keto group of Isatin with different aromatic primary amines. They were characterized by means of spectral data and subsequently subjected to the *in vitro* antibacterial activities against gram positive and gram negative strains of microbes. It was observed that the compound with electron withdrawing substituents exhibited good antibacterial activities against almost all the micro organisms.

## 1. Introduction

Microwave-assisted organic synthesis is widely used as a source of heating in drug synthesis. Drug molecules can be built in a fraction of the time by this method. As a result, this technique has rapidly gained acceptance as a valuable tool for accelerating drug discovery and development processes. A microwave is a form of electromagnetic energy, which falls at the lower end of the electromagnetic spectrum and is defined in a measurement of frequency as 300 to 300,000 Megahertz. The microwave region of the electromagnetic spectrum lies between infrared and radio frequencies. The basic mechanism of microwave assisted synthesis involves agitation of polar molecules or ions that oscillate under the effect of an oscillating electric or magnetic field. In the presence of an oscillating field, particles try to orient themselves or be in phase with the field. Only materials that absorb microwave radiation are relevant to microwave chemistry. These materials can be categorized according to the three main mechanisms of heating such as dipolar polarization, conduction mechanism, and interfacial polarization. The technique offers simple, clean, fast, efficient, and economical for the synthesis of a large number of drug molecules, having provided the momentum for many medicinal chemists to switch from traditional heating method to microwave heating method [1, 2]. Thus it follows green chemistry approach. The role of green chemistry is essential in ensuring that the next generation of chemicals, materials, and energy is more sustainable than the current generation. Worldwide demand for environmentally friendly chemical processes and products requires the development of novel and cost-effective approaches to prevent pollution of the environment [3]. The important area of green chemistry is the elimination of solvents during chemical processes or the replacement

TABLE 1: Comparative study on yield and reaction time of the synthesized compounds 2(a–j) by conventional and microwave assisted method.

| Compound code | Conventional method | | | Microwave assisted method | | |
|---|---|---|---|---|---|---|
| | Time (h) | Energy (Temp. °C) | Yield (%) | Time (min.) | Energy (power. watt) | Yield (%) |
| 2a | 3 | 98–100 | 64 | 5 | 210 | 84 |
| 2b | 3 | 98–100 | 62 | 7 | 210 | 82 |
| 2c | 3 | 98–100 | 56 | 6 | 210 | 85 |
| 2d | 2 | 98–100 | 58 | 10 | 210 | 75 |
| 2e | 2 | 98–100 | 66 | 7 | 210 | 85 |
| 2f | 2 | 98–100 | 53 | 9 | 210 | 74 |
| 2g | 2 | 98–100 | 57 | 8 | 210 | 77 |
| 2h | 3 | 98–100 | 53 | 10 | 210 | 76 |
| 2i | 3 | 98–100 | 56 | 10 | 210 | 72 |
| 2j | 3 | 98–100 | 60 | 8 | 210 | 84 |

TABLE 2: TLC report and melting point data of the synthesized compounds 2(a–j).

| Compound code | R | Tm (°C) | $R_f$ |
|---|---|---|---|
| 2a | phenyl | 144-145 | 0.56 |
| 2b | 2-nitrophenyl | 120-121 | 0.67 |
| 2c | 3-nitrophenyl | 140-141 | 0.62 |
| 2d | 4-nitrophenyl | 150-151 | 0.68 |
| 2e | 3-chlorophenyl | 135-136 | 0.66 |
| 2f | 4-chlorophenyl | 140-141 | 0.67 |
| 2g | 4-bromophenyl | 230-231 | 0.62 |
| 2h | 4-fluorophenyl | 141-142 | 0.56 |
| 2i | 3-Cl-4-F-phenyl | 129-130 | 0.54 |
| 2j | 2,6-dichlorophenyl | 241-142 | 0.55 |

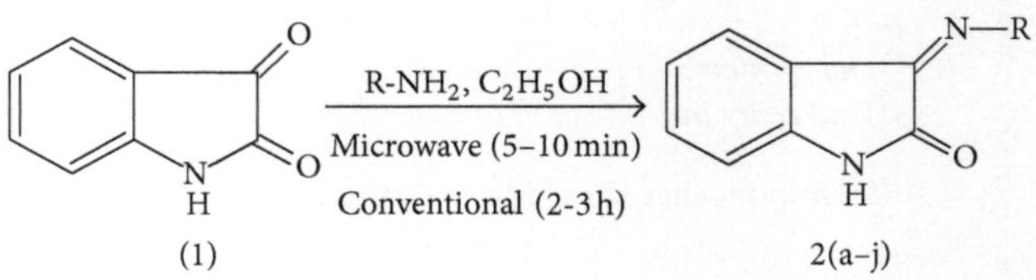


SCHEME 1

of hazardous solvents with environmental friendly solvents. Heterocyclic compounds are the most commonly used pharmacophore in the development of drugs and pharmaceutical substances. Due to their drug-likeness and structural diversity, these are routinely employed in high-throughput screening at early stages of drug discovery processes. Therefore, the goal of the present study is to carry out the synthesis of some heterocyclic compounds under green chemistry approach [4]. The above facts prompted us to synthesize some Schiff base of Isatin derivatives by using microwave heating method [5–7]. Azomethine group (–C=N–) containing compounds typically known as Schiff bases has been synthesized by the condensation of primary amines with active carbonyls [8]. Schiff bases form a significant class of compounds in medicinal chemistry due to various biological activities such as antibacterial [9], antifungal [10], antiviral [11], anti-HIV [12], and anticonvulsant [13] activities. The structures of the synthesized compounds were confirmed by means of their physical and spectral data. The synthesized compounds were evaluated for their possible antibacterial activities by cup plate method.

## 2. Experimental Method

*2.1. Materials.* The chemicals and solvents used for the experimental work were of commercial grade. All the melting points were taken in open capillaries and are uncorrected (Table 2). Followup of the reactions and checking the purity of the compounds were made by TLC on precoated Silica gel-aluminum plates (Type 60 F254, Merck, Darmstadt, Germany) and were visualized by exposure to UV-light (254 nm) or iodine vapor for few seconds. The IR spectra of the compounds were recorded on FT-IR Spectrophotometer, model IR Affinity-1 (SHIMADZU), using KBr powder and the values are expressed in $cm^{-1}$. $^1H$ NMR spectra of selected compounds were recorded on multinuclear FT NMR Spectrometer, model Advance-II (Bruker), (at 400 MHz) using tetramethylsilane as an internal standard. The multiplicities of the signals are denoted with the symbols s, d, t, and m for singlet, doublet, triplet, and multiplet, respectively. The microwave irradiated synthesis was performed in scientific microwave oven, Catalyst System (operating between 140–700 W). All the reactions were carried out at power level-1, which corresponds to 140 W.

*2.2. Conventional Synthesis of Schiff Bases of Isatin.* Equimolar (0.01 mol) quantity of Isatin and substituted anilines were dissolved in ethanol (10 mL) and refluxed for 3 h in presence of glacial acetic acid. In between TLC was checked to confirm the completion of reaction. After completion of reaction, the reaction mixture was kept overnight to get the solid product. The product was filtered, dried, and recrystallized from ethanol (Scheme 1).

*2.3. Microwave Synthesis of Schiff Bases of Isatin.* The required quantities of above reactants are subjected to microwave irradiation for 5–10 min at power level 1 (140 watts). In between TLC was checked to confirm the completion of

TABLE 3: Antibacterial activities of the synthesized compounds 2(a–j).

| Compounds | Diameter of zone of inhibition (millimeters) | | | | |
|---|---|---|---|---|---|
| | *P. vulgaris* | *P. aeruginosa* | *E. coli* | *S. aureus* | *S. epidermidis* |
| 2a | 12.23 ± 0.57 | 15.02 ± 1.00 | 12.00 ± 1.00 | 13.32 ± 0.57 | 12.56 ± 0.57 |
| 2b | 16.33 ± 0.55 | 17.10 ± 1.00 | 22.00 ± 1.00 | 23.00 ± 1.00 | 15.33 ± 0.57 |
| 2c | 18.33 ± 0.55 | 18.11 ± 1.00 | 21.00 ± 1.00 | 19.00 ± 1.00 | 14.00 ± 1.00 |
| 2d | 18.66 ± 0.52 | 16.00 ± 1.00 | 18.66 ± 0.57 | 19.34 ± 0.57 | 17.00 ± 1.00 |
| 2e | 15.33 ± 0.57 | 17.00 ± 1.00 | 16.33 ± 0.57 | 16.32 ± 0.57 | 14.33 ± 0.57 |
| 2f | 21.32 ± 0.53 | 21.00 ± 1.00 | 21.00 ± 1.00 | 19.34 ± 0.57 | 15.33 ± 0.57 |
| 2g | 17.33 ± 0.57 | 18.00 ± 1.00 | 20.00 ± 1.00 | 19.23 ± 0.57 | 12.33 ± 0.57 |
| 2h | 16.33 ± 0.56 | 20.10 ± 1.00 | 20.00 ± 1.00 | 22.20 ± 1.00 | 16.33 ± 0.57 |
| 2i | 12.33 ± 0.57 | 14.10 ± 1.00 | 12.33 ± 0.57 | 16.32 ± 0.57 | 11.66 ± 0.57 |
| 2j | 14.34 ± 0.53 | 14.10 ± 1.00 | 14.33 ± 0.57 | 17.35 ± 0.57 | 16.00 ± 1.00 |
| Control | — | — | — | — | — |
| Standard | 23.33 ± 0.57 | 20.33 ± 0 .57 | 31.33 ± 0.57 | 21.66 ± 0.57 | 23.33 ± 0.57 |

Results were expressed as mean ± S.D. ($n = 3$); "—" indicates no zone of inhibition.

reaction. After completion of reaction, the reaction mixture was kept overnight to get the solid product. The product was filtered, dried, and recrystallized from ethanol (Scheme 1).

## 3. Results and Discussion

The synthetic protocol followed was outlined in Scheme 1. The synthesized compounds were characterized by their physical and spectral studies. As compared to conventional heating, microwave heating provides high yield with pure product in less reaction time (Table 1). The IR spectra of all the synthesized compounds were recorded using KBr (Merck). All the compounds have exhibited -NH stretching in the region of 3400–3200 $cm^{-1}$. IR spectra of synthesized compounds showed the presence of characteristic absorption peaks around 3000–3100 $cm^{-1}$ (C–H, Ar), 2800–2780 $cm^{-1}$ (-C=N–), 1640–1602 $cm^{-1}$ (C=N), 1580–1520 $cm^{-1}$ (C=C Ar), and 1346–1330 $cm^{-1}$ (C–N). The halogenated derivatives showed the IR absorption bands in the regions and 1400–1000 $cm^{-1}$, 800–600 $cm^{-1}$, and 600–500 $cm^{-1}$ which correspond to (C–F str.), (C–Cl str.), (C–Br str.), respectively. The N–O stretching vibrations in nitro group occur near 1550–1475 $cm^{-1}$ (asymmetrical) and 1365–1290 $cm^{-1}$ (symmetrical), with the band at 1550 $cm^{-1}$ being the stronger. The $^1$HNMR spectra above synthesized compounds have shown singlet in the region of $\delta$ 9.35 to 9.98 corresponding to secondary amino group (–NH). The aromatic protons resonate as multiplet in the region of $\delta$ 7.03 to 7.90. The synthesized compounds were evaluated for their *in-vitro* antibacterial activities by cup plate method against both gram positive and gram negative bacteria. The results of the study were summarized in Table 3 including the activity of standard. The compound 2f showed most promising effect whereas compound 2a and 2i were found to be least effective against *P. vulgaris*. Similarly the compound 2f was found to have maximum zone of inhibition and compounds 2i and 2j were found to have minimum zone of inhibition against *P. aeruginosa*. The compound 2b exhibited most promising effect whereas compound 2a was found to have mild effect against *E. coli* and *S. aureus*. The compound 2d showed maximum effect against *S. epidermidis* whereas compound 2i was found to have mild effect against this species. Most of the tested compounds exhibited antibacterial activities against all the test organisms but the activity was less than that of the standard drug tetracycline in this test concentration. Thus taking into account the antibacterial activities of the tested compounds most of the compounds displayed poor activities against *S. epidermidis* and promising activities against *P. vulgaris*, *P. aeruginosa*, *E. coli*, and *S. aureus*.

The photographs of zone of inhibition of standard, control, and some of the tested compounds were given in Figure 1. Figure 1(a) showed the zone of inhibition of the standard drug against *E. coli* whereas there is no zone of inhibition in case of control. Similarly Figure 1(b) displayed the zone of inhibition of the standard drug against *S. aureus* whereas there is no zone of inhibition in case of control. Figure 1(c) showed the zone of inhibition of test compound 2 h (L) which was more than that of 2a (R) against *S. aureus*. Figure 1(d) showed the zone of inhibition of test compounds 2i (L) and 2f (R) against *P. vulgaris*. In case of Figure 1(e) it was clearly visible that the diameter of zone of inhibition of compound 2c was more than that of compound 2d against *E. coli*. Figure 1(f) showed the diameter of zone of inhibition of compounds 2c and 2d against *P. aeruginosa*.

## 4. Biological Evaluation

*4.1. Antibacterial Study.* The synthesized compounds were screened *in-vitro* for their antibacterial activities against *Staphylococcus aureus* (MTCC-87), *Escherichia coli* (MTCC-40), *Staphylococcus epidermidis* (MTCC-2639), *Pseudomonas aeruginosa* (MTCC-424), and *Proteus vulgaris* (MTCC 426) using cup plate method [14, 15]. The compounds were tested at 500 $\mu$g concentration in dimethyl sulphoxide (DMSO), using nutrient agar as the medium. After 24 h of incubation

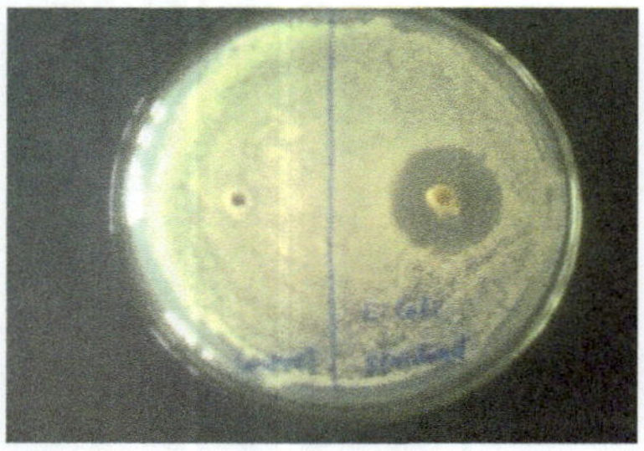

(a) Zone of inhibition of control and standard against *E. coli*

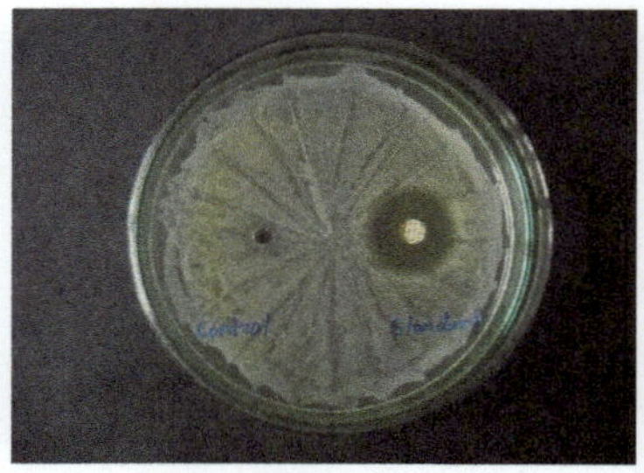

(b) Zone of inhibition of control and standard against *S. aureus*

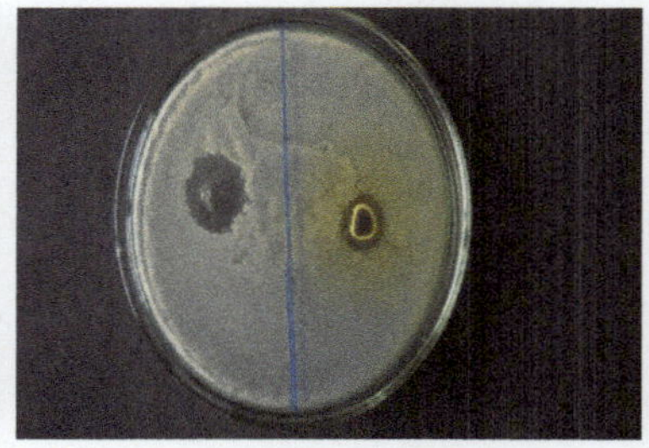

(c) Zone of inhibition of test compounds 2 h (L) and 2a (R) against *S. aureus*

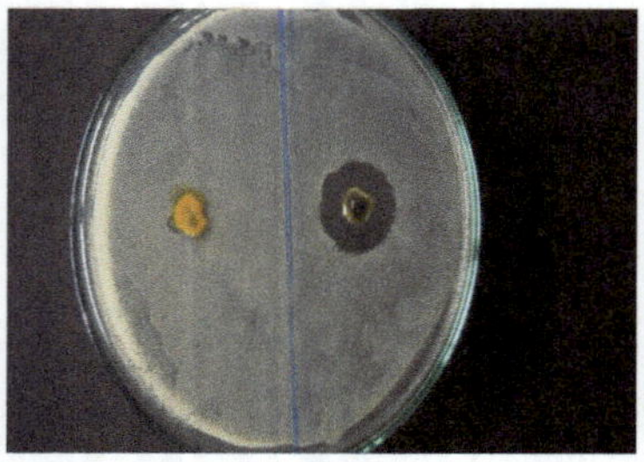

(d) Zone of inhibition of test compounds 2i (L) and 2f (R) against *P. vulgaris*

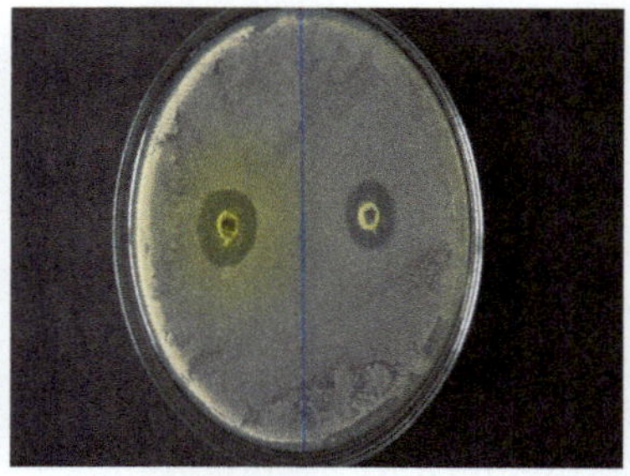

(e) Zone of inhibition of test compounds 2c (L) and 2d (R) against *E. coli*

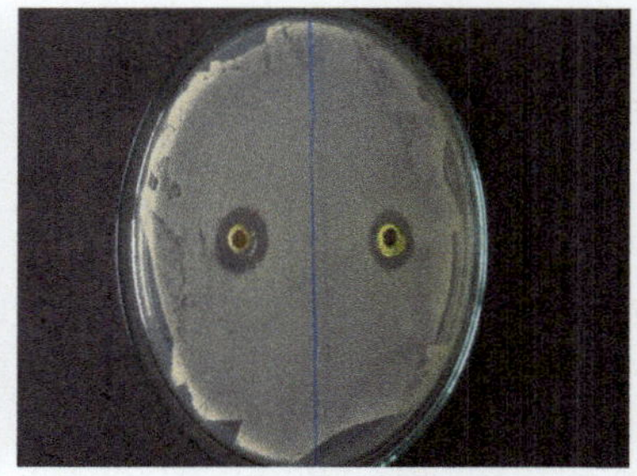

(f) Zone of inhibition of test compounds 2c (L) and 2d (R) against *P. aeruginosa*

FIGURE 1: Photographs displaying zone of inhibition of control, standard, and various test compounds against different test organisms (R: right, L: left).

at 37°C, the zone of inhibition formed were measured in millimeters against standard drug tetracycline and the data were presented in Table 3.

## 5. Conclusion

We have synthesized some Schiff base of Isatin derivatives by conventional and microwave irradiation method. With the help of microwave synthesis, the yield of product was increased from 60% upto 85% as compared to conventional method. By microwave irradiation the reactions were completed within 5–10 minutes and the products were obtained in good to high yields, which reduced the time, waste, and formation of byproduct. The microwave assisted synthesis is simple ecofriendly and can be used as an alternative to the existing conventional heating method. From the results of antibacterial studies it was concluded that the tested compounds exhibited significant antibacterial activities against both gram positive and gram negative organisms. Among the tested compounds, compound substituted with electron withdrawing group in Isatin residue preferably at paraposition showed promising antibacterial activities; this may be attributed to their enhanced electronic character which favors greater penetration through microbial membrane.

## Acknowledgments

The authors are very much thankful to Roland Institute of Pharmaceutical Sciences, Berhampur, Odisha, India, for providing necessary facilities to carry out the research work. The authors are also thankful to SAIF, Panjab University, Chandigarh, for spectral analysis of synthesized compounds.

## References

[1] M. Charde, A. Shukla, V. Bukhariya, J. Mehta, and R. Chakole, "A review on: a significance of microwave assist technique in green chemistry," *International Journal of Phytopharmacy*, vol. 2, no. 2, pp. 39–50, 2012.

[2] S. Ravichandran and E. Karthikeyan, "Microwave synthesis—a potential tool for green chemistry," *International Journal of ChemTech Research*, vol. 3, no. 1, pp. 466–470, 2011.

[3] S. Sharma, S. Gangal, and A. Rauf, "Green chemistry approach to the sustainable advancement to the synthesis of heterocyclic chemistry," *Rasayan Journal of Chemistry*, vol. 1, no. 4, pp. 693–717, 2008.

[4] M. Kidwai, "Green chemistry trends toward sustainability," *Pure and Applied Chemistry*, vol. 78, no. 11, pp. 1983–1992, 2006.

[5] H. M. Mamun, M. A. Foysal, and M. Mahabub, "Microwave-assisted efficient synthesis of isatins and spiro-thiadiazolines under green chemistry protocol," *Journal of Scientific Research*, vol. 2, no. 2, pp. 322–329, 2010.

[6] A. Kulshrestha and S. Baluja, "Microwave promoted synthesis of some Schiff bases," *Archives of Applied Science Research*, vol. 2, no. 3, pp. 221–224, 2010.

[7] R. Somani, S. Pawar, S. Nikam, P. Shirodkar, and V. Kadam, "Microwave assisted synthesis and antimicrobial activity of some Schiff's bases," *International Journal of ChemTech Research*, vol. 2, no. 2, p. 860, 2010.

[8] S. K. Sridhar, S. N. Pandeya, J. P. Stables, and A. Ramesh, "Anticonvulsant activity of hydrazones, Schiff and Mannich

bases of isatin derivatives," *European Journal of Pharmaceutical Sciences*, vol. 16, no. 3, pp. 129–132, 2002.

[9] S. N. Pandeya, D. Sriram, G. Nath, and E. De Clercq, "Synthesis and antimicrobial activity of Schiff and Mannich bases of isatin and its derivatives with pyrimidine," *Farmaco*, vol. 54, no. 9, pp. 624–628, 1999.

[10] A. Jarrahpour, D. Khalili, E. De Clercq, C. Salmi, and J. M. Brunel, "Synthesis, antibacterial, antifungal and antiviral activity evaluation of some new bis-Schiff bases of isatin and their derivatives," *Molecules*, vol. 12, no. 8, pp. 1720–1730, 2007.

[11] S. S. Chhajed and M. S. Padwal, "Antimicrobial evaluation of some novel Schiff and Mannich bases of isatin and its derivatives with quinolin," *International Journal of ChemTech Research*, vol. 2, no. 1, pp. 209–213, 2010.

[12] T. R. Bal, B. Anand, P. Yogeeswari, and D. Sriram, "Synthesis and evaluation of anti-HIV activity of isatin $\beta$-thiosemicarbazone derivatives," *Bioorganic and Medicinal Chemistry Letters*, vol. 15, no. 20, pp. 4451–4455, 2005.

[13] M. Verma, S. N. Pandeya, K. N. Singh, and J. P. Stables, "Anticonvulsant activity of Schiff bases of isatin derivatives," *Acta Pharmaceutica*, vol. 54, no. 1, pp. 49–56, 2004.

[14] S. H. Gillespie, *Medical Microbiology-Illustrated*, Butterworth Heinemann, London, UK, 1994.

[15] U. K. Singh, S. N. Pandeya, A. Singh, B. K. Srivastava, and M. Pandey, "Synthesis and antimicrobial activity of Schiff's and N-Mannich bases of isatin and its derivatives with 4-Amino-N-carbamimidoyl benzene sulfonamide," *International Journal of Pharmaceutical Sciences and Drug Research*, vol. 2, no. 2, pp. 151–154, 2010.

# Condensation Energy for a Two-Gap Superconducting State in Nanoparticles

**Sergei P. Kruchinin,[1] Hiroyuki Kawabe,[2] Hidemi Nagao,[3] and Yasuhiro Nakazawa[4]**

[1] *Bogolyubov Institute for Theoretical Physics, National Academy of Science of Ukraine, Kiev 03680, Ukraine*
[2] *Department of Social Work, Faculty of Social Work, Kinjo University, 1200 Kasama, Hakusan, Ishikawa 924-8511, Japan*
[3] *Division of Mathematical and Physical Science, Graduate School of Natural Science and Technology, Kanazawa University, Kakuma, Kanazawa 920-1192, Japan*
[4] *Department of Chemistry, Graduate School of Science, Osaka University, Toyonaka, Osaka 560-0043, Japan*

Correspondence should be addressed to Sergei P. Kruchinin; skruchin@i.com.ua

Academic Editor: Raphael Schneider

Recent technology and experiments have fabricated high-quality superconducting $MgB_2$ nanoparticles. We investigate properties of two-gap superconductivity in nanosized systems by using a two-sublevel model. In the present work, we analyze the results obtained for superconducting granules in the case of multiband superconductivity. We discuss the finite size effect in multiband superconductors. A definition of the critical level spacing of two-gap superconductivity is also presented, and we discuss the condensation energy and the parity gap of two-gap superconductivity in relation to the size dependence of those properties with two bulk gaps and the effective pair scattering process between two sublevels.

## 1. Introduction

Recent advances in nanoscience have demonstrated that fundamentally new physical phenomena are found when systems are reduced in size to dimensions that become comparable to the fundamental microscopic length scales of a material under study. Superconductivity is a macroscopic quantum phenomenon, and therefore, it is especially interesting to see how this quantum state is influenced when the samples are reduced to nanometer sizes. In such systems, new states of matter can be engineered that do not occur in bulk materials.

The properties of superconducting materials with ultrasmall sizes differ from those in bulk [1]. The characteristics of a superconducting granule depend on the discrete energy spectrum, the parity of the number of electrons, and quantum fluctuations [2]. The dependence of superconducting properties on the parity of the number of electrons participating in the Cooper coupling was called the "parity effect" [3]. The analogous phenomenon is known in nuclear physics. The parity effect was experimentally observed in small superconducting aluminum granules [4, 5].

After the discovery of multiband superconductors such as $MgB_2$ with $T_c = 39$ K [6] and iron pnictides with $T_c = 45$ K [7], the physicists were faced with the question about the influence of the number of bands on the properties of superconducting granules. This problem was discussed in recent works [8–11]. Here, we plan to generalize the results of calculations of the thermodynamics of superconducting granules to the case of a two-band model of superconductivity.

Recent discovery of superconductivity of $MgB_2$ has much attracted great interest with multigap superconductivity [6]. Although multigap superconductivity had been discussed theoretically [12] in 1958, the multigap superconductivity has been observed experimentally in the 1980s [13].

$MgB_2$ is the first material whose effects are so dominant and implications are so thoroughly explored. Magnesium diboride is reported to be an anisotropic superconductor with conventional Bardeen-Cooper-Schrieffer electron-phonon coupling [7, 14]. The band structure of $MgB_2$ calculated in several works since the discovery of the superconductivity is similar to that of graphite and is formed by by the $\sigma$ and $\pi$ zones. Magnesium diboride has two superconducting gaps, 4 meV and 7.5 meV, due to the $\pi$ and $\sigma$ electron bands. The two-gap structure was established in a number of experiments, and the two-gap superconductivity has been also discussed by many groups.

In the recent years, great efforts have been devoted to the fabrication of $MgB_2$ nanostructures that could play a crucial role in the field of applied superconductivity [7]. The ideal candidate is one-dimensional (1D) nanostructures including nanotubes, nanowires, and nanoparticles. As for $MgB_2$, nanoparticles of approximately 20–100 nm in size are available (see Figure 1) [8].

For ultra-small superconducting grains, the experiments [4, 5] by Black et al. have also generated much interest in the size dependence of superconductivity. Properties of ultra-small superconducting grains have been theoretically investigated by many groups [10, 11, 15–20]. Such ultra-small grains were considered by Anderson [1]. The standard BCS theory gives a good description of the phenomenon of superconductivity in large samples. For ultra-small Al grains [4, 5], the bulk gap has been discussed in relation to physical properties of ultra-small grains such as the parity gap [3], condensation energy [18], electron correlation [16], and as well as the size dependence of the level spacing [17] of samples. In this paper, we investigate the properties of two-gap superconductivity in a nanosized system by using a two-sublevel model in the case of strong interaction. We discuss the condensation energy and the parity gap of two-gap superconductivity in relation to the size dependence of those properties in the case of two bulk gaps and the effective pair scattering process between two sublevels.

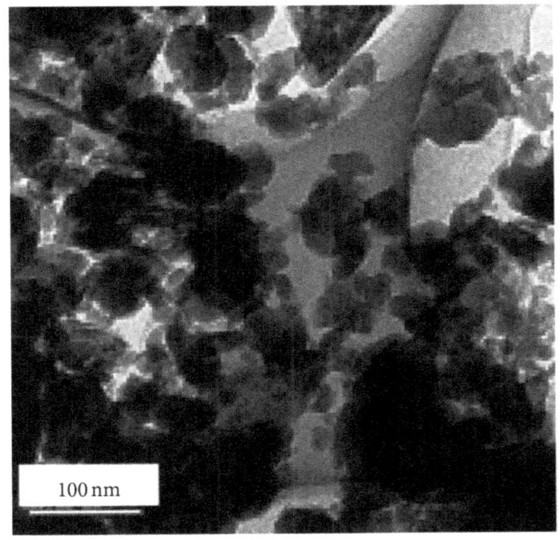


FIGURE 1: The micrograph of 20–100 nm nanoparticles of $MgB_2$.

## 2. The Model and the Solution

In the case of a two-gap superconductor, we can consider a model with two sublevels corresponding to two independent bands. We consider a pairing Hamiltonian with two sublevels $\nu = 1, 2$ corresponding to two bands 1 and 2 written as

$$H = H_0 + H_{\text{int}}, \quad (1)$$

where

$$H_0 = \sum_{j,\sigma,\nu} \left[\varepsilon_{\nu j} - \mu\right] a^{\dagger}_{\nu j\sigma} a_{\nu j\sigma}, \quad (2)$$

$$\begin{aligned} H_{\text{int}} = &- \sum_{\nu} g_\nu \sum_{j,j' \in I_\nu} a^{\dagger}_{\nu j\uparrow} a^{\dagger}_{\nu j\downarrow} a_{\nu j'\downarrow} a_{\nu j'\uparrow} \\ &+ g_{12} \sum_{j\in I_1, k\in I_2} a^{\dagger}_{1j\uparrow} a^{\dagger}_{1j\downarrow} a_{2k\downarrow} a_{2k\uparrow} + \text{H.c.}, \end{aligned} \quad (3)$$

where $a^{\dagger}_{\nu j\sigma}(a_{\nu j\sigma})$ is the creation (annihilation) operator in sublevels $\nu$ with spin $\sigma$ and the energies $\varepsilon_{\nu j}$. Here, the operators for each sublevel satisfy the anticommutation relations, the operators between sublevels are independent, and $\mu$ is the chemical potential. The sums of $j$ and $k$ in (3) are over the set $I_1$ of $N_{1I_1}$ states corresponding to a half-filled band 1 with fixed width $2\omega_{1D}$ and the set $I_2$ of $N_{2I_2}$ states for band 2, respectively. The second term in (1) is the interaction Hamiltonian, $g_1$ and $g_2$ are the effective electron-electron interaction constants including the e-p interaction for each sublevel, respectively, and $g_{12}$ is the effective interaction constant that corresponds to the pair scattering process including the phonon scattering between two bands. Note that, in the bulk limit of (1), we can derive a similar coupled gap equation presented in [9].

In this study, we assume that the Debye energies for two sublevels are the same $\omega_{1D} = \omega_{2D} = \omega_D$. Within this assumption, $N_{1I_1}$ and $N_{2I_2}$ are relatively estimated by the density of state (DOS) for two bands as $N_{1I_1}/N_{2I_2} = \rho_1/\rho_2$, where $\rho_1$ and $\rho_2$ are DOSs for two bands, respectively. The interaction constants $g_1$ and $g_2$ can be written as $d_1\lambda_1$ and $d_2\lambda_2$, respectively, $d_1 = 2\omega_D/N_{1I_1}$ and $d_2 = 2\omega_D/N_{2I_2}$ mean the mean energy level spacing, and $\lambda_1$ and $\lambda_2$ are the dimensionless parameters for two sublevels. We consider the intersublevel interaction constant $g_{12} = \sqrt{d_1 d_2}\lambda_{12}$. In summary, we obtain a relation of $\rho_1/\rho_2 = N_{1I_1}/N_{2I_2} = d_2/d_1$. It is convenient to apply the path integral approach to the treatment of fluctuations of the order parameters. In works [16, 17], the problem of ultra-small superconducting grains was considered within the path integral method

$$Z(\mu, T) = \int D^2\Delta(\tau)\, e^{-S[\Delta]}, \quad (4)$$

where $T$ is the temperature and $S[\Delta]$ is the action.

We now adapt the developed technique of path integration in the case of two-gap superconductivity. In what follows, we will use this method to study the influence of fluctuations on the two-gap superconductivity in ultra-small superconducting grains. This approach gives an exact expression for the canonical partition function of a superconductor [3]. Within the path integral approach, we obtain the canonical partition function for the Hamiltonian of (1) in the following form:

$$Z(\mu,T) = \int D\Delta_1 D\Delta_1^* D\Delta_2 D\Delta_2^* e^{-S[\Delta_1,\Delta_2]}, \quad (5)$$

where the action $S[\Delta_1,\Delta_2]$ is defined as

$$\begin{aligned} &S[\Delta_1,\Delta_2] \\ &= -\sum_j \left[\mathrm{Tr}\ln G_{1j}^{-1} - \frac{\xi_{1j}}{T}\right] - \sum_k \left[\mathrm{Tr}\ln G_{2k}^{-1} - \frac{\xi_{2k}}{T}\right] \\ &+ \int_0^{1/T} d\tau \frac{1}{g_1 g_2 - g_{12}^2} \\ &\times \left[g_2|\Delta_1(\tau)|^2 + g_1|\Delta_2(\tau)|^2 \right. \\ &\left. + g_{12}\left(\Delta_1(\tau)\Delta_2(\tau)^* + \Delta_1(\tau)^*\Delta_2(\tau)\right)\right]. \end{aligned} \quad (6)$$

Here, $\Delta_1$ and $\Delta_2$ are bulk gaps for sublevels 1 and 2, respectively, $\xi_{\nu j} = \varepsilon_{\nu j} - \mu$, and $G_{\nu j}^{-1}(\tau,\tau') = [-d/d\tau - \xi_{\nu j}\sigma^z - \Delta_\nu(\tau)\sigma^+ - \Delta_\nu^*(\tau)\sigma^-]\delta(\tau-\tau')$, where $\sigma^\pm = \sigma^x \pm i\sigma^y$ and $\sigma^{x,y,z}$ are the Pauli matrices. $G_1^{-1}$ and $G_2^{-1}$ satisfy antiperiodic boundary conditions.

In the case of stronger interaction, $\Delta_1 \gg d_1$ and $\Delta_2 \gg d_2$ can be considered in the mean-field approximation for the order parameters within the path integral approach. Substituting the time-independent order parameters into the action of (6), we have

$$\begin{aligned} \Omega(\mu) = &\sum_{j,\nu}\left(\xi_{\nu j} - \epsilon_{\nu j}\right) + \frac{1}{g_1 g_2 - g_{12}^2} \\ &\times \left[g_2\Delta_1^2 + g_1\Delta_2^2 + g_{12}\left(\Delta_1^*\Delta_2 + \Delta_1\Delta_2^*\right)\right], \end{aligned} \quad (7)$$

where $\epsilon_{1j} = (\xi_{1j}^2 + \Delta_1^2)^{1/2}$ and $\epsilon_{2k} = (\xi_{2k}^2 + \Delta_2^2)^{1/2}$. In (7), the values of $\Delta_1$ and $\Delta_2$ must be chosen in a way that minimizes $\Omega$. From the minimization of $\Omega$, we obtain the coupled gap equation at zero temperature for a two-gap system:

$$\begin{pmatrix}\Delta_1 \\ \Delta_2\end{pmatrix} = \begin{pmatrix} g_1\sum_j \frac{1}{2\epsilon_{1j}} & -g_{12}\sum_k \frac{1}{2\epsilon_{2k}} \\ -g_{12}\sum_j \frac{1}{2\epsilon_{1j}} & g_2\sum_k \frac{1}{2\epsilon_{2k}} \end{pmatrix} \begin{pmatrix}\Delta_1 \\ \Delta_2\end{pmatrix}. \quad (8)$$

From this equation, we formally obtain an expression for a bulk gap at zero temperature in the case of two-gap superconductivity:

$$\widetilde{\Delta}_1 = \omega \sinh^{-1}\left(\frac{1}{\eta_1}\right), \qquad \widetilde{\Delta}_2 = \omega \sinh^{-1}\left(\frac{1}{\eta_2}\right), \quad (9)$$

where $1/\eta_\nu = (\lambda_{\nu'} + \alpha_\pm^{(-1)^{\nu'}}[\eta_1,\eta_2]\lambda_{12})/(\lambda_1\lambda_2 - \lambda_{12}^2)$ $(\nu \neq \nu')$ and $\alpha_\pm[\eta_1,\eta_2] = \pm\sinh(1/\eta_1)/\sinh(1/\eta_2)$. For two-band superconductivity, we can consider two cases for the phases of the gaps: $\mathrm{sgn}(\widetilde{\Delta}_1) = \mathrm{sgn}(\widetilde{\Delta}_2)$ and $\mathrm{sgn}(\widetilde{\Delta}_1) = -\mathrm{sgn}(\widetilde{\Delta}_2)$. For the same phase, $\alpha_+$ is used in (8), and we use $\alpha_-$ for the opposite phases. Note that $\widetilde{\Delta}_1 = -\widetilde{\Delta}_2$ in the limit of strong intersublevel coupling $\lambda_{12}$, that is, for the opposite phases. At $\lambda_{12} = 0$, we find the same results of two bulk gaps derived from the conventional BCS theory for two independent sublevels.

In nanosized single-band superconductivity, the condensation energy can be defined as $E_{N,b}^C(\lambda) = E_{N,b}^G(0) - E_{N,b}^G(\lambda) - n\lambda d$, where $E_{N,b}^G$ is the ground state energy of an $N$-electron system in the interaction band, $b$ is the number of electrons on single occupied levels, and $\lambda$ and $n$ are the dimensionless coupling parameter and the number of pair occupied levels, respectively. In the case of a nanosized two-band system, the condensation energy can be written in a similar way as follows: $E_{N_1,b_1;N_2,b_2}^C(\lambda_1,\lambda_2,\lambda_{12}) = E_{N_1,b_1;N_2,b_2}^G(0,0,0) - E_{N_1,b_1;N_2,b_2}^G(\lambda_1,\lambda_2,\lambda_{12}) - \sum_\nu n_\nu\lambda_\nu d_\nu$, where $E_{N_1,b_1;N_2,b_2}^G(\lambda_1,\lambda_2,\lambda_{12})$ means the ground state energy of the $(N_1 + N_2)$-electron system. From (5) and (7), the condensation energy of a two-sublevel system can be expressed in terms of the condensation energies of independent single level systems:

$$\begin{aligned} &E_{N_1,b_1;N_2,b_2}^C(\lambda_1,\lambda_2,\lambda_{12}) \\ &= E_{N_1,b_1}^C(\lambda_1) + E_{N_2,b_2}^C(\lambda_2) \\ &- \frac{\lambda_{12}^2}{\lambda_1\lambda_2 - \lambda_{12}^2}\left(\frac{\Delta_1^2}{d_1\lambda_1} + \frac{\Delta_2^2}{d_2\lambda_2} + \frac{2(\Delta_1^*\Delta_2 + \Delta_1\Delta_2^*)}{\sqrt{d_1 d_2}\lambda_{12}}\right), \end{aligned} \quad (10)$$

where $E_{N_1,b_1}^C(\lambda_1)$ and $E_{N_2,b_2}^C(\lambda_2)$ correspond to the condensation energies for single band cases and the third term means the cross term between two bands. At the same phases of $\Delta_1$ and $\Delta_2$, the condensation energy of (10) decreases; that is, there appears the instability by the coupling constant $\lambda_{12}$. On the other hand, at the opposite phases, the condensation energy becomes larger, because $\Delta_1^*\Delta_2 + \Delta_1\Delta_2^* < 0$. We can expect that the condensation energy of two-gap superconductivity becomes more stable than that of two independent systems due to the intersublevel coupling $\lambda_{12}$ and the opposite phases.

To discuss the critical level spacing for a two-gap system, we start from the coupled gap equation of (8). For the case of the critical level spacing of a two-gap system, we have

$$\begin{aligned} 1 = &\lambda_1\sum_j \frac{1}{2|\widetilde{\xi}_{1j}|} + \lambda_2\sum_k \frac{1}{2|\widetilde{\xi}_{2k}|} \\ &- \left(\lambda_1\lambda_2 - \lambda_{12}^2\right)\sum_j \frac{1}{2|\widetilde{\xi}_{1j}|}\sum_k \frac{1}{2|\widetilde{\xi}_{2k}|}, \end{aligned} \quad (11)$$

where $\widetilde{\xi}_i = \xi_i/d_i$ for sublevels $i = 1, 2$. For the odd or even grain, (11) can be approximately solved by using the digamma function: for the odd case, the critical level spacing becomes

$$d_{1c}^o = \omega_D e^\gamma \exp\left[-\frac{1}{\lambda}\right], \qquad d_{2c}^o = \frac{d_2}{d_1} d_{1c}^o, \quad (12)$$

and, for the even case,

$$d_{1c}^{e} = 4\omega_D e^{\gamma} \exp\left[-\frac{1}{\lambda}\right], \qquad d_{2c}^{e} = \frac{d_2}{d_1} d_{1c}^{e}. \tag{13}$$

Here, we use

$$\frac{1}{\lambda} = \frac{1}{2x}\left[\lambda_1 + \lambda_2 - ax + \sqrt{(\lambda_1 - \lambda_2 - ax)^2 + 4\lambda_{12}^2}\right] \tag{14}$$

with $x = \lambda_1\lambda_2 - \lambda_{12}^2$, $a = \log(d_1/d_2)$. From these expressions, we find the relations

$$d_{1c}^{e} = 4d_{1c}^{o}, \qquad d_{2c}^{e} = 4d_{2c}^{o}, \tag{15}$$

$$d_{1/2c}^{o} \approx \frac{e^{\gamma}}{2} \exp\left[\frac{1}{\eta_{1/2}} - \frac{1}{\lambda}\right] \widetilde{\Delta}_{1/2}. \tag{16}$$

In the case of $|\lambda_1 - \lambda_2| \gg \lambda_{12}$, (16) can be approximately rewritten as

$$d_{1/2c}^{o} \approx \frac{e^{\gamma}}{2} \exp\left[\frac{\lambda_2 - \lambda_1 + 2\alpha\lambda_{12}}{\lambda_1\lambda_2 - \lambda_{12}^2}\right] \widetilde{\Delta}_{1/2}. \tag{17}$$

On the other hand, in the limit of $|\lambda_1 - \lambda_2| = \lambda_{12}$, we have

$$d_{1/2c}^{o} \approx \frac{e^{\gamma}}{2} \exp\left[\frac{(1+\alpha)\lambda_{12}}{\lambda_1\lambda_2 - \lambda_{12}^2}\right] \widetilde{\Delta}_{1/2}. \tag{18}$$

For the case of $\lambda_{12} = 0$, (16) can be rewritten as $d_{1/2c}^{o} \approx \exp[\gamma]/2 \exp[1/\lambda_1 - 1/\lambda_2]\widetilde{\Delta}_{1/2}$. Therefore, when the coupling constants $\lambda_1$ and $\lambda_2$ become the same value, we have a relation similar to that for a single level system: $d_{1/2c}^{o} \approx 0.89\widetilde{\Delta}_{1/2}$. These results suggest that the critical level spacing strongly depends on $\lambda_{12}$ and the difference between the effective interaction constants for the sublevels. The relation in (15) is the same relation as in the conventional nanosized BCS theory.

In the case of two sublevel spacings, the chemical potential lies halfway between the highest occupied and the lowest unoccupied levels with smaller level spacing in the half-filled case as shown in Figure 2(a). We assume that $d_1 < d_2$ and that the numbers of occupied levels corresponding to each sublevel are $n_1$ and $n_2$, respectively. Then, the total number of electrons becomes $N = 2n_1 + 2n_2$. When we consider $N = 2n_1 + 2n_2 + 1$, the chemical potential lies on the level $\varepsilon_{1n_1+1}$, as shown in Figure 2(b). Figure 2(c) shows the position of the chemical potential in the case of $N = 2n_1 + 2n_2 + 2$. The parity gap of nanosized two-gap superconductivity is written as

$$\Delta_p^1 = E_{2n_1+1+2n_2,1}^{G} - \frac{1}{2}\left(E_{2n_1+2n_2,0}^{G} + E_{2(n_1+1)+2n_2,0}^{G}\right). \tag{19}$$

From (7) and the ground state energy $E_{N,b}^{G} = \Omega_{\mu_N} + \mu_N N$, we obtain

$$\Delta_p^1 = \Delta_1 - \frac{d_1}{4}\left(\frac{\rho_1}{\rho_2} - 1\right). \tag{20}$$

In view of Figures 2(c), 2(d), and 2(e), we can obtain another formula for a parity gap:

$$\Delta_p^2 = E_{2(n_1+1)+2n_2+1,1}^{G} - \frac{1}{2}\left(E_{2(n_1+1)+2n_2,0}^{G} + E_{2(n_1+1)+2(n_2+1),0}^{G}\right), \tag{21}$$

which yields

$$\Delta_p^2 = \Delta_2 - \frac{d_2}{4}\left(\frac{3\rho_2}{\rho_1} - 1\right). \tag{22}$$

The present results suggest two kinds of the dependence of the parity gap on the level spacing. We note that the parity gap does not depend upon the effective interaction $\lambda_{12}$, and the structure around the Fermi level plays an important role, by contributing to the size dependence of the parity gap.

## 3. Conclusion

Thus, we have investigated the properties of nanosized two-gap superconductivity by using a two-sublevel model in the framework of the mean-field approximation. In view of the discussion for the condensation energy in nanosized two-gap superconductivity, the phases of the gaps are very important to stabilize the superconductivity. At the same phases, the two-gap superconductivity is instable by the coupling constant $\lambda_{12}$. On the other hand, at the opposite phases, the superconductivity becomes stable. We can expect that the condensation energy of two-gap superconductivity becomes more stable than that of two independent systems due to the intersublevel coupling $\lambda_{12}$ and to the opposite phases.

The parity effect can be observed only at sufficiently low temperatures. It becomes especially significant in ultra-small granules about 10 Å in size. The experimental investigation of superconductivity in such granules becomes possible now due to the developed elegant experimental technique [6, 7]. The relations concerning the parity effect and the energy condensation in the case of multiband superconductivity, which are obtained in the present work, can be verified in the same experiments as in [6, 7] but with the use of nanogranules made of $MgB_2$.

In summary, a model corresponding to nanosized two-gap superconductivity has been presented, and the expression of the partition function of a nanosized system has been analytically derived by using the path integral approach. A definition of the critical level spacing of the two-gap superconductivity has been also presented, and we discuss the condensation energy and the parity gap of the two-gap superconductivity in relation to the size dependence of those properties with two bulk gaps and the effective pair scattering process between two sublevels. The results of this work can be tested in the tunneling experiments with $MgB_2$ nanoparticles [2, 4]. In work [19], the limits of the application of calculations to one-band nanosuperconductors were considered in detail. In the next article, we will plan to study such limits for two-band nanosuperconductors, as well as the influence of fluctuations on the two-gap superconductivity in ultra-small superconducting grains.

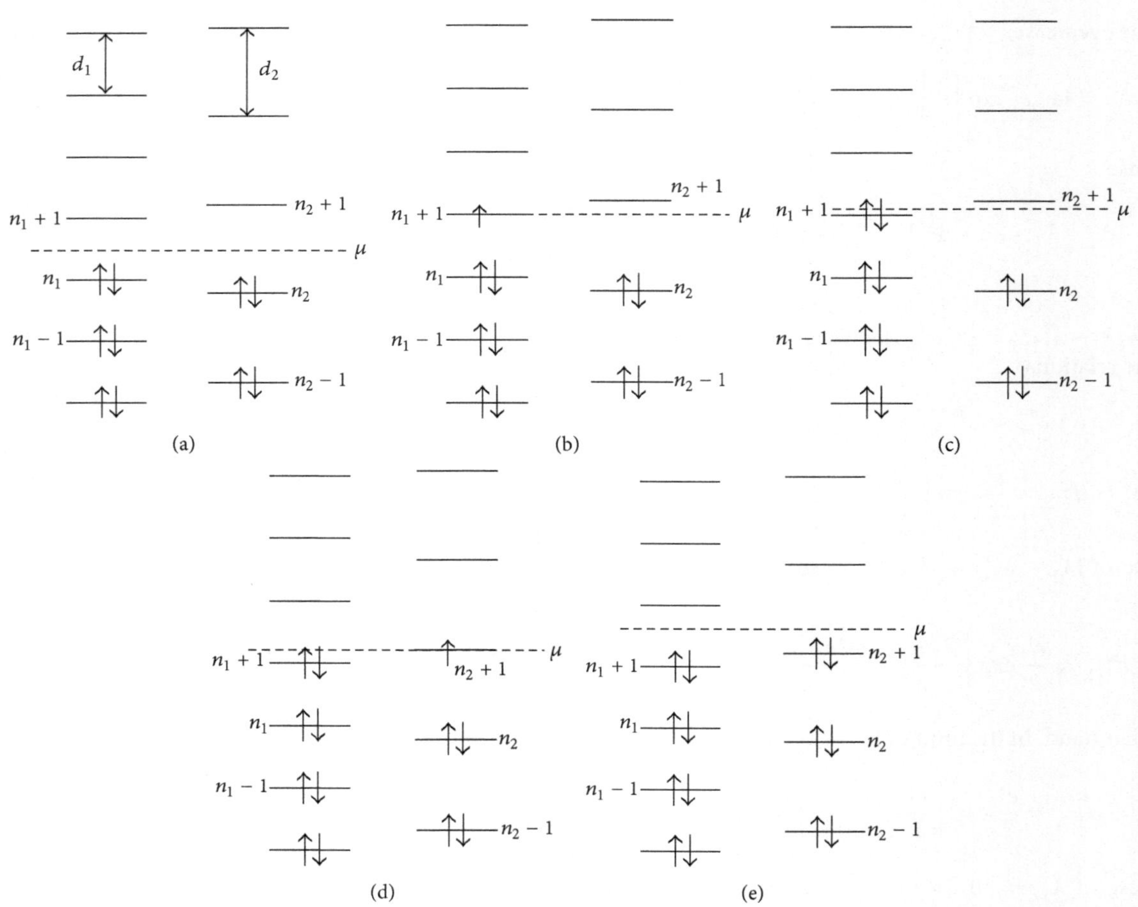


Figure 2: Positioning of the chemical potential relative to the electronic energy levels in a two-gap superconducting grain. Solid and dotted lines mean two sublevels. (a) Half-filled system with $2n_1 + 2n_2$ electrons, (b) $2n_1 + 1 + 2n_2$-electron system, (c) $2(n_1 + 1) + 2n_2$-electron system, (d) $2(n_1 + 1) + 2n_2 + 1$-electron system, and (e) $2(n_1 + 1) + 2(n_2 + 1)$-electron system.


## Acknowledgments

The authors thank Professors Jun Akimitsu and Karl Bennemann for their continued encouragement and helpful discussions. S. P. Kruchinin thanks the Research Center for Structural Thermodynamics, Osaka University, for hosting the author as a guest professor to accomplish this work.


## References


[1] P. W. Anderson, "Theory of dirty superconductors," *Journal of Physics and Chemistry of Solids*, vol. 11, no. 1-2, pp. 28–30, 1959.

[2] J. von Delft and D. C. Ralph, "Spectroscopy of discrete energy levels in ultrasmall metallic grains," *Physics Report*, vol. 345, no. 2-3, pp. 61–173, 2001.

[3] K. A. Matveev and A. I. Larkin, "Parity effect in ground state energies of ultrasmall superconducting grains," *Physical Review Letters*, vol. 78, no. 19, pp. 3749–3752, 1997.

[4] C. T. Black, D. C. Ralph, and M. Tinkham, "Spectroscopy of the superconducting gap in individual nanometer-scale aluminum particles," *Physical Review Letters*, vol. 76, no. 4, pp. 688–691, 1996.

[5] D. C. Ralph, C. T. Black, and M. Tinkham, "Spectroscopic measurements of discrete electronic states in single metal particles," *Physical Review Letters*, vol. 74, no. 16, pp. 3241–3244, 1995.

[6] P. Shirage, K. Miyazawa, K. Kihou et al., "Synthesis of ErFeAsO-based superconductors by the hydrogen doping method," *Europhysics Letters*, vol. 92, no. 5, p. 57011, 2010.

[7] Q. Yang, J. Sha, X. Ma, Y. Ji, and D. Yang, "Aligned single crystal $MgB_2$ nanowires," *Superconductor Science and Technology*, vol. 17, no. 8, p. L31, 2004.

[8] C. Cui, D. Liu, Y. Shen et al., "Nanoparticles of the superconductor $MgB_2$: structural characterization and in situ study of synthesis kinetics," *Acta Materialia*, vol. 52, no. 20, pp. 5757–5760, 2004.

[9] H. Nagao, S. P. Kruchinin, A. M. Yaremko, and K. Yamaguchi, "Multiband superconductivity," *International Journal of Modern Physics B*, vol. 16, no. 23, pp. 3419–3428, 2002.

[10] E. A. Yuzbashyan, A. A. Baytin, and B. L. Altshuller, "Finite-size corrections for the pairing Hamiltonian," *Physical Review B*, vol. 71, no. 9, Article ID 094505, 10 pages, 2002.

[11] A. M. Garsia-Garsia, J. D. Urbina, E. Yuzbashyan, K. Rihter, and B. Altshuler, "Bardeen-Cooper-Schrieffer theory of finite-size superconducting metallic grains," *Physical Review Letters*, vol. 100, no. 18, Article ID 187001, 2008.

[12] H. Suhl, B. T. Matthias, and R. Walker, "Bardeen-Cooper-Schrieffer theory of superconductivity in the case of overlapping bands," *Physical Review Letters*, vol. 3, no. 12, pp. 552–554, 1959.

[13] G. Binnig, A. Baratoff, H. E. Hoenig, and J. G. Bednorz, "Two-band superconductivity in Nb-doped $SrTiO_3$," *Physical Review Letters*, vol. 45, no. 16, pp. 1352–1355, 1980.

[14] J. Nagamatsu, N. Nakagawa, T. Muranaka, Y. Zenitani, and J. Akimitsu, "Superconductivity at 39 K in magnesium diboride," *Nature*, vol. 410, no. 6824, pp. 63–64, 2001.

[15] B. Jankó, A. Smith, and V. Ambegaokar, "BCS superconductivity with fixed number parity," *Physical Review B*, vol. 50, no. 2, pp. 1152–1161, 1994.

[16] J. von Delft, A. D. Zaikin, D. S. Golubev, and W. Tichy, "Parity-affected superconductivity in ultrasmall metallic grains," *Physical Review Letters*, vol. 77, no. 15, pp. 3189–3192, 1996.

[17] R. A. Smith and V. Ambegaokar, "Effect of level statistics on superconductivity in ultrasmall metallic grains," *Physical Review Letters*, vol. 77, no. 24, pp. 4962–4965, 1996.

[18] V. N. Gladilin, V. M. Fomin, and J. T. Devreese, "Magnetic susceptibility of ultrasmall superconductor grains," *Physical Review B*, vol. 70, no. 14, Article ID 144506, 2004.

[19] V. N. Gladilin, V. M. Fomin, and J. T. Devreese, "Shape of nanosize superconducting grains: does it influence pairing characteristics?" *Solid State Communications*, vol. 121, no. 9-10, pp. 519–523, 2002.

[20] J. Dukelsky, S. Pittel, and G. Sierra, "Colloquium: exactly solvable Richardson-Gaudin models for many-body quantum systems," *Reviews of Modern Physics*, vol. 76, no. 3, pp. 643–662, 2004.

# Experimental Investigation into the Forced Convective Heat Transfer of Aqueous $Fe_3O_4$ Nanofluids under Transition Region

**Jie Ma,[1,2] Yinchen Xu,[1] Wenlie Li,[1] Jiantao Zhao,[1] Shuping Zhang,[1,2] and Sergey Basov[1]**

[1] *College of Science, University of Shanghai for Science and Technology, Jungong Road 516, Shanghai 200093, China*
[2] *Green Biochemical and Ecochemical Engineering Lab, University of Shanghai for Science and Technology, Jungong Road 516, Shanghai 200093, China*

Correspondence should be addressed to Jie Ma; majie0203ch@hotmail.com

Academic Editor: Yang Xu

The forced convective heat transfer (FCHT) properties of nanofluids, made of $Fe_3O_4$ nanomaterials and deionized water, are firstly measured by a self-made forced convective heat transfer apparatus. The nanofluid flows through a horizontal copper tube in the transition region with Reynolds numbers in the range of 2500–5000. Some parameters including Reynolds number, axial distance, and mass concentration are also investigated. The preliminary results are firstly presented that the heat transfer coefficients of $Fe_3O_4$ nanofluids systematically decrease with increasing concentration of nanoparticles under transition region which contradicts the initial expectation.

## 1. Introduction

Due to restrictions of their thermal properties, traditional fluids such as water, engine oil, and ethylene glycol are inadequate for high heat flux application. Some nanofluids, made of nanoparticles and a based liquid, have been attracting much attention for they can improve the convective heat transfer and thermal conductivity of the based liquids [1, 2]. Some nanomaterials, such as Cu, Fe, $TiO_2$, $Al_2O_3$, CuO, $SiO_2$, and carbon tube as a main ingredient of nanofluid, have been investigated [3, 4]. Wen and Ding and Lai et al. confirmed that the heat transfer coefficient of nanofluid is increased with flow rate and nanoparticle volume fraction [2, 5]. Anoop et al. demonstrated that the heat transfer coefficient of nanofluid is enhanced with the decrease of size of nanoparticles [6]. Heris et al. found that heat transfer coefficient of nanofluid were obvious difference with different kind of nanoparticles [7]. However, some researches about heat transfer of nanofluids are often difficult to reproduce and even contrary with other relative previous research [8].

There are three modes of flow of a fluid: a fluid with Reynolds numbers below 2000 is named as a laminar flow; a turbulent flow usually refers to the fluid with Reynolds number higher than 10000; the fluid with Reynolds numbers in range from 2000 to 10000 is often called as transition region fluid. As yet, most researches about nanofluids are focused on the laminar or turbulent flow. Contrarily, the properties of nanofluid in transition region are commonly ignored or abandoned because the performances of fluid in the region are affected by many factors, and many some uncertain or irregular results are obtained even in the same conditions [9]. However, in nature a large number of significant fluids perform in transition region, such as blood passing through large arteries, which are characterized by Reynolds numbers [10]. So it is necessary to investigate fluid under the transition region for providing useful information both in academic and technological aspects.

$Fe_3O_4$ nanofluids, as important magnetic fluids, have been widely investigated and applied into many fields. However, according to our investigation, the study is still insufficient about the heat transfer of this nanofluid [11–13], and no report is found concerning the forced convective heat transfer (FCHT) of water-based magnetic $Fe_3O_4$ nanofluids under full developed transition region. Syam Sundar et al. [14] have estimated the FCHT properties of aqueous magnetic $Fe_3O_4$ nanofluids under turbulent flow (Reynolds number > 10000) and found that the heat transfer coefficient of the fluids was enhanced by adding the nanoparticles. Li and Xuan [15] have investigated the influence of an external magnetic field to the convective heat transfer of aqueous magnetic fluid flow over a fine wire under the laminar flow region.

In this study, $Fe_3O_4$ @ phenol formaldehyde resin (PFR) water-based nanofluid is constructed via a facile hydrothermal approach. The obtained black nanofluids show superparamagnetic and high stability. The FCHT properties of the water-based $Fe_3O_4$ and $Fe_3O_4$ @ (PFR) nanofluids are preliminary discussed under transition region. Some important parameters, such as particles concentration, Reynolds number, and the axial profile distance, are discussed. It is found that the heat transfer coefficients of the nanofluids systematically decrease with increasing particle concentration. This is in contradiction to the initial expectation. This investigation may provide a threshold to further exploit the peculiar heat transfer phenomena under transition region.

## 2. Experimental

*2.1. Reagents and Materials.* Iron(III) chloride hexahydrate ($FeCl_3 \cdot 6H_2O$, AR), Iron(II) sulfate heptahydrate ($FeSO_4 \cdot 7H_2O$, AR), Hexamethylene tetramine ($(CH_2)_6N_4$, AR), Sodium hydroxide (NaOH, AR), Phenol ($C_6H_5OH$, AR), Ethylene glycol ($C_2H_4(OH)_2$, AR), and Ethanol ($C_2H_5OH$, AR) were all purchased from China National Medicines Corporation Ltd. and used as received without further purification. De-ionized water used in all experiment is self-made.

*2.2. Preparation of Nanofluids.* In a typical procedure, 15 mmol $FeCl_3 \cdot 6H_2O$ and 12 mmol $FeSO_4 \cdot 7H_2O$ were dissolved in 100 mL de-ionized water with 50 mL ethylene glycol in a three-necked bottle. The mixture was quickly stirred with a mechanical whisk at 65°C. After 30 minutes, the aqueous NaOH (1 M) 50 mL were added into the bottle and the black suspension was formed at once. The black suspension had been constantly stirred for 60 minutes, and then it is transferred into a 500 mL autoclave with Teflon lining. After that, 150 mL de-ionized water, 0.50 g $C_6H_5OH$, and 1.00 g $C_6H_{12}N_4$ were added into the autoclave. The autoclave was sealed and kept in a controllable temperature oven at presetting 120°C for 4 h.The autoclave was cooled naturally after the heating process. The black precipitate was washed by ethanol for several times, respectively. Finally, $Fe_3O_4$ @ PFR nanocomposites were redispersed into deionized water to form nanofluids. The naked $Fe_3O_4$ nanofluid was prepared via two steps: the naked $Fe_3O_4$ nanoparticles were dispersed into water by a mechanical whisk, and then the pH value of the mixture was adjusted to 9-10 by 1 M aqueous NaOH solution for preparing stable fluid.

*2.3. The Measurement of the Forced Convective Heat Transfer Behavior of the Nanofluids*

*2.3.1. Experimental Setup.* The forced convective heat transfer properties of nanofluids were measured by self-made simplified apparatus referred to the literature [16]. The schematic diagram of experimental apparatus is shown in Figure 1.

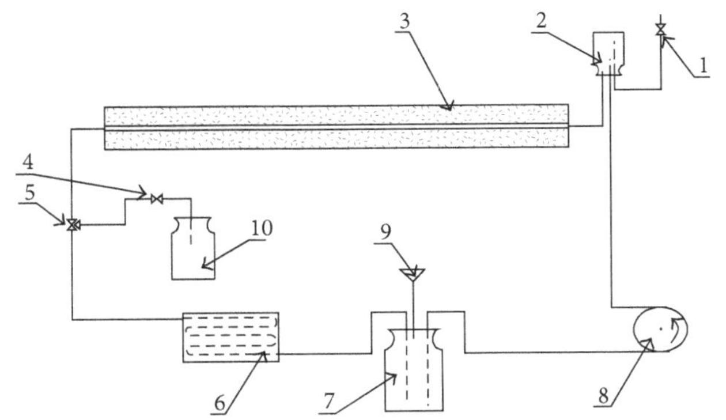


FIGURE 1: The schematic diagram of experimental apparatus for detecting the forced convective heat transfer of fluid. (1) Steam trap, (2) Water tank, (3) Test section, (4) Swith, (5) Vacuum valve, (6) Air cooler, (7) Reservoir tank, (8) Pump, (9) Funnel, and (10) Collection tank.

The test chamber consisted of 110 cm annular horizontal straight copper tube with 5 mm inner diameter and 1 mm thickness. Seven thermocouples are dispersedly instated on the tube, among them, five thermocouples are mounted on the test section at axial positions to measure the wall temperature distribution, and remaining two thermocouples are, respectively, inserted at the inlet and exit of the test section to measure the temperature of nanofluids there. Nanofluids flow inside the copper tube and pass through the test section for heat exchanger. The test chamber is covered by a heat-insulating shield with 10 cm thickness for blocking the heat lost. The flow rate of fluid is controlled by a peristaltic type pump. The supply heat is regulated by the resistance wires which are intertwined on the copper tube. The air cooler is used to cool fluids out of the test unit.

*2.3.2. Relative Problem Calculation Formula.* The de-ionized water was used as fluid to test the reliability and accuracy of the experimental system. The results were compared with the predictions of the following Gnielinski equation [17] under fully developed section in transition region:

$$\mathrm{Nu} = \frac{(f/8)(\mathrm{Re}-1000)P_r}{1+12.7\sqrt{f/8}\left(P_r^{2/3}-1\right)} \quad 2300 \le \mathrm{Re} \le 10^6. \tag{1}$$

The resistance coefficient $f$ and Reynolds number are calculated as follows:

$$f = (1.82\ \lg \mathrm{Re} - 1.64)^{-2}, \tag{2}$$

$$\mathrm{Re} = \frac{\rho \cdot v \cdot D}{\mu}. \tag{3}$$

The forced convective heat transfer coefficient ($h$) in this experiment is calculated as follows:

$$\overline{h_{nf}} = \frac{q}{T_w - T_f} = \frac{C_{pnf} \cdot \rho_{nf} \cdot \overline{v} \cdot A \cdot (T_{b2} - T_{b1})}{\pi DL \cdot \left(T_w - \overline{T_b}\right)}. \tag{4}$$

$C_{pnf}$ in this experiment is calculated, respectively, as follows:

$$C_{pnf} = \frac{m_s \cdot C_{ps} + m_w \cdot C_{pw}}{m_{nf}}. \tag{5}$$

The $\rho_{nf}$ value is given as $1\,\mathrm{g\,cm^{-1}}$ for the concentration of nanoparticles is very low, and the influence can be ignored to the dense of fluid; $D$ and $L$ values are 0.5 and 110 cm in turn; So (4) can be simplified into (6) as follow:

$$\overline{h_{nf}} = \frac{C_{pnf} \cdot \overline{v} \cdot (T_{b2} - T_{b1})}{880 \cdot \left(T_w - \overline{T_b}\right)}. \tag{6}$$

## 3. Results and Discussion

The crystal phase of as-obtained $Fe_3O_4$ @ PFR nanocomposites is characterized by X-ray power diffraction, and a typical XRD pattern of them is shown in Figure 2. The pattern reveals the obtained nanocrystalline product consisted of F-centered cube structure. All diffraction peaks are easily indexed as cube phase $Fe_3O_4$ given in JCPDS (no. 65-3107) with lattice constants $a = 8.383(9)$ Å. The average crystallite size of samples is estimated about 20 nm by the Scherrer equation autocalculated in Jade software. Concurrently, a widen peak appeared in the region between 20° and 30° of two-theta diffraction angle due to amorphous PFR. It suggests that the $Fe_3O_4$ @ PFR nanocomposite can be facilely synthesized via the two-step method. It is found that $Fe_3O_4$ @ PFR nanofluids show higher stability than naked $Fe_3O_4$ nanofluids after they have been statically laid aside in sealing bottle for 7 days. It should be ascribed to water-solute PFR polymer, which can catch some $H_2O$ molecular by hydrogen bonds and increase the dispersity of $Fe_3O_4$ nanoparticles.

For inspecting the reliability and accuracy of the self-made experimental system, the pure water as work fluid measured the Nusselt number change with Reynolds number in the range of 2400–3300. The measurements values, which are shown in Figure 3, are compared with the calculated values via the classical Gnielinski equation [17]. From the figure, the deviations between measured and calculated values are less than 12% over the Reynolds number range, which is located in the allowable range of deviation in transition region [9, 10]. That is to say, the self-made apparatus is reliable to measure the forced convent heat transfer of as-obtained nanofluids.

The forced convective heat transfer coefficient ($\overline{h_{nf}}$) of a fluid rises with Reynolds number (Re) increasing, but the increased amplitude changes with the type of fluid. Reynolds number of fluid can be calculated by (3) according to the change of flow speed, so some examinations are measured at different flow speed of fluid controlled by a special pump. The relevancies between $\overline{h_{nf}}$ and Re of some fluids are inspected and corresponding results curves are shown in Figure 4, where $\overline{h_{nf}}$ of nanofluid is calculated by (6). The heat capacities of as-obtained nanofluids, which are lower than that of pure water, are calculated by empirical equation (5). The results clearly indicate that pure water shows the highest $\overline{h_{nf}}$ of all tested samples in transition region at a

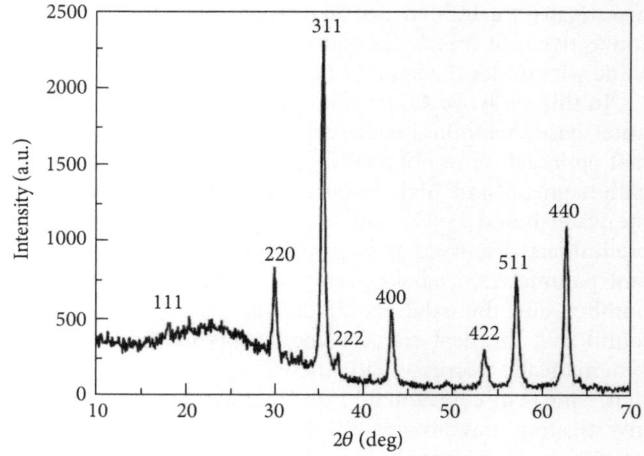


FIGURE 2: The typical XRD Pattern of the as-obtained $Fe_3O_4$ nanoparticles coated by PFR.

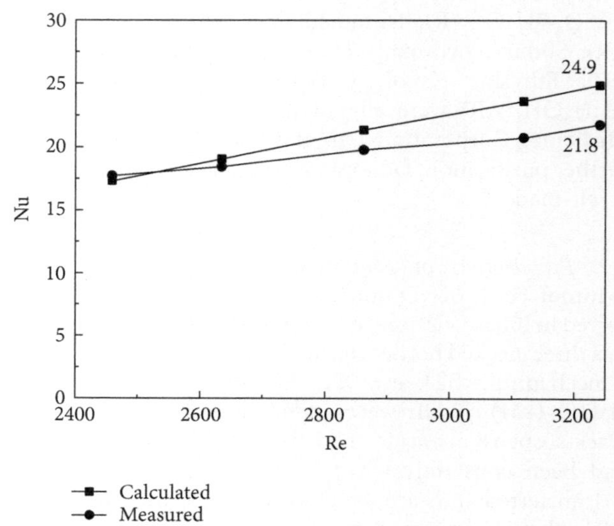


FIGURE 3: Comparisons between the calculated value by Gnielinski equation and measured value by the self-made apparatus in the deionized water system.

same Reynolds number. That is to say, the forced convective heat transfer capacity of water cannot be enhanced by adding tiny $Fe_3O_4$ nanoparticles in current investigation conditions. $Fe_3O_4$ @ PFR nanofluids demonstrate superior $\overline{h_{nf}}$ to the naked $Fe_3O_4$ nanofluids shown in Figure 4(b). The heat transfer coefficient of the naked $Fe_3O_4$ nanofluids is declined with the concentration increase of $Fe_3O_4$ nanoparticles. The above experimental results suggest that it is disadvantaged to raise the forced convective heat transfer efficiency of nanofluid by the increase concentration of the nanoparticles in the present research system. These results are contradictive with the initial expectation and some reports that the FCHT properties of the fluid may be improved by adding some nanoparticles [14, 15]. However, we also found that the result

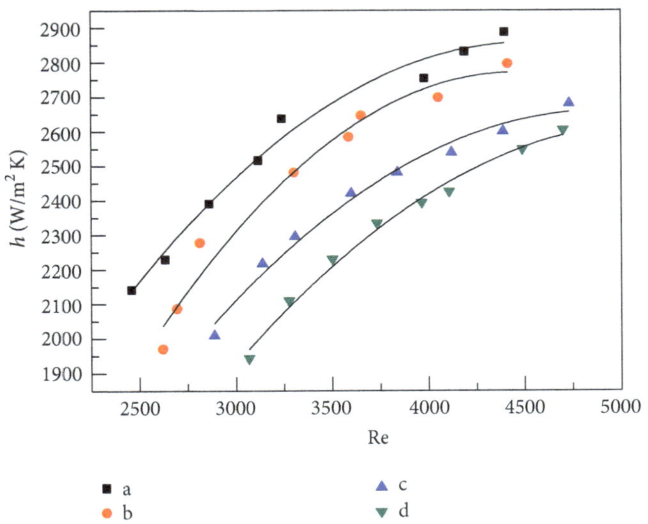


FIGURE 4: Dependence of heat transfer coefficient of samples on Re. (a) Deionized water, (b) 0.16% (m/m) $Fe_3O_4$ @ PFR nanofluid, (c) 0.05% $Fe_3O_4$ nanofluid, and (d) 0.24% $Fe_3O_4$ nanofluid.

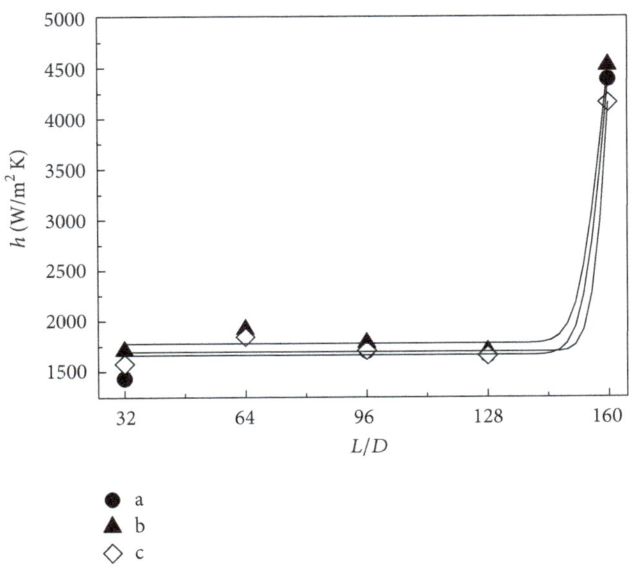


FIGURE 5: The effect of axial profile length/diameter ($L/D$) to the forced convective heat transfer coefficient ($h$) of some typical samples (Re = 2650 ± 50). (a) 0.16% (m/m) $Fe_3O_4$ @ PFR nanofluid, (b) 0.05% $Fe_3O_4$ nanofluid, and (c) 0.24% $Fe_3O_4$ nanofluid.

is similar with the research results about CuO and $TiO_2$ nanofluid and $Al_2O_3$ nanofluid under natural convective heat transfer conditions [18, 19]. The reasons of deterioration should be attributed to the effects coming from particle-fluid slip, sedimentation of nanoparticles [18], and the change of fluid parameters with nanoparticles self-characters [19].

The curve shown in Figure 5 indicates the relationship of the local heat transfer coefficients against the axial distance/diameter ($L/D$) from the test starting point to exit of the test section at Reynolds numbers around 2650 ± 50. The experiments are given at the fully developed section and exit section because of the ratio between axial distance and inner diameter of tube beyond 60. From the trend of the curve, it can be found that the influence from the entrance sector can be neglected at the fully developed section for the heat transfer coefficient keeps identical value at different positions in the $L/D$ range from 32 to 140. This result suggests that the heat transfer coefficient keep stable in the fully developed sector for the temperature of fluid at different position are almost same in the fully developed sector. On the contrary, when the fluid flows into exit region, the heat transfer coefficients of all kinds of fluid are boosted. This phenomenon should be attributed to the obvious difference in temperature which existed in the exit sector and the outer cooling-sector fluid. Compared to the $\overline{h_{nf}}$ values at exit sector of two $Fe_3O_4$ nanofluids with different concentration, it indicates that the increased concentration of nanoparticles is disadvantage to improve the forced heat transfer coefficient under transition region with present investigated conditions. These results are coincident with the results shown in Figure 4.

## 4. Conclusions

This paper is concerned with the forced convective heat transfer of the $Fe_3O_4$ @ PFR nanofluids. Experiments were carried out under transition region conditions. The following conclusions were obtained.

(1) The pure water shows the highest $\overline{h_{nf}}$ of all tested samples in transition region at a same Reynolds number. This is in contradiction to the initial expectation. This may be attributed to the complex factors that affected the forced convective heat transfer properties of nanofluids.

(2) The $Fe_3O_4$ @ PFR nanofluids can enhance the convective heat transfer with the increases of the Reynolds number in transition region. However, the increased concentration of $Fe_3O_4$ nanoparticles is disadvantage to raise nanofluid the forced convective heat transfer efficiency in transition region.

(3) The enhancement is particularly significant in the outlet region and gets steady in the developed region.

Some important factors are investigated and some valuable results are obtained, but many other factors, such as the viscosity, the concentration of nanomaterials, and test temperature are still waiting for investigation. The part of research is on the way.

## Nomenclature

$h$: Heat transfer coefficient
$q$: Heat flux
$T_w$: Tube wall temperature
$T_b$ Average bulk temperature
$C_p$: Specific heat capacity

$w$: Water
$nf$: Nanofluid
$f$: Resistance coefficient
Re: Reynolds number
$L$: Tube length
$A$: Tube cross-section area
$v$: Fluid velocity
$\mu$: Viscosity
$\rho$: Density
$k$: Coefficient of thermal conductivity
Nu: Nusselt number
$D$: Tube diameter.

## Acknowledgments

The authors gratefully acknowledge the support of the National Natural Science Foundation of China (no. 20906061), and the Shanghai International Cooperation Foundation (no. 073458014) for financial support.

## References

[1] S. P. Jang and S. U. S. Choi, "Role of Brownian motion in the enhanced thermal conductivity of nanofluids," *Applied Physics Letters*, vol. 84, no. 21, pp. 4316–4318, 2004.

[2] D. Wen and Y. Ding, "Experimental investigation into convective heat transfer of nanofluids at the entrance region under laminar flow conditions," *International Journal of Heat and Mass Transfer*, vol. 47, no. 24, pp. 5181–5188, 2004.

[3] T. K. Hong, H. S. Yang, and C. J. Choi, "Study of the enhanced thermal conductivity of Fe nanofluids," *Journal of Applied Physics*, vol. 97, no. 6, Article ID 064311, 4 pages, 2005.

[4] H. Masuda, A. Ebata, K. Teramae, and N. Hishinuma, "Alteration of thermal conductivity and viscosity of liquid by dispersing ultra-fine particles (Dispersion of G-$Al_2O_3$, $SiO_2$ and $TiO_2$ ultra-fine particles)," in *Netsu Bussei*, vol. 4, pp. 227–233, 1993.

[5] W. Y. Lai, S. Vinod, P. E. Phelan, and R. Prasher, "Convective heat transfer for water-based alumina nanofluids in a single 1.02-mm tube," *Journal of Heat Transfer*, vol. 131, no. 11, pp. 1–9, 2009.

[6] K. B. Anoop, T. Sundararajan, and S. K. Das, "Effect of particle size on the convective heat transfer in nanofluid in the developing region," *International Journal of Heat and Mass Transfer*, vol. 52, no. 9-10, pp. 2189–2195, 2009.

[7] S. Z. Heris, S. G. Etemad, and M. N. Esfahany, "Experimental investigation of oxide nanofluids laminar flow convective heat transfer," *International Communications in Heat and Mass Transfer*, vol. 33, no. 4, pp. 529–535, 2006.

[8] D. Wen and Y. Ding, "Experimental investigation into the pool boiling heat transfer of aqueous based $\gamma$-alumina nanofluids," *Journal of Nanoparticle Research*, vol. 7, no. 2-3, pp. 265–274, 2005.

[9] W. Yu-ting, L. Bin, M. Chong-fang, and G. Hang, "Convective heat transfer in the laminar-turbulent transition region with molten salt in a circular tube," *Experimental Thermal and Fluid Science*, vol. 33, no. 7, pp. 1128–1132, 2009.

[10] J. P. Abraham, E. M. Sparrow, and W. J. Minkowycz, "Internal-flow Nusselt numbers for the low-Reynolds-number end of the laminar-to-turbulent transition regime," *International Journal of Heat and Mass Transfer*, vol. 54, no. 1–3, pp. 584–588, 2011.

[11] K. Parekh and H. S. Lee, "Magnetic field induced enhancement in thermal conductivity of magnetite nanofluid," *Journal of Applied Physics*, vol. 107, no. 9, Article ID 09A310, 2010.

[12] J. Philip, P. D. Shima, and B. Raj, "Enhancement of thermal conductivity in magnetite based nanofluid due to chainlike structures," *Applied Physics Letters*, vol. 91, no. 20, Article ID 203108, 3 pages, 2007.

[13] W. Yu, H. Xie, L. Chen, and Y. Li, "Enhancement of thermal conductivity of kerosene-based $Fe_3O_4$ nanofluids prepared via phase-transfer method," *Colloids and Surfaces A*, vol. 355, no. 1-3, pp. 109–113, 2010.

[14] L. Syam Sundar, M. T. Naik, K. V. Sharma, M. K. Singh, and T. C. Siva Reddy, "Experimental investigation of forced convection heat transfer and friction factor in a tube with $Fe_3O_4$ magnetic nanofluid," *Experimental Thermal and Fluid Science*, vol. 37, pp. 65–71, 2012.

[15] Q. Li and Y. Xuan, "Experimental investigation on heat transfer characteristics of magnetic fluid flow around a fine wire under the influence of an external magnetic field," *Experimental Thermal and Fluid Science*, vol. 33, no. 4, pp. 591–596, 2009.

[16] Y. Ding, H. Alias, D. Wen, and R. A. Williams, "Heat transfer of aqueous suspensions of carbon nanotubes (CNT nanofluids)," *International Journal of Heat and Mass Transfer*, vol. 49, no. 1-2, pp. 240–250, 2006.

[17] J. P. Abraham, E. M. Sparrow, and J. C. K. Tong, "Heat transfer in all pipe flow regimes: laminar, transitional/intermittent, and turbulent," *International Journal of Heat and Mass Transfer*, vol. 52, no. 3-4, pp. 557–563, 2009.

[18] N. Putra, W. Roetzel, and S. K. Das, "Natural convection of nano-fluids," *Heat and Mass Transfer*, vol. 39, no. 8-9, pp. 775–784, 2003.

[19] D. Wen and Y. Ding, "Formulation of nanofluids for natural convective heat transfer applications," *International Journal of Heat and Fluid Flow*, vol. 26, no. 6, pp. 855–864, 2005.

# 24

# Studying the Effects of Adding Silica Sand Nanoparticles on Epoxy Based Composites

**Tahir Ahmad,[1] Othman Mamat,[2] and Rafiq Ahmad[1]**

[1] *Department of Metallurgy and Materials Engineering, College of Engineering and Emerging Technologies, University of the Punjab, Lahore, Pakistan*
[2] *Mechanical Engineering Department, Universiti Teknologi PETRONAS, Bandar Seri Iskandar, 31750 Tronoh, Perak, Malaysia*

Correspondence should be addressed to Tahir Ahmad; tahirengg4051@yahoo.com

Academic Editor: Amir Kajbafvala

The research about the preparation of submicron inorganic particles, once conducted in the past decade, is now leading to prepare polymer matrix composite (PMC) reinforced with nanofillers. The objective of present research is to study the modified effects of reinforcement dispersion of nanoparticle silica in epoxy resin on the physical properties, mechanical and thermal behaviour, and the microstructure of resultant composites. Stirrer mixing associated with manual mixing of silica sand nanoparticles (developed in our earlier research) (Ahmad and Mamat, 2012) into epoxy was followed by curing being the adopted technique to develop the subject nanocomposites. Experimental values showed that 15 wt.% addition of silica sand nanoparticles improves Young's modulus of the composites; however, a reduction in tensile strength was also observed. Number of holes and cavities produced due to improper mixing turn out to be the main cause of effected mechanical properties. Addition of silica sand nanoparticles causes a reduction in degree of crystallinity of the nanocomposites as being observed in differential scanning calorimetry (DSC) analysis.

## 1. Introduction

Nowadays industrial and academic research laboratories are focusing much of their efforts to develop and improve physical, mechanical, and electrical properties of polymer nanocomposites. Their interest to develop such composites is mainly because of the fact that nanoparticles present a high surface-to-volume ratio which may induct unique properties to these nanocomposites as compared to macro-scale composites [1]. Andritsch et al. [2] prepared nanocomposites filled with $SiO_2$, $Al_2O_3$, and AIN with an average particles size from 22 nm to 10 nm and studied the nanoparticles dispersion and DC breakdown strength. They observed the highest DC breakdown strength for 0.5 wt.% for two component system. It was also noticed that the DC breakdown strength increased with 10 wt.% as compared to 2 and 5 wt.% and again decreased with 15 wt.% nano-fillers. Yasmin et al. [3] studied the dispersion of clay nanoparticles in epoxy-based composites using a three roll mill with the concentration of 1–10 wt.%. This technique was found highly efficient and environmentally friendly due to excellent dispersion of nanoparticles within a short time. Elastic modulus was improved up to 80% with 10 wt.% addition of clay nanoparticles in epoxy. Absence of any improvement in tensile strength of the nanocomposites over pure epoxy was due to the clustering of nanoparticles and/or to the occasional occurrence of nano- to microsize voids in the microstructure. Li et al. [4] prepared the Epoxy nanocomposites with exfoliated nanoclay and a novel ultrafine full-vulcanized powdered rubber. The morphology of the epoxy/rubber/clay nanocomposites has been studied using X-ray diffraction (XRD) and transmission electron microscopy (TEM). Their study showed that the unmodified clay was fully exfoliated and uniformly dispersed in the resulting nanocomposite and found that the impact strength of this special epoxy/rubber/clay nanocomposite increased up 107% over the neat epoxy resin. Thermal analyses showed that thermal stability of the nanocomposite was much better than that of epoxy nanocomposite based on organically modified clay. Damian et al. [5] synthesized epoxy-based nanocomposites reinforced with multiwalled

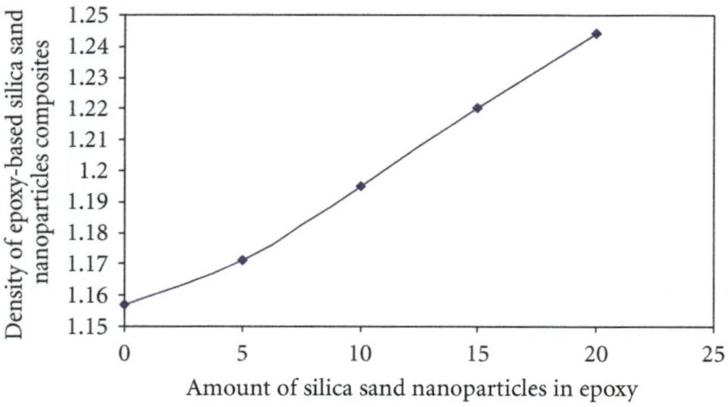


FIGURE 1: Density analyses of epoxy-$SiO_2$ composites.

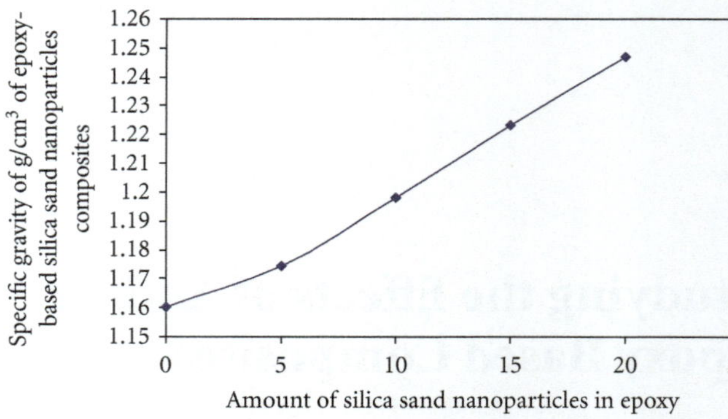


FIGURE 2: Specific gravity analyses of epoxy-$SiO_2$ composites.

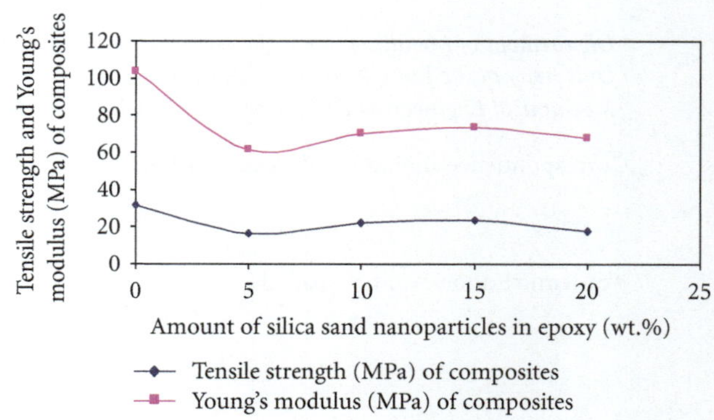


FIGURE 3: Tensile strength and Young's modulus of epoxy-$SiO_2$ composites.

carbon nano-tubes (MWNT). They functionalized MWNTs in order to increase the dispersion in the polymer matrix and thus to ensure good adhesion between the two components. Their study showed that the structure of amine MWNT-B100 exhibit compatibility with the epoxy matrix. However, the higher thermostability was achieved by the composite based on MWNTCOOH as observed from TGA. Frache et al. [6] prepared the epoxy based nanocomposites reinforced with home-made hydrotalcites (HTlc) and montmorillonite-(MMT-)type layered silicates and studied the morphology, dispersion and thermal analysis of the nanocomposites. Their research showed possibility to obtain nano-structured materials, characterized by the resin intercalation into HTlc galleries, by using a proper exchange of the pristine hydrotalcite with stearate anions. The thermal analysis indicated that nanocomposites based on HTlc were prone to decompose, both in air and nitrogen, while in the case of the nanocomposite based on the organophilic MMT a slight improvement was found in air. Jordan et al. [7] studied the effect of addition of micron and nanosized silica inclusions on polyurethane matrix and found it as an increment in density as the filler content increased. He investigated that the composite density further can be increased with micron-sized inclusions as compared to the density for the nanosized inclusions. In above studies, no body tried to utilize silica sand nanoparticles as reinforcement agents in epoxy based composites. Exclusively, the objective of present research is to fill this gap and study the effect of addition of silica sand nanoparticles on epoxy based composites.

## 2. Experimental Work

The processing and characterization of silica sand nanoparticles have already been discussed in our earlier papers [8]. Before mixing with epoxy resin (E-51), silica sand nanoparticles were dried at 120°C to remove any moisture for 5 hours. The silica sand nanoparticles were mixed with epoxy using stirrer and manual mixing for several hours. The resultant mixture was degassed and cured in steel mould at 120°C to form dog-bone-shaped samples (ASTM D638-97). ASTM D790-03 standard was used to develop flexural test specimens. The Universal Tensile Testing machine (5 KN) LLOYD Instruments, LR 5 K was engaged to determine the tensile and flexural properties of the composites with a cross-head speed of 50 mm/min. ASTM D 792-08 standard was used to determine the densities of epoxy based composites. Differential scanning calorimeter (DSC) in nitrogen atmosphere was incorporated to analyse the thermal properties of the epoxy based composites. A 5 mg weight sample was prepared to obtain the scans between 40 to 210°C wit a heating rate of 10°C/min.

## 3. Results and Discussion

*3.1. Processing of Tronoh Silica Sand.* The silica sand originated naturally from Tronoh Perak, Malaysia was washed with water to remove any clay particles followed by drying it in an oven at 120°C. The dried silica sand was meshed about <600 $\mu$m using sieve analysis. The <600 $\mu$m size of natural silica sand was ground to nanoparticles using dry ball milling with zirconia balls as grinding media with different steps of sieve analysis. This high-energy milling was known as one of the "top-down" nanoparticles type approach, which generally relies on physical methods for their production. Different particles sizes went through a precise sieve analysis (Sieve Shaker (EFL2000). The analysis of silica sand nanoparticles had been verified using the ZetaSizer, Nano ZS (ZEN 3600) (MALVERN) analyzer. The nanoparticles results were also checked microscopically through FESEM.

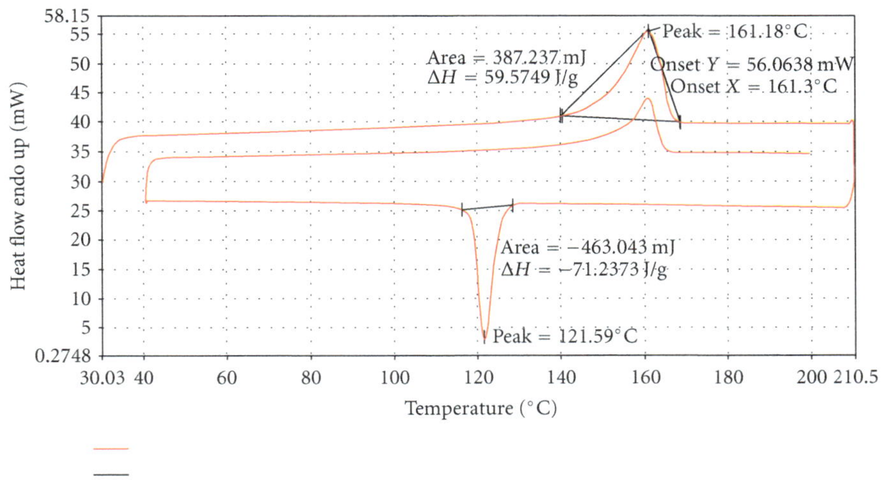


FIGURE 4: DSC curves of crystallization of epoxy-$SiO_2$ composites.

The chemical composition of silica sand nanoparticles was analyzed adopting XRF technique [8].

### 3.2. Characterization of the Epoxy-Silica Sand Nanoparticles Composites

*3.2.1. Density and Specific Gravity Determination of Epoxy-Silica Nanoparticles Composites.* The silica sand being denser as compared to epoxy may cause an enhancement in density of epoxy when added as reinforcement as shown in Figure 1. For every 5 wt.% increase in weight of silica sand nanoparticles, the density of epoxy raised with an average incremental value of 1.8%. Temperature recorded during testing procedure remained as 22.6°C. The inclusions of silica nanoparticles when added to polymer tended to create a linkage of bonding between $SiO_2$ and polymer which some time results in higher density of the composites.

Results obtained above may actually be expanded to get specific gravity values for each sample. The density and specific gravity were determined using D 792-08 ASTM standard. Specific gravity, or relative density, increases in accordance to sample density as well as wt% increment of silica sand nanoparticles. An increase between 2%–6% can be seen by comparing silica sand nanoparticles added epoxy with its pure form as control element. For every 5% increase in weight of silica sand nanoparticles, it can be found that specific gravity increased by an average value of 1.8%, which is the same compared to density measurement. The results of specific gravities are shown in Figure 2.

*3.2.2. Mechanical Properties of the Epoxy-Silica Sand Nanoparticles Composites.* The properties of the composites depend upon the method of dispersion of particles, aggregation of particles, and interaction between particles and polymers. The thickness and density of the interphase developed between two constituents depend on the interaction between particles and polymer [7]. Interparticle distance also plays an important role in the interphase behaviour and properties of the composites. Tensile properties of the composites are mostly affected by the materials, method, specimen condition and preparation, and also by percentage of the reinforcement [9]. For the comparative testing of materials, the preparation of samples and adopted method are very important to get relatively same properties of the composites. Silica nanoparticles, if not properly mixed with the epoxy, cause voids and increase level of porosity inside the system. This is the reason why pure epoxy exhibited highest value of tensile strength and Young's modulus, and why this strength got reduced when silica nanoparticles between 15 wt.% and 20 wt.% dispersed in it. The mechanical properties of the epoxy-silica sand nanoparticles composites are shown in Figure 3. As wt.% of silica sand nanoparticles increases, tensile strength as well as Young's modulus also rises. These numbers conversely drop when 20 wt% of silica sand nanoparticles is introduced to system. Nature of interface is important in producing high-performance epoxy-nanosilica composites [10]. Better tensile properties are achieved when the samples have good interface between matrix and filler because the applied load mostly transferred through the filler via interface.

*3.2.3. Thermal Analysis of the Epoxy-Silica Sand Nanoparticles Composites.* The thermal analysis of epoxy-$SiO_2$ nanocomposites was performed using DSC technique as shown in Table 1. Neat epoxy showed the values of crystallization temperature between 111.12 and 110.32°C; however, for 20 wt.% of silica sand nanoparticles addition to epoxy, the values remain between 121.59 and 119.97. The increase in weight percentage of silica sand nanoparticles to epoxy increases the crystallization temperature; however, it reduces the percentage crystallinity of the nanocomposites as shown in Figure 4 DSC curves. Melting endotherms showed a slight difference for all samples as observed in DSC thermograms and the same results were also found by other researcher using different kinds of nanoparticles [10]. Built-in software

TABLE 1: DSC analysis of epoxy-silica sand nanoparticles composites.

| Composites | Crystallization $T_c$. (onset) (°C) | Crystallization $T_c$. (peak) (°C) | Melting $T_m$.(onset) (°C) | Melting $T_m$.(peak) (°C) | $\Delta H$ (J/g) | Crystallinity (%) |
|---|---|---|---|---|---|---|
| Pure epoxy | 110.32 | 111.12 | 160.20 | 161.14 | 250.774 | 86.8 |
| 5 wt.% silica sand + epoxy | 112.34 | 113.40 | 160.33 | 161.40 | 240.523 | 83.5 |
| 10 wt.% silica sand + epoxy | 115.44 | 116.20 | 160.45 | 161.21 | 239.804 | 80.3 |
| 15 wt.% silica sand + epoxy | 117.95 | 118.60 | 179.64 | 161.46 | 237.495 | 79.5 |
| 20 wt.% silica sand + epoxy | 119.97 | 121.59 | 138.30 | 145.55 | 230.444 | 77.6 |

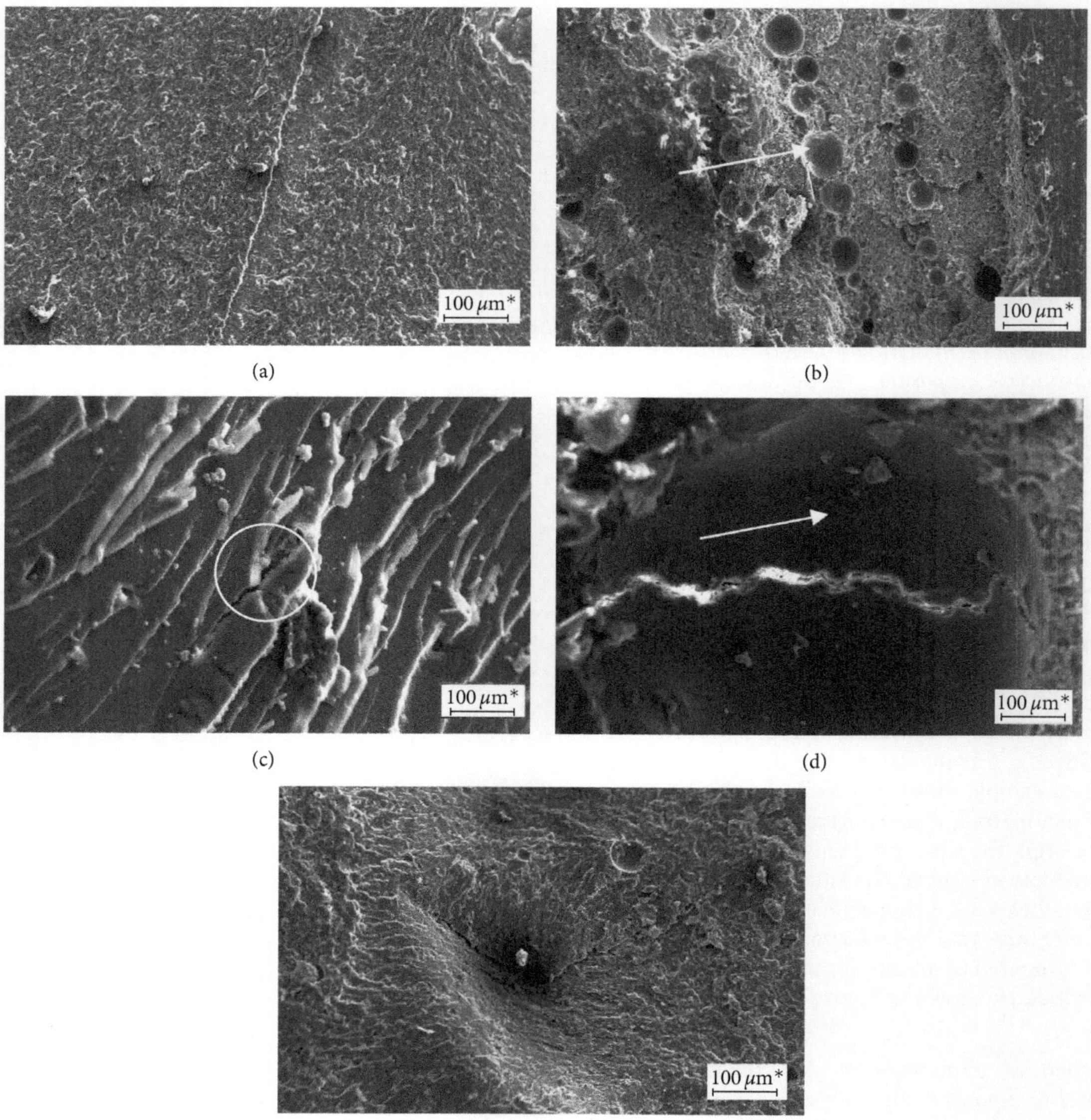


FIGURE 5: Fracture surface analysis of pure epoxy, epoxy-$SiO_2$ nanoparticles composites with (a) pure, (b) 5 wt.% $SiO_2$, (c) 10 wt.% $SiO_2$, (d) 15 wt.% $SiO_2$, (e) 20 wt.% $SiO_2$.

as shown in (1) below was used to determine the heat of fusion ($\Delta H$) from the scans as follows:

$$X_{DSC} = \frac{\Delta H_{sample}}{293}. \quad (1)$$

The reduction in degree of crystallinity was mainly due to reducing trend of heat of fusion from 5 wt.% to 20 wt.% of silica sand nanoparticles in epoxy. The same results were also discussed by some other researchers [11, 12].

*3.2.4. Fracture Surfaces SEM Analysis of Tensile Samples of Epoxy-$SiO_2$ Composites.* For pure epoxy, Figure 5(a) shows a brittle fracture which is due to quite surface roughness, and river line pattern in fracture surface [10]. A lot of holes and cavities can be seen in Figure 5(b), which could be due to other small molecular chemical compounds adhered to the silica nanoparticles. This behaviour confirms the reason, as discussed earlier, why tensile strength drops drastically between pure epoxy and 5% silica nanoparticles added to it. In Figure 5(c), tiny crack lines are visible and the arrow indicates the slow propagating crack area. Circled crack propagation in Figure 5(c) justifies that the crack direction is from top to bottom in the micrograph. Figures 5(d) and 5(e) signify crack propagation as a result from cavities and holes formed, due to improper mixture process. The white zones in SEM analysis show that the deformation and cohesive failure of the composites is because of the silica sand nanoparticles surrounds a large part of the polymer matrix. Decreasing trend of mechanical properties is due to agglomeration of silica sand nanoparticles in epoxy as shown in SEM results images.

## 4. Conclusion

Epoxy-$SiO_2$ nanocomposites were successfully produced using Tronoh silica sand nanoparticles as reinforcement. The density of the epoxy-$SiO_2$ nanocomposites was increased from 1.157 g/cm$^3$ to 1.244 g/cm$^3$ with an average increment of 1.796%. The mechanical properties of the nanocomposites decrease with the addition of silica sand nanoparticles due to improper mixing of the nanoparticles. SEM micrograph showed white and agglomerate zones on fracture surfaces which indicate that silica sand nanoparticles are not homogeneously dispersed in epoxy. A reduction in degree of crystallinity in epoxy based nanocomposites was observed with increasing trend of silica sand nanoparticles addition in epoxy as shown by DSC analysis.

## Acknowledgments

The authors would like to thank the Universiti Teknologi PETRONAS Perak, Malaysia and University of the Punjab Lahore Pakistan for providing the necessary support to complete this research.

## References

[1] J. Gonzalez-Benito and D. Olmos, "Efficient dispersion of nanoparticles in the thermoplastic polymers," *Society of Plastics Engineers, Plastics Research Online*, 2010.

[2] T. Andritsch, R. Kochetov, Y. T. Gebrekiros, U. Lafont, P. H. F. Morshuis, and J. J. Smit, "Synthesis and dielectric properties of epoxy based nanocomposites," in *Proceedings of the Annual Report Conference on Electrical Insulation and Dielectric Phenomena (CEIDP '09)*, pp. 523–526, October 2009.

[3] A. Yasmin, J. L. Abot, and I. M. Daniel, "Processing of clay/epoxy nanocomposites by shear mixing," *Scripta Materialia*, vol. 49, no. 1, pp. 81–86, 2003.

[4] B. Li, X. Zhang, J. Gao et al., "Epoxy based nanocomposites with fully exfoliated unmodified clay: mechanical and thermal properties," *Journal of Nanoscience and Nanotechnology*, vol. 10, no. 9, pp. 5864–5868, 2010.

[5] C. M. Damian, A. M. Pandele, C. Andronescu, A. Ghebaur, S. A. Garea, and H. Iovu, "Epoxy-based nanocomposites reinforced with new amino functionalized multi-walled carbon nanotubes," *Fullerenes Nanotubes and Carbon Nanostructures*, vol. 19, no. 3, pp. 197–209, 2011.

[6] A. Frache, O. Monticelli, M. Nocchetti, G. Tartaglione, and U. Costantino, "Thermal properties of epoxy resin nanocomposites based on hydrotalcites," *Polymer Degradation and Stability*, vol. 96, no. 1, pp. 164–169, 2011.

[7] J. Jordan, K. I. Jacob, R. Tannenbaum, M. A. Sharaf, and I. Jasiuk, "Experimental trends in polymer nanocomposites—a review," *Materials Science and Engineering A*, vol. 393, no. 1-2, pp. 1–11, 2005.

[8] T. Ahmad and O. Mamat, "Tronoh silica sand nanoparticle production and applications design for composites," *Defect and Diffusion Forum*, vol. 330, pp. 39–47, 2012.

[9] American Standard of Testing Material (ASTM), *Standard Test Method for Tensile Properties of ReinForced Thermosetting Plastics Using Straight Sided Specimens*, ASTM International, West Conshohocken, Pa, USA, 2004.

[10] Y. Zheng, K. Chonung, G. Wang, P. Wei, and P. Jiang, "Epoxy/nano-silica composites: curing kinetics, glass transition temperatures, dielectric, and thermal-mechanical performances," *Journal of Applied Polymer Science*, vol. 111, no. 2, pp. 917–927, 2009.

[11] K. Chrissafis, K. M. Paraskevopoulos, E. Pavlidou, and D. Bikiaris, "Thermal degradation mechanism of HDPE nanocomposites containing fumed silica nanoparticles," *Thermochimica Acta*, vol. 485, no. 1-2, pp. 65–71, 2009.

[12] M. Joshi and B. S. Butola, "Studies on nonisothermal crystallization of HDPE/POSS nanocomposites," *Polymer*, vol. 45, no. 14, pp. 4953–4968, 2004.

# 25

# Aqueous Medium Synthesis Route for Randomly Stacked Molybdenum Disulfide

**Pravas Kumar Panigrahi and Amita Pathak**

*Department of Chemistry, Indian Institute of Technology, Kharagpur, West Bengal 721302, India*

Correspondence should be addressed to Amita Pathak; ami@chem.iitkgp.ernet.in

Academic Editor: Dongsheng Wen

Synthesis of poorly crystalline, randomly oriented rag-like structures of molybdenum disulfide has been reported starting from aqueous solutions of ammonium molybdate, and thioacetamide in presence of sodium dodecyl sulfate via calcination of the amorphous precipitates, obtained through acidification of the *in situ* generated intermediate of ammonium tetrathiomolybdate. X-ray photoelectron spectroscopy, UV-visible spectroscopy, and X-ray diffraction of the calcined samples reveal the formation of single-phase $MoS_2$, while the amorphous precipitates have been found to be a mixture of $Mo_2S_5$, $MoS_3$, and a trace amount of $H_2MoS_4$. Highly folded and disordered layers of rag-like $MoS_2$ have been confirmed through high-resolution transmission electron microscopy. The electrical conductivity for the cold pressed pellet of the $MoS_2$ sample is found to be significantly higher than that of 2H-$MoS_2$ and increases further on annealing.

## 1. Introduction

Molybdenum disulfide ($MoS_2$) is a hexagonal layer structured material. The $MoS_2$ layers, each of which consists of a layer of molybdenum atoms sandwiched between two layers of sulfur atoms, are connected by weak van der Waals force. Due to the weak interactions between the sheets of sulfur atoms, $MoS_2$ exhibits excellent lubrication property [1, 2]. It also finds applications as electrode in high-density lithium batteries [3], solar cells [4] in form of thin films, and as intercalation host to produce new materials [5, 6]. From a catalytic point of view, $MoS_2$ possesses outstanding properties towards methanation of CO with hydrogen [7] and hydrodesulfurization reaction for removal of sulfur compounds from crude oil [8, 9]. It has been found that the catalytic activity of $MoS_2$ strongly depends on the number of coordinatively unsaturated sites [10]. Thus, the catalytic activity of $MoS_2$ is influenced by the particle size as well as the number of layers, demonstrating that the morphology is also an important parameter for such applications.

Till date, $MoS_2$ powders with different morphologies have been synthesized through a wide range of methods, such as thermal decomposition of ammonium tetrathiomolybdate or amorphous $MoS_3$ [11, 12], reaction of stoichiometrically mixed molybdenum and sulfur powders in vacuum at high temperature [13], and gas-phase reactions of molybdenum oxides with $H_2S$ under reducing atmosphere [14, 15]. There are also reports on the use of other high-energy sources like laser [16], electron beam [17], and $\gamma$-radiation [18] for the synthesis of nanotubes, fullerene-like, and other curved nanostructures of $MoS_2$. As an alternative to these methods, which require high temperatures/pressures and/or sophisticated instrumentations, solution-based synthesis routes have been developed [19–21] and have gained popularity due to their simplicity and convenience. For example, hydrothermal/solvothermal processes have been carried out at different temperatures (100–300°C) and incubation periods (12 h–3 days) by various research groups [22–25] to obtain fine powders of $MoS_2$ with a variety of crystallinity. Among these reports, Li et al. [23] observed $MoS_2$ nanoparticles with short range stacking of layers, when a solution containing $(NH_4)_2Mo_3S_{13}$ and hydrazine was treated hydrothermally at 180°C for 16 h. Sonochemical method [26] has also been explored to produce $MoS_2$. Alternatively, Bezverkhy et al. [27] have synthesized highly dispersed $MoS_2$ by using $(NH_4)_2MoS_4$ and $N_2H_4{\cdot}H_2O$ via aqueous medium precipitation method. The same group has also reported the formation of single layered $MoS_2$ with short fringes, when processed in presence of an organic surfactant in the reaction mixture [28]. Similarly, Hu et al. [29] have prepared the ball-like $MoS_2$ nanoparticles on calcining the amorphous molybdenum sulfide ($MoS_x$), obtained through precipitation from the solution of $Na_2MoO_4$ and thioacetamide in water and alcohol. Recently, Wu et al. [30] have reported the preparation of poorly crystalline $MoS_2$ nanospheres through calcination of $MoS_3$ precursor obtained through precipitation in an aqueous medium in presence of polyethylene glycol.

Considering the technical simplicity and versatility of the precipitation synthesis approaches, we have described here an aqueous solution-based precipitation route for the preparation of poorly crystalline $MoS_2$ with randomly stacked layers using commonly available starting reagents. Unlike most of the previous reports, the composition of the precursor powders has been studied by XRD, XPS, and thermal analysis. The present paper also includes the optical and electrical studies of the calcined powders, which have not been widely reported for the poorly crystalline $MoS_2$ nanostructures, prepared and studied so far in the literature.

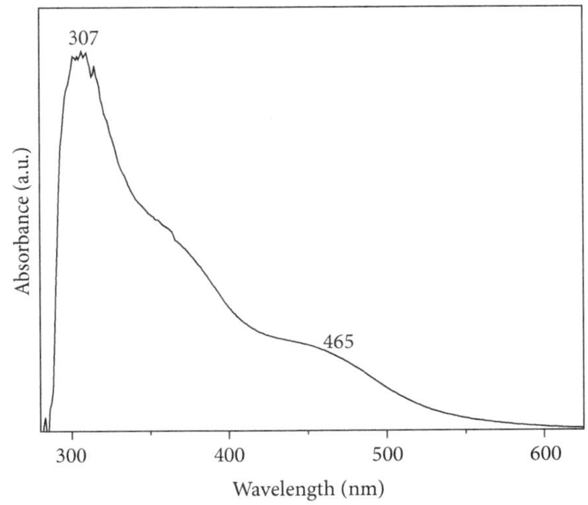


FIGURE 1: Optical absorption spectrum of the precursor solution containing $MoS_4^{2-}$ ions.

## 2. Experimental

*2.1. Synthesis.* The preparation of $MoS_2$ was carried out through modification of the method described in literature [31]. All the chemicals were purchased and used as received. To start, 30 mL of 0.008 M ammonium molybdate ($(NH_4)_6Mo_7O_{24}\cdot 4H_2O$, Merck India, 98%) solution was taken, and sodium dodecyl sulfate (SDS) of 10 times of cmc (*critical micelle concentration*) was added to it under constant stirring to obtain a clear solution. Then, 9.60 mL of 0.23 M sodium dithionite ($Na_2S_2O_4$, BDH, England, 98% pure) solution and 45 mL of 0.20 M thioacetamide ($CH_3CSNH_2$, Spectrochem India, 99%) solution were added into the former solution and were thoroughly mixed together by stirring. The solution mixture was heated (~90°C) over a water bath to obtain a clear reddish yellow color solution. Acidification of this solution with concentrated HCl (pH < 1) led to a dark brown colored precipitate. The precipitate was isolated using a centrifuge and was washed with water for several times. Drying of the precipitate gave rise to brownish black powders, which were calcined at 400°C for 2 h under argon atmosphere to obtain the black powders of $MoS_2$.

*2.2. Characterization.* The optical absorption spectrum of the reddish yellow colored solution of the intermediate as well as the ethanol dispersion of the calcined powders were studied at room temperature using Shimadzu UV-1601 UV-visible spectrophotometer. The phase and structure analyses of the virgin and the calcined powders were carried out on X'pert Pro 3040/60 high-resolution X-ray diffractometer using Cu K$\alpha$ radiation ($\lambda$ = 1.5418 Å). X-ray photoelectron spectra (XPS), using VG Microtech ESCA2000 Multilab apparatus, was obtained for the virgin and the calcined powders for determining the oxidation states of Mo and S in the respective samples. Thermal analysis of the dried precipitate was performed at the heating rate of 6°C per min under dynamic argon using Perkin-Elmer Pyris Diamond TG/Differential thermal analyzer and alumina as the reference. Transmission electron microscope of model JEOL2010 (UHR) was used to study the microstructure of the calcined powders. The FTIR spectrum of the precursor sample was taken on a Perkin Elmer Fourier transform analyzer (FTIR spectrometer RXI). Electrical conductivity measurement was carried out on pressed pellet of the as-prepared calcined powders of $MoS_2$ using a computer-controlled HIOKI LCR HITESTER at an AC signal of 1.0 V. The surfaces of the pellets were coated with silver paint for electrical contact. Finally, the electrical connection of the pellet to the apparatus was made through nickel electrode and put in a temperature controlled furnace. The hydrodynamic size of the synthesized $MoS_2$ sample was measured by using 90 Plus Particle Size Analyzer (made by Brookhaven Instruments). The BET (Brunauer-Emmett-Teller) specific surface area of the calcined sample was measured through nitrogen adsorption isotherms at 77 K by using Quantachrome Instruments (Autosorb-1, Model no. ASI-C-9).

## 3. Results and Discussions

*3.1. Analysis of the Precursor Precipitate.* To understand the chemistry involved in the formation of precursor precipitate, the composition of the reddish yellow colored precursor solution obtained on heating the solution mixture over a water bath was characterized by UV-visible spectroscopy (Figure 1). This optical absorption spectrum shows two absorption peaks centered at ~307 and 465 nm corresponding to the tetrathiomolybdate ions ($MoS_4^{2-}$) [32, 33]. These ions possibly get formed as an intermediate on heating the solution of heptamolybdate ($Mo_7O_{24}^{6-}$) and thioacetamide through

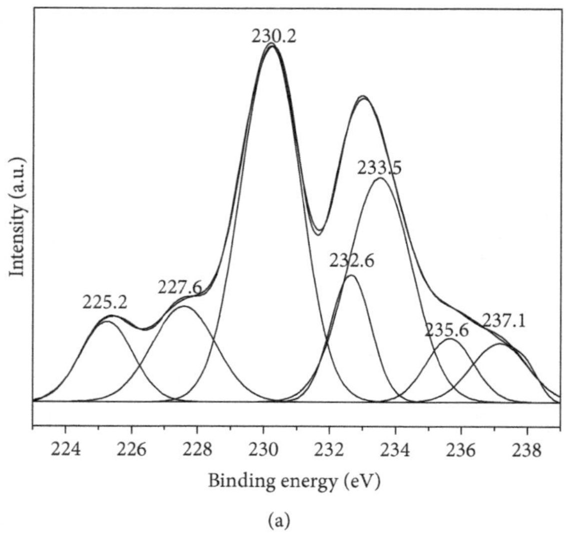


FIGURE 2: XPS spectra (a) for Mo, after and before curve peak fitting, and (b) for S in the as-precipitated precursor sample obtained after acidification of ATM solution.

exchange of oxygen atoms in molybdate ions by sulfur. The possible reaction involved can be written as follows:

$$Mo_7O_{24}{}^{6-} + 28CH_3CSNH_2 + 4H_2O \longrightarrow 7MoS_4{}^{2-} + 8H^+ + 28CH_3CONH_2 \quad (1)$$

Acidification of the solution containing tetrathiomolybdate ions with concentrated HCl led to brownish black colored precipitate of the sulfides. The composition of this as-precipitated precursor was investigated through analyzing the oxidation states of Mo and S in the samples by X-ray photoelectron spectroscopy. The survey spectra (not shown) showed the peaks for carbon, an impurity in the vacuum system, and oxygen in addition to the elements of interest (i.e., Mo and S). The deconvoluted and curve-fitted spectra shown in Figure 2(a) corroborated to seven peaks centered at 225.2, 227.6, 230.2, 232.6, 233.5, 235.6, and 237.1 eV corresponding to Mo and S species present in the sample. The peak at 227.6 eV may be attributed to the 2s binding energy of S in elemental state [34]. The peak around 225.2 eV, which has a lower shift of 2.4 eV with respect to 2s binding energy of free S, originates from the sulfur bonded to Mo. The rest of the peaks can readily be fitted into three groups, 230.2 and 233.5 eV, 232.6 and 235.6 eV, 233.5 and 237.1 eV, representing the 3d doublets, $3d_{5/2}$ and $3d_{3/2}$, respectively, of various oxidation states of Mo.

It is reported that, for any oxidation state of Mo, the 3d doublet (i.e., $3d_{5/2}$ and $3d_{3/2}$) has an energy separation of ~3.1 eV and a characteristic $3d_{5/2} : 3d_{3/2}$ peak intensity ratio of 3 : 2 [35]. Furthermore, the $3d_{5/2}$ and $3d_{3/2}$ binding energies of all the oxidation states of Mo show a higher shift in comparison to the respective binding energy values for elemental Mo, that is, 228 and 231.1 eV [36]. The peak pair (230.2 eV and 233.5 eV), which has a shift from 3d doublet of elemental Mo, higher than that of Mo (IV) in $MoS_2$ (1.6 eV) and lower than that for Mo (VI) in $MoS_3$ (4.2 eV), can be, respectively, assigned to $3d_{5/2}$ and $3d_{3/2}$ (i.e., the 3d doublet) binding energies of Mo (V) species in $Mo_2S_5$. The 3d doublet binding energies at 232.6 and 235.6 eV in Figure 2(a) may appear due to Mo (VI) species in $MoS_3$. However, the higher energy shift of this peak pair from that of elemental Mo than the expected value of 4.2 eV and close proximity of the binding energies of the respective 3d doublets (i.e., $3d_{5/2}$ and $3d_{3/2}$) of $MoS_3$ and $MoO_3$ led us to believe that the XPS peak pair at 232.6 and 235.6 eV may be partially contributed by 3d doublet for the Mo (VI) species in $MoO_3$. This trace amount of $MoO_3$ may be formed during the handling of the sample in ambient atmosphere. The lone peak at 237.1 eV matches well with the respective $3d_{3/2}$ binding energy of Mo (VI) species in $H_2MoS_4$ [35], whereas its $3d_{5/2}$ counterpart of 3d doublet, with binding energy at 233.8 eV, may have merged with the broad peak at 233.5 eV. The insoluble compound, $H_2MoS_4$, which may be resulted from one of the acidification subreactions, eventually produces $MoS_3$ with liberation of $H_2S$ on heating above 190°C.

Figure 2(b) shows the XPS spectra for S in the as-precipitated virgin precursors obtained after acidification of ATM solution. The figure indicates the presence of three major peaks with their maxima at 162.9, 166.8, and 168.9 eV. The broad peak at 162.9 eV has a lower binding energy value compared to that of the S 2p doublet (i.e., 164.1 eV and 165.2 eV for S $2p_{3/2}$ and S $2p_{1/2}$, resp.) in elemental sulfur [35]. However, this value is higher than the expected XPS peak positions for 2p doublet for S in $MoS_2$ (i.e., 161.7 eV and 162.7 eV for S $2p_{3/2}$ and S $2p_{1/2}$, resp.). In consistency with the previous literature, the peak at 162.9 eV is thus assigned to the binding energy of S 2p in $MoS_3$ [37]. On the other hand, the XPS peaks of S located at around 166.8 and 168.9 eV show higher binding energy values compared to that of S in zero oxidation state. These peaks possibly appeared due to the presence of bisulfite (with S in +4 oxidation state) and bisulfate (with S in +6 oxidation state) residue in the precipitate. The bisulfite and bisulfate ions are possibly produced from aqueous solution of sodium dithionite in presence of oxygen [38]. Nevertheless, it is difficult to ascertain the identity of $Mo_2S_5$ and $H_2MoS_4$ from the XPS analysis of sulfur because of nonavailability of relevant XPS data in the literature. However, on the basis of Mo 3d peak analysis, it can be predicted that the brownish black uncalcined samples, which get formed during the acidification of the intermediate ammonium tetrathiomolybdate (ATM) solution, could be a mixture of amorphous $MoS_3$, $Mo_2S_5$, elemental S, and a small amount of $H_2MoS_4$. The plausible set of reactions (2)–(4) can be written as follows [39–41]:

$$MoS_4^{2-} + 2H^+ \longrightarrow MoS_3\downarrow + H_2S\uparrow \quad (2)$$

$$2MoS_4^{2-} + 4H^+ \longrightarrow Mo_2S_5\downarrow + 2H_2S\uparrow + S\downarrow \quad (3)$$

$$MoS_4^{2-} + 2H^+ \longrightarrow H_2MoS_4 \longrightarrow MoS_3\downarrow + H_2S\uparrow \quad (4)$$

The simultaneous thermogravimetric (TG), differential thermogravimetric (DTG), and differential thermal analysis (DTA) of the as-precipitated brownish black powders, shown in Figure 3, supported the XPS results. The TG curve shows an initial weight loss of ~2.8% below 150°C, possibly corresponding to the removal of adsorbed water from the sample. The temperature region between 150 and 600°C in the TG curve can be divided into four distinct stages corresponding to various reactions involved in the decomposition process. The weight loss of ~22.7% associated with the endothermic effect in the stage-I region (150–240°C) could be taken as a collective effect due to loss of occluded vapors of $H_2S$, HCl, and the by-product elemental sulfur formed due to reaction (3) along with the conversion of $MoS_3$ to $Mo_2S_5$ [42] as per the following reaction:

$$2MoS_3 \xrightarrow{190-220^\circ C} Mo_2S_5 + S\uparrow \quad (5)$$

The stage-II (240–390°C) involves the endothermic decomposition of $Mo_2S_5$ to amorphous $MoS_2$ and an additional exothermic effect [41] for the transformation $MoS_2$ (amorphous) → $MoS_2$ (crystalline) in the sample. As a result, the overall thermal effect in DTA curve for stage-II is a small exothermic, which may be superimposed on a small endothermic peak. The weight loss (~9%) in the TG curve, in the corresponding region, matches well with the expected loss of sulfur during the conversion as follows:

$$Mo_2S_5 \xrightarrow{\triangle} S\uparrow + MoS_2\,(\text{amorphous}) \longrightarrow MoS_2\,(\text{Crystalline}) \quad (6)$$

The stage-III (390–430°C) and IV (430–600°C), exothermic and endothermic effects, respectively, represent the oxidation of $MoS_2$ to $MoO_2$ and $MoO_2$ to $Mo_2O_3$, respectively [42]. The formation of $MoO_2$ may occur due to the oxygen impurities present in argon (carrier gas for the instrument) and subsequent conversion to $Mo_2O_3$ at higher temperature due to local deficiency of oxygen.

Finally, the degree of crystallinity of the dried as-precipitated precursor powders was studied by X-ray diffraction (XRD), which shows a broad and diffused diffraction spectrum (Figure 4) indicating the sample to be predominantly

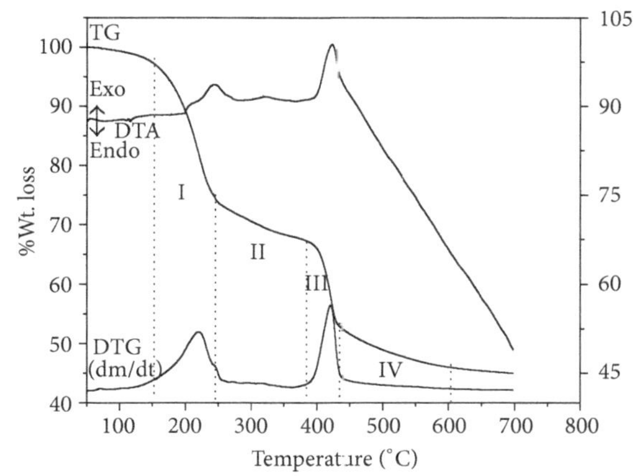


FIGURE 3: Thermograms for the decomposition of the as-precipitated precursor sample.

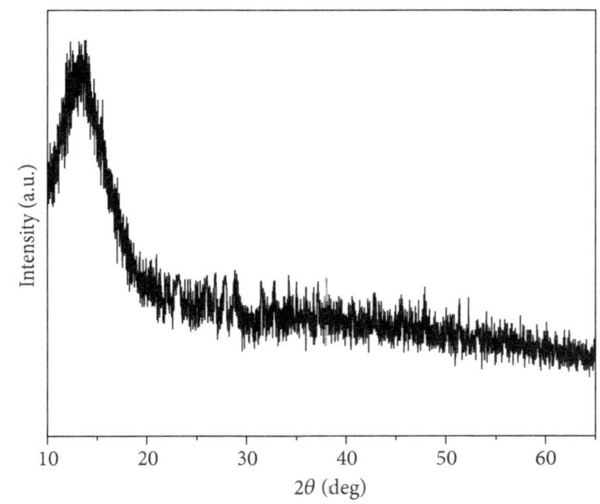


FIGURE 4: XRD pattern of the as-precipitated precursor powders.

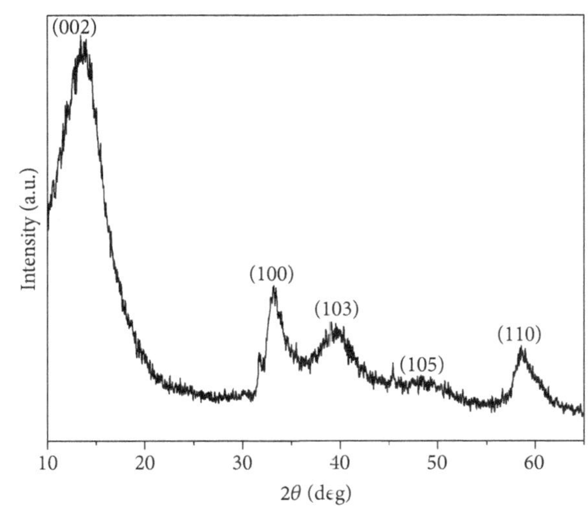


FIGURE 5: XRD pattern of the $MoS_2$ sample after calcination at 400°C for 2 h.

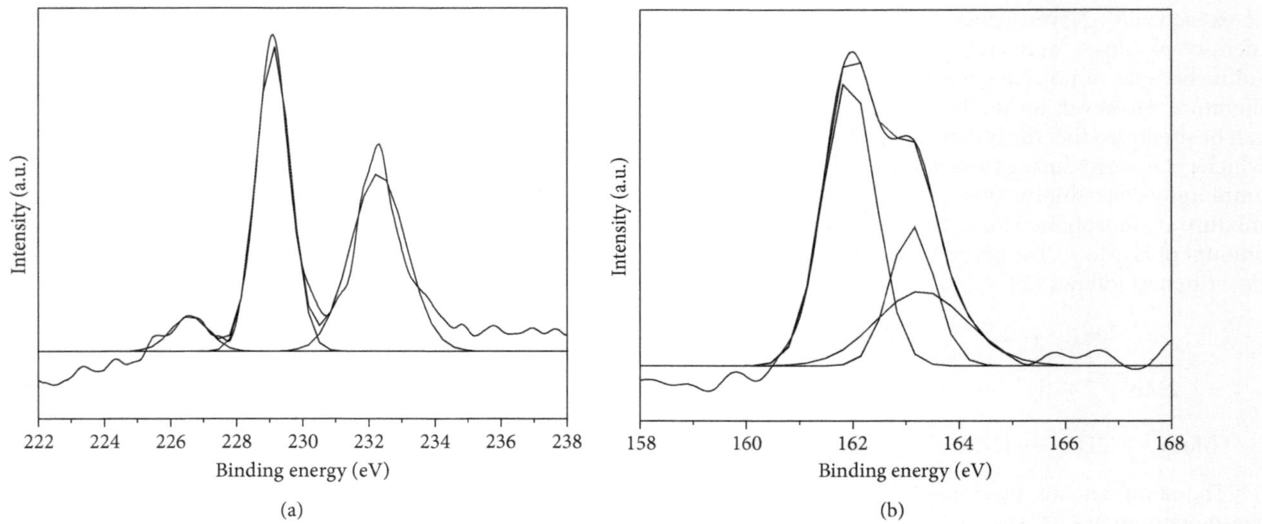


FIGURE 6: XPS spectra, before and after curve peak fitting, (a) for Mo and (b) for S in $MoS_2$ sample, obtained after calcination at 400°C for 2 h.

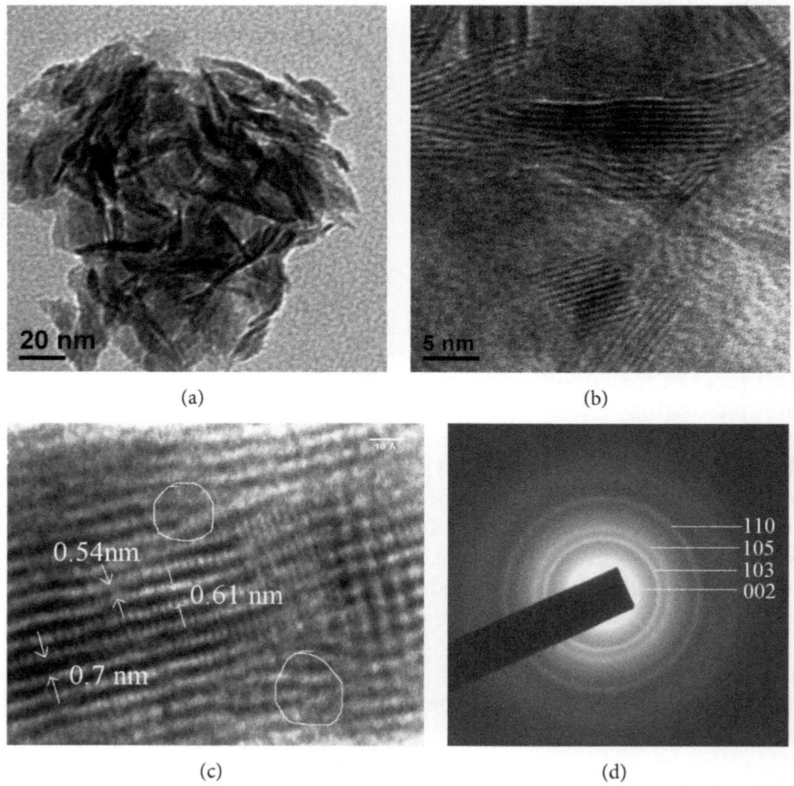


FIGURE 7: Morphological analysis of the as-obtained $MoS_2$ product obtained after calcination at 400°C for 2 h; (a) low-resolution image; (b), (c) high resolution images showing clear view of lattice fringes; (d) electron diffraction pattern of a selected area.

amorphous. The solitary peak centered at ~$2\theta = 13.4°$ may be due to some amount of crystalline materials present in the sample. Of the sulfide compounds of molybdenum, $MoS_3$ are reported to be amorphous [12, 43]. However, $Mo_2S_5$, whose crystallinity nature is unknown given to the unavailability of any information in JCPDS database, may be responsible for the broad peak in Figure 4.

*3.2. Analysis of the Calcined Sample.* XRD pattern of the sample, obtained on calcinations of the as-precipitated powders at 400°C in argon atmosphere for 2 h, shows broad but distinct XRD peaks (Figure 5). All the XRD peaks are indexed to the respective $d_{hkl}$ values of the hexagonal cell of bulk $MoS_2$ with lattice parameters $a$ = 3.159 Å and $c$ = 12.162 Å (according to JCPDS Card no. 37-1492). The $d_{002}$ line for

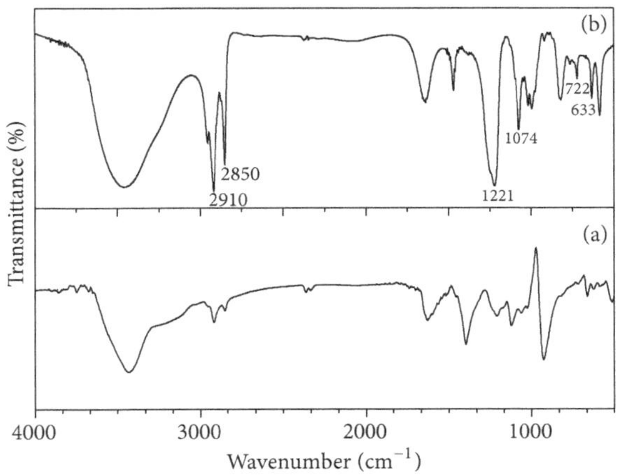


FIGURE 8: FTIR spectra of (a) the as-precipitated virgin precursor powders and (b) pure SDS.

the calcined powders is observed at $2\theta = 13.70°$, which indicates a shift towards lower angle with respect to bulk 2H-$MoS_2$. The observed shift may be attributed to lattice expansion of ~6.2% along $c$-axis due to induction of crystal defects or strains owing to the curvature of the layers [30]. This prediction is substantiated by larger *ca.* "$c$" value of the sample ($a = 3.105$ Å and $c = 12.911$ Å) compared to that reported for 2H-$MoS_2$. The diffraction pattern is comparable with the scan reported by Wildervanck and Jellinek for $MoS_2$ prepared through the decomposition of $MoS_3$ [12]. The average crystallite size is found to be about 33 Å in $c$-direction (corresponding to approximately six $MoS_2$ layers) and 78 Å in $a$-direction. The calculations of dimensions and number of layers have been carried out in the standard manner [44] by using full width at half maximum (FWHM) of the well-defined diffraction peaks. Furthermore, the broadening of the $d_{hkl}$ lines gives an indication of the formation of poorly crystalline $MoS_2$ powders.

Finally, the purity and composition of the calcined powders were analyzed by XPS. The peak pair (229.1 and 232.3 eV) in Figure 6(a) represents the 3d doublet (i.e., $3d_{5/2}$ and $3d_{3/2}$, resp.) of the Mo (IV) species in $MoS_2$, while the peak centered at 226.6 eV can be assigned to S 2s binding energy in $MoS_2$. Figure 6(b) shows two peaks centered at 161.9 and 163.1 eV corresponding to the S 2p doublet, that is, $2p_{3/2}$ and $2p_{1/2}$, respectively in $MoS_2$ [35]. The small peak of insignificant intensities beyond 163.1 eV may be attributed to some impurities associated with sulfur. Thus, the XPS studies indicate the formation of pure $MoS_2$ through heat treatment of the precursor precipitate, obtained on acidification of ATM solution.

The detailed microstructure of the calcined $MoS_2$ sample is analyzed by transmission electron microscopy. Low-resolution transmission electron micrograph (TEM) of the sample (Figure 7(a)) indicated the formation of typical lamellar structures of $MoS_2$ analogous to that observed by Chianelli et al. [45], who termed them as rag-like structures. Peng et al. also reported similar type of rag structures, which were synthesized under drastic conditions of hydrothermal reaction [24]. At higher magnification, these structures (Figure 7(b)) reveal the presence of different number of lattices ranging from 3 to 11. The low stacking and random orientation of the lattices, characteristics of rag-like structures, are also supported by the broad XRD peaks (Figure 5). At still higher resolution, the image (Figure 7(c)) indicates the existence of edge dislocations (in the encircled domains of the image) in the microstructure of the sample. This type of defects generally induced nonuniformity in d-spacing between the lattices [46], which is justified by the measured fringe widths, that is, 0.54, 0.61, and 0.7 nm, at different location of the image. This observation further demonstrates the formation of highly folded and disorder stacking character of rag-like structures [12]. However, the average width of the lattice fringes measured from the image is found to be ~0.62 nm, which is in close agreement with the characteristic $d_{002}$ plane of the 2H-$MoS_2$ crystal. The selected area electron diffraction (SAED) patterns of the sample (Figure 7(d)) show diffused diffraction rings, which are indexed by matching the d-values of the $MoS_2$ crystal in the JCPDS card 37-1492 with those calculated from measurement of the ring radii. The observed diffraction rings can be attributed to coherent diffraction by aggregate of partially crystalline nanoparticles.

Finally, the hydrodynamic size of synthesized $MoS_2$ powders was measured, which is found to be around 299.3 nm. Since the measured size for the particles is higher than that observed from TEM image, it can be indicated that there is presence of aggregation in the sample [47].

Based on the previous literature, it has been proposed that the formation of the poorly crystalline rag structures of $MoS_2$ may be due to the presence of some amount of SDS residue in the precipitated precursor powders, which is evident from its FTIR spectra (Figure 8(a)). The appearance of typical infrared peaks at 2919 and 2850 cm$^{-1}$ (asymmetric and symmetric –$CH_2$ vibration), 1221 cm$^{-1}$ (C–C skeletal vibration), 1074 and 633 cm$^{-1}$ ($SO_4^{2-}$), and 722 cm$^{-1}$ (C–O stretching vibration) gives enough indication of the existence of SDS in the sample [48]. This is also confirmed by comparing the observed FTIR peaks with those obtained from pure SDS (Figure 8(b)). According to the previous reports, the precipitates, obtained through low-temperature reactions in the presence of long chain surfactants, such as cetyltrimethylammonium bromide (CTAB) or polyethylene glycol (PEG), were found to be contaminated with surfactants even after extensive washing [22, 30]. These adsorbed surfactants were believed to be decomposed under heat treatment, resulting in the abnormal expansion of interlayer gaps. Drawing similarity with the report, it can thus be predicted that the thermal decomposition of the carbonaceous compound (i.e., SDS) adsorbed on the precursor precipitate obtained through the present preparation procedure prevents the regular stacking of the layers of molybdenum sulfide and induces faulted stacking in the structures of the synthesized $MoS_2$.

Further verification of the role of SDS in the formation of poorly crystalline $MoS_2$ powders with highly folded and disorder stacking layered structure has been performed by

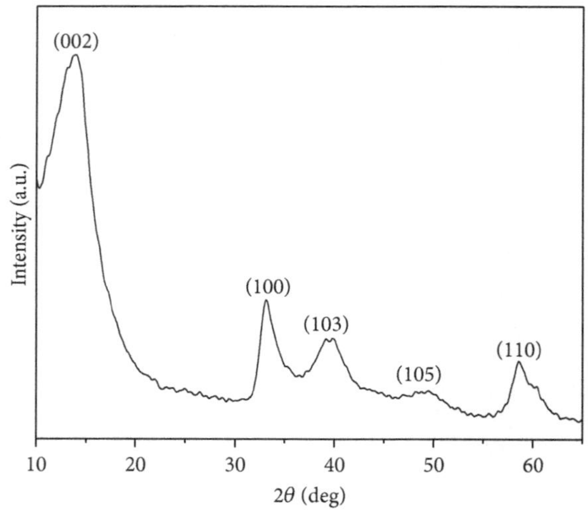


(a)

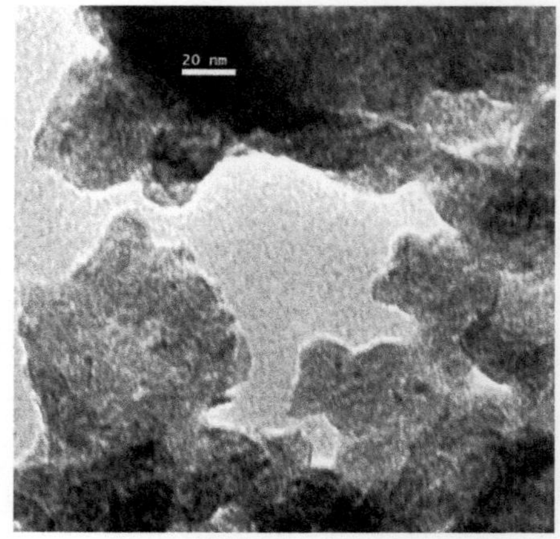


(b)

Figure 9: (a) XRD pattern and (b) TEM image of $MoS_2$ sample obtained on calcination (at 400°C for 2 h) of the precipitate prepared in absence of SDS.

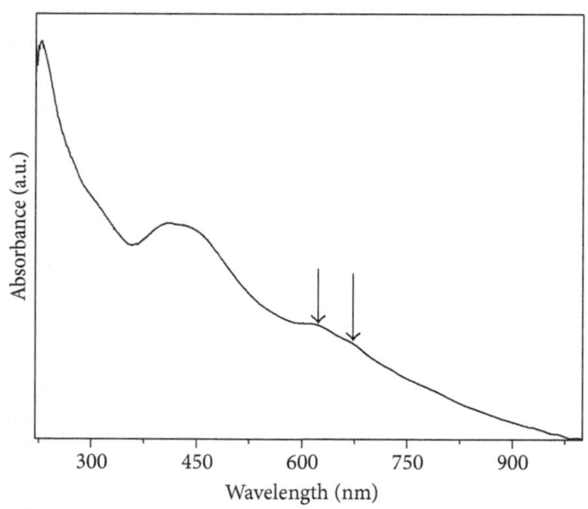


Figure 10: Optical absorption spectrum of the as-obtained $MoS_2$ sample.

carrying out the XRD and microscopy analysis of the $MoS_2$ sample obtained in the absence of SDS (Figure 9). The XRD pattern in Figure 9(a) displays X-ray diffraction pattern similar to that shown in Figure 5. However, the $2\theta$ values for the (002) diffraction peak position for $MoS_2$ samples prepared in absence of SDS (Figure 9(a)) and in presence of SDS (Figure 5) are found to be at $2\theta \approx 14.0$ and 13.4, respectively, compared to that for bulk 2H-$MoS_2$ ($2\theta \approx 14.40°$) (according to standard data in JCPDS Card no. 37-1492). This relatively higher downward shift of the (002) peak in case of sample prepared in presence of SDS indicates the larger lattice expansion along $c$-axis, which can be ascribed to the folding and randomness in the stacking of the $MoS_2$ layers [30]. This finding is also supported by the TEM images of the samples. The TEM image of the $MoS_2$ sample obtained in absence of SDS indicates the formation of agglomerations of some irregularly shaped coarse particles in contrast to that represented in Figure 7 (for $MoS_2$ sample obtained in the presence of SDS).

*3.3. Study of the Properties of the Synthesized $MoS_2$ Sample.* In order to study the optical property of the calcined powders, UV-visible absorption was taken at room temperature after dispersion in ethanol. The spectra in Figure 10 depict two weak bands centered at ~675 and 617 nm (marked by arrows), and a strong and very broad band with the maxima centered at ~400 nm. The first two peaks correspond to exciton A and B of $MoS_2$, respectively. These exciton energy values are found to be red shifted in comparison to the respective exciton energy values assigned to the bulk 2H-$MoS_2$ (654.3 and 593.3 nm) but are comparable to the respective values for nested *IF*-like structures (667 and 616.3 nm) at 175 K [49]. Furthermore, the observed UV-visible spectral pattern is found similar to that reported by Liao et al. [50] for $MoS_2$ fibers synthesized through a low-temperature soft chemical route. It can thus be inferred that the curved nanostructures of the synthesized sample may be responsible for the observed optical spectrum of the synthesized $MoS_2$ sample.

For studying the semiconducting behavior of the prepared $MoS_2$ samples, two pellets were prepared by pressing $MoS_2$ powders. One of the pellets was heated at 400°C for 4 h in argon atmosphere before carrying out the electrical measurement. The plots of log of conductivity ($\log \sigma$) versus inverse of temperature ($1/T$) at different frequencies (such as 100, 200, 500, and 1000 kHz) are shown in Figures 11(a) and 11(b) for the $MoS_2$ pellets, before and after heat treatment, respectively. Both of the samples exhibit typical

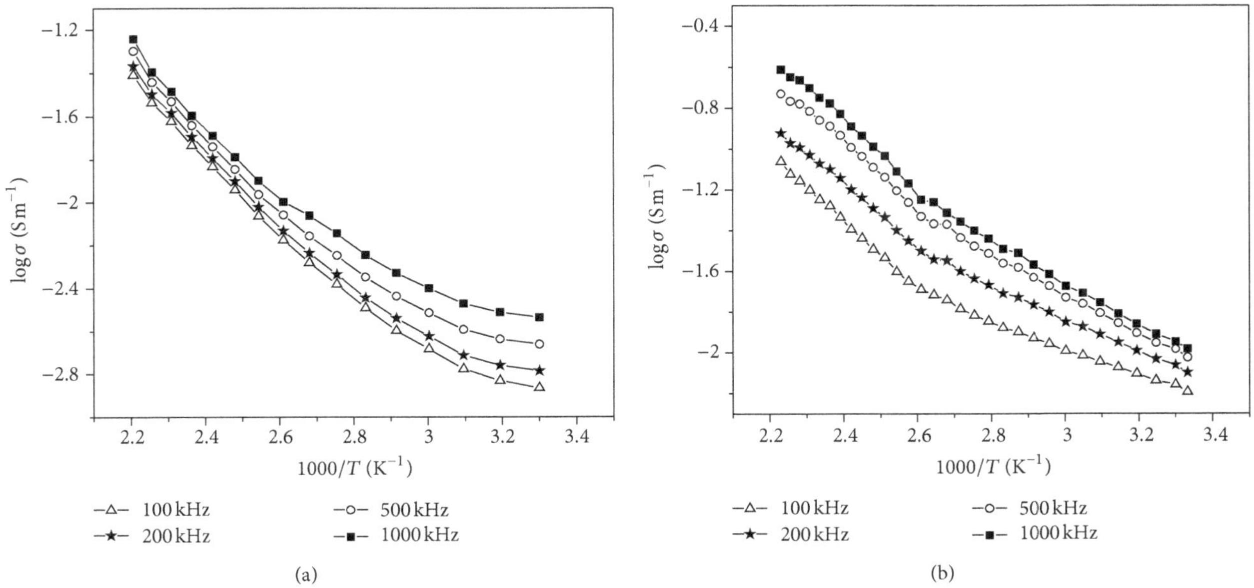


FIGURE 11: Variation of conductivity with temperature at different frequencies for pressed pellets of $MoS_2$; (a) before heat treatment and (b) after heat treatment.

semiconductor behaviors, that is, increase in the electrical conductivity with rise in temperature. In the low frequency range (below 1 kHz), the electrical conductivity for both heat-treated and virgin $MoS_2$ pellets is found to be frequency independent and comparable to the DC conductivity values. The conductivity (in both, heat-treated and virgin pellets) is, however, observed to increase beyond the frequency range of 1 kHz. The frequency dependence of conductivity can be given by the following equation [51]:

$$\sigma = \sigma_{dc} + \sigma_{ac} = \sigma_{dc} + A\omega^{n}, \quad (7)$$

where $A$ and $n$ are materials parameters ($0 < n < 1$), $\sigma_{dc}$ is "DC conductivity" or frequency independent part, $\sigma_{ac}$ and $\omega$ are AC conductivity and angular frequency, respectively.

Further, from Figure 11, the plots of log $\sigma$ versus $(1/T)$ are observed to undergo a change in slope beyond 373 K for both the samples. Relatively less significant rise in electrical conductivities of the samples at temperatures below 373 K is probably due to $\sigma_{ac}$ part being the major contributor to the total measured conductivity for the sample, owing to the fact that AC conductivity of semiconductors is only weakly dependent on temperature compared to $\sigma_{dc}$ [52]. At intermediate and higher temperature range, the $\sigma_{dc}$ part possibly contributes significantly, and thus the samples shows a noticeable rise in conductivity with temperature. Similar nonlinear dependence of log conductivity on $(1/T)$ for $MoS_2$ has been reported by Sienicki [53], who showed that log $\sigma$ for $MoS_2$ varied exponentially with reciprocal of temperature $(1/T)$ at higher temperatures.

In addition, the conductivity measured at room temperature (303 K) for the heat-treated pellet ($1.13 \times 10^{-2}$ S m$^{-1}$) is found to be 16 times higher than that of the virgin pellet ($6.88 \times 10^{-4}$ S m$^{-1}$). This significant improvement in $\sigma_{ac}$ may be due to the densification of the heat-treated pellet through grain growth at higher temperature. The activation energies ($E_a$) of the samples were also calculated from the slope of the log $\sigma$ versus $1/T$ curve at a frequency of 1000 kHz using Arrhenius equation. Contrary to the trend of $\sigma$, the $E_a$ values for heat-treated pellet (0.259 eV) and virgin pellet (0.258 eV) were found to be comparable.

Finally, the BET surface area of the calcined sample has been determined by using BET surface area measurement instrument, which is found to be 35 m$^2$/g. In spite of the low surface area value, based on some previous studies [10, 54, 55] and the morphological analysis, the synthesized $MoS_2$ sample in the present technique, however, may be predicted to exhibit better catalytic activity. According to the studies by Tanaka and Okuhara [54], the difference in catalytic activity of layered transition metal chalcogenides such as $MoS_2$ is not accounted for by BET surface area or pore structure but depends on the morphology, in particular on the ratio of basal and edge planes. This fact was also vindicated by the experimental observation by Iwata et al. [10] and Uzcanga et al. [55]. For example, Iwata et al. [10] claimed that the poorly crystalline $MoS_2$ sample with highly bent multilayered structure provides larger number of catalytically active sites than the well-crystallized sample in spite of having surface area as low as 10 m$^2$/g. This improvement in catalytic property was ascribed to the high content of edges and defects induced in the $MoS_2$ layers of the synthesized sample. Uzcanga et al. [55] also anticipated enhanced catalytic performance for the sonochemically prepared $MoS_2$ powders compared to that of the conventionally prepared sample, even though the former sample displayed lower surface area.

## 4. Conclusions

We have reported a simple aqueous solution-based precipitation route for the preparation of rag-like structures of $MoS_2$ using some easily available laboratory reagents. As all of the reagents used are fairly soluble in water, there is less possibility of having impurities in the final products. The acidification of ATM solution with concentrated HCl resulted in the formation of a mixture of amorphous $MoS_3$ and $Mo_2S_5$ with small amount of elemental sulfur and $H_2MoS_4$, as evident from XPS and thermal analysis. However, XRD patterns indicated the formation of pure but poorly crystalline $MoS_2$ phase on calcining the as-precipitated samples at 400°C for 2 h in argon atmosphere. The calcined powders were found to be composed of significantly folded and randomly oriented S-Mo-S layers of $MoS_2$ as visible from high-resolution TEM images. This observation was supported by the red shifting of the exciton peak positions in UV-visible spectrum of solid $MoS_2$. The unique morphologies of the prepared $MoS_2$ sample may be applied as a potential catalyst for hydrodesulfurization reactions. The electrical conductivity of $MoS_2$ sample was found to increase on annealing.

## Acknowledgments

The authors acknowledge the Department of Science and Technology, Government of India, for the financial support to carry out this work. The authors also acknowledge their sincere thanks to Professor P. V. Satyam and Professor Shikha Varma of Institute of Physics, Bhubaneswar, for sample characterization by HRTEM and XPS, respectively.

## References

[1] L. Rapoport, N. Fleischer, and R. Tenne, "Applications of $WS_2$ ($MoS_2$) inorganic nanotubes and fullerene-like nanoparticles for solid lubrication and for structural nanocomposites," *Journal of Materials Chemistry*, vol. 15, no. 18, pp. 1782–1788, 2005.

[2] M. H. Cho, J. Ju, S. J. Kim, and H. Jang, "Tribological properties of solid lubricants (graphite, $Sb_2S_3$, $MoS_2$) for automotive brake friction materials," *Wear*, vol. 260, no. 7-8, pp. 855–860, 2006.

[3] N. Imanishi, K. Kanamura, and Z. I. Takehara, "Synthesis of $MoS_2$ thin film by chemical vapor deposition method and discharge characteristics as a cathode of the lithium secondary battery," *Journal of the Electrochemical Society*, vol. 139, no. 8, pp. 2082–2087, 1992.

[4] E. Gourmelon, O. Lignier, H. Hadouda et al., "$MS_2$ (M = W, Mo) photosensitive thin films for solar cells," *Solar Energy Materials and Solar Cells*, vol. 46, no. 2, pp. 115–121, 1997.

[5] W. M. R. Divigalpitiya, R. F. Frindt, and S. R. Morrison, "Inclusion systems of organic molecules in restacked single-layer molybdenum disulfide," *Science*, vol. 246, no. 4928, pp. 369–371, 1989.

[6] A. Zak, Y. Feldman, V. Lyakhovitskaya et al., "Alkali metal intercalation of $MS_2$(M = W, Mo) fullerene-like nanoparticles and bulk (2H) powders," *Journal of the American Chemical Society*, vol. 124, no. 17, pp. 4747–4758, 2002.

[7] J. Chen, S. L. Li, Q. Xu, and K. Tanaka, "Synthesis of open-ended $MoS_2$ nanotubes and the application as the catalyst of methanation," *Chemical Communications*, no. 16, pp. 1722–1723, 2002.

[8] E. Devers, P. Afanasiev, B. Jouguet, and M. Vrinat, "Hydrothermal syntheses and catalytic properties of dispersed molybdenum sulfides," *Catalysis Letters*, vol. 82, no. 1-2, pp. 13–17, 2002.

[9] B. Yoosuk, J. H. Kim, C. Song, C. Ngamcharussrivichai, and P. Prasassarakich, "Highly active $MoS_2$, $CoMoS_2$ and $NiMoS_2$ unsupported catalysts prepared by hydrothermal synthesis for hydrodesulfurization of 4,6-dimethyldibenzothiophene," *Catalysis Today*, vol. 130, no. 1, pp. 14–23, 2008.

[10] Y. Iwata, K. Sato, T. Yoneda et al., "Catalytic functionality of unsupported molybdenum sulfide catalysts prepared with different methods," *Catalysis Today*, vol. 45, no. 1–4, pp. 353–359, 1998.

[11] M. Nath, A. Govindaraj, and C. N. R. Rao, "Simple synthesis of $MoS_2$ and $WS_2$ nanotubes," *Advanced Materials*, vol. 13, no. 4, pp. 283–286, 2001.

[12] J. C. Wildervanck and F. Jellinek, "Preparation and crystallinity of molybdenum and tungsten sulfides," *Zeitschrift Für Anorganische und Allgemeine Chemie*, vol. 328, no. 5-6, pp. 309–318, 1964.

[13] R. N. Viswanath and S. Ramasamy, "Preparation and isochronal sintering behavior of molybdenum disulfide compound," *Journal of Materials Science*, vol. 25, no. 12, pp. 5029–5035, 1990.

[14] Y. Feldman, G. L. Frey, M. Homyonfer et al., "Bulk synthesis of inorganic fullerene-like $MS_2$ (M = Mo, W) from the respective trioxides and the reaction mechanism," *Journal of the American Chemical Society*, vol. 118, no. 23, pp. 5362–5367, 1996.

[15] Y. Feldman, E. Wasserman, D. J. Srolovitz, and R. Tenne, "High-rate, gas-phase growth of $MoS_2$ nested inorganic fullerenes and nanotubes," *Science*, vol. 267, no. 5195, pp. 222–225, 1995.

[16] P. A. Parilla, A. C. Dillon, K. M. Jones et al., "The first true inorganic fullerenes?" *Nature*, vol. 397, article 114, 1999.

[17] M. J. Yacamán, H. López, P. Santiago, D. H. Galván, I. L. Garzón, and A. Reyes, "Studies of $MoS_2$ structures produced by electron irradiation," *Applied Physics Letters*, vol. 69, no. 8, pp. 1065–1067, 1996.

[18] G. Chu, G. Bian, Y. Fu, and Z. Zhang, "Preparation and structural characterization of nano-sized amorphous powders of $MoS_2$ by $\gamma$-irradiation method," *Materials Letters*, vol. 43, no. 3, pp. 81–86, 2000.

[19] M. Potoczek, K. Przybylski, and M. Rekas, "Defect structure and electrical properties of molybdenum disulphide," *Journal of Physics and Chemistry of Solids*, vol. 67, no. 12, pp. 2528–2535, 2006.

[20] A. J. Grant, T. M. Griffiths, G. D. Pitt, and A. D. Yoffe, "The electrical properties and the magnitude of the indirect gap in the semiconducting transition metal dichalcogenide layer crystals," *Journal of Physics C*, vol. 8, no. 1, article 004, pp. L17–L23, 1975.

[21] E. Benavente, M. A. Santa Ana, and G. González, "Electrical conductivity of $MoS_2$ based organic-inorganic nanocomposites," *Physica Status Solidi B*, vol. 241, no. 10, pp. 2444–2447, 2004.

[22] N. Berntsen, T. Gutjahr, L. Loeffler, J. R. Gomm, R. Seshadri, and W. Tremel, "A solvothermal route to high-surface-area nanostructured $MoS_2$," *Chemistry of Materials*, vol. 15, no. 23, pp. 4498–4502, 2003.

[23] Q. Li, M. Li, Z. Chen, and C. Li, "Simple solution route to uniform $MoS_2$ particles with randomly stacked layers," *Materials Research Bulletin*, vol. 39, no. 7-8, pp. 981–986, 2004.

[24] Y. Peng, Z. Meng, C. Zhong et al., "Hydrothermal synthesis of $MoS_2$ and its pressure related crystallization," *Journal of Solid State Chemistry*, vol. 159, no. 1, pp. 170–173, 2001.

[25] J. H. Zhan, Z. D. Zhang, X. F. Qian, C. Wang, Y. Xie, and Y. T. Qian, "Solvothermal synthesis of nanocrystalline $MoS_2$ from $MoO_3$ and elemental sulfur," *Journal of Solid State Chemistry*, vol. 141, no. 1, pp. 270–273, 1998.

[26] M. M. Mdleleni, T. Hyeon, and K. S. Suslick, "Sonochemical synthesis of nanostructured molybdenum sulfide," *Journal of the American Chemical Society*, vol. 120, no. 24, pp. 6189–6190, 1998.

[27] I. Bezverkhy, P. Afanasiev, and M. Lacroix, "Aqueous preparation of highly dispersed molybdenum sulfide," *Inorganic Chemistry*, vol. 39, no. 24, pp. 5416–5417, 2000.

[28] P. Afanasiev, G. F. Xia, G. Berhault, B. Jouguet, and M. Lacroix, "Surfactant-assisted synthesis of highly dispersed molybdenum sulfide," *Chemistry of Materials*, vol. 11, no. 11, pp. 3216–3219, 1999.

[29] K. H. Hu, Y. R. Wang, X. G. Hu, and H. Z. Wo, "Preparation and characterization of ball-like $MoS_2$ nanoparticles," *Materials Science and Technology*, vol. 23, no. 2, pp. 242–246, 2007.

[30] Z. Wu, D. Wang, and A. Sun, "Surfactant-assisted fabrication of $MoS_2$ nanospheres," *Journal of Materials Science*, vol. 45, no. 1, pp. 182–187, 2010.

[31] P. Pramanik and S. Bhattacharya, "Deposition of molybdenum chalcogenide thin films by the chemical deposition technique and the effect of bath parameters on these thin films," *Materials Research Bulletin*, vol. 25, no. 1, pp. 15–23, 1990.

[32] J. W. McDonald, G. D. Friesen, L. D. Rosenhein, and W. E. Newton, "Syntheses and characterization of ammonium and tetraalkylammonium thiomolybdates and thiotungstates," *Inorganica Chimica Acta*, vol. 72, pp. 205–210, 1983.

[33] H. W. Hou, X. Q. Xin, and S. Shi, "Mo(W, V)-Cu(Ag)-S(Se) cluster compounds," *Coordination Chemistry Reviews*, vol. 153, pp. 25–56, 1996.

[34] J. Grigas, E. Talik, and V. Lazauskas, "X-ray photoelectron spectra and electron structure of $Bi_2S_3$ crystals," *Physica Status Solidi B*, vol. 232, no. 2, pp. 220–230, 2002.

[35] H. W. Wang, P. Skeldon, and G. E. Thompson, "XPS studies of $MoS_2$ formation from ammonium tetrathiomolybdate solutions," *Surface and Coatings Technology*, vol. 91, no. 3, pp. 200–207, 1997.

[36] C. D. Wagner, W. M. Riggs, L. E. Davis, and J. F. Moulder, *Handbook of X-Ray Photoelectron Spectroscopy*, Perkin-Elmer, Minneapolis, Minn, USA, 1979.

[37] Y. Peng, Z. Meng, C. Zhong, J. Lu, Z. Yang, and Y. Qian, "Tube- and ball-like amorphous $MoS_2$ prepared by a solvothermal method," *Materials Chemistry and Physics*, vol. 73, no. 2-3, pp. 327–329, 2002.

[38] C. E. Housecroft and A. G. Sharpe, *Inorganic Chemistry*, Pearson, Upper Saddle River, NJ, USA, 3rd edition, 2008.

[39] V. Yatirajam, U. Ahuja, and L. R. Kakkar, "Use of sulfide precipitation in separation and determination of molybdenum (V)," *Talanta*, vol. 23, no. 11-12, pp. 819–822, 1976.

[40] H. W. Wang, P. Skeldon, G. E. Thompson, and G. C. Wood, "Synthesis of molybdenum disulphide by acidification of ammonium tetrathiomolybdate solutions," *Journal of Materials Science Letters*, vol. 15, no. 6, pp. 494–496, 1996.

[41] H. W. Wang, P. Skeldon, G. E. Thompson, and G. C. Wood, "Synthesis and characterization of molybdenum disulphide formed from ammonium tetrathiomolybdate," *Journal of Materials Science*, vol. 32, no. 2, pp. 497–502, 1997.

[42] H. W. Wang, P. Skeldon, and G. E. Thompson, "Thermogravimetric-differential thermal analysis of the solid-state decomposition of ammonium tetrathiomolybdate during heating in argon," *Journal of Materials Science*, vol. 33, no. 12, pp. 3079–3083, 1998.

[43] P. Ratnasamy, L. Rodrique, and A. J. Leonard, "Structural and textural studies in molybdenum sulfide systems," *Journal of Physical Chemistry*, vol. 77, no. 18, pp. 2242–2245, 1973.

[44] H. P. Klug and L. E. Alexander, *X-Ray Diffraction Procedures for Polycrystalline and Amorphous Materials*, John Wiley & Sons, New York, NY, USA, 1974.

[45] R. R. Chianelli, E. B. Prestridge, T. A. Pecoraro, and J. P. Deneufville, "Molybdenum disulfide in the poorly crystalline "rag" structure," *Science*, vol. 203, no. 4385, pp. 1105–1107, 1979.

[46] A. Plancon, "New modeling of X-ray diffraction by disordered lamellar structures, such as phyllosilicates," *American Mineralogist*, vol. 87, no. 11-12, pp. 1672–1677, 2002.

[47] P. Venkatesan, N. Puvvada, R. Dash et al., "The potential of celecoxib-loaded hydroxyapatite-chitosan nanocomposite for the treatment of colon cancer," *Biomaterials*, vol. 32, no. 15, pp. 3794–3806, 2011.

[48] J. Coates, *Interpretation of Infrared Spectra, A Practical Approach, Encyclopedia of Analytical Chemistry*, John Wiley & Sons, Chichester, UK, 2000.

[49] G. L. Frey, S. Elani, M. Homyonfer, Y. Feldman, and R. Tenne, "Optical-absorption spectra of inorganic fullerenelike $MS_2$ (M = Mo, W)," *Physical Review B*, vol. 57, no. 11, pp. 6666–6671, 1998.

[50] H. Liao, Y. Wang, S. Zhang, and Y. T. Qian,, "A solution low-temperature route to $MoS_2$ fiber," *Chemistry of Materials*, vol. 13, no. 1, pp. 6–8, 2001.

[51] M. Siekierski, W. Wieczorek, and J. Przyluski, "AC conductivity studies of composite polymeric electrolytes," *Electrochimica Acta*, vol. 43, no. 10-11, pp. 1339–1342, 1998.

[52] S. R. Elliot, "AC conduction in amorphous chalcogenide and pnictide semiconductors," *Advances in Physics*, vol. 36, no. 2, pp. 135–217, 1987.

[53] W. Sienicki, "Statistical methods of investigations on the electrical properties of molybdenum disulphide intercalates," *Materials Chemistry and Physics*, vol. 72, no. 3, pp. 347–351, 2001.

[54] K. I. Tanaka and T. Okuhara, "Regulation of intermediates on sulfided nickel and molybdenum disulfide catalysts," *Catalysis Reviews*, vol. 15, no. 1, pp. 249–292, 1977.

[55] I. Uzcanga, I. Bezverkhyy, P. Afanasiev, C. Scott, and M. Vrinat, "Sonochemical preparation of $MoS_2$ in aqueous solution: replication of the cavitation bubbles in an inorganic material morphology," *Chemistry of Materials*, vol. 17, no. 14, pp. 3575–3577, 2005.

# Studying the Enrichment of Ice Cream with Alginate Nanoparticles Including Fe and Zn Salts

**Armoon Sharifi,[1] Leila Golestan,[1] and Mazyar Sharifzadeh Baei[2]**

[1] *Department of Food Science, Ayatollah Amoli Branch, Islamic Azad University, Amol, Iran*
[2] *Islamic Azad University, Ayatollah Amoli Branch, Amol, Iran*

Correspondence should be addressed to Armoon Sharifi; Armoon.sharifi@yahoo.com

Academic Editor: John Zhanhu Guo

The aim of this research was developing alginate nanoparticles as a carrier for food enrichment. In this research, Fe/Zn-loaded alginate nanoparticles were prepared and characterized as point size, morphology, FTIR, loading efficacy (LE), and release properties and used in ice cream structure. After this stage, absorption of the salts was measured and sensory and rheological evaluations were taken for samples. Results showed that alginate nanoparticles have average size between 90 and 135 nm. Also, the shape of the nanoparticles is regular and smooth without aggregation phenomena. FTIR certified that Zn/Fe loaded into alginate nanoparticles. Also, loading efficacy of Zn/Fe was 70–85% and release profile of nanoparticles showed a steady state. Alginate nanoparticles could decrease the loss of Fe/Zn in comparison control. Furthermore, these nanoparticles have no side effects on sensory and rheological properties. Hence, this nanoparticle can be suggestive for the enrichment of ice cream and probably other foods.

## 1. Introduction

One of the most popular aspects of nanotechnology is designing of new vehicle for delivery and transition of materials. Today, researchers achieve new results about probability application of drug delivery systems for food enrichment. Regine et al. studied calcium absorption from fortified ice cream formulations compared with calcium absorption from milk and designed a calcium-fortified ice cream formulation that is lower in fat than regular ice cream and could provide a useful source of additional dietary calcium [1]. Huynh et al. also used solvent-free $\beta$-carotene nanoparticles for food fortification [2]. Chan et al. encapsulated herbal aqueous extract through absorption with ca-alginate hydrogel beads and found the potential of using absorption process and hydrogel material for the encapsulation of herbal aqueous extract [3]. Goto et al. used soybean ferritin gene for iron fortification of rice seed [4]. There was little investigation in fortification with the nanomaterials field. In recent decades, researchers used polysaccharide polymers such as alginate, chitosan, and dextran due to their biodegradable properties. Alginate has different application in food science such as food additive. Furthermore, alginate was certified by WHO/FDA expert food committee on food additives at the 39th meeting in 1992 [5]. Alginate is applied to numerous kinds of food, such as ice cream and jelly, for gelling properties. Machado et al. prepared calcium alginate nanoparticles Using water-in-oil (W/O) nanoemulsions [6]. Zhang et al. prepared chitosan/alginate nanoparticles for oral delivery of insulin and found that these nanoparticles can be good vectors showing a good efficiency in oral administration of the protein of peptide drugs [7]. Ahmad et al. used alginate nanoparticles for drug delivery of ATDs (rifampicin, isoniazid, pyrazinamide, and ethambutol) in mice against tuberculosis [8]. Zohri et al. used Nisin-loaded chitosan/alginate nanoparticles as an antimicrobial agent against *Staphylococcus aureus* in raw and pasteurized milk samples and achieved desirable results [9]. It seems that with attention to the alginate in the stabilizing of ice cream, it can be a good candidate for drug delivery in ice cream. Hence, alginate nanoparticles were developed for food fortification and used in ice cream.

## 2. Method and Materials

*2.1. Materials.* Dialysis membrane bag Mw 100 kD, Spectrum Laboratories, and zinc chloride salt (Zncl2), and ferric salt

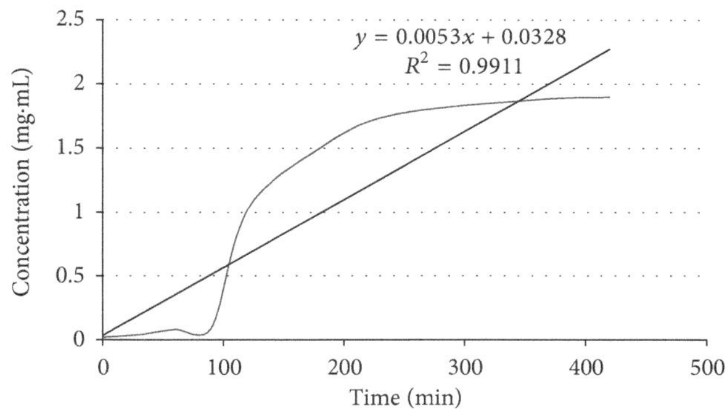


FIGURE 1: Release profile of Fe/Zn from nanoparticles.

(Fecl3), ultrafilter Amicon cutoff 100 kD, and alginate were purchased from Sigma-Aldrich.

*2.2. Preparation of Fe/Zn-Loaded Alginate Nanoparticles.* Alginate nanoparticles were prepared using a desolvation method. Alginate powder was added to distilled water. Ethanol solution was added continuously or intermittently into 1% alginate solution at pH 7 under stirring at 700 rpm at room temperature until the solution became just turbid. In continuous addition method ethanol was added continuously in the solution with rate addition about 1.0 to 2.0 mL per min, and for intermittent method, 2 mL of ethanol was added for every 5 min interval. After this 5 mgr of zinc salt and $Fe^{+3}$ salt, solutions continuously were added to the solution and then stirred for 10 min and nanoparticles centrifuged at 11000 rpm for 10 min and separated.

*2.3. Nanoparticle Characterization*

*2.3.1. Release Profile and Loading Efficacy Determination of Nanoparticles.* Release studies for Fe/Zn salts were done at 0, 15, 30, 60, 90, and 120 min and thermograms of release were drawn at different times by dialysis membrane bag methods (Figure 1) [9]. For this purpose, the 1000 KD (MWCO) membrane was selected as the MWCO being sufficiently large to allow the passage of Fe/Zn salts. For the experiment, 5 mL of a 1 mg/mL solution of phosphate-buffered saline (PBS) at pH 7.4 (release media) was poured into the inner tube of the dialyzer. The dialyzer tube was placed into a 50 mL glass cylinder containing release media, which was continually stirred at 300 rpm using a small magnetic stir bar to prevent the formation of an unstirred water layer at the membrane/outer solution interface. Diffusion to the outer solution at 37°C was assessed by sampling the contents of the outer solution at periodic intervals of 0, 15, 30, 60, and 120 min [8, 10]. For loading studies, samples were centrifuged in Amicon Ultra-15 (Ultracel-100 k Millipore Co., USA) with a molecular weight cutoff to remove free molecules of polymer as well. The alginate nanoparticles were prepared separately to be used as blank solution in subsequent atomic absorption analysis. Samples were centrifuged in Amicon Ultra-15 (Ultracel-100 K) for 20 min at 5000 rpm, and then the absorbance of the solutions in the tubes was measured.

The amount of iron in nanoparticle samples was determined on the basis of Tautkus et al., and the loading efficacy was evaluated as follows (Tautkus et al. 2003, [5]):

$$\begin{aligned} \text{LE } (\%) = {} & \frac{\text{Fe}}{\text{Zn}} \text{ salts Total} \\ & - \frac{\text{Fe}}{\text{Zn}} \frac{\text{supernatant}}{\text{Fe/Zn}} \text{ salts Total} \end{aligned} \tag{1}$$

*2.3.2. Size Disruption of Nanoparticles.* For this purpose, Zetasizer 3000 HS (Malvern Instruments, UK) was used. Nanoparticles were diluted and measured by the device. Mean and standard deviations of the free nanoparticles and Fe/Zn-loaded nanoparticles were measured and presented for comparison.

*2.3.3. Morphology of Nanoparticles.* The nanoparticles morphology such as shape and incidence of aggregation phenomena was studied by SEM (scanning electron microscopy, JSM-6610, 30 kv). For this purpose, the samples of the nanoparticle suspensions (5–10 ll) were mounted on metal stubs, plating coated under vacuum, and then observed.

*2.3.4. FTIR Evaluation.* FTIR spectra were obtained using 8400S (Shimadzu Co., Japan) FTIR spectrometer. Samples were dried in a vacuum desiccator, mixed with micronized KBr powder, and compressed into discs using a manual tablet press [11].

*2.4. Ice Cream Making Producing.* For ice cream making, first all of the consequents of mixture include milk, sugar, cream, dried milk, and flavoring agent (Vanilla) calculated by Pierson square method and weighted. Then, milk and cream are mixed well for 5 min with electrical mixer. This mixture was poured in steel bottle and heated till 40°C slowly. The other materials added to ice cream after this stage. The mixture hashed and then set at 80°C for 25 seconds and stored at refrigerator for 4 hour. After this, Fe/Zn-loaded nanoparticles added to ice cream. There were three-trial sample and one blank sample for experimental analysis.

## 3. Ice Cream Analysis

*3.1. Fe/Zn Absorbance Amount.* Seven people (women and men with 20–30 ages) were studied and the enrichment ice that was used cream varied from 10 mg Fe to 14 mg Zn at the morning before food eating. The amount of Fe/Zn calculated at urine samples and the average of the results were reported.

*3.2. Sensory Evaluation.* Ice cream samples were organoleptically evaluated for texture, flavour, body/texture, and overall acceptability. For sensory evaluation, 15 trained panelist were selected and samples included ice cream as blank and ice cream including nanoparticles in three replications were evaluated with a 5-point hedonic test.

*3.3. Rheological Properties.* For this purpose, Yield stress assay was done for ice cream, and it was used from Bohlin

Viscometer with thermal circulator. Bob spindle opted on the basis viscosity of mixture. The mixture after ageing time was poured in C30 cup and its temperature reached 10°C and yield stress was between 14.2 and 501.7 (Figure 4).

*3.4. Statistical Analysis.* Statistical analysis was conducted using Spss (version 16) to determine the means and standard deviation of the rheological data and sensory data. An ANOVA was conducted to determine differences between means.

## 4. Results and Discussion

*4.1. Release Profile and Loading Efficacy Determination of Nanoparticles Evaluation.* There are numerous possible applications of alginate polymer drug delivery systems using both encapsulation with matrices and encapsulation with alginate surfactant oil in water emulsions as delivery agents. As a designed carrier for protein or peptides delivery, the alginate-chitosan microspheres have to experience the gastrointestinal tract when they are orally administrated. Hence, alginate nanoparticle has benefit for oral delivery [7]. As seen in the profile, nanoparticles first release at initial time that named burst release. These phenomena occur due to swelling properties and alginate nanoparticles; this result was in agreement with Gazori and Zohri results [9, 10, 12]. After this nanoparticle reaches a steady-state stage and this situation continues till all of Zn/Fe are released. This type of release profile is suitable due to gradual absorption for oral delivery of encapsulates [7]. Also loading study showed that nanoparticles have 70–85% from Fe/Zn salts.

*4.2. Size Disruption of Nanoparticles Evaluations.* It can be observed that the mean size of nanoparticles is 98 nm and the mean of PDI (polydisparity index) of the nanoparticles was 0.1 PDI and is related to homogeneity of the particles. It showed that these nanoparticles have good stability and can be good candidates for drug delivery.

*4.3. Morphology of Nanoparticles.* SEM was used for certification of alginate nanoparticles and provided morphological information on the Fe/Zn-loaded alginate nanoparticles. Spherical, distinct, regular, and smooth shapes were seen for the nanoparticles through the SEM images (Figure 2). The particle size range of the nanoparticles was 90–200 nm.

*4.4. FTIR Evaluation.* The FTIR spectra of alginate powder and nanoparticles are shown in Figure 3. The spectra showed a band at approximately 3757 $cm^{-1}$, 2855 $cm^{-1}$, and 1739 $cm^{-1}$ which are assigned to Amide I, Amide II, and Amide III, respectively. The most prominent FT Raman band centered at 880 $cm^{-1}$ is mainly due to the O/H deformation mode, while the band at 1400 $cm^{-1}$ is ascribed to deformation of the CH2 groups. The C/O/C and C/OH stretching modes give rise to several close-lying bands in the spectral regions of 1250/1290 and 1000/1025 $cm^{-1}$, respectively, but in nanoparticles spectra two additional band in 3390 $cm^{-1}$ and 1636 $cm^{-1}$ were observed that this changes related to loading of Fe/Zn functional groups in alginate egg structure and appearance of new bands, Zohri et al.'s result was in agreement with this also [10].

TABLE 1: Sensory properties of ice cream with Zn/Fe-loaded nanoparticles.

| Variables correlated with sensory properties | $A$ | $B$ | $A \times B$ |
|---|---|---|---|
| Flavor | $70.08^{**}$ | $154.14^{**}$ | $16.05^{**}$ |
| Aroma | $613.00^{**}$ | $904.00^{**}$ | $121.00^{**}$ |
| Oral texture | $99.90^{**}$ | $61.18^{**}$ | $7.18^{**}$ |

Means ± standard errors of the measured parameters are presented. The common superscripts in each column is indicative of statistical insignificance that is $P > 0.05$.

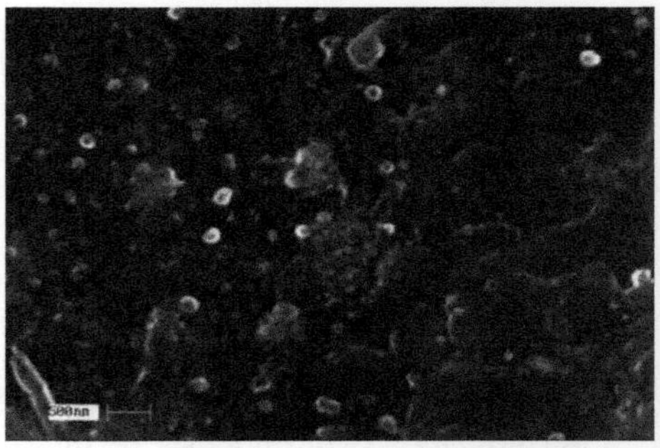

FIGURE 2: Scanning electron microscope image of Zn/Fe-loaded nanoparticles.

*4.5. Fe/Zn Absorbance Amount.* Results of urine assay showed that more than 90% of Zn/Fe was absorbed. This is because chelating properties of alginate structure can be loaded minerals and it can be prevent its lost. Generally, the aim of drug delivery systems is to prevent interaction between encapsulate and the other ingredients and also deliver proposed material to the appropriate place at a suitable time [13].

*4.6. Sensory Evaluation.* Table 1 showed that there were any significant difference between control samples and ice cream with Fe/Zn-loaded nanoparticles. One of the reasons for this result can be related to the size of nanoparticles. The average size of the nanoparticles is between 90 and 135 nm. This size is smaller than to be tasted by tasting buddies. There were not any differences between samples but there was significant relation between characteristic of ice cream at 1% and 5% levels [13, 14]. Zohri et al. certified that alginate can play debittering role and improve the sensory properties of UF feta cheese [15].

*4.7. Rheological Properties Evaluation.* The changes in viscosity and yield stress against shear rating in ice cream samples that include nanoparticles were similar to control samples and it decreased with increase of shear rate. This occurs due to the accumulation of molecules in low shear rates, and this causes increasing in viscosity. When shear rate increases, the molecules set one side and this causes an increasing in friction

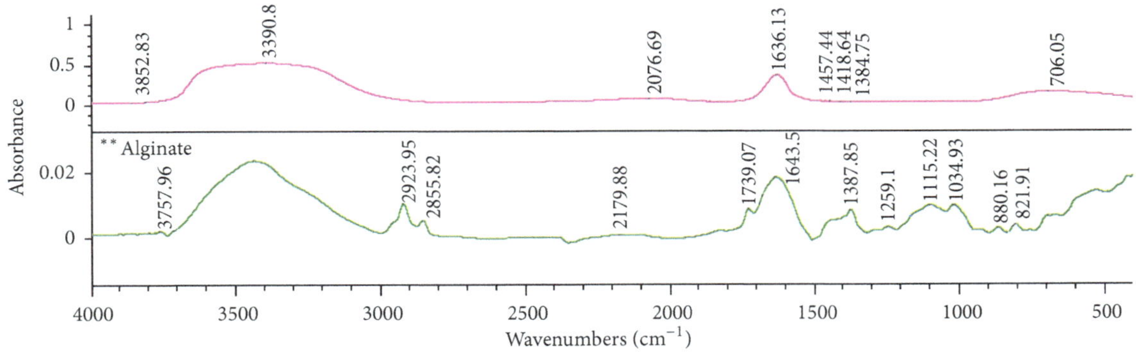


FIGURE 3: FTIR spectra of the alginate powder and nanoparticles.

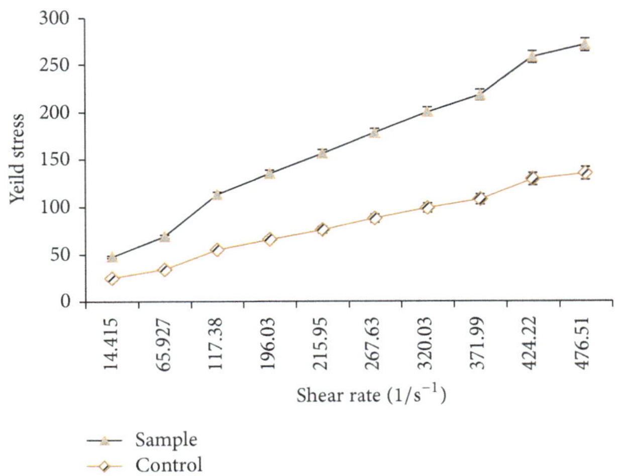


FIGURE 4: Rheological properties of control sample and Fe/Zn-loaded nanoparticles.

and the viscosity decreases. The other reason it can be related to the size of nanoparticles and dispersing of nanoparticles in ice cream structures. Hence, these nanoparticles have no rules in rheological properties [16]. This result was in agreement with Mandar et al.'s [13].

## References


[1] R. M. van der Hee, M. Silvia, S. Marieke et al., "Calcium absorption from fortified ice cream formulations compared with calcium absorption from milk," *Journal of the American Dietetic Association*, vol. 109, no. 5, pp. 830–835, 2009.

[2] P. T. Huynh and P. Takhistov, *free beta carotene nanoparticles for food fortification [Ph.D. thesis]*, Libraryof Rutger University, 2010.

[3] E. S. Chan, Z. H. Yim, S. H. Phan, R. F. Mansa, and P. Ravindra, "Encapsulation of herbal aqueous extract through absorption with ca-alginate hydrogel beads," *Food and Bioproducts Processing*, vol. 88, no. 2-3, pp. 195–201, 2010.

[4] F. Goto, T. Yoshihara, N. Shigemoto, S. Toki, and F. Takaiwa, "Iron fortification of rice seed by the soybean ferritin gene," *Nature Biotechnology*, vol. 17, no. 3, pp. 282–286, 1999.

[5] W. R. Gombotz and S. F. Wee, "Protein release from alginate matrices," *Advanced Drug Delivery Reviews*, vol. 31, no. 3, pp. 267–285, 1998.

[6] A. H. E. . Machado, D. Lundberg, A. J. Ribeiro et al., "Preparation of calcium alginate nanoparticles using water-in-oil (W/O) nanoemulsions," *Langmuir*, vol. 28, no. 9, pp. 4131–4141, 2012.

[7] N. Zhang, J. Li, W. Jiang et al., "Effective protection and controlled release of insulin by cationic $\beta$-cyclodextrin polymers from alginate/chitosan nanoparticles," *International Journal of Pharmaceutics*, vol. 393, no. 1-2, pp. 213–219, 2010.

[8] Z. Ahmad, R. Pandey, S. Sharma, and G. K. Khuller, "Alginate nanoparticles as antituberculosis drug carriers: formulation development, pharmacokinetics and therapeutic potential," *The Indian Journal of Chest Diseases & Allied Sciences*, vol. 48, no. 3, pp. 171–176, 2006.

[9] M. Zohri, A. Nomani, T. Gazori et al., "Characterization of chitosan/alginate self-assembled nanoparticles as a protein carrier," *Journal of Dispersion Science and Technology*, vol. 32, no. 4, pp. 576–582, 2011.

[10] M. Zohri, M. S. Alavidjeh, I. Haririan et al., "A comparative study between the antibacterial effect of nisin and nisin-loaded chitosan/alginate nanoparticles on the growth of Staphylococcus aureus in raw and pasteurized milk samples," *Probiotics and Antimicrobial Proteins*, vol. 2, no. 4, pp. 258–266, 2010.

[11] T. Gazori, M. R. Khoshayand, E. Azizi, P. Yazdizade, A. Nomani, and I. Haririan, "Evaluation of Alginate/Chitosan nanoparticles as antisense delivery vector: formulation, optimization and in vitro characterization," *Carbohydrate Polymers*, vol. 77, no. 3, pp. 599–606, 2009.

[12] T. Gazori, I. Haririan, S. Fouladdel, A. Namazi, A. Nomani, and E. Azizi, "Inhibition of EGFR expression with chitosan/alginate nanoparticles encapsulating antisense oligonucleotides in T47D cell line using RT-PCR and immunocytochemistry," *Carbohydrate Polymers*, vol. 80, no. 4, pp. 1042–1047, 2010.

[13] R. P. Mandar, M.-G. San, and M. Fernanda, "Characterization of ergocalciferol loaded solid lipid nanoparticles," *Journal of Food Science*, vol. 77, pp. 8–13, 2012.

[14] G. R. Shaviklo, G. Thorkelsson, K. Sveinsdottir, and F. Rafipour, "Chemical properties and sensory quality of ice cream fortified with fish protein," *Journal of the Science of Food and Agriculture*, vol. 91, no. 7, pp. 1199–1204, 2011.

[15] M. Zohri, M. Shafiee Alavidjeh, S. S. Mirdamadi et al., "Nisin-loaded chitosan/alginate nanoparticles: a hopeful hybrid biopreservative," *Journal of Food Safety*, vol. 33, no. 1, pp. 40–49, 2013.

[16] B. Sarmento, S. Martins, A. Ribeiro, F. Veiga, R. Neufeld, and D. Ferreira, "Development and comparison of different nanoparticulate polyelectrolyte complexes as insulin carriers," *International Journal of Peptide Research and Therapeutics*, vol. 12, no. 2, pp. 131–138, 2006.

# Permissions

The contributors of this book come from diverse backgrounds, making this book a truly international effort. This book will bring forth new frontiers with its revolutionizing research information and detailed analysis of the nascent developments around the world.

We would like to thank all the contributing authors for lending their expertise to make the book truly unique. They have played a crucial role in the development of this book. Without their invaluable contributions this book wouldn't have been possible. They have made vital efforts to compile up to date information on the varied aspects of this subject to make this book a valuable addition to the collection of many professionals and students.

This book was conceptualized with the vision of imparting up-to-date information and advanced data in this field. To ensure the same, a matchless editorial board was set up. Every individual on the board went through rigorous rounds of assessment to prove their worth. After which they invested a large part of their time researching and compiling the most relevant data for our readers. Conferences and sessions were held from time to time between the editorial board and the contributing authors to present the data in the most comprehensible form. The editorial team has worked tirelessly to provide valuable and valid information to help people across the globe.

Every chapter published in this book has been scrutinized by our experts. Their significance has been extensively debated. The topics covered herein carry significant findings which will fuel the growth of the discipline. They may even be implemented as practical applications or may be referred to as a beginning point for another development. Chapters in this book were first published by Hindawi Publishing Corporation; hereby published with permission under the Creative Commons Attribution License or equivalent.

The editorial board has been involved in producing this book since its inception. They have spent rigorous hours researching and exploring the diverse topics which have resulted in the successful publishing of this book. They have passed on their knowledge of decades through this book. To expedite this challenging task, the publisher supported the team at every step. A small team of assistant editors was also appointed to further simplify the editing procedure and attain best results for the readers.

Our editorial team has been hand-picked from every corner of the world. Their multi-ethnicity adds dynamic inputs to the discussions which result in innovative outcomes. These outcomes are then further discussed with the researchers and contributors who give their valuable feedback and opinion regarding the same. The feedback is then collaborated with the researches and they are edited in a comprehensive manner to aid the understanding of the subject.

Apart from the editorial board, the designing team has also invested a significant amount of their time in understanding the subject and creating the most relevant covers. They scrutinized every image to scout for the most suitable representation of the subject and create an appropriate cover for the book.

The publishing team has been involved in this book since its early stages. They were actively engaged in every process, be it collecting the data, connecting with the contributors or procuring relevant information. The team has been an ardent support to the editorial, designing and production team. Their endless efforts to recruit the best for this project, has resulted in the accomplishment of this book. They are a veteran in the field of academics and their pool of knowledge is as vast as their experience in printing. Their expertise and guidance has proved useful at every step. Their uncompromising quality standards have made this book an exceptional effort. Their encouragement from time to time has been an inspiration for everyone.

The publisher and the editorial board hope that this book will prove to be a valuable piece of knowledge for researchers, students, practitioners and scholars across the globe.

# List of Contributors

**Satyajit Tripathy, Sabyasachi Das, Sandeep Kumar Dash, Sourav Chattopadhyay and Somenath Roy**
Immunology and Microbiology Laboratory, Department of Human Physiology with Community Health, Vidyasagar University, Midnapore West Bengal 721102, India

**Yu-Wen Chen and Der-Shing Lee**
Department of Chemical Engineering, National Central University, Chung-Li 320, Taiwan

**Amit Kumar Mittal, Abhishek Kaler, Aparna Vasant Mulay and Uttam Chand Banerjee**
Department of Pharmaceutical Technology, National Institute of Pharmaceutical Education and Research, Sector-67, SAS Nagar, Punjab 160062, India

**B. Karthikeyan and B. Loganathan**
Department of Chemistry, Annamalai University, Annamalai Nagar, Tamil Nadu 608 002, India

**Duhita G. Sant, Tejal R. Gujarathi, Shrikant R. Harne, Balu A. Chopade and Karishma R. Pardesi**
Department of Microbiology, University of Pune, Pune, Maharashtra 411007, India

**Sougata Ghosh**
Institute of Bioinformatics and Biotechnology, University of Pune, Pune, Maharashtra 411007, India

**Rohini Kitture and Sangeeta Kale**
Department of Applied Physics, Defence Institute of Advanced Technology, Girinagar, Pune 411025, India

**Xiaoping Jin, Jürgen Parisi and Joanna Kolny-Olesiak**
Energy and Semiconductor Research Laboratory, Department of Physics, Carl von Ossietzky University of Oldenburg, Carl von Ossietzky Straße 9–11, 26111 Oldenburg, Germany

**Véronic Landry, Pierre Blanchet and Gabrielle Boivin**
FPInnovations, 319 Franquet, Quebec, QC, Canada G1P 4R4

**T. Prakash and R. Jayaprakash**
Nanotechnology Laboratory, Department of Physics, Sri Ramakrishna Mission Vidyalaya College of Arts and Science, Coimbatore, Tamil Nadu 641020, India

**G. Neri**
Department of Electronic Engineering, Chemistry and Industrial Engineering, University of Messina, 98166 Messina, Italy

**Sanjay Kumar**
Department of Physics, B. R. Ambedkar Bihar University, Muzaffarpur, Bihar 842001, India

**A. K. Singh**
Nanomaterials and Sensors Laboratory, Defence Institute of Advanced Technology, Girinagar, Pune, Maharashtra 411025, India
DRDO Centre for Piezoceramics and Devices, ARDE, Pashan, Pune, Maharashtra 411021, India

**Umesh T. Nakate**
Nanomaterials and Sensors Laboratory, Defence Institute of Advanced Technology, Girinagar, Pune, Maharashtra 411025, India

**Pravas Kumar Panigrahi and Amita Pathak**
Department of Chemistry, Indian Institute of Technology Kharagpur, Kharagpur, West Bengal 721302, India

**Nydeia W. Bolden, Vijaya K. Rangari and Shaik Jeelani**
Department of Materials Science and Engineering, Tuskegee University, Tuskegee, AL 36088, USA

**Seyhan Boyoglu**
Department of Pediatrics, Emory-Children's Center, Emory University, Atlanta, GA 30322, USA

**Shree R. Singh**
Center for NanoBiotechnology Research, Alabama State University, Montgomery, AL 36104, USA

**M. Th. Makhlouf, B. M. Abu-Zied and T. H. Mansoure**
Department of Chemistry, Faculty of Science, Assiut University, Assiut 71516, Egypt

**Kuan-Jen Chen**
The Instrument Center, National Cheng Kung University, Tainan 70101, Taiwan

**Fei-Yi Hung, Truan-Sheng Lui and Cheng-Hung Chen**
Department of Materials Science and Engineering, Institute of Nanotechnology and Microsystems Engineering, Center for Micro/Nano Science and Engineering, National Cheng Kung University, Tainan 70101, Taiwan

**Sheng-Po Chang**
Institute of Microelectronics and Department of Electrical Engineering, Advanced Optoelectronic Technology Center, National Cheng Kung University, Tainan 701, Taiwan

**Ram Prasad and Vyshnava Satyanarayana Swamy**
Amity Institute of Microbial Technology, Amity University Uttar Pradesh, Sector 125, Noida 201303, India

**V. L. Chandraboss, B. Karthikeyan, J. Kamalakkannan, S. Prabha and S. Senthilvelan**
Department of Chemistry, Annamalai University, Annamalai Nagar, Tamil Nadu 608 002, India

**M. Virginia Roldán, Nora Pellegri and Oscar de Sanctis**
Laboratorio de Materiales Cer´amicos, FCEIyA, IFIR, Universidad Nacional de Rosario (UNR), Avenida Pellegrini 250, S2000BTP Rosario, Argentina

**Shailesh K. Dhoke, Narayani Rajgopalan and A. S. Khanna**
Department of Metallurgical Engineering and Materials Science, Indian Institute of Technology Bombay, Powai, Mumbai 400076, India

**K. Pradeesh, Nageswara Rao Kotla, Shahab Ahmad, Vindesh K. Dwivedi and G. Vijaya Prakash**
Nanophotonics Lab, Department of Physics, Indian Institute of Technology Delhi, Hauz Khas, New Delhi 110016, India

**Pranjal Chandra**
Amity Institute of Biotechnology, Amity University Uttar Pradesh, Sector 125, Noida 201303, India
Department of Chemistry, Pusan National University, Busan 609-735, Republic of Korea
Institute of BioPhysio Sensor Technology, Pusan National University, Busan 609-735, Republic of Korea

**Jai Singh**
Faculty of Nanotechnology and Advanced Materials, HMC, and Graphene Research Institute, Sejong University, Seoul 143-747, Republic of Korea

**Amardeep Singh**
Department of Chemistry, Pusan National University, Busan 609-735, Republic of Korea

**Ananya Srivastava**
Department of Chemistry, DDU Gorakhpur University, Gorakhpur 273009, India

**Rajendra N. Goyal**
Institute of BioPhysio Sensor Technology, Pusan National University, Busan 609-735, Republic of Korea
Department of Chemistry, Indian Institute of Technology Roorkee, Roorkee 247667, India

**Yoon Bo Shim**
Department of Chemistry, Pusan National University, Busan 609-735, Republic of Korea
Institute of BioPhysio Sensor Technology, Pusan National University, Busan 609-735, Republic of Korea

**Jalal Albadi**
College of Science, Behbahan Khatam Alanbia University of Technology, Behbahan 736254, Iran

**Abdolhossein Razeghi**
Catalysis and Reaction Engineering, Tehran University, Tehran, Iran

**Hossein Abbaszadeh and Azam Mansournezhad**
Department of Chemistry, Gachsaran Branch, Islamic Azad University, Gachsaran, Iran

**Jnyanaranjan Panda, V. Jagannath Patro and Biswa Mohan Sahoo**
Department of Pharmaceutical Chemistry, Roland Institute of Pharmaceutical Sciences and Berhampur, Odisha 760010, India

**Jitendriya Mishra**
University Institute of Pharmaceutical Sciences, UGC Centre of Advanced Study (CAS), Panjab University, Chandigarh 160014, India

**Sergei P. Kruchinin**
Bogolyubov Institute for Theoretical Physics, National Academy of Science of Ukraine, Kiev 03680, Ukraine

**Hiroyuki Kawabe**
Department of Social Work, Faculty of Social Work, Kinjo University, 1200 Kasama, Hakusan, Ishikawa 924-8511, Japan

**Hidemi Nagao**
Division of Mathematical and Physical Science, Graduate School of Natural Science and Technology, Kanazawa University, Kakuma, Kanazawa 920-1192, Japan

**Yasuhiro Nakazawa**
Department of Chemistry, Graduate School of Science, Osaka University, Toyonaka, Osaka 560-0043, Japan

**Jie Ma and Shuping Zhang**
College of Science, University of Shanghai for Science and Technology, Jungong Road 516, Shanghai 200093, China. Green Biochemical and Ecochemical Engineering Lab, University of Shanghai for Science and Technology, Jungong Road 516, Shanghai 200093, China

**Yinchen Xu, Wenlie Li, Jiantao Zhao and Sergey Basov**
College of Science, University of Shanghai for Science and Technology, Jungong Road 516, Shanghai 200093, China

**Tahir Ahmad and Rafiq Ahmad**
Department of Metallurgy and Materials Engineering, College of Engineering and Emerging Technologies, University of the Punjab, Lahore, Pakistan

**Othman Mamat**
Mechanical Engineering Department, Universiti Teknologi PETRONAS, Bandar Seri Iskandar, 31750 Tronoh, Perak, Malaysia

**Pravas Kumar Panigrahi and Amita Pathak**
Department of Chemistry, Indian Institute of Technology, Kharagpur, West Bengal 721302, India

**Armoon Sharifi and Leila Golestan**
Department of Food Science, Ayatollah Amoli Branch, Islamic Azad University, Amol, Iran

**Mazyar Sharifzadeh Baei**
Islamic Azad University, Ayatollah Amoli Branch, Amol, Iran